COVER ILLUSTRATION

This north-looking oblique view of the lunar highlands reveals the heavily-cratered nature of the ancient (4.0 to 4.6 b.y.) lunar crust. The thousands of coincident impacts evidenced by these overlapping craters have degraded and erased records of early crustal formation and original surface morphology while transforming the rocks into complex breccias. During this intense bombardment, the pristine crustal rocks were repeatedly excavated and redistributed over the lunar surface in the form of crater ejecta, developing into the brecciated impact formations which now cover most of the globe to depths measured in kilometers. Very careful study of the individual breccia fragments is required to decipher pre-impact history of the lunar crust.

The large central crater is Heaviside (163 km diameter, near 13.9°S, 167.6°E). The flat-floored, hexagonal crater in the foreground, Ibn Hayyan, contains a complex fracture pattern that indicates either an unusual impact process or post-impact modification by igneous activity, which would further complicate interpretations of early crustal history.

This metric photo was taken from an altitude of 121 km by the crew of Apollo 17. NASA photo #AS17-837.

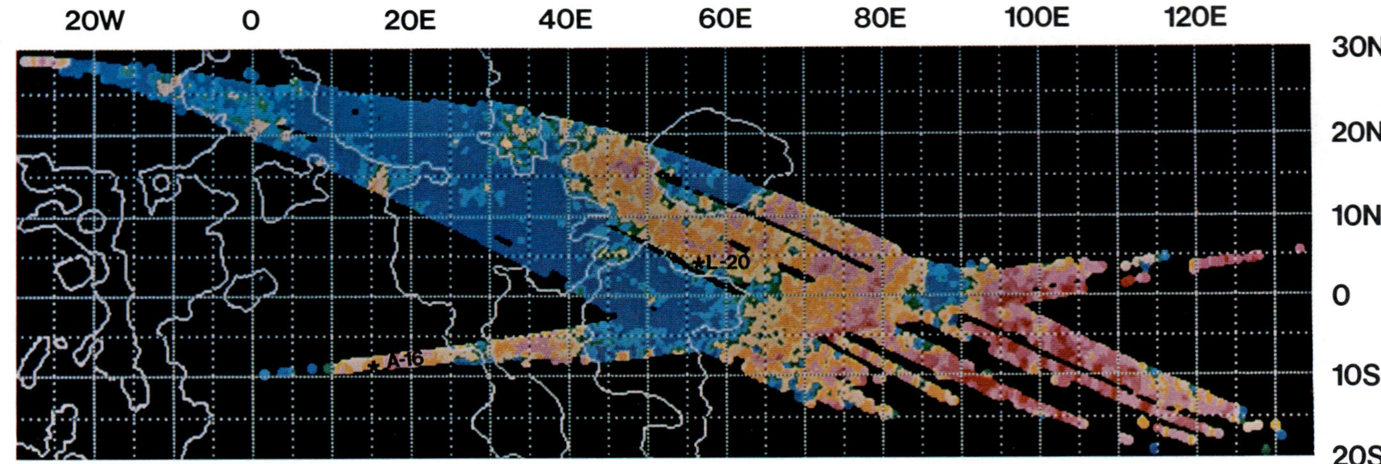

Plate 1. This Mg/Al image of Apollo 15 and 16 orbital X-ray fluorescence data has been constructed to emphasize variations of magnesium/aluminum ratios in terra soils on the moon. An Mg/Al threshold value was set to display all mare basalts in blue and to assign a full range of colors to variations in highland regions (see the color key and orbital vs. sample data comparisons on the opposite page.) The image illustrates the chemical dissimilarity between the nearside and farside terra and also the wide variety of compositions within each of these petrographic provinces. If the more anorthositic (lower Mg/Al) eastern farside characterizes the differentiated terra crust, the distribution of this crustal material on the nearside surface is limited to a few local areas, e.g., the area immediately east of the Apollo 16 landing site and the western rim of the Smythii Basin (80E to 90E). One might expect that unaltered highland crustal material excavated from depth during the formation of the Smythii Basin would be exposed on the basin rim. The chemical resemblance of the western rim to farside Mg/Al values indicates that the farside province may extend several hundred kilometers into the nearside eastern limb. However, the average soil compositions at the only two highland landing sites, Apollo 16 and Luna 20, are not comparable to the farside for areas as extensive as those within the field of view of the X-ray spectrometer (diam. 3 lunar degrees). For additional information on the X-ray experiment, see Adler *et al.* (1973) *Proc. Lunar Sci. Conf. 4th,* p. 2783–2791. For details of data handling methods and imaging techniques, see Andre *et al.* (1977) *Science* **197,** 986–989 and Bielefeld *et al.* (1977) *Proc. Lunar Sci. Conf. 8th,* p. 901–908. (**Constance G. Andre*** and **Isidore Adler,**** **National Air and Space Museum* ***University of Maryland*.)

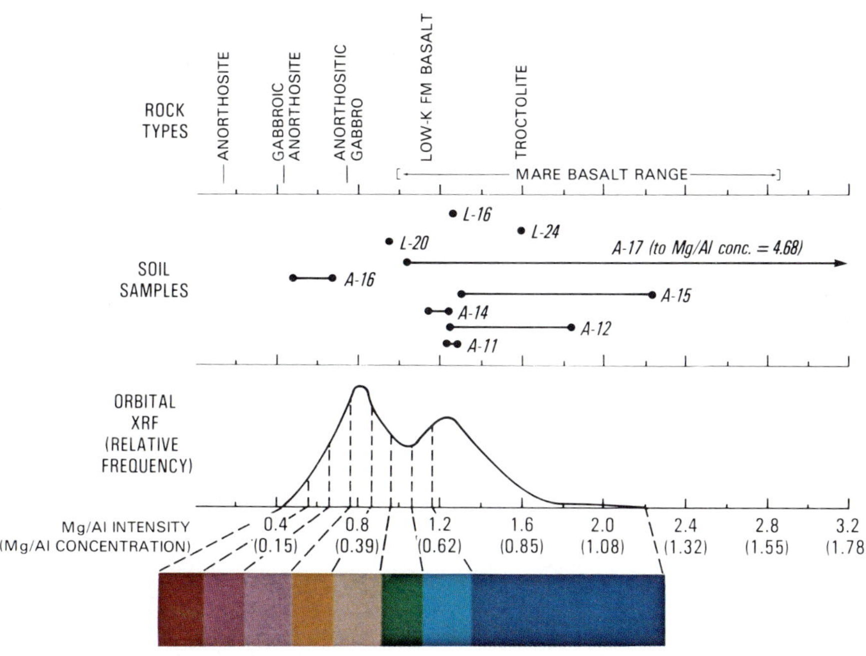

Plate 2 (above). Color key for the orbital X-ray Mg/Al image on the facing page. Even increments of Mg/Al intensity (and concentration) ratios for "highland" areas were used to define the color bar. The relative frequency of occurrence for the orbital Mg/Al values is shown in the histogram above the scale. The range and frequency of the values represent predominantly nearside areas within the X-ray coverage. The horizontal bars above the histogram represent Mg/Al ranges for soil samples from each Apollo or Luna mission corresponding to the same scale used for the orbital data. The uppermost section shows how several lunar rock types relate to the soil ranges and to orbital values. Note that the Mg/Al ratio is a sensitive indicator for distinguishing the anorthosite, gabbroic anorthosite, anorthositic gabbro, low-K Fra Mauro basalts, and troctolite members of the terra suite.

GEOCHIMICA ET COSMOCHIMICA ACTA

Supplement 12

PROCEEDINGS OF THE CONFERENCE ON THE LUNAR HIGHLANDS CRUST

Houston, Texas, November 14–16, 1979

GEOCHIMICA ET COSMOCHIMICA ACTA
Journal of The Geochemical Society and The Meteoritical Society
Supplement 12

PROCEEDINGS OF THE CONFERENCE ON THE LUNAR HIGHLANDS CRUST

Houston, Texas, November 14–16, 1979

Compiled by the
Lunar and Planetary Institute
Houston, Texas

Pergamon Press
New York • Oxford • Toronto • Sydney • Frankfurt • Paris

Pergamon Press Offices:

U.S.A.	Pergamon Press Inc., Maxwell House, Fairview Park, Elmsford, New York 10523, U.S.A.
U.K.	Pergamon Press Ltd., Headington Hill Hall, Oxford OX3 0BW, England
CANADA	Pergamon of Canada Ltd., 150 Consumers Road, Willowdale, Ontario M2J 1P9, Canada
AUSTRALIA	Pergamon Press (Aust) Pty. Ltd., P.O. Box 544, Potts Point, NSW 2011, Australia
FRANCE	Pergamon Press SARL, 24 rue des Ecoles, 75240 Paris, Cedex 05, France
FEDERAL REPUBLIC OF GERMANY	Pergamon Press GmbH, 6242 Kronberg/Taunus, Pferdstrasse 1, Federal Republic of Germany

---

First edition 1980

Type set by Precision Typographers,
printed by Publishers Production International,
and bound by Arnold's Bindery
in the United States of America

Library of Congress Cataloging in Publication Data

Conference on the Lunar Highlands Crust, Houston,
Tex., 1979.
Proceedings of the Conference on the Lunar
Highlands Crust, Houston, Texas, November 14–16,
1979.

(Geochimica et cosmochimica acta: Supplement; 12)
1. Lunar petrology—Congresses. 2. Moon—
Surface—Congresses. I. Lunar and Planetary
Institute. II. Series.
OB592.C68 1980 559.9′1 80-18706
ISBN 0-08-026304-6

CONTENTS

Preface

The understanding of processes responsible for the formation of ancient planetary crusts is a primary objective of planetary science. Inasmuch as evidence of the earliest history of the Earth has been erased and the ancient cratered terrains of Mercury and Mars have not yet been sampled, the samples of the lunar highlands can be said to hold the key to our understanding of early planetary crustal formation. Unfortunately, an intense period of bombardment of the lunar crust has scrambled the record of the first 600 million years. Nevertheless, some pristine-ancient crustal rocks have been identified and have preserved a record of the ~4.6 b.y. melting events associated with crustal formation.

The LPI Topical Conference on the Lunar Highlands Crust was held in Houston, November 14–16, 1979. This three-day meeting was attended by 108 scientists from the U.S., Germany, Canada, and Great Britain who presented papers on the following topics:

1. Regional characteristics of the lunar highlands crust
2. Petrology of the lunar highlands
3. Chemistry and chronology of the highlands crust
4. Physical processes of crustal evolution
5. Magma oceans and crustal formation

These proceedings contain 23 papers in which authors have reported the results of their studies of highlands samples and have attempted to place in perspective with our knowledge of material from other lunar sites and of the crusts of other planets.

Publication of these proceedings, which is sponsored by the Lunar and Planetary Institute on behalf of Principal Investigators in the Lunar and Planetary Programs, has occurred in parallel with *Proceedings of the Eleventh Lunar and Planetary Science Conference,* which include many papers relevant to the lunar highlands.

On behalf of the Institute, I thank the authors, editors and reviewers for their cooperation in accomplishing the tight publication schedule. In particular, I thank Jim Papike, who, as Science Editor, succeeded admirably in keeping everyone's productivity at a very high level indeed and who, together with the Associate Editors, deserves much of the credit for the scientific quality of this volume. My personal appreciation is extended also to the members of the Institute editorial staff who shared every stage in the development of this book, and in particular to Ms. Paula Criswell, Technical Editor, who has borne the major responsibility for technical preparation of these *Proceedings*.

Lunar and Planetary Institute
Houston, Texas
May, 1980

Russell B. Merrill

Papike, J.J. and Merrill, R.B., eds.
Proc. Conf. Lunar Highlands Crust (1980), p. 1-49
Printed in the United States of America

Mission objectives for geological exploration of the Apollo 16 landing site[1]

W. R. Muehlberger

The University of Texas at Austin, Austin, Texas 78712

Friedrich Hörz

Geology Branch, NASA Johnson Space Center, Houston, Texas 77058

J. R. Sevier

Lunar and Planetary Institute, Houston, Texas 77058

G. E. Ulrich

U.S. Geological Survey, Flagstaff, Arizona 86001

Summary—This document is a committee report which outlines scientific objectives and specific rationales developed for the exploration of the Apollo 16 site, our most important mission to the lunar highlands. Although premission hypotheses erroneously favored a volcanic origin for the Cayley plains as well as Descartes Mountains, the major mission objectives were successfully accomplished: to delineate the nature and origin of two major physiographic units of the lunar central highlands. The units are composed of fragmental impact deposits; present debate focuses on specific source areas of the samples and depositional mechanism(s).

The premission exploration strategy was to use impact craters of various diameters and associated depths of penetration as stratigraphic probes. Specific sampling procedures, operational constraints, and suites of samples that were collected for specific local objectives, are described for lunar scientists who are unfamiliar with mission planning in general and more specifically with Apollo 16. A large number of samples taken to reconstruct local stratigraphic-structural relationships are either insufficiently studied or have not been studied at all. While such stratigraphic-structural relationships are extremely complex, or even chaotic, at Apollo 16, additional insight into local geologic processes could be obtained by future investigation of these special sample suites.

The report concludes with a summary of currently viable hypotheses concerning the origin of Cayley and Descartes materials. It is hoped that future studies of samples and other data may provide

[1]This is an invited committee report written by some members of the "Apollo 16 Traverse Planning Team." The respective roles of the above individuals for Apollo 16 were: Muehlberger: Principal Investigator, Field Geology Team; Hörz: Mission Science Crew Instructor; Sevier: Chairman, Traverse Planning Team; Ulrich: Co-Investigator and PI representative, Field Geology Team. Final editing was done by Muehlberger and Hörz. Contributions of P. Butler, J. W. Head, C. A. Hodges, R. O. Pepin and W. C. Phinney are gratefully acknowledged.

discrimination of these model interpretations, which principally focus on local versus distant source areas for the majority of the Apollo 16 sample suite.

INTRODUCTION

Because the Apollo 16 mission to the Descartes landing site in the central lunar highlands remains our only bona fide highland mission, interpretation of local and regional geologic processes at this site will play a central role in our general understanding of lunar crustal evolution.

This paper differs from most other contributions in this volume because it is an effort to look at the past by reexamining our original overall and local scientific objectives for the Apollo 16 landing site. Why did we go there? Why and how did we pick specific sampling stations? What was considered important field evidence in reconstructing a geologic, structural-stratigraphic framework for better understanding of local and regional geologic processes that may be recorded in the collected samples, photographs and verbal descriptions by astronauts J. Young and C. Duke? Much of the general lunar exploration philosophy and the specific plans for Apollo 16 evolved in formal and informal meetings and is contained in a vast body of gray literature or, unfortunately, is not documented at all. As a consequence the Lunar and Planetary Sample Team (LAPST) convened a subcommittee[1] consisting of individuals intimately familiar with—if not largely responsible for—most of the detailed Apollo 16 planning, in the hope that some of this information could be unearthed and disseminated for the benefit of lunar scientists, especially newcomers to the program. It was also hoped that revival of old mission rationales would result in an assessment of how individual samples and especially sample suites were investigated as compared with the intended objectives.

The Apollo missions sampled (usually only once) most of the major stratigraphic units recognized from telescope geology on the near side of the Moon (Fig. 1).

Apollo 11—Relatively old, dark mare basalt
Apollo 12—Relatively young, light mare basalt
Apollo 14—Ejecta from Imbrium Basin contained in the Fra Mauro formation
Apollo 15—Lunar crust from the rim of the Imbrium Basin and mare basalts
Apollo 16—Constructional units of the lunar highlands (Cayley, Descartes)
Apollo 17—Pre-Imbrian lunar crust from the rim of the Serenitatis Basin and pyroclastic materials

The Apollo 16 site was selected because it was possible to sample two major, morphologically distinct units of the central lunar highlands, covering about 11% of the entire lunar nearside: (a) the Cayley plains, representative of widespread

[1]The committee members were J. R. Sevier (Chairman), J. W. Head, F. Hörz, C. M. Meyer, J. J. Papike, W. R. Muehlberger, R. O. Pepin and G. E. Ulrich.

Fig. 1. Distribution of Cayley plains material on the lunar front side according to Wilhelms and McCauley (1971). The locations of the Apollo landing sites are also shown.

highlands plains units and thus considered the result of a significant, moonwide geologic process and (b) the Descartes mountains, a more localized physiographic unit.

PREMISSION ACTIVITIES

Regional and local geology

The principal geologic objectives of the Apollo 16 mission were the investigation of two major physiographic units, the "Cayley plains" and the "Descartes Mountains" (Fig. 2). On the basis of their topographic expression both units were interpreted as volcanic formations (Milton, 1968; Wilhelms and McCauley, 1971;

Fig. 2. Regional view of the Apollo 16 landing site. Note relatively subdued, flat morphology of plains materials in contrast to rugged appearance of the Descartes Mountains. The extremely bright ejecta aprons (North and South Ray craters), within reach of a Rover traverse, provided excellent potential for stratigraphic probing and for detailed investigation of relatively recent lunar regolith processes.

Milton and Hodges, 1972; Elston *et al.*, 1972; Trask and McCauley, 1972; Head and Goetz, 1972).

Due to their subdued, relatively flat topography and ponded nature in local topographic depressions, the Cayley plains were interpreted as volcanic deposits, rich in Si and Al because of their substantially higher albedo relative to mare basalts. They were interpreted either as genuine flows—supported by well documented embayment relationships with other terra features—or as pyroclastic deposits, as inferred from seemingly thin, veneer-like surface deposits mimicking the underlying topography in certain areas. Although Eggleton and Marshall (1962) had interpreted the plains circumferential to Imbrium sculpture as a "smooth" facies of Imbrium ejecta, this concept was abandoned because it seemingly could not account for the moonwide extent of the plains. Therefore, local to regional volcanic sources seemed to be required for their formation.

In contrast to the widespread Cayley plains, the topography of the Descartes Mountains is virtually unique on the moon. No other deposits of identical morphology have been recognized, although similar hilly and furrowed materials have been mapped in several places (Wilhelms and McCauley, 1971). The Descartes formation had to be at least 1000 m thick, based on morphology and topography. No genetic relationship to any local or regional cratering event was apparent. Thus the Descartes Mountains were interpreted as volcanic "constructs" with an origin analogous to terrestrial volcanic extrusions, perhaps along fissures.

The general stratigraphic relationships of both the Cayley plains and Descartes mountains were in general well established, although locally puzzling relationships of superposition occurred, leading to the view that the Apollo 16 plains and Descartes Mountains could have been emplaced almost contemporaneously, with the Descartes materials deposited last, and with volcanic activity possibly continuing into very recent (Copernican) lunar time. Crater density statistics on the Cayley plains were consistent with superposition relationships and placed their formation between the Imbrium basin event and extrusion of the basaltic mare fill. Lower crater densities were observed on Descartes, which in part supported the younger age relative to the plains, but it was recognized that the rugged topography rendered a "crater age" rather uncertain. Analyses of superposition relations led to locally contradictory deposition sequences. Milton and Hodges (1972) even described gradational transitions between Cayley and Descartes.

All of the above observations are still valid today, although the actual mission results mandate different interpretations. Why and where did the preferred premission interpretations go wrong? Of course, no precise and especially single answer can be given, but contributing factors of an historical and a philosophical nature may be identified for the benefit of future mission planning.

The "Descartes" site was always considered an attractive landing site to explore the central highlands. As early as June 1969, it was considered as a candidate for Apollo 19 by the site selection committee, and by October 1969 it was being considered for as early as Apollo 16. (Apollo 11 had successfully landed in July 1969 and NASA was concentrating on developing the capability to land very close to a pre-determined point for Apollo 12. This was extremely important in that specific sampling objectives could be precisely determined and planned in advance, rather than being "generalized" as they were on Apollo 11. Apollo 12 demonstrated this capability in December 1969 by landing within 100 m of the pre-planned location.) Subsequently, the cancellation of Apollo 18–20 and the failure of Apollo 13 to land at "Fra Mauro" contributed to a serious reassessment of site priorities. Nevertheless, Descartes remained a high priority site for an early mission, in spite of the fact that photographic coverage of the Descartes area was deemed insufficient to assure spacecraft- and astronaut safety; Apollo 14 was to take appropriate pictures, especially the lacking stereophotographs using a high resolution topographic camera, complemented by handheld Hasselblad cameras equipped with 80- and 500 mm lenses. The topographic camera failed; the Hasselblad photography was successful. Final acceptance of "Descartes" as the Apollo 16 site by the Apollo Site Selection Board occurred in June of 1971. Detailed, photogeologic study and other preparations were already well underway at this time.

The Apollo 11 and 12 landings had occurred in relatively simple geologic settings. Apollo 14 had returned complex impact products seemingly related to the formation of the Imbrium basin. Similarily complex breccias were expected from Apollo 15. It was apparent that petrogenetic processes in the lunar crust were severely masked by impact processes, if not modified beyond recognition. Therefore it was desirable to select for Apollo 16 a landing site on volcanic highland formations that appeared to be little modified by large scale impact processes and that had seemingly little structural complexity; ideally, a geologic setting analogous to the mare basalts was desired. The ponded Cayley plains constituted the only conceivable analogues.

In the context of the highlands volcanics hypothesis, the Apollo 16 site offered an additional inducement: the presence of a second physiographic unit, the Descartes mountains, thought to have formed by volcanic process different from the emplacement of the Cayley plains. It probably represented a different eruptive style and/or composition occurring somewhat later in highlands history. With advanced spacecraft technology and greatly augmented surface mobility provided by the lunar roving vehicle, site selection philosophies evolved from primarily single objective missions (Apollo 11–14) to dual and even multiple purpose missions for Apollo 15–17. It was thus possible on Apollo 16 to add exploration of the Descartes Mountain unit to the primary mission objectives.

As indicated above, photographic coverage for detailed site studies was fair at best; in particular no stereoscopic imagery was available. Scientific site selection criteria were based primarily on earth based telescope photography and observations, in part complemented by Ranger VIII photos (Milton, 1968). Specific mission preparations also included Orbiter IV imagery and the Apollo 14 Hasselblad photos (Milton and Hodges, 1972). Furthermore a multidisciplinary approach, employing other remote sensed information was not feasible at this time. Except for albedo maps, complementary data sets to assist and greatly refine photogeologic interpretations were not fully developed (e.g., Head and Goetz, 1972). The Apollo 16 experience is one of the strongest arguments for comprehensive site characterization using data sets from a wide variety of techniques in future mission planning (see also Hinners, 1972).

Philosophically, the uniformitarian principle is the basis to formulate physically sound hypotheses for geologic exploration. It leads to working hypotheses, against which actual findings may be tested. Ten years after Apollo 11 and in view of missions to other planetary bodies, notably Voyager's spectacular findings, we begin to realize how limited this terrestrial experience may be and how difficult extrapolation to other planetary bodies is. Although terrestrial geologic processes and preserved features are indeed numerous and varied, processes and preserved features that have no terrestrial analogue(s) may occur on many planetary objects. Premission interpretations of the Cayley plains were consistent with terrestrial volcanic hypotheses; terrestrial analogues to ponded impact deposits are not recognized. Although an impact origin of Cayley materials is now established, the actual mechanism of deposition is still the subject of lively debate.

Upon landing on the lunar surface, astronaut John Young remarked: "There you are, our mysterious and unknown Descartes highland plains; Apollo 16 is going to change your image." Little did he know how true his prediction was. Not only the image of the actual landing site, but indeed our entire knowledge and conception of lunar highland processes rests largely on the enormous scientific value of Apollo 16. We can explore and detail lunar highland processes, because the landing site was selected on a physiographic unit of widespread occurrence and thus of global significance, regardless of premission inferences as to specific formation mechanisms. The basic mission rationale was to collect and return data to be used in studying the nature and origin of the Cayley plains and Descartes mountains, objectives which were successfully accomplished, albeit by now recognized as extremely complex.

Surface exploration plans

The actual landing site at 15°30′47″E longitude and 8°59′34″S latitude was selected because it satisfied a crucial planning element: in case of Rover failure, the major mission objectives could be achieved by walking traverses. Furthermore, the landing site offered excellent opportunities for stratigraphic probing of the plains in the form of two exceedingly young impact features, North- and South-Ray Craters.

Our sampling procedures and traverse design were such that, by sampling craters of various sizes and in various locations, we would be able to reconstruct the stratigraphy of the Cayley, determine lateral variations within it and compare this to the Descartes. This strategy rests on impact cratering mechanics, i.e., systematics in the excavation and ejection sequence result in an overturned flap of inverted stratigraphy. The "Exploration Packages" shown in Fig. 3 evolved as groups of stations with specific focuses and aims.

EVA (Extravehicular Activity) 1:
The initial short traverse was within easy range of the Lunar Module (LM) to Flag (290 m diameter) and Spook (370 m diameter) craters, on whose rims we could sample Cayley excavated from depths of up to 60 m. From the samples we expected to be able to determine what lateral continuity there was in the Cayley (samples common to both Spook and Flag) and the characteristics of the deepest layer at Spook (samples unique to it alone). Further, rays from South Ray crater

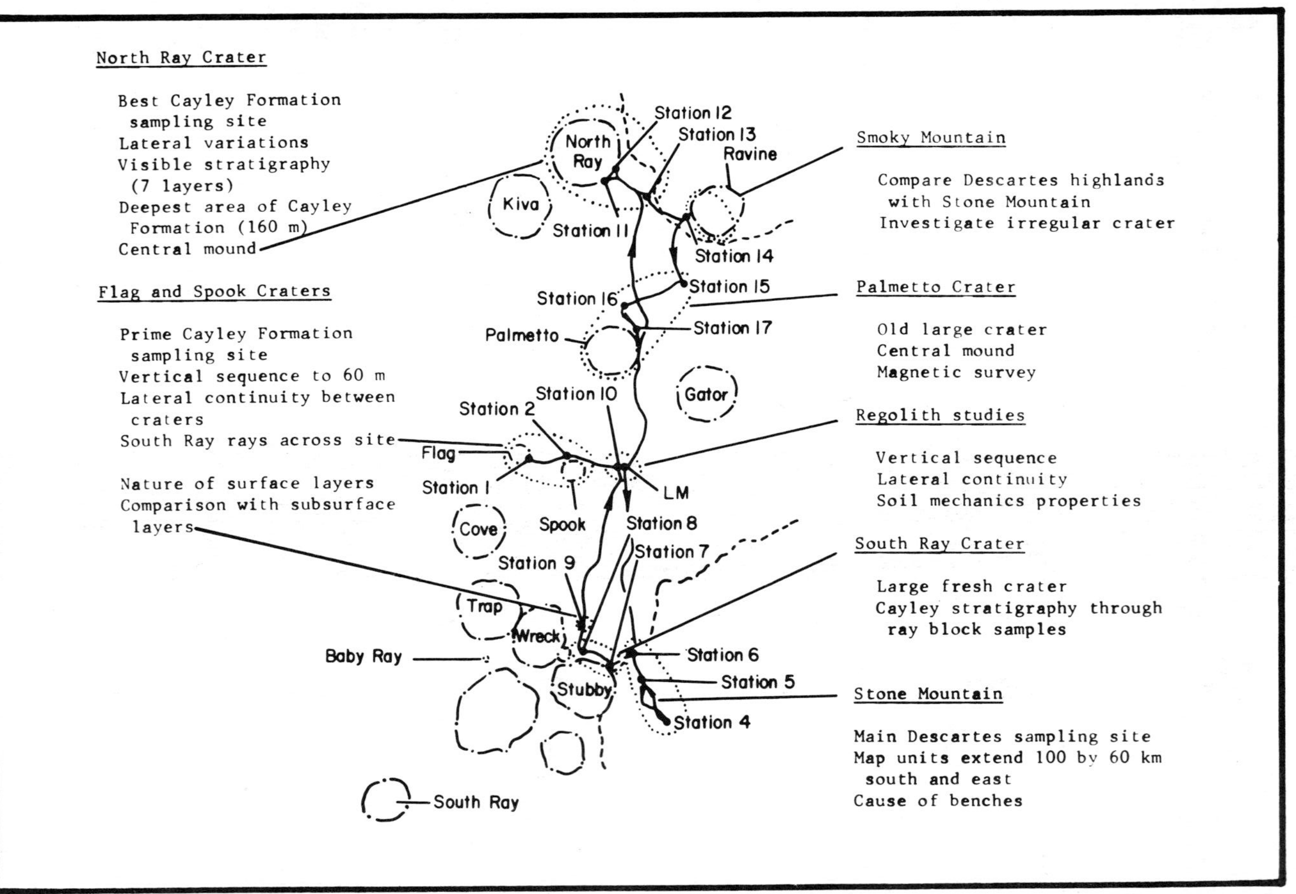

Fig. 3. Premission traverse design showing major objectives of individual EVA's and of specific stations.

traversed this area and thus we could obtain samples from this fresh, 570 m diameter crater, too remote to visit. Thus, following emplacement and activation of the Apollo Lunar Surface Experiments Package (ALSEP), a good and representative sample of Cayley materials could be obtained, the primary mission objective, during a single EVA with approximately 3 hours of geologic exploration. This objective could be met, if necessary, by walking, in case the Rover vehicle failed.

EVA 2:
The second traverse went due south to Stone Mountain, to ensure the earliest possible sampling of Descartes materials. Station 4 was high on Stone Mountain at Cinco a., a cluster of five craters. Station 5 was placed at an intermediate elevation and Station 6 at the base of the mountains to help determine the amount and/or rates of downhill transport. Thus EVA 2 concentrated largely on Descartes materials. The later parts of this traverse were intended to sample ray materials from South Ray crater (Station 7, 8) and to acquire a set of special soil and other samples from increasing regolith depths (Station 9) to study small scale lunar surface processes.

EVA 3:
This traverse was designed to explore North Ray Crater and to spend two (2!) hours sampling and studying this kilometer-size crater (Stations 11 and 12), essentially the same size as Meteor Crater, Arizona. This was an unprecedented amount of time to spend at a single place on the Moon ("How can it possibly take more than 30 minutes to sample, photograph and describe a crater?"). In premission, high-sun photographs with about 5 m resolution, North Ray crater appeared to have 7 different layers in the wall, a dark central mound, and a giant dark boulder on the southeast rim. Because of its diameter, the North Ray event should have penetrated to a depth of $\approx$ 160 m into the Cayley plains, possibly into the roots of the adjacent Smoky Mountain. Sampling along the rim, stereo sets of telephoto and polarized-light photographs, photographic panoramas of the crater interior, plus a sampling stop near the outer edge of the ejecta blanket (Station 13) were included to assure that samples from all depths within the crater were collected and photographically documented to relate these samples to the stratigraphy in the crater walls.

The return leg was designed to investigate an irregular crater at the foot of Smoky Mountain (Station 14) for comparison with Descartes material from EVA 2. Two additional stops (Stations 15 and 16) were scheduled on the rims of relatively small, fresh craters to better evaluate lateral variation of the Cayley plains over distances of a few km. Last, an old, subdued crater, Palmetto, approximately 1 km in diameter, was scheduled (Station 17) for comparison with the North Ray Crater findings.

Again some remarks concerning major elements of planning philosophy are given for historical as well as future reference. All scientific planning had to be compatible with numerous constraints imposed by astronaut safety, the fundamental limitation of space exploration. The single most limiting factor was "consumables" such as oxygen and coolants carried by the crew. These dictated not only total range of traverse away from the LM, but also the time available at any specific station. Although mobility was enormously increased by the lunar roving vehicle, the consumables had to be budgeted such that at any given time the astronauts could make it back to the LM on foot if the Rover failed; this "walk back" envelope defined a rigid time-distance framework for all activities. For example, it virtually mandated that it would be most prudent to visit (when consumables reserves were the maximum) the most distant stations first, without intervening stops for observations, etc. (thus the straight dashes to Stations 1, 4 and 11/12). Once at a given station, a maximum stay time could be defined and was rigorously applied. Unscheduled stops—at the crew's discretion—had unavoidable impact on subsequent activities and station times. Detailed check lists were attached to the astronauts' cuffs including 1) navigational information, 2) mandatory station activites related to Rover performance, status of "consumables," aiming of TV antenna etc., 3) scientific station tasks and sampling activities. These "cuff lists" allowed the crew to dispense the former two as efficiently as possible, thereby gaining time for geologic observations and sampling, sometimes restricted to as little as 5 or 10 minutes. Successful sampling, photographic documentation and detailed verbal descriptions attest to the enormous discipline of astronauts Young and Duke.

Another constraint, unusual for terrestrial geologic field work, was the fact that only a limited number of specimens could be sampled, no more than ≈ 80 kg in total mass. Within this limit, the most representative sample suite had to be selected, indeed a formidable task by any standard, additionally complicated by the lack of any iterative option(s) such as going back to a station. This weight limitation is in part reflected by the number of short stops at intermediate ranges to at least ensure representative geographic distribution of the samples; stay times at the most distant, major stations (1, 4 and 11/12), were maximized to accomplish the most detailed and purposeful sampling.

Thus, based on limitations for mobility, time and weight, even for a nominal mission, the above EVA plans and procedures appeared to be the most promising to fulfill the principal mission objectives, and hopefully to obtain a reasonable cross-section of the Cayley plains and their relationship to the Descartes formation.

THE MISSION

The LM landed at almost precisely the preplanned point (Fig. 4). Thus, initially no modifications to the traverses seemed necessary. The accidental breaking of the cable to the heat flow experiment during the early part of EVA 1 caused implementation of contingency plans to the successive EVAs. EVA 2 was conducted under a revised schedule including abbreviated station times and deletion of low priority tasks. Early return to the LM area was necessary to possibly attempt repair of the heat flow cable. The decision to abandon such repair efforts was made late in EVA 2 and therefore additional sampling stations near the LM

Fig. 4. Apollo 16 landing site and actual traverses accomplished. Note the major modifications of EVA 3.

resulted. As a consequence, the LM area of Apollo 16 is more intensively sampled than any other area of the Moon!

A major change in the preplanned EVA 3 traverse design resulted from the decision to lift off at the preplanned time even though the lunar landing had been delayed for 6 hours because of control problems with the command module which might have affected the trans-earth injection burn. EVA 3 was reduced from 7 to 5 hours, which permitted only stops at North Ray Crater in the final traverse. To keep the maximum time available for sampling, some of the preplanned photographic tasks were deleted at these stations.

POST MISSION EVALUATION

Geologic setting of samples

By the end of EVA 1 it became obvious that most of the samples described and collected were impact breccias, rather than volcanic rocks. This pattern prevailed in EVAs 2 and 3 as well as in detailed work on the returned samples. Thus the premission volcanic hypothesis is now replaced by impact related hypotheses for the formation of both the Cayley plains as well as the Descartes Mountains. Details of depositional processes and history are still unsolved and various interpretations seem viable. Also, surprisingly, Descartes and Cayley appear remarkably similar, if not identical, in lithological character. Some specific Apollo 16 features are discussed briefly below; a much more detailed account is given by AFGIT (in press).

Four distinct geologic areas with associated sample materials can be established: (1) the central area on the Cayley plains, (2) Stone Mountain, (3) North Ray Crater and (4) South Ray Crater ejecta (see Figs. 3 and 4).

1. The central Cayley area

The smoothest area on the plains where the LM landed was not as level as had been expected (Schaber, AFGIT, 1980). Indeed it is a highly irregular, densely cratered terrain with several tens of meters of surface relief. The regolith is 10–15 m thick as inferred from a bench inside Buster Crater at Station 2 and a small unsampled, very blocky crater some 600 m south of the LM which apparently penetrated into very competent strata. The rocks collected include impact breccias of variable colors, clast contents and matrix characteristics; genuine (impact) melt rocks are rare. In general, Schaber (AFGIT, in press) concludes that the majority of the ''central area'' Cayley samples are from relatively shallow depth generally less than 70 m deep.

2. Stone Mountain (Stations 4, 5, and 6)

The samples collected highest on Stone Mountain were on the hummocky ejecta, not far below the rim, of Cinco a. crater at Station 4 (Sanchez, AFGIT, in press). This was the primary location for sampling the Descartes highlands. Although high-sun orbital photographs revealed no bright ray material in this area, two rocks returned have exposure ages of 2 m.y., the generally accepted age of formation for South Ray Crater, and the crew photographed a field of fragments that were interpreted to be South Ray ejecta. The best candidate for Descartes material may be the double core tube (64002/64001), which is as yet unopened. The influence of South Ray Crater needs to be determined so that the Descartes samples may be identified. At Station 5, lower on Stone Mountain, an area in a small crater, shielded from South Ray Crater ejecta, was sampled in an attempt to avoid the South Ray contamination recognized at Station 4. Samples 65015 (metaclastic crystalline rock) and 65315 (light-matrix breccia) may be good examples to test for Descartes highlands characteristics at this locality.

3. North Ray Crater (Stations 11 and 13)
This km-wide, 230 m-deep crater appears to lie on the boundary between Smoky Mountain and the plains. The 50 m.y. old impact may have penetrated a hill of mass-wasted Smoky Mountain material overlying the edge of the plains or possibly a downfaulted block of mountain-derived material (Ulrich, AFGIT, in press). The dominant rock type on the crater rim and in its walls is a very friable, melt-poor, light-matrix breccia. It is a coarsely cataclastic to finely brecciated anorthosite and produces some of the oldest ages (ca. 4.26 b.y.) at the site. Some materials have been identified as ferroan anorthosites (Dowty *et al.*, 1974). A minor part of the boulder population at Station 11 and the only boulder present at Station 13, 0.75 km to the southeast, are dark-matrix breccias which are richer in glass and obviously reached higher temperatures than the melt poor rocks. Dark matrix boulders are generally angular and unfilleted. Such blocks are believed to be derived from the floor of North Ray Crater and may represent a resistant stratum or irregular pod of the material that is more common in the plains samples. These rocks were not molten at the time of their excavation although some glass-filled fractures within the breccias (such as 67915) could have formed at that time.

4. South Ray Crater (Stations 8 and 9)
Station 8 was selected within a blocky ray area to sample ejecta approximately 4 km from its source at South Ray Crater (Reed, AFGIT, in press). Several exposure ages of approximately 2 m.y. (e.g., Arvidson *et al.*, 1975) on samples of meter-size boulders appear to confirm the presence of South Ray ejecta. Interestingly, the soil particles in this area do not record similar exposure ages (McKay and Heiken, 1973). Thus the ray patterns visible on orbital photographs appear to be caused by larger rock fragments, not by ejected soils. Boulder chips which should represent Cayley material ejected from South Ray Crater include 68115 and 68815 (dark-matrix breccias) and 68415 and 68416 (fine-grained holocrystalline rocks from two corners of a meter-size boulder which appears to have a fairly homogeneous matrix throughout). Station 9 was selected to sample an inter-ray area and the single core tube, 69001, will provide a good comparison with the double core tube at Station 8 to examine differences in plains regolith between nonray- and ray-covered areas from South Ray Crater.

South Ray Crater should be an important reference site for characterizing the Cayley unit. Samples from Stations 2, 4, 8, 9 and the LM area (LM, 10, 10a) offer strong possibilities of recovering South Ray ejecta material. Angular fragments, absence of fillets, and 2 m.y. exposure ages are obvious clues to recognizing such samples. It is possible that a few samples from Baby Ray Crater, still younger than South Ray, are in the collection.

Despite these four geographically and morphologically distinct settings for the Apollo 16 sample collection and despite distinctly different small scale geologic environments based on surface characteristics and proximity to various size impact craters, the samples recovered appear grossly similar in lithologic type(s) and relative abundance, even within individual sampling stations where statistically significant samples could be collected (Wilshire *et al.*, 1973; Wilshire *et al.*, AFGIT, in press). Despite an enormous textural variety of rocks ranging from igneous to metamorphic, and especially including a large variety of breccia types, there is no obvious sample provenance relating to structural and stratigraphic implications (Fig. 5). Their chemistries are also grossly similar. The evidence for processing by meteorite impact is overwhelming. This implies that the samples were dislodged from their original source area. Because no distinct sample provenance appears to exist, the various lithologies appear to have resided in close juxtaposition beneath the Apollo 16 landing site prior to being excavated by local

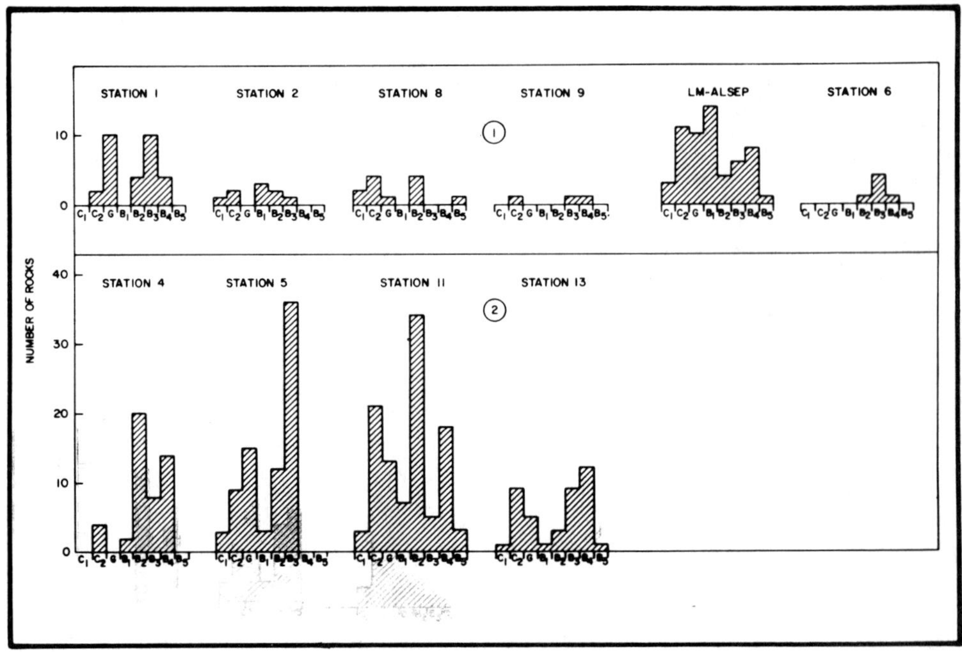

Fig. 5. Relative frequencies of various lithologies encountered at each station according to handspecimen description and macroscopic-textural criteria of 468 samples (> 1 g in mass) by Wilshire *et al.* (AFGIT, in press). Note that B_1–B_2 and B_4–B_5 are closely related pairs; their discrimination here is by clast color and is not necessarily a genetic difference. Rocks in group 1 are from "Cayley" plains stations, those of group 2 from the Descartes Mountains (stations 4 and 5) and North Ray Crater (stations 11 and 13). C_1: igneous rocks; C_2: metaclastic rocks; G = glasses; B_1: light matrix-light clast breccias; B_2: light matrix-dark clast breccias; B_3: intermediate color breccias with light and dark clasts; B_4: dark matrix-light clast breccias; B_5: dark matrix-dark clast breccias (Wilshire *et al.*, 1973).

impacts. Complex cratering histories and the close spatial association of varied lithologies make reliable reconstruction of a local stratigraphy extremely difficult.

Specific sample suites

In the following, an attempt is made to recapitulate some sampling procedures and to emphasize the scientific rationale for specific samples and especially sample suites. Ultimately such information together with sample data must be combined in efforts to reconstruct the complex, possibly even chaotic, stratigraphic-structural relationships at the Apollo 16 site. On occasion these complexities appear beyond challenge and thus a tendency exists to describe individual case histories of specific samples. If the case histories of specific sample *suites,* taken from specific geologic settings were established, additional stratigraphic information might emerge. The purpose of this section is to call attention to the existence of an original sampling rationale for such suites.[1]

Rake samples

Considering time limitations as well as restricted overall sample weight, rake samples were considered best for collecting the widest variety of rocks most representative of an individual station. Furthermore there exists the possibility that some "exotic" material from more distant locations and perhaps even remote source areas outside the landing site will be collected. A large soil sample (500–1000 g) was considered an integral part of the rake sample procedure, so that a statistically significant number of "rocks" of various sizes may be compared with each other (i.e., "walnuts," 1–10 mm samples and <1 mm fines, etc.). Thus rake and (large) soil samples must be considered complementary and should be studied concurrently.

Rake samples were scheduled at Stations 1, 4, 8, 11/12, 14, and 17. The crew was encouraged to collect additional rake samples at their discretion because the rake procedure was considered the best sampling approach for any area if there was insufficient time for documented collection of hand specimens.

While each preplanned rake sample had a specific purpose, additional objectives emerge when these samples are viewed in the context of the overall landing site:

1. Lateral Continuity of Cayley

a) Stations 8, LM, 17 and 11/12 represent a South to North profile with sample stations about 2–3 km apart.

---

[1]For clarity in this section, the premission rationale for each topic is described first; actual mission activities and/or results from subsequent sample analyses follow. Details concerning scientific objectives, sampling procedures, sampling locales, lunar surface documentation, etc. are presented by Hörz *et al.* (1972). A detailed listing of all samples is presented in Appendix II, documenting the extent of sample allocations and thereby an approximate measure of analytical work performed as of early 1980. Based on this compilation, *samples judged by the authors to be insufficiently studied are annotated with an asterisk in the body of this report* .

b) Stations 14, 1, 11/12 and 8 were the most easterly, westerly, northerly and southerly sampling stations on Cayley, thus defining the largest areal spread on this unit.

c) Station 17 was in the middle of a sampling triangle formed by Stations 1, 11/12 and 14, giving an additional tie point for lateral correlations of the Cayley plains.

2. Vertical Extent of Cayley

According to systematic depth/diameter relationships of impact craters, the depths of excavation are predictable for a variety of the landing site craters. The samples recovered from Station 1 (Flag), 8 (South Ray), 11/12 (North Ray), 17 (Palmetto) and 14 (Ravine) should contain progressively more deep-seated materials, as they were located on the rims of these craters, specifically next to small, subsequent cratering events (e.g., Plum, End and Cat Craters) that punched into the overturned flaps of the larger craters. Station 8 was expected to provide samples from shallow Cayley horizons in the form of South Ray Crater ejecta. The rake sampling program was designed to yield a depth profile of approximately 150–200 m, thus permitting stratigraphic interpretation.

3. Crater Ages

Systematic determination of exposure ages on a statistically representative suite of walnut samples will, hopefully, yield formation ages of the main and/or subsidiary craters. Four to six craters could be sampled and their respective formation ages would help constrain the absolute meteorite flux. Crater production and degradation models could be based on absolute chronology, at least for post Copernican timescales.

A total of 14 rake samples were actually retrieved from Stations 1, 4, 5, 8, 11, 13 and the LM/ALSEP area; many of them were improvised at the crew's discretion, because (a) they were under considerable time pressure and (b) they could not discern distinct stratigraphic/lithologic units and were forced to retrieve a "best representative" sample suite via the rake procedure. All preplanned stations were sampled; shortening of EVA 3 forced deletion of important rake sample locations (Stations 14 and 17).

The following rake samples were collected and their premission rationale is still in large part valid; "specimen" refers to samples >1 g in mass.

Station 1: Most westerly Cayley; prime Cayley site

"walnuts:"	61515-61577	(33 specimens)*
"soils:"	61500-61510	(1 specimen)

Station 4: Prime Descartes site

a) "walnuts:"	64535-64589	(30 specimens)*
"soils:"	64500-64525	(11 specimens)
b) "walnuts:"	64810-64837	(13 specimens)*
"soils:"	64800-64804	—

Station 5: At "contact" of Cayley/Descartes (crew struggled in vain to find contact)

a) "walnuts:" 65310-65366 (22 specimens)*
"soils:" no corresponding soil

b) "walnuts:" 65510-65588 (39 specimens)*
"soils:" 65500-65504 —

c) "walnuts:" 65715-65795 (41 specimens)*
"soils:" 65700-65704 —

d) "walnuts:" 65920-65927 (3 specimens)*
"soils:" 65900-65916 (7 specimens)

Station 8: Most southerly Cayley

"walnuts:" 68515-68537 (13 specimens)*
"soils:" 68500-68510 (1 specimen)

Station 11: Most northerly Cayley; North Ray stratigraphy

a) "walnuts:" 67515-67576 (32 specimens)*
"soils:" 67510-67514 — *

b) "walnuts:" 67615-67697 (38 specimens)*
"soils:" 67600-67610 (1 specimen)

c) "walnuts:" 67710-67776 (32 specimens)*
"soils:" 67700-67708 (4 specimens)

Station 13: Additional North Ray ejecta(?) and "normal" Cayley surface

"walnuts:" 63515-63598 (39 specimens)*
"soils:" 63500-63515 (6 specimens)

LM/ALSEP: "Typical" Cayley

a) "walnuts:" 60515-60536 (11 specimens)*
"soils:" 60500-60510 —

b) "walnuts:" 60615-60679 (35 specimens)*
"soils:" 60600-60604 —

The best "Descartes" sampling locations were Stations 4 and 5, especially 5. According to the Apollo 16 Lunar Sample Information Catalog, a total of only 4 documented samples >50 g are available from Station 4 and 7 "handspecimens" from Station 5, as compared with 54 and 112, respectively, "walnut" specimens > 1 g, contained in the rake samples. These rake samples must play a crucial role in determining the nature of Descartes as well as Cayley.

The premission rationale for individual stations in the Cayley Formation remains valid, although unraveling the local stratigraphy appears to be significantly more difficult than anticipated. The extent of lateral and vertical continuity of

Cayley could in principle be attacked via the rake samples in accord with the premission arguments and rationale. Study of the "walnut" samples could furthermore be augmented by investigation of the numerous 4–10 mm chips in the associated large soil samples, if not by the < 4 mm fractions of the soils themselves.

Core samples

Acquisition of regolith materials by means of single and double drive tubes and the deep drill core was considered paramount for the understanding of small scale regolith and other surface processes, particularly in a preserved microstratigraphic context. Furthermore, lithologic contents of the cores also reflect local geologic environments and sample provenance. Originally double drive tubes were planned at Stations 4, 8, 10 and 14; a single drive tube was to be taken at Station 9 and immediately placed into a special vacuum container (CSVS = Core Sample Vacuum Container) to preserve its pristine character. The deep drill location was in the ALSEP vicinity, as it also served for emplacement of the heat flow experiment probes.

The basic philosophy which the crew followed was to select within a specific station a location that was free of any obvious, recent, small scale (m and larger) crater and to stay away from large boulders. The reasons for these specific selection criteria were 1) to ensure that a "typical" regolith was recovered that represented a significant, undisturbed time history of regolith evolution, and 2) to minimize the risk of encountering buried rocks and blocks.

The following cores were actually taken:

Station 4: as planned; double drive
Station 8: as planned; double drive
Station 9: as planned; single drive
LM/ALSEP/10: 2 double drive tubes plus deep drill at the corners of a nearly equilateral triangle, ≈ 50 m on a side.

Station 4: cores 64001/64002*

The prime objective was to obtain pure "Descartes" regolith. Secondary objectives were related to small scale mass wasting processes on the slopes of Stone Mountain. Station 4 was preferred over 5 because of increased probability that materials from topographically and stratigraphically higher portions of the Stone Mountain massif would be encountered. In addition, core location and retrieval activities were to be tied to penetrometer measurements to address potential small and large scale, lateral and vertical continuity of the core profile. The core was to be taken in the middle of a diamond pattern of penetrometer readings.

This core still represents probably the best "Descartes" material; however, there may be some "contamination" from South Ray ejecta scattered across Station 4. Materials recovered from the soil mechanics trench (64420) should be considered complementary to the core. This core (64001/64002)* is still unopened as of early 1980, despite its great potential for characterizing the Descartes mountains unit.

Station 8: Cores 68001/68002*

The major core objective at Station 8 was to retrieve fine-grained South Ray Crater ejecta, expected to be extremely young and immature and thus significant in studying small scale regolith processes. Station 8 was as close as possible to South Ray and was scheduled on a prominent high albedo ray of this extremely youthful-appearing crater. In addition to ray material, it was also deemed necessary to retrieve the "older" regolith that formed the substrate on which the South Ray ejecta were deposited. Ideally, the contact of "young" South Ray ejecta and "old" regolith surface could be retrieved in one double drive tube. If, however, the South Ray deposits were judged in the field to be too thick, then single drive tubes from locations on and off the ray deposits were scheduled. Retrieval of the young deposits was unquestionably the prime objective and highest priority at Station 8; collection of old substrate regolith was desirable. The planned activities at Station 9 positively ensured collection of the latter materials.

The crew could not detect any albedo and/or color changes indicative of fine grained ray deposits. They became, however, increasingly more confident in associating boulder trains and clusters with South Ray ejecta. As a consequence Station 8 was established in one of these boulder areas. The crew was confident that the boulders were indeed from South Ray and that Station 8 was properly located.

The specific drive tube location was beyond the boulder area and 1–2 m from the edge of a 10–15 m diameter crater. Thus the core stratigraphy may include materials from this local impact event, but a significant contribution from boulder erosion products is not expected. This core is still unopened as of early 1980. However, detailed study of local soils from Station 8 (and others) failed to demonstrate unequivocally the addition of fine-grained "young" South Ray material to local soils (e.g., McKay and Heiken, 1973).

Station 9: Core 69001* (CSVC)

This station was designed to accommodate a variety of "special samples," such as a "turnover" boulder and soil samples from specific, shallow depths. The overall station criterion was to look for "typical" Cayley devoid of any South Ray ejecta (other than perhaps the "turnover boulder" and here it was recommended that the crew look for a highly rounded boulder in contrast to the angular South Ray fragments).

The actual stop satisfied the above criterion in that no obvious, fine-grained South Ray material of high albedo was present. Station 9 appears to be a typical "undisturbed" Cayley regolith. The specific core location was smooth and free of large rocks (> 5 cm). This core (69001)* was placed in the CSVC, where it still remains (early 1980). The core must be treated as part of a very detailed regolith sampling effort, which includes samples 69003* ("beta cloth"), 69004 ("velvet cloth"), 69920 ("skim"-sample), 69940 and 69960 ("scoop"-samples), i.e., samples taken from successively greater regolith depth, as described below.

LM/ALSEP; Cores 60013/60014*; Cores 60009/60010; deep drill cores 60001 through 60007

The overall objective of the LM/ALSEP core materials was to characterize Cayley regolith. The location of the deep drill was entirely controlled by the location of the heat flow experiment (HFE). The geologic rationale was to sample a deep section of typical Cayley regolith.

Actual location was about 175 m SW of LM and 25 m S of ALSEP site. The area is generally flat, although numerous 2–6 m diameter craters are visible.

A drive tube close to the LM was required to support the penetrometer experiment (60013/60014). Small and large scale vertical and horizontal variation between two Cayley profiles (these drive tubes and the deep drill) in close proximity appeared interesting for investigating regolith evolution.

Sampling location for cores 60013/60014 was 120 m WSW of LM. Compared to the other drive tube locations, this site is relatively devoid of boulders and rocks, and has a remarkably smooth appearance. No obvious craters are visible that could have affected the regolith stratigraphy in relatively recent history.

Drive tubes 60009/60010 were originally to be taken at Station 14; however, this station had to be abandoned and thus the tubes were used in the LM area at a location which was in some meaningful geometrical relation to drive tubes 60013/14 and the deep drill, already taken. The actual sampling site was ≈ 100 m SW of the LM. The area was devoid of any noticeable North- or South Ray Crater ejecta.

The two drive tubes together with the deep drill stems taken in the LM/ALSEP area represent a unique spatial concentration of core materials. As a group, these cores should permit small and large scale stratigraphic correlations of the LM/ALSEP regolith profile. As can be seen from the LM/ALSEP station map (AFGIT, in press) the core locations delineate a triangle of 50–60 m side length; it is possible to look for lateral regolith continuity along a N–S cross section. Further, the penetrometer measurement locations were purposely designed to complement the core correlations over small (1–2 m) and larger distances (> 10 m).

Radial sampling of small craters

Time permitting, the crew was encouraged to proceed with a "radial sampling" procedure at small crater structures. Station 17 was specifically assigned for this task and planned activities at Stations 11/12 and 13 represented short cut versions of radial sampling. The objective was to collect a statistically significant number of samples from the ejecta blankets of these craters. Applying the principle of inverted target stratigraphy in these deposits (Shoemaker, 1963; Gault *et al.*, 1968), an attempt to reconstruct the original target stratigraphy could be made, especially for the upper few 10's of m in the Cayley plains. Deeper stratigraphic samples were expected from the larger craters such as Flag (Station 1), Spook (Station 2) and North Ray (11–12), structures too large to attempt comprehensive radial sampling.

Station 17 had to be abandoned because of the shortened EVA 3. Poorly defined radial samples, however, may be contained in Station 1, 2, 3 and the combination of Station 11/13 materials regarding the stratigraphic content of the Flag, Buster and North Ray crater targets, respectively. Some stratigraphic information must also be present in the ejecta boulders from South Ray, characterized by angular blocks in the surface photographs and very young ($\approx$ 2 m.y.) exposure ages (Arvidson *et al.*, 1975).

Formation ages of specific craters

Ejecta from a variety of photogeologically "young" craters could be collected with ease. Exposure age determinations on such ejecta may yield the time of crater formation. These ages, together with the crater's overall state of degradation, would provide important information on the magnitude of the meteoroid flux during recent lunar history. Large ($>$ 1 m) boulders appeared to be the best collection targets, because of the increased probability of relatively simple surface residence.

Many ejecta fragments—both large and small—were collected from a variety of craters and the following summary emerges: exposure ages of South Ray and North Ray crater ejecta tend to cluster around 2 m.y. and 50 m.y., respectively, thus indicating the most likely times of formation for these two craters. A comprehensive summary of Apollo 16 exposure ages is presented by Arvidson *et al.* (1975).

"Buster" is a crater which is relatively well sampled. Six samples were collected (62235, 62236, 62237, 62255, 62275 and 62295) of which three were dated by the Kr method (62235: 160 m.y.; 62295: 235 m.y.; 62255: 1.7 m.y.; see Arvidson *et al.*, 1975). Additional studies on the remaining samples may reveal that the older ages are related to the formation of Buster Crater, perhaps even Spook Crater. Inasmuch as Spook Crater is a degraded, partially filled structure, a possible date of formation would have far-reaching consequences for reconstructing temporal variability of the lunar meteoroid flux and associated implications for the crater degradation model of Soderblom (1970), now widely applied via the D_L parameter (e.g., Boyce, 1976) and other photogeologic dating techniques.

Special regolith samples

The interaction of the space environment with the lunar surface is the subject of a variety of investigations. Such investigations are concerned with the particle track record of solar and galactic radiation, the chemical characteristics of the solar wind, radioactive isotopes produced by solar flares and cosmic radiation, noble gases, micrometeoroids, and various sputtering phenomena. Although many of these studies are actually concerned with studying the sun, quantitative understanding of these processes contributes substantially to our knowledge about small scale lunar surface processes such as the erosion of rocks, the rate of lateral and vertical regolith mixing, the obliteration of impact craters, the migration of volatile and semivolatile materials and the influx of meteoritic materials. In addition, careful documentation of the lunar surface will also have important applications for remote sensing of compositional data.

As a consequence special sampling procedures, including specialized tools and/or containers, were designed for the Apollo 16 mission. These procedures as well as scientific objectives were detailed by Hörz *et al.* (1972). The prime objectives of the procedures were to return pristine soil and rock samples for specialized studies under well documented conditions.

Pristine regolith materials

Two special Contact Soil Sampling Devices (CSSD) were fabricated to collect the uppermost regolith layer at Station 9. The CSSD is an aluminum box into which a collector plate is inserted. The collector plates were draped with nylon fabrics; one fabric had properties similar to Beta cloth and the draped plate was floating freely in the container. Therefore, no load other than its own weight was applied during sampling. Based on premission tests this CSSD collected (ideally) a layer of regolith approximately 100 μm thick. The fabric on the second CSSD consisted of a velvet-like nylon fabric, and the plate was springloaded. A regolith layer approximately 0.5 mm thick was expected to stick to the velvet cloth. Samples 69003* (Beta cloth) and 69004 (velvet) were collected at Station 9, approximately 2.5 km from the landing point (LM).

Adjacent to the two CSSD's, two other regolith samples were taken, to incorporate successively deeper soil materials. The skim sample (69920) was retrieved by carefully skimming across the regolith surface with the regular scoop. It is estimated that the maximum penetration was 5 mm. The fourth sample was a standard scoop sample (69940) with an estimated penetration of 3 cm. A drive tube (69001) taken in the vicinity provides material from an even greater depth (estimated penetration: 27 cm).

After collection of these materials, a soil sample (69960) was taken from underneath a boulder that was overturned by the crew. The soil was shielded from the space environment since the time the boulder was deposited; therefore it is valuable material for comparison between exposed and shielded soils.

Sample 69004, the velvet draped sampling device, was opened in 1978; none of the other samples has been distributed and investigated in the context of the premission sampling rationale and objectives. Considerable crew time was invested during training, special hardware was built and an entire station was designed around this comprehensive soil sampling routine, because it was deemed highly important to have soils from shallow, carefully controlled depth(s). Perhaps the efforts on the velvet sampler will kindle new interest.

Pristine rock surfaces

To avoid abrasion and other factors resulting in possible degradation of rock surfaces during transit, two ''padded bags'' were used for samples 67215* and 67235* taken at Station 11.

When the samples arrived in the LRL, it was obvious that the two bags were well protected by soil-filled documented sample bags within the larger sample collection pouch (SCB). Sample 67215 seems to be a relatively friable white matrix breccia; it appears that sample 67235 is a more competent, although highly fractured rock. Both samples appear unsuitable for their intended purpose.

Fillet samples

The combination of a variety of special investigations (particle tracks, gamma ray spectroscopy of short-lived isotopes, noble gases, micrometeorite craters) may eventually lead to a quantitative understanding of lunar erosion processes and such studies require samples of special surface geometries or histories, if not both. Such samples may be soil "fillets" banked against a boulder, especially when compared with a rock chip taken from the top of the same boulder. Such fillet samples may also yield information about the transportation mechanisms and rates of regolith materials. Fillet sampling procedures were carried out at Stations 1, 8, and 11.

Station 1: A boulder on the south rim of Plum Crater possessed a well-developed soil fillet (61280). The associated boulder chip (61295) is a moderately friable breccia. The rake soil (61500) may be a suitable reference soil, although other soil samples are also available (61240, 61220).

Station 8: This station was located in a boulder-strewn field related to the South Ray cratering event. Boulder A consisted of a highly recrystallized breccia (68115). Hardly any fillet had developed around the boulder, although a sample was taken (68120); thus, the erosion of "hard" rocks could be studied. A suitable reference soil is 68500.

Station 11: A group of white boulders attracted the attention of the crew. Although a fillet sample was not necessarily planned, the variety of materials collected (67455, 67475, 67460) satisfies the requirements for such a sample set. The white boulders are friable breccias that are probably very easily eroded. The actual fillet soil is 67460; the two breccia chips are 67455 and 67475; a suitable reference soil could be 67480.

Although some of the above samples were studied extensively (e.g., chip 67455), none of the above sample suites was pursued with the specific "fillet" objective in mind.

Permanently shadowed soil

Migration and redistribution of volatile and semivolatile elements on the lunar surface were discovered in a variety of investigations on regolith materials. Neither the activation mechanism nor the mode of migration is well established. By sampling soil materials from permanently shadowed areas, it was hoped that a concentration of these elements would be found, because the shadowed areas may act as cold traps. Therefore, the soil from the surface of a permanently shadowed area should be especially enriched in volatiles and semivolatiles, and a soil sample from a greater depth at the same location may serve as a reference sample.

A hole more than 1 m deep and approximately 50 cm wide was observed at the south end of Shadow Rock at Station 13. Soil materials (63320, 63340) were taken in the manner described above. Because of the shadowed condition, no precise surface photographs are available, but the comments of the crew about the geometry of the "gopher hole" suggest that the soil materials were well shielded from the sun. After the samples were received in the LRL, samples 63320 and 63340 were placed in specially sealed containers. The closest soil sample (63500) from normal, exposed regolith was taken approximately 15m southeast of Shadow Rock in connection with a rake sample.

These samples were studied as intended in the original mission objectives. The most extensive account is by Eberhardt *et al.* (1976). According to their data the shadowed soils are enriched in volatiles, e.g., 63321 and 63341 are enriched by 38% and 28%, respectively, in $^{40}Ar_{tr}$ relative to the control soil 63501. Thus volatiles do concentrate in "cold traps."

East-west split

Various rare gases and other volatiles make up what can be called a lunar atmosphere. Most interest has focused on argon. Most of these gases are present in an ionized state and may therefore be accelerated and redistributed by the solar wind electric field along trajectories that trend approximately north-south. At the completion of the trajectory, the gases are implanted into the regolith materials. The angle of incidence seems to be rather small. If soil materials are taken between two boulders, the sides of which strike in an east-west direction, the interboulder area should be devoid of such reimplanted gas species. Comparison with an unshielded soil sample collected in the vicinity should aid in understanding the implanation of rare gases by this ion pumping mechanism from the atmosphere.

Such a set of soil samples (67940, 67960) was taken at Station 11 between House Rock and Outhouse Rock. These soils were studied in detail along their original objectives. The most extensive account is by Eberhardt *et al.* (1976). They state that there is no compelling evidence that the shielded soil was deficient in reaccelerated $^{40}Ar_{tr}$, although good geometrical shielding conditions existed. The experiment is, however, inconclusive because of the distinct possibility that a previously exposed soil component containing trapped atmospheric Ar was transported laterally into the E-W split area.

POST MISSION GEOLOGY OF THE LANDING SITE

Cataclastic, fragmental, melt-bearing and commonly highly polymict breccias, together with genuine impact-melt rocks are overwhelming evidence that both the Cayley plains as well as the Descartes Mountains are made up of (fragmental) impact deposits. Furthermore, present evidence suggests that Cayley and Descartes materials may not be distinctive lithologically; however, the Descartes samples may not have been exhaustively studied.

Most of the information to follow is extracted from AFGIT (in press) which contains numerous post mission details including sample descriptions, their geologic setting, distribution of North- and South-Ray ejecta, crew descriptions and analysis of surface photographs, station maps, topography and attempts to reconstruct local cross sections and stratigraphy, including critiques on other stra-

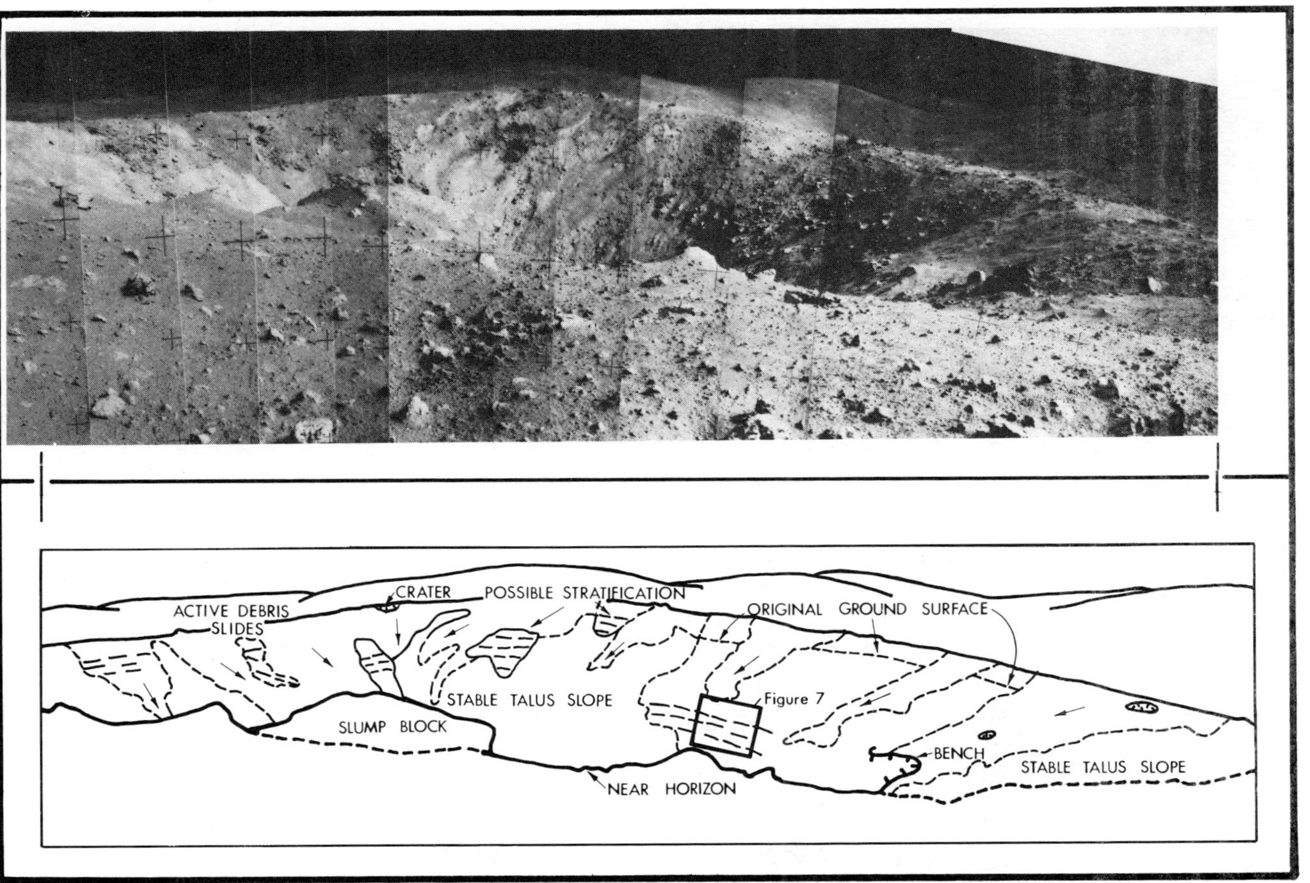

Fig. 6. Panoramic view of North Ray Crater, approximately 1 km in diameter.

Fig. 7. Telephotograph of large light-matrix breccia blocks on northeast wall of North Ray Crater (intentionally underexposed to enhance textures in shadows).

tigraphic models. Only a short synopsis of the latter efforts is presented here to assist formulation of diagnostic tests for various hypotheses during future sample studies.

North Ray Crater

The sample suites collected along the south rim of North Ray Crater at Station 11 and 750 m southeast of the crater rim at Station 13 should represent the range of compositional variation present in the entire crater wall. This crater is 1000 m across and 230 m deep and penetrates a low ridge on the west flank of 400 m high Smoky Mountain, 4.5 km north of the landing point (Fig. 6).

Prior to the mission, North Ray Crater was deemed the best potential source for samples of the Cayley Plains to a depth of nearly 200 m. Photographs obtained during the mission support the idea that some materials from the adjoining Des-

---

Fig. 8. Abundance of rock types inferred from analysis of surface photographs (8a) and actually collected (8b) on the rim and ejecta blanket of North Ray Crater (AFGIT, in press); rock classification after Wilshire *et al.* (1973).

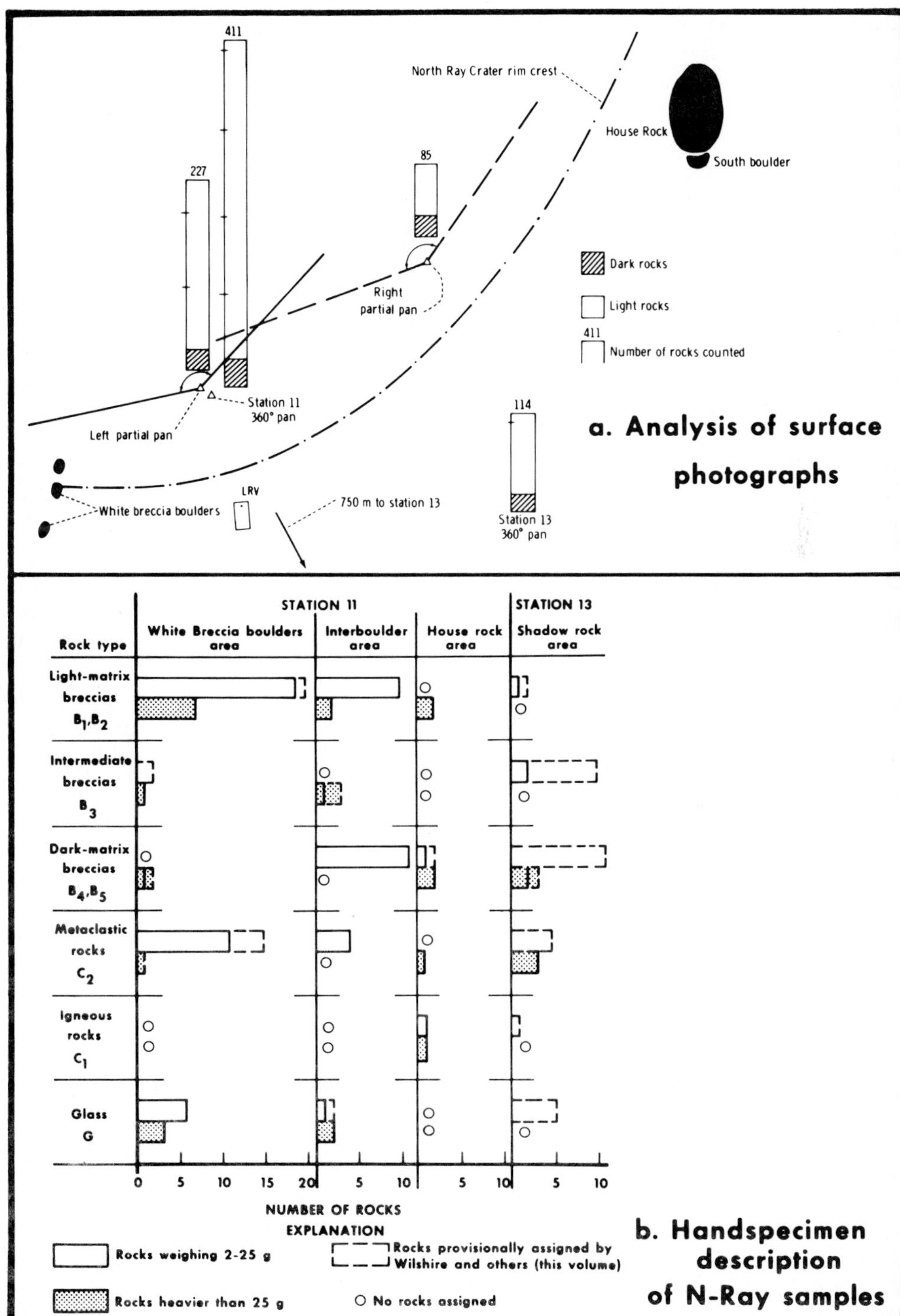

411
227
85
114
North Ray Crater rim crest
House Rock
South boulder
Right partial pan
Dark rocks
Light rocks
Number of rocks counted
Station 11 360° pan
Left partial pan
White breccia boulders
LRV
750 m to station 13
Station 13 360° pan
a. Analysis of surface photographs
STATION 11
STATION 13
Rock type
White Breccia boulders area
Interboulder area
House rock area
Shadow rock area
Light-matrix breccias B_1, B_2
Intermediate breccias B_3
Dark-matrix breccias B_4, B_5
Metaclastic rocks C_2
Igneous rocks C_1
Glass G
0 5 10 15 20 5 10 5 10 5 10
NUMBER OF ROCKS
EXPLANATION
Rocks weighing 2-25 g
Rocks heavier than 25 g
Rocks provisionally assigned by Wilshire and others (this volume)
No rocks assigned
b. Handspecimen description of N-Ray samples

cartes Mountains may be present in the crater wall. Therefore the rocks and soils are ambiguous as to which are plains vs. mountain samples, but it is safe to assume that they are all North Ray ejecta. Important data derived from lunar surface and orbital photography is summarized in AFGIT (1973), Hodges *et al.* (1973), Ulrich (1973), and AFGIT (in press).

A grossly layered sequence is visible only in the upper part of the east wall of the crater (Fig. 6). It is either obscured by debris slides or is absent in the remainder of the crater. The dominant rock type is a very friable, light-matrix breccia (Figs. 6, 7) which generally makes small to moderate-sized boulders that are partly buried by their own debris, forming massive fillets. Dark-matrix breccias and glass-coated rocks increase in proportion toward House Rock, itself a giant (12 m high, 25 m long) dark-matrix breccia block. The dark-matrix breccias are coherent and unfilleted; their resistance to fragmentation by micrometeorite impacts makes it likely that they have been preferentially preserved in the rake samples (Fig. 8). The streak of dark-colored material extending from House Rock to the large dark blocks on the floor of the crater suggests that this enormous boulder came from the basal unit. The major unit (70 m thick) in the east wall of the crater (Figs. 7, 9) appears to be outcrop of light-matrix breccias that incorporated irregular, dark blocks. The top unit probably includes additional light-matrix breccias beneath the crater ejecta and regolith. Talus covers most of the lower parts of the crater thus preventing an accurate assessment of rock types that might be present; however, the absence of ledges or blocks projecting through the talus suggests that friable light-matrix breccias are the most likely rock type under the talus. Polarimetric studies show an absence of crystalline rocks in the area of Figs. 6 and 7 and the presence of only breccias or glass. Fig. 9 shows geologic information derived from orbital and surface photographs together with a true-scale topographic cross-section of North Ray Crater. It summarizes the actual field observations and forms the basis for Fig. 10, which illustrates various stratigraphic interpretations. The common basis of such interpretative attempts is to establish a stratigraphic order from photographic and petrographic observations that could potentially apply to much, if not all, of the entire Apollo 16 landing site.

South Ray Crater

The most detailed descriptive and interpretative accounts of South Ray Crater are by Ulrich (1973) and Ulrich *et al.*, and Reed, both in AFGIT (in press). It is 680 m across and about 130 m deep. Although the crater was not visited, its extensive ejecta and ray systems seem to exert a dominant influence over the entire landing site, particularly the southern half (Fig. 11). Many samples that display exposure ages of $\approx$ 2 m.y. are thought to be derived from South Ray; these samples have various petrologic characteristics. The absence of fine-grained South Ray ejecta in the Apollo 16 collection is conspicuous (McKay and Heiken, 1973). Because of its apparently significant contribution to the sample collection,

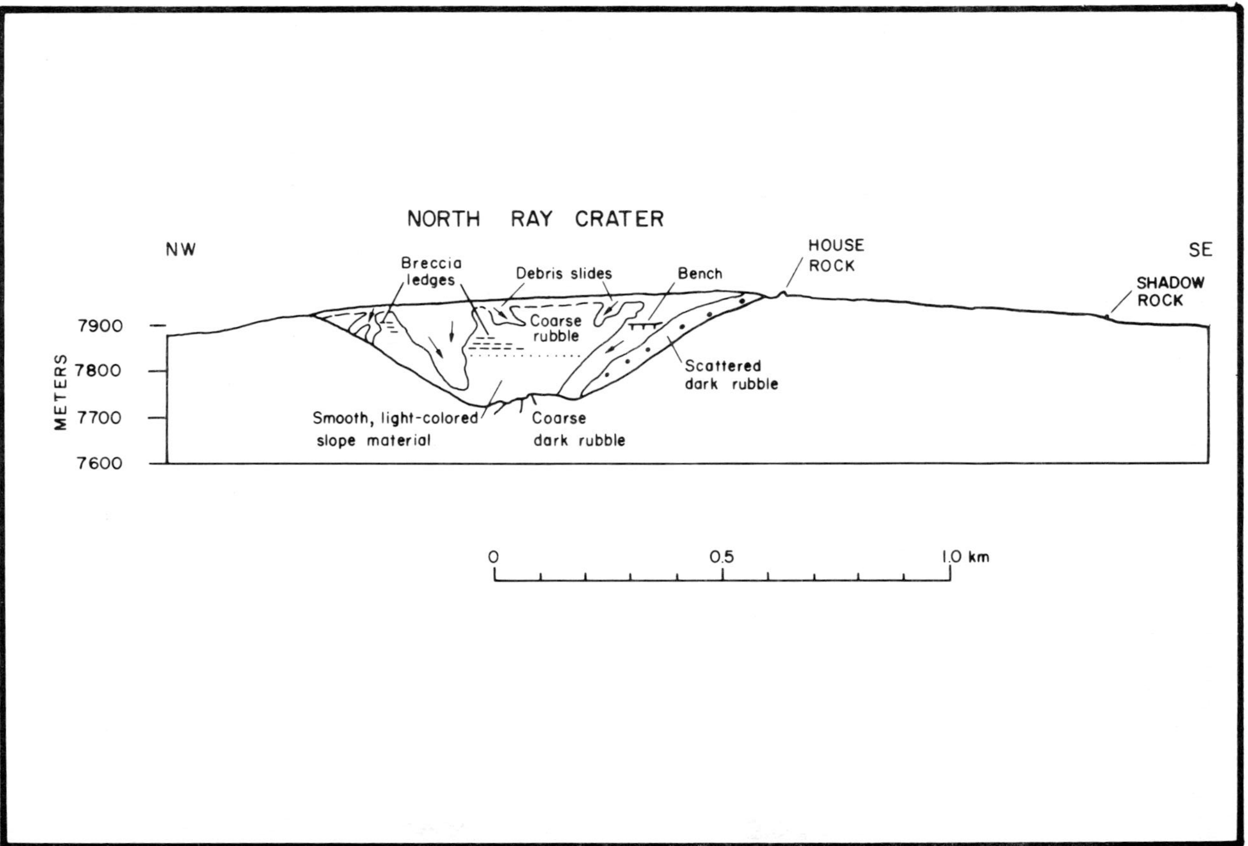

Fig. 9. Diagrammatic cross-section of North Ray Crater derived from field observations and surface photographs combined with actual topographic cross-section of North Ray crater (Hodges *et al.*, 1973), representing the factual "field evidence."

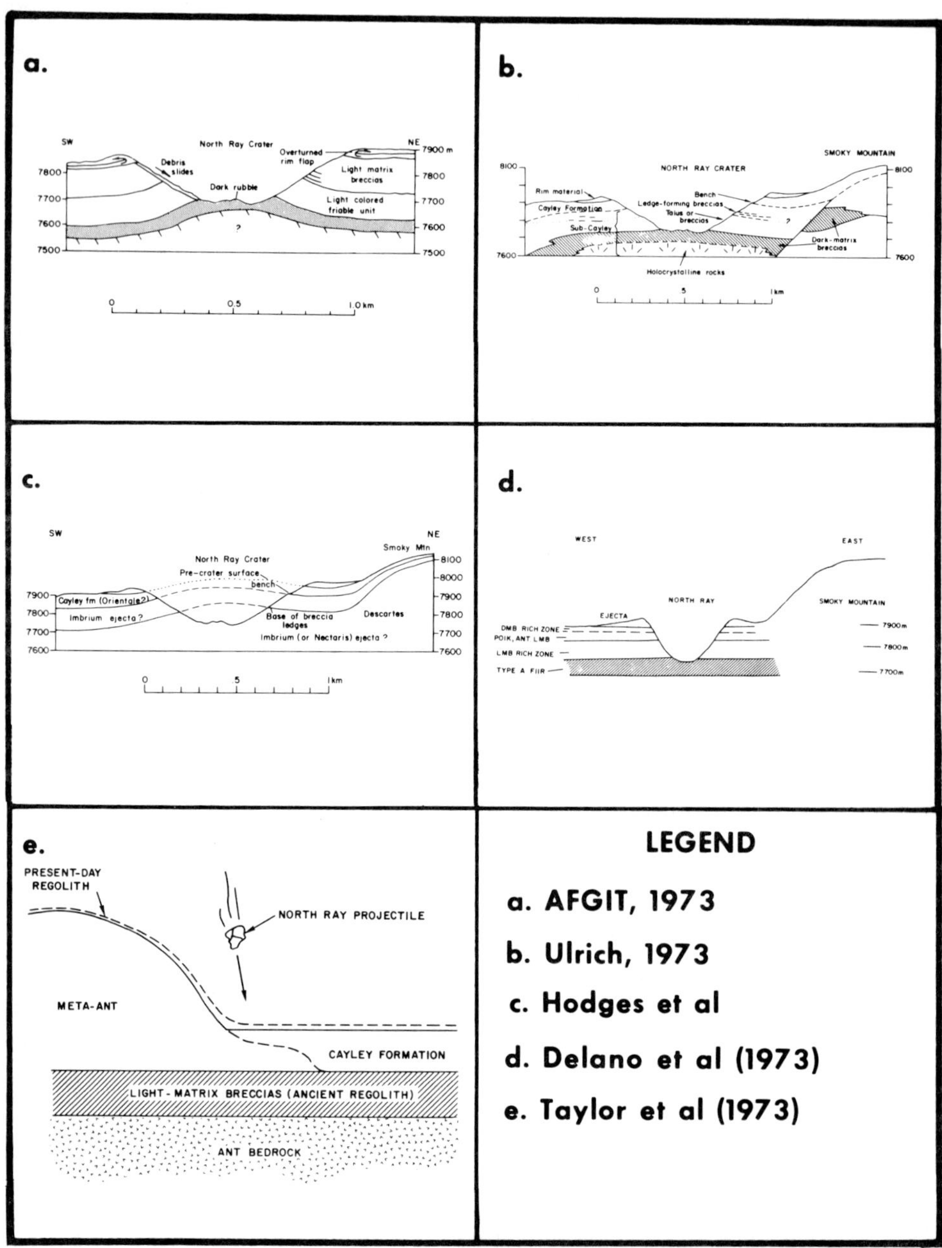

Fig. 10. Proposed cross-sections and/or stratigraphic implications of North Ray Crater: a) AFGIT, 1973; b) Ulrich, 1973; c) Hodges *et al.*, 1973; d) Delano *et al.*, 1973; e) Taylor *et al.*, 1973.

Note gross similarities of all stratigraphic models placing light-matrix breccia materials in the upper section(s) and the competent dark-matrix breccias close to the crater bottom; the reader is referred to the original references for details of arguments, interpretations, and implications.

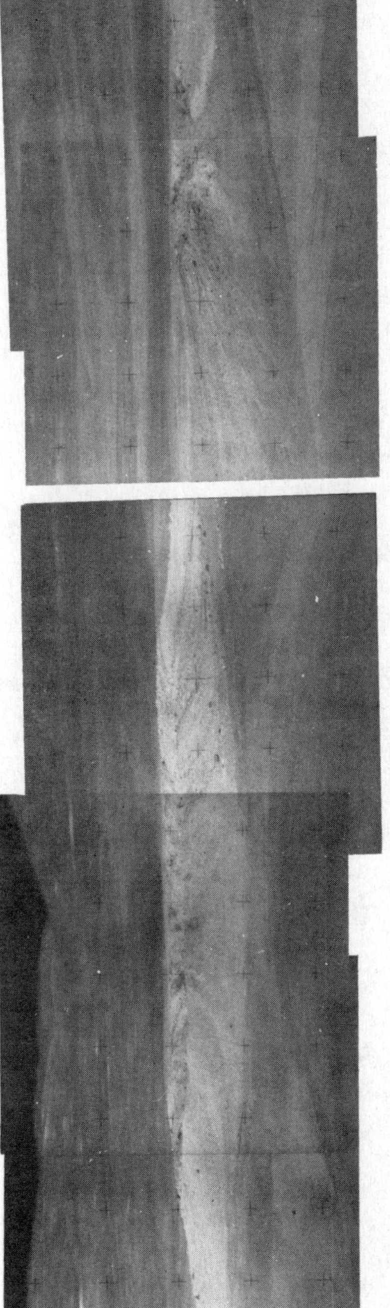

Fig. 11. Panoramic telephoto views of South and Baby Ray Craters (taken from station 4, high on Stone Mountain, looking west).

South Ray's stratigraphic section was interpreted using surface and orbital photographs, analysis of boulder sizes and shapes (angular versus rounded), and petrologic observations.

In contrast to North Ray, it appears relatively certain that the upper stratigraphic sections of South Ray are composed of compact, dark rocks, visible in the crater walls and especially manifested as numerous, large (1–10 meters) blocks dominating the crater rim (Fig. 11). This stratum may be 60 m thick. The lower walls of South Ray Crater appear to be made up of light gray materials. Interestingly and importantly, the near surface layer of Baby Ray Crater, a 130 m diameter structure, is also composed of the competent looking, dark lithologies so prominent at South Ray, leading to the conclusion that a continuous stratum of this material exists between South Ray and Baby Ray Craters (Figs. 11 and 12).

In view of the sample evidence, the darker stratum in this area is interpreted as the source for dark matrix breccias; the deeper and lighter unit may be an impact melt and the source for most of the light-grey igneous lithologies, e.g., 68415, 61416, etc. The presence of light matrix breccias in this stratigraphic column cannot be excluded (Fig. 12).

Mound craters

Irregular mounds on the floor of several ≈ 1 km sized craters in the Apollo 16 site were tentatively interpreted by Milton and Hodges (1972) as indicative of a stratigraphic discontinuity caused by local, competent strata at depth, based on the cratering experiments of Quaide and Oberbeck (1968). Postmission analysis (Ulrich, AFGIT, in press) reveals that such central mounds are confined exclusively to craters 0.5 to 1.8 km across, occur only within the Cayley plains, are clearly recognizable in craters of different degradational states and, importantly, do *not* occur on *all* craters of the above size class. It is therefore concluded that competent strata of highly localized lateral extent are present. Some of the more prominent "mound" craters include North Ray, Kiva, Palmetto, Haystack NE, Haystack SW and South Ray; Gator Crater is an example of a normal, bowl-shaped crater lacking a central mound. Using crater-diameter and mound height measurements and the geometric relationships of Quaide and Oberbeck's experimental craters, the depth of this competent stratum averages ≈ 150 m, although the variation (110 to 190 m) is probably more significant.

In summary, detailed synthesis of photogeologic and sample evidence leads to the conclusion that extremely local, rather than regional, stratigraphic sequences may exist. There is, as yet, no strong evidence of stratigraphic continuity across the entire landing site. In contrast, the two best determined sections (North Ray and South Ray) appear demonstrably different. The intermittent occurrence of central mound craters lends additional support for the existence of localized,

Fig. 12. Proposed cross-sections and stratigraphic inferences for South Ray Crater: a) AFGIT, 1973; b) Ulrich and Reed, AFGIT, in press; c) Delano *et al.*, 1973.

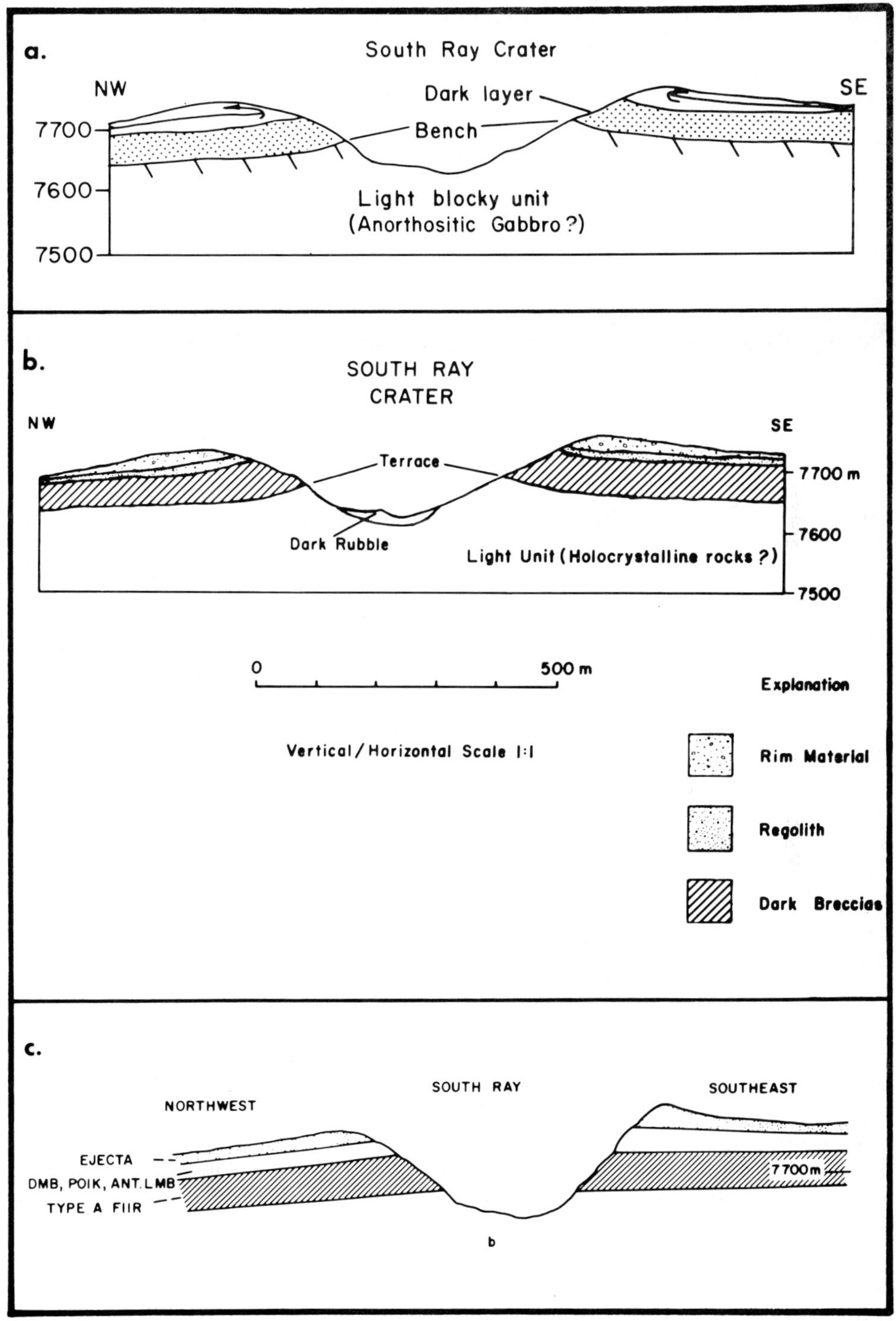
a.
South Ray Crater
NW
SE
Dark layer
Bench
7700
7600
7500
Light blocky unit
(Anorthositic Gabbro ?)
b.
SOUTH RAY
CRATER
NW
SE
Terrace
7700 m
7600
7500
Dark Rubble
Light Unit (Holocrystalline rocks ?)
0
500 m
Vertical / Horizontal Scale 1:1
Explanation
Rim Material
Regolith
Dark Breccias
c.
SOUTH RAY
NORTHWEST
SOUTHEAST
EJECTA
DMB, POIK, ANT. LMB
TYPE A FIIR
7700m
b

competent subsurface materials of limited extent, i.e., scales of ≈ km. It is therefore concluded that the Cayley plains consist predominantly of fragmented impact deposits with interspersed pods and lenses of more resistant materials such as dark matrix breccias and/or still more competent, igneous-textured, impact melts.

"POST MISSION" REGIONAL GEOLOGY

Two contrasting schools of thought exist about the origin of the Apollo 16 impact deposits: one view maintains that the samples are predominantly of local origin, while the other view suggests more distant, basin-related source areas. The landing site is ≈ 600 km W of the Nectaris Basin center, 1600 km SE of the Imbrium Basin and ≈ 3500 km from the Orientale center to the WSW.

Predominantly "local" origin

Using laboratory cratering experiments and photogeologic studies, Oberbeck and Morrison (1974) developed diagnostic criteria to identify secondary craters and crater chains. Adding cratering theory and orbital geochemistry Oberbeck *et al.* (1974, 1975) formulated the so-called "secondary cratering" hypothesis, summarized by Oberbeck (1975). In this view the highly subdued plains topography as well as their generally ponded nature is essentially caused by mass wasting processes from local topographic highs.

During large scale cratering, primary ejecta are expelled along ballistic trajectories and reimpact the local and distant crater environs with sufficient terminal velocities to generate secondary craters. The mass excavated from the local substrate may be many times that of the incoming projectile. Angles of impact from the local horizontal are typically very shallow (15°–25°) for secondary impacts. As a consequence, secondary crater ejecta have preferred momenta, radially away from the primary crater. Ejecta of secondary craters, if in close juxtaposition, interfere with each other and combine into a massive debris surge of intimately mixed "local" substrate as well as "primary" crater material.

This "secondary cratering" hypothesis addresses primarily the origin of highlands plains, such as the Cayley, their principal characteristics reflecting the last depositional phase via a debris surge from relatively large impact events; downslope mass wasting by both primary and secondary craters, however, is a continuous process as well as is the reexcavation and mixing of the ponded materials.

Applying these concepts specifically to the Apollo 16 landing site, Morrison and Oberbeck (1975) suggested a maximum of ≈ 15% of primary Imbrium material in the local Cayley plains; potential contributions from Orientale are significantly smaller. However, a substantial fraction of the substrate encountered by the Imbrium ejecta could have been derived previously from the nearby Nectaris basin. Thus, while the "secondary cratering" hypothesis predominantly mixes and reworks "local" materials, it is not in a position to delineate the origin of the substrate, which itself may have varied crater sources—both local as well as regional, if not global. The important points are (1) that relatively small amounts of primary crater materials are required to produce the subdued plains and (2) that "local" materials dominate volumetrically the combined effects of distant craters and basins.

Oberbeck and co-workers do not specifically address the formation and source area of the Descartes Mountains; however, materials mass-wasted downslope should be part of the plains samples. Implicit in the secondary cratering hypothesis—a relatively low velocity impact regime—is that most of the thermal energy manifested in breccias and melt rocks derives from primary craters.

Somewhat different arguments for a predominantly local sample origin were presented by Head (1974) after studying specifically the local and regional cratering history of the Apollo 16 area, especially the effects of relatively large craters ($\geq$ 30 km diameter). In his view the ponded, level and subdued Cayley morphology largely reflects the original flatness of relatively large crater floors caused by ejecta fallback and a coherent melt sheet. Although modification by secondary cratering processes is permitted in this view, the major morphological features are related to major local cratering events at the Apollo 16 site. Importantly, the petrogenesis of the returned samples is inferred to be directly related to this cratering history.

As illustrated in Figs. 13 and 14a, Head (1974) delineates the following sequence: the oldest recognizable crater, unnamed crater A of $\approx$ 150 km diameter, is almost centered on the landing site and is thought to predate the Nectaris basin impact. Based on interpreted superposition relations, unnamed craters C and B appear to bracket deposition of the Descartes materials, which are interpreted as Nectaris ejecta, possibly ponded to unusually large thicknesses. Thus crater C is older and B is younger than Nectaris, implying that unnamed crater B, $\approx$ 60 km diameter, had a profound effect on the Apollo 16 landing site. Craters Dolland B and "unnamed D" are superposed on crater "B", and because radial furrows indicative of the "Imbrium Sculpture" cut their rims, they are interpreted to be older than Imbrium, the latter being the last, large scale cratering event that may have affected the area, possibly depositing as much as 50m Imbrium ejecta. Contributions from Orientale were judged negligible. Detailed consequences for sample provenance and petrogenesis related to cratering mechanics are discussed by Head (1974), the major implications being illustrated in Fig. 14a.

In summary, two different approaches, those by Oberbeck and Head, advocate local sample provenance. The major difference between the two hypotheses is essentially one of scale and emphasis: the "secondary cratering" hypothesis attempts to address the general principles of plains formation, while the photogeologic study of Head is specifically aimed at delineating the local and regional cratering history of the Apollo 16 site.

The conclusions reached are not mutually exclusive, but complementary, inasmuch as Head predominantly outlines the specific history of "primary" crater effects, i.e., the "substrate" history; this substrate is repeatedly ploughed over and reworked by Oberbeck *et al.*'s "secondary cratering" regime not only by

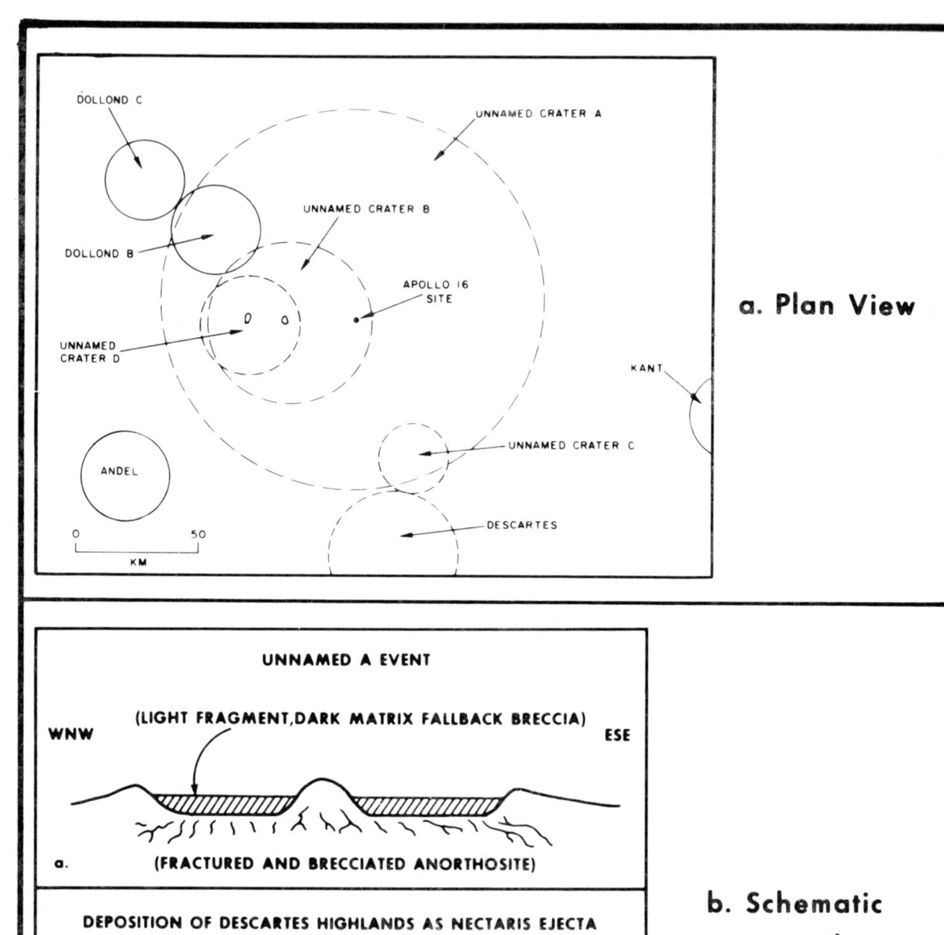
DOLLOND C
UNNAMED CRATER A
UNNAMED CRATER B
DOLLOND B
APOLLO 16 SITE
UNNAMED CRATER D
KANT
UNNAMED CRATER C
ANDEL
DESCARTES
0 50
KM
a. Plan View
UNNAMED A EVENT
(LIGHT FRAGMENT, DARK MATRIX FALLBACK BRECCIA)
WNW
ESE
a.
(FRACTURED AND BRECCIATED ANORTHOSITE)
DEPOSITION OF DESCARTES HIGHLANDS AS NECTARIS EJECTA
b.
UNNAMED B CRATERING EVENT
WALL
UNNAMED B
EJECTA
FALLBACK BRECCIA
c.
POST UNNAMED B EVENTS (DOLLOND B,C)
DOLLOND B
APOLLO 16 SITE
MAJOR BASIN CONTRIBUTIONS TOO SMALL TO BE SHOWN
d.
b. Schematic cross sections

the craters identified by Head, but by many additional ones in the broader Apollo 16 region, including the Nectaris, Imbrium, and Orientale basin events. There may be a difference in the makeup of the Descartes materials, which seem to be in Head's view pure Nectaris ejecta, while the secondary cratering hypothesis would predict more of a mixture of "local" and Nectaris components, if the Descartes Mountains were indeed formed by the deposition of Nectaris material.

Predominantly basin-related origin

Hodges *et al.* (1973), Moore *et al.* (1974) and Hodges and Muehlberger (AFGIT, in press), Ulrich and Reed (AFGIT, in press) propose that the upper 200 to 500 m of the Apollo 16 Cayley plains are primarily from the Imbrium Basin, with a possible though minor contribution from the Orientale basin (Fig. 14b and c). This interpretation rests to a large degree on photogeologic studies of various ejecta morphologies circumferential to Orientale, the youngest and best preserved basin. In particular, Moore *et al.* (1974) delineate systematic changes in ejecta morphology with radial distance and emphasize various erosional and depositional features, such as flow fronts and channels, ponding of ejecta, draping of preexisting relief, etc. Moore *et al.* (1974) also synthesize observational data on various sized lunar craters and conclude that the continuous ejecta, essentially composed of primary crater material, extend some 1.5 radii beyond a crater's rim; beyond that range the ejecta become increasingly more discontinuous and the effects of secondary impacts will be progressively more pronounced. Theoretical calculations furthermore show that substantial masses of primary Orientale ejecta may be ballistically distributed moonwide.

Applying these basic observations to the more degraded Imbrium ejecta blanket, Moore *et al.* (1974) conclude that as much as 500 m of Imbrium ejecta could have been deposited at the Apollo 16 site. Hodges *et al.* (1973), Eggleton (AFGIT, in press) and Hodges and Muehlberger (AFGIT, in press) detail many additional analogies of the Orientale and Imbrium ejecta facies and depositional/erosional features, in particular "Imbrium Sculpture" close to Apollo 16. Therefore they also conclude that the Apollo 16 Cayley plains are predominantly of Imbrium derivation, in part ejected as already highly reworked impact deposits that formed the upper parts of the Imbrium target or brecciated, shocked and melted solely by the Imbrium impact itself.

Hodges and Muehlberger (AFGIT, in press) specifically disagree with Head (1974) on the origin of the Descartes mountains and ascribe formation of this unit also to the deposition of Imbrium ejecta, supporting evidence being derived from 1) gradational transitions between Cayley plains and Descartes mountains, 2)

Fig. 13. Location and formation sequence of major, primary impact craters that significantly affected the geologic history of the Apollo 16 landing site according to Head (1974): a) Plan view; b) Schematic cross-sections illustrating proposed sequence of events (see also Fig. 2).

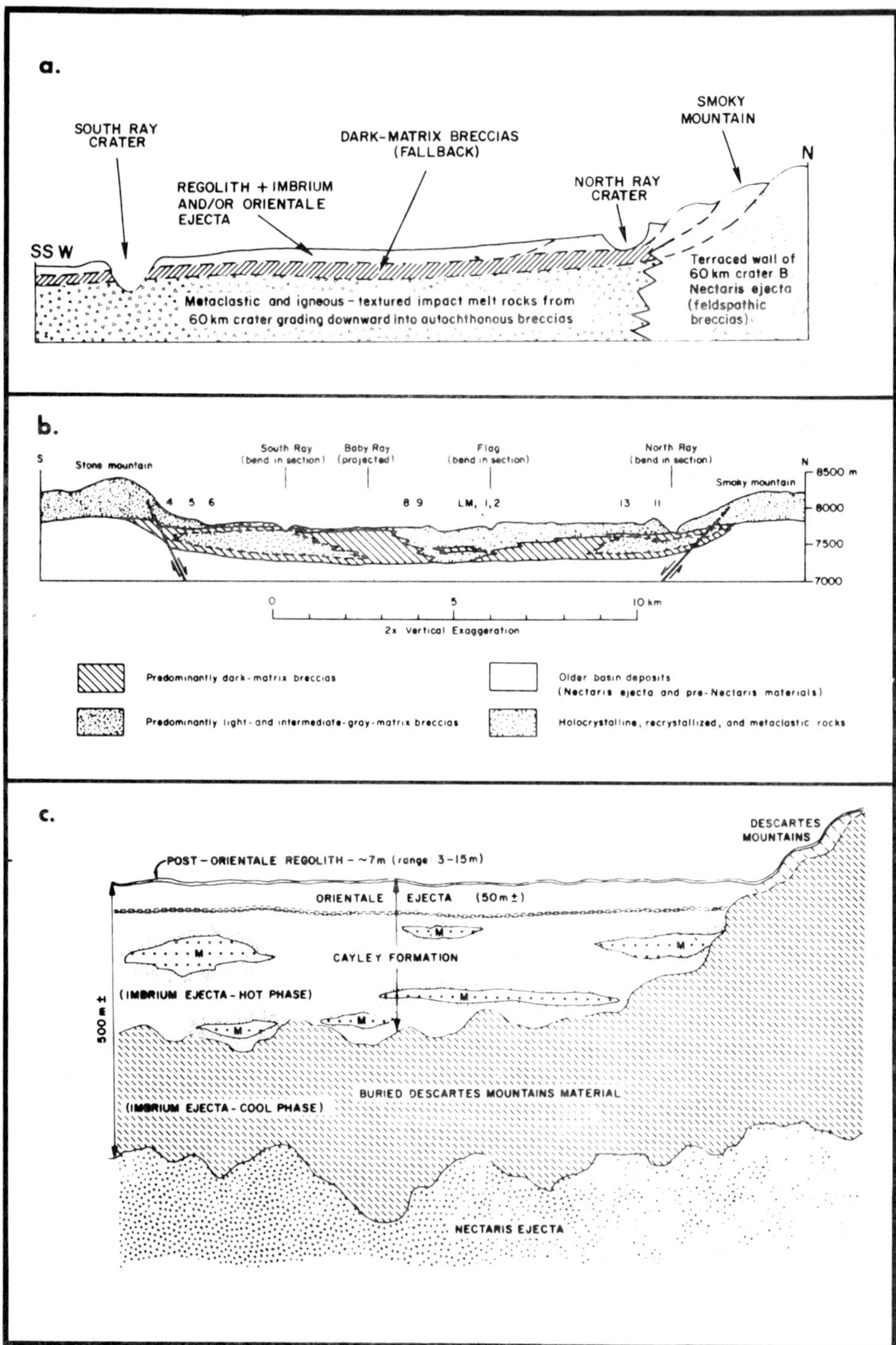
a.
SMOKY MOUNTAIN
SOUTH RAY CRATER
DARK-MATRIX BRECCIAS (FALLBACK)
N
NORTH RAY CRATER
REGOLITH + IMBRIUM AND/OR ORIENTALE EJECTA
SSW
Terraced wall of 60 km crater B Nectaris ejecta (feldspathic breccias)
Metaclastic and igneous-textured impact melt rocks from 60 km crater grading downward into autochthonous breccias
b.
S
Stone mountain
South Ray (bend in section)
Baby Ray (projected)
Flag (bend in section)
North Ray (bend in section)
N
8500 m
Smoky mountain
4 5 6
8 9
LM, 1, 2
13
11
8000
7500
7000
0
5
10 km
2x Vertical Exaggeration
Predominantly dark-matrix breccias
Predominantly light- and intermediate-gray-matrix breccias
Older basin deposits (Nectaris ejecta and pre-Nectaris materials)
Holocrystalline, recrystallized, and metaclastic rocks
c.
DESCARTES MOUNTAINS
POST-ORIENTALE REGOLITH - ~7m (range 3-15m)
ORIENTALE EJECTA (50m ±)
M
CAYLEY FORMATION
(IMBRIUM EJECTA - HOT PHASE)
500 m ±
BURIED DESCARTES MOUNTAINS MATERIAL
(IMBRIUM EJECTA - COOL PHASE)
NECTARIS EJECTA

some Imbrium-related sculpturing in the Descartes mountains, 3) the apparent deflection of the Descartes ridges by the Kant Plateau and 4) the generally very rugged, thus youthful, appearance of Descartes. These observations are considered incompatible with a Nectaris age; also, similarly sculptured terrain is nowhere observed around the Nectaris basin. Nevertheless, Nectaris material could well be present at depth (Hodges and Muehlberger, AFGIT, in press) underlying the Imbrium ejecta.

There is also an interpretative difference concerning "unnamed crater B" of Head (1974). In Head's view, this crater is younger than deposition of Descartes materials. Hodges and Muehlberger (AFGIT, in press) argue that it must predate Descartes, and they even question the existence of such a crater altogether because no photogeologic evidence for an ejecta blanket from unnamed crater B seems to exist in the Descartes mountains (see Fig. 2).

The latest and most detailed account favoring an Imbrium Basin sample origin is given by Hodges and Muehlberger (AFGIT, in press); their interpretations are illustrated in Fig. 14c. They show Imbrium ejecta that are segregated into hot, partly molten deposits that ponded preferentially in local topographic lows, now forming the Cayley plains, and colder, dry ejecta that are responsible for formation of the Descartes mountains. The drained hot material accumulated into localized pools of melt that cooled sufficiently slowly to produce igneous textures and which had significant thermal aureoles to produce igneous textures and the thermally metamorphosed lithologies. Similarities in absolute ages at Apollos 14 and 16 further suggest an Imbrium derivation.

Hodges and Muehlberger (AFGIT, in press) draw particular attention to the small scale crater degradation data and their implications regarding absolute and relative times of deposition of Cayley units related to the Imbrium impact (Boyce

---

Fig. 14. Schematic histories and geologic cross-sections of the Apollo 16 landing site: a) Head, 1974; b) Ulrich and Reed, AFGIT, in press; c) Hodges and Muehlberger, AFGIT, in press.

Although of variable scale and detail, Figs. 14b and c are conceptually very similar. Dimensions and locations of various subsurface materials are inferred from topographic relations and areal sample distribution in Fig. b). Fig. c) is schematic and illustrates the basic concepts.

Note that Fig. 14 a) has North Ray penetrate Smoky Mountain material, which is stratigraphically distinct from the plains materials; all of the basic stratigraphic elements are associated with local cratering events with only a minor addition of Imbrium ejecta and negligible Orientale contributions; these latter materials are concentrated, however, in the near surface horizons of the landing site. In contrast, models b and c portray both the Descartes and plains materials as massive deposits of Imbrium ejecta; specifically the melts are of Imbrium derivation and aided in mobilizing the Cayley so that it could pool in low areas and embay the less fluid, cold- and thus ridgy- Descartes facies; therefore relatively more melt is indicated for the plains. Note also that interpretations b and c depict a close spatial association of melt pools (= igneous rocks) and thermally metamorphosed (= competent dark matrix breccias), the latter resulting from the melt pools' thermal aureoles. Any crater into Cayley with a floor mound presumably interacted with such localized melt pools.

et al., 1974; Chao *et al.*, 1975). The "crater age" of the plains is younger than the Imbrium event and essentially synchronous with the Orientale formation. Thus there is the distinct possibility that Orientale ejecta, sufficiently thick to mask and "rejuvenate" the Imbrium deposits, are present at Apollo 16 (and over large areas of the lunar globe), although such rejuvenation may also be accomplished by "seismic shaking" according to Schultz and Gault (1975). Chao *et al.* (1975) postulate a significant contribution of Orientale ejecta to the Apollo 16 collection, whereas Hodges and Muehlberger (AFGIT, in press) consider it a possibility, pending clarification of the moonwide rejuvenation mechanism of the Cayley surfaces.

Thus a synthesis of presently viable hypotheses concerning the origin of the Cayley plains and Descartes Mountains demonstrates that a wide range of contrasting proposals exist. Although all hypotheses agree that local materials and ejecta from the Nectaris-, Imbrium- and Orientale impact could be contained in the Apollo 16 collection, there is considerable debate and uncertainty on the relative significance of these possible source locales and it is here where interpretations diverge widely. These contrasting views entail far-reaching and specific consequences for the petrogenesis of the samples and especially for chronological sequences of events. It appears possible to broadly distinguish between "local" sources and processes versus "distant" sources from a single dominating event by future analysis of the Apollo 16 rocks.

"Well, it's back to the drawing boards—or wherever geologists go" (T. K. Mattingly, Apollo 16 astronaut, April 1972). It is largely the challenge of sample students to decide whether we have to draw big or small circles around the Apollo 16 landing site. The answer is undoubtedly contained in the Apollo 16 materials but will not come easily and, unquestionably, will require continued, concerted, multidisciplinary studies of the entire sample collection. Both the "Traverse Planning Team" as well as the Apollo 16 crew will continue to provide any assistance possible.

Acknowledgment—The Lunar and Planetary Institute is operated by the Universities Space Research Association under Contract No. NSR 09-051-001 with the National Aeronautics and Space Administration. This paper constitutes the Lunar and Planetary Institute Contribution No. 403.

REFERENCES

AFGIT (Apollo Field Geology Investigation Team) (1973) Apollo 16 exploration of Descartes: A geologic summary. *Science* **179,** 62–69.

AFGIT Geology of the Apollo 16 area, central lunar highlands (G. E. Ulrich, C. A. Hodges and W. R. Muehlberger, eds.). *U.S. Geol. Survey Open File Report No. 79-1091,* 1128 pp. (To be published as *U.S. Geol. Survey Prof. Paper No. 1048*). In press.

Arvidson R., Crozaz G., Drozd R. Y., Hohenberg C. M., and Morgan C. Y. (1975) Cosmic ray exposure ages of features and events at the Apollo landing sites. *The Moon* **13,** 259–276.

Boyce J. M. (1976) Ages of flow units in the lunar nearside maria based on Lunar Orbiter IV photographs. *Proc. Lunar Sci. Conf. 7th,* p. 2717–2738.

Boyce J. M., Dial A. L., and Soderblom L. A. (1974) Ages of the lunar nearside light plains and maria. *Proc. Lunar Sci. Conf. 5th,* p. 11–23.

Chao E. C. T., Hodges C. A., Boyce J. M., and Soderblom L. A. (1975) Origin of lunar light plains. *J. Res. U.S. Geol. Survey* **3,** 379–392.

Delano J. W., Bence A. E., Papike J. J., and Cameron K. L. (1973) Petrology of the 2–4 mm soil fraction from the Descartes region of the Moon and stratigraphic implications. *Proc. Lunar Sci. Conf. 5th,* p. 537–551.

Dowty E., Prinz M., and Keil K. (1974) Ferroan anorthosite: A widespread and distinctive lunar rock type. *Earth Planet. Sci. Lett.* **24,** 15–25.

Eberhardt P., Eugster O., Geiss E., Grögler N., Guggisberg S., and Mörgeli M. (1976) Noble gases in the Apollo 16 special soils from the east-west split and the permanently shadowed area. *Proc. Lunar Sci. Conf. 7th,* p. 563–585.

Eggleton R. E. and Marshall C. H. (1962) Notes on the Apenninian Series and pre-Imbrian stratigraphy in the vicinity of Mare Imbrium and Mare Nubium. In *Astrogeol. Studies Ann. Prog. Rept., February 1961–August 1961, Pt. A,* p. 132–137. U.S. Geol. Survey open-file report.

Elston D. P., Boudette E. L., Schafer J. P., Muehlberger W. R., and Sevier J. R. (1972) Apollo 16 field trips. *Geotimes* **17,** 27–30.

Gault D. E., Quaide W. L., and Oberbeck V. R. (1968) Impact cratering mechanics and structures. In *Shock Metamorphism of Natural Materials* (B. M. French and N. M. Short, eds.), p. 87–89. Mono, Baltimore.

Head J. W. (1974) Stratigraphy of the Descartes region (Apollo 16): Implications for the origin of samples. *The Moon* **11,** 77–99.

Head J. W. and Goetz A. F. H. (1972) Descartes region—evidence for Copernican-age volcanism. *J. Geophys. Res.* **77,** 1368–1374.

Hinners N. W. (1972) Apollo 16 site selection. *Apollo 16 Prelim. Sci. Rep.* NASA SP-315, p. 1–1 to 1–3.

Hodges C. A., Muehlberger W. R., and Ulrich G. E. (1973) Geologic setting of Apollo 16. *Proc. Lunar Sci. Conf. 5th,* p. 1–25.

Hörz F., Carrier W. D. III, Young J. W., Duke C. M., Nagle J. S., and Fryxell R. (1972) Apollo 16 special samples. *Apollo 16 Prelim. Sci. Rep.* NASA SP-315, p. 7–24 to 7–54.

McKay D. S. and Heiken G. H. (1973) The South Ray Crater age paradox. *Proc. Lunar Sci. Conf. 5th,* p. 41–47.

Milton D. J. (1968) Geologic map of the Theophilus Quadrangle of the Moon. U.S. Geol. Survey Misc. Geol. Inv. Map I-546 (LAC 78).

Milton D. J. and Hodges C. A. (1972) Geologic maps of the Descartes region of the Moon, Apollo 16 premission maps. U.S. Geol. Survey Misc. Geol. Inv. Map I-748 (2 sheets).

Moore H. J., Hodges C. A., and Scott D. H. (1974) Multiringed basins—illustrated by Orientale and associated features. *Proc. Lunar Sci. Conf. 5th,* p. 71–100.

Morrison R. H. and Oberbeck V. R. (1975) Geomorphology of crater and basin deposits: Emplacement of the Fra Mauro formation. *Proc. Lunar Sci. Conf. 6th,* p. 2503–2530.

Oberbeck V. R. (1975) The role of ballistic erosion and sedimentation on lunar stratigraphy. *Rev. Geophys. Space Phys.* **13,** 337–362.

Oberbeck V. R., Hörz F., Morrison R. H., Quaide W. L., and Gault D. E. (1975) On the origin of the lunar smooth plains. *The Moon* **12,** 19–54.

Oberbeck V. R. and Morrison R. H. (1974) Laboratory simulation of the herringbone pattern associated with lunar secondary crater chains. *The Moon* **9,** 415–455.

Oberbeck V. R. Morrison R. H., Hörz F., Quaide W. L., and Gault E. (1974) Smooth plains and continuous deposits of craters and basins. *Proc. Lunar Sci. Conf. 5th,* p. 111–126.

Quaide W. L. and Oberbeck V. R. (1968) Thickness determinations of the lunar surface layer from impact craters. *J. Geophys. Res.* **73,** 5247–5270.

Schultz P. H. and Gault D. E. (1975) Seismic effects from major basin formations on the Moon and Mercury. *The Moon* **12,** 159–177.

Shoemaker E. M. (1963) Impact mechanics at Meteor Crater, Arizona. In *The Moon, Meteorites, and Comets* (B. M. Middlehurst and G. P. Kuiper, eds.), p. 301–336. Univ. Chicago Press, Chicago.

Soderblom L. A. (1970) A model for lunar impact erosion applied to the lunar surface. *J. Geophys. Res.* **75,** 2655–2661.

Taylor G. J., Drake M. J., Hallam M. E., Marvin U. B., and Wood J. A. (1973) Apollo 16 stratigraphy: The ANT-hills, the Cayley plains, and a pre-Imbrian regolith. *Proc. Lunar Sci. Conf. 4th,* p. 553–568.

Trask N. J. and McCauley J. F. (1972) Differentiation and volcanism in the lunar highlands: Photogeologic evidence and Apollo 16 implications. *Earth Planet. Sci. Lett.* **14,** 201–206.

Ulrich G. E. (1973) A geologic model for North Ray Crater and stratigraphic implications for the Descartes region. *Proc. Lunar Sci. Conf. 4th,* p. 27–39.

Wilhelms D. E. and McCauley J. F. (1971) Geologic map of the nearside of the Moon. U.S. Geol. Survey Misc. Geol. Inv. Map I-703.

Wilshire H. G., Stuart-Alexander D. G., and Jackson E. D. (1973) Apollo 16 rocks: Petrology and classification. *J. Geophys. Res.* **78,** 2379–2392.

Appendix I: Apollo 16 Premission Documents

A large variety of technologically and scientifically oriented documents was used in planning and executing the Apollo missions. Among those were many internal NASA documents which are no longer available from the issuing agency. However, many of these Apollo 16 pre- and post-mission documents have been collected in the Library/Information Center, Lunar and Planetary Institute. These documents are indispensable for the historian. They illustrate well the daily interaction of scientists and engineers in their continuous efforts to optimize scientific return within a set of rigorous, technological constraints.

Some of the Apollo 16 pre-mission documents in this collection are:

Apollo Site Selection Board, Minutes (1966–1972). (This board made the final decision(s) where to land.)

Group For Lunar Exploration, Minutes (1967–1970). (This committee made scientific evaluations of the candidate landing sites and recommended to the Site Selection Board.)

Science Working Panel, Minutes (1970–1972). (This panel detailed and prioritized all surface activities related to scientific exploration at a specific landing site.)

Apollo 16, Apollo Lunar Surface Experiments Package: ALSEP Training Handout Apollo 16 Crew Debriefing. NASA Manned Spacecraft Center, Houston, TX; November 1971.

Lunar Surface Scientific Equipment for Apollo Mission 16; Lunar Surface Project Office, Manned Spacecraft Center, Houston, TX, January 10, 1972.

Apollo 16 Final Flight Plan, Manned Spacecraft Center, Houston, TX, March 6, 1972.

Apollo 16 Traverse Planning Kit, Manned Spacecraft Center, Houston, TX, March 9, 1972.

Apollo Mission J-2 (Apollo 16) Mission Science Planning Document, Manned Spacecraft Center, Houston, TX, March 15, 1972.

Apollo 16 Science Handbook, Manned Spacecraft Center, Houston, TX, April 7, 1972.

Apollo 16 Lunar Surface Procedures. Final. Manned Spacecraft Center, Houston, TX, March 16, 1972.

Traverse Briefings for Apollo 16 Crew; The Traverse Planning Team, March 15, 1972.

Apollo 16, Mission Requirements Document, Manned Spacecraft Center, Houston, TX.

Appendix II: Sample Allocation Status (February 1980)

Comprehensive synthesis of analytical work performed on the Apollo 16 collection clearly exceeds the scope of this report. On the other hand, we wish to convey some measure as to the potential thoroughness, or lack thereof, of investigations performed on each sample mentioned in the report. An approximate estimate concerning the degree of sample characterization may be obtained from the total number of subsplits allocated to various investigators. We therefore consulted the curator's latest allocation list and established the following groupings:

(a) 0-2 allocations: sample essentially unstudied
(b) 3-5 allocations: sample insufficiently studied
(c) 6-20 allocations: sample well studied, but not necessarily comprehensively
(d) >20 allocations: sample studied comprehensively

The following table characterizes each sample in terms of the above allocation groups; samples are listed in numerical sequence. "Walnuts" are samples typically larger than 1 g collected together with normal soil sample or rake sample procedures; "rocks" are individually collected, bagged or documented samples typically >100 g.

Number	Type	Group	Number	Type	Group
60001	core	c	60629	walnut	a
60002	core	d	60635	walnut	a
60003	core	d	60636	walnut	b
60004	core	d	60637	walnut	a
60005	core	d	60638	walnut	a
60006	core	d	60639	walnut	c
60007	core	d	60645	walnut	a
60009	core	d	60646	walnut	a
60010	core	d	60647	walnut	a
60013	core	a	60648	walnut	a
60014	core	a	60649	walnut	a
60500	soil	a	60655	walnut	a
60501	<1 mm	d	60656	walnut	a
60502	1–2 mm	b	60657	walnut	a
60503	2–4 mm	c	60658	walnut	a
60504	4–10 mm	a	60659	walnut	a
60510	rake soil	a	60665	walnut	a
60515	walnut	a	60666	walnut	b
60516	walnut	a	60667	walnut	a
60517	walnut	a	60668	walnut	a
60518	walnut	a	60669	walnut	a
60519	walnut	a	60675	walnut	a
60525	walnut	b	60676	walnut	a
60526	walnut	b	60677	walnut	a
60527	walnut	a	60678	walnut	a
60528	walnut	a	60679	walnut	a
60529	walnut	a	61015	rock	d
60535	walnut	a	61016	rock	d
60600	rake soil	a	61135	rock	c
60601	<1 mm	d	61175	rock	d
60602	1–2 mm	c	61195	rock	d
60603	2–4 mm	a	61220	soil	b
60604	4–10 mm	b	61221	<1 mm	d
60615	walnut	a	61222	1–2 mm	b
60616	walnut	a	61223	2–4 mm	a
60617	walnut	a	61224	4–10 mm	c
60618	walnut	a	61225	walnut	a
60619	walnut	a	61226	walnut	a
60625	walnut	b	61240	soil	a
60626	walnut	b	61241	<1 mm	d
60627	walnut	a	61242	1–2 mm	c
60628	walnut	a	61243	2–4 mm	b

Number	Type	Group	Number	Type	Group
61244	4–10 mm	a	62235	rock	d
61245	walnut	a	62236	rock	c
61246	walnut	a	62237	rock	d
61247	walnut	a	62255	rock	d
61248	walnut	a	62275	rock	c
61249	walnut	a	62295	rock	d
61280	soil	a	63320	soil	c
61281	<1 mm	d	63321	<1 mm	d
61282	1–2 mm	b	63322	1–2 mm	a
61283	2–4 mm	a	63323	2–4 mm	a
61284	4–10 mm	a	63324	4–10 mm	a
61295	rock	d	63340	soil	c
61500	rake soil	c	63341	<1 mm	d
61501	<1 mm	d	63342	1–2 mm	b
61502	1–2 mm	b	63343	2–4 mm	a
61503	2–4 mm	a	63344	4–10 mm	a
61504	4–10 mm	a	63500	rake soil	c
61505	walnut	a	63501	<1 mm	d
61515	walnut	a	63502	1–2 mm	c
61516	walnut	b	63503	2–4 mm	d
61517	walnut	a	63504	4–10 mm	a
61518	walnut	a	63505	walnut	b
61519	walnut	a	63506	walnut	a
61525	walnut	b	63507	walnut	a
61526	walnut	a	63508	walnut	a
61527	walnut	a	63509	walnut	a
61528	walnut	a	63515	walnut	a
61529	walnut	a	63525	walnut	a
61535	walnut	a	63526	walnut	a
61536	walnut	a	63527	walnut	a
61537	walnut	a	63528	walnut	a
61538	walnut	a	63529	walnut	a
61539	walnut	a	63535	walnut	a
61545	walnut	a	63536	walnut	a
61546	walnut	a	63537	walnut	a
61547	walnut	a	63538	walnut	a
61548	walnut	a	63539	walnut	a
61549	walnut	a	63545	walnut	c
61555	walnut	a	63546	walnut	a
61556	walnut	a	63547	walnut	a
61557	walnut	a	63548	walnut	a
61558	walnut	a	63549	walnut	c
61559	walnut	a	63555	walnut	a
61565	walnut	a	63556	walnut	b
61566	walnut	a	63557	walnut	c
61567	walnut	a	63558	walnut	a
61568	walnut	a	63559	walnut	a
61569	walnut	a	63565	walnut	a
61575	walnut	a	63566	walnut	a
61576	walnut	a	63567	walnut	a
61577	walnut	a	63568	walnut	a

Number	Type	Group	Number	Type	Group
63569	walnut	a	64557	walnut	a
63575	walnut	a	64558	walnut	a
63576	walnut	a	64559	walnut	a
62577	walnut	a	64565	walnut	a
63578	walnut	a	64566	walnut	a
63579	walnut	a	64567	walnut	c
63585	walnut	b	64568	walnut	a
63586	walnut	a	64569	walnut	b
63587	walnut	a	64575	walnut	a
63588	walnut	a	64576	walnut	a
63589	walnut	b	64577	walnut	a
63595	walnut	a	64578	walnut	a
63596	walnut	a	64579	walnut	a
63597	walnut	a	64585	walnut	a
63598	walnut	b	64586	walnut	a
64001	core	a	64587	walnut	a
64002	core	a	63588	walnut	a
64420	soil	a	64589	walnut	a
64421	<1 mm	a	64800	rake soil	a
64422	1–2 mm	b	64801	<1 mm	d
64423	2–4 mm	d	64802	1–2 mm	b
64424	4–10 mm	a	64803	2–4 mm	a
64425	walnut	a	64804	4–10 mm	a
64500	rake soil	a	64815	walnut	b
64501	<1 mm	d	64816	walnut	a
64502	1–2 mm	b	64817	walnut	a
64503	2–4 mm	b	64818	walnut	b
64504	4–10 mm	a	64819	walnut	b
64505	walnut	a	64825	walnut	a
64506	walnut	a	64826	walnut	b
64507	walnut	a	64827	walnut	a
64508	walnut	a	64828	walnut	a
64509	walnut	a	64829	walnut	a
64515	walnut	a	64835	walnut	a
64516	walnut	a	64836	walnut	a
64517	walnut	a	64837	walnut	a
64518	walnut	a	65015	rock	d
64519	walnut	a	65315	rock	d
64525	walnut	a	65325	walnut	b
64535	walnut	b	65326	walnut	a
64536	walnut	c	65327	walnut	a
64537	walnut	c	65328	walnut	a
64538	walnut	a	65329	walnut	a
64539	walnut	a	65335	walnut	a
64545	walnut	a	65336	walnut	a
64546	walnut	a	65337	walnut	b
64547	walnut	a	65338	walnut	a
64548	walnut	b	65339	walnut	a
64549	walnut	a	65345	walnut	a
64555	walnut	a	65346	walnut	a
64556	walnut	a	65347	walnut	a

Number	Type	Group
65348	walnut	a
65349	walnut	a
65355	walnut	a
65356	walnut	a
65357	walnut	a
65358	walnut	a
65359	walnut	a
65365	walnut	b
65366	walnut	a
65500	rake soil	c
65501	<1 mm	d
65502	1–2 mm	a
65503	2–4 mm	a
65504	4–10 mm	a
65510	rake soil	a
65511	<1 mm	b
65512	1–2 mm	a
65513	2–4 mm	a
65514	4–10 mm	d
65515	walnut	a
65516	walnut	a
65517	walnut	a
65518	walnut	a
65519	walnut	a
65525	walnut	a
65526	walnut	a
65527	walnut	a
65528	walnut	a
65529	walnut	a
65535	walnut	a
65536	walnut	a
65537	walnut	a
65538	walnut	a
65539	walnut	a
65545	walnut	a
65546	walnut	a
65547	walnut	a
65548	walnut	a
65549	walnut	a
65555	walnut	a
65556	walnut	a
65557	walnut	a
65558	walnut	a
65559	walnut	a
65565	walnut	a
65566	walnut	a
65567	walnut	a
65568	walnut	a
65569	walnut	a
65575	walnut	c
65576	walnut	a

Number	Type	Group
65577	walnut	a
65578	walnut	a
65579	walnut	a
65585	walnut	a
65586	walnut	a
65587	walnut	a
65588	walnut	a
65700	rake soil	a
65701	<1 mm	d
65702	1–2 mm	d
65703	2–4 mm	a
65704	4–10 mm	a
65715	walnut	a
65716	walnut	a
65717	walnut	a
65718	walnut	a
65719	walnut	a
65725	walnut	a
65726	walnut	a
65727	walnut	a
65728	walnut	a
65729	walnut	a
65735	walnut	a
65636	walnut	a
65737	walnut	a
65738	walnut	a
65739	walnut	a
65745	walnut	a
65746	walnut	a
65747	walnut	a
65748	walnut	a
65749	walnut	a
65755	walnut	a
65756	walnut	a
65757	walnut	a
65758	walnut	a
65759	walnut	a
65765	walnut	a
65766	walnut	a
65767	walnut	a
65768	walnut	a
65769	walnut	a
65775	walnut	a
65776	walnut	a
65777	walnut	a
65778	walnut	a
65779	walnut	a
65785	walnut	a
65786	walnut	a
65787	walnut	a
65788	walnut	a

Number	Type	Group
65789	walnut	a
65795	walnut	a
65900	rake soil	a
65901	<1 mm	c
65902	1–2 mm	c
65903	2–4 mm	a
65904	4–10 mm	a
65905	walnut	a
56906	walnut	a
65907	walnut	a
65908	walnut	a
65909	walnut	a
65915	walnut	a
65916	walnut	a
65925	walnut	a
65926	walnut	a
65927	walnut	a
67215	rock	a
67235	rock	a
67455	rock	d
67460	soil	b
67461	<1 mm	d
67462	1–2 mm	b
67463	2–4 mm	b
67464	4–10 mm	a
67475	rock	d
67480	soil	c
67481	<1 mm	d
67482	1–2 mm	b
67483	2–4 mm	c
67484	4–10 mm	a
67485	walnut	a
67486	walnut	a
67487	walnut	a
67488	walnut	a
67489	walnut	a
67495	walnut	a
67510	rake soil	a
67511	<1 mm	a
67512	1–2 mm	a
67513	2–4 mm	a
67514	4–10 mm	a
67515	walnut	a
67516	walnut	a
67517	walnut	a
67518	walnut	a
67519	walnut	a
67525	walnut	a
67526	walnut	a
67527	walnut	a
67528	walnut	a

Number	Type	Group
67529	walnut	a
67535	walnut	a
67536	walnut	a
67537	walnut	a
67538	walnut	a
67539	walnut	a
67545	walnut	a
67546	walnut	a
67547	walnut	a
67548	walnut	a
67549	walnut	a
67555	walnut	a
67556	walnut	a
67557	walnut	a
67558	walnut	a
67559	walnut	c
67565	walnut	b
67566	walnut	a
67567	walnut	a
67568	walnut	a
67569	walnut	a
67575	walnut	a
67576	walnut	a
67600	rake soil	a
67601	<1 mm	d
67602	1–2 mm	c
67603	2–4 mm	d
67604	4–10 mm	a
67615	walnut	b
67616	walnut	b
67617	walnut	a
67618	walnut	b
67619	walnut	a
67625	walnut	a
67626	walnut	a
67627	walnut	d
67629	walnut	b
67635	walnut	b
67636	walnut	b
67637	walnut	b
67638	walnut	a
67639	walnut	a
67645	walnut	a
67646	walnut	a
67647	walnut	a
67648	walnut	a
67649	walnut	a
67655	walnut	a
67656	walnut	a
67657	walnut	b
67658	walnut	a

Number	Type	Group	Number	Type	Group
67659	walnut	a	67757	walnut	a
67665	walnut	a	67758	walnut	a
67666	walnut	a	67759	walnut	a
67667	walnut	b	67765	walnut	a
67668	walnut	a	67766	walnut	a
67669	walnut	a	67767	walnut	a
67675	walnut	a	67768	walnut	a
67676	walnut	a	67769	walnut	a
67685	walnut	a	67775	walnut	a
67686	walnut	a	67776	walnut	a
67687	walnut	a	67915	rock	d
67688	walnut	a	67940	soil	a
67695	walnut	a	67941	<1 mm	d
67696	walnut	a	67942	1–2 mm	b
67697	walnut	a	67943	2–4 mm	a
67700	rake soil	b	67944	4–10 mm	b
67701	<1 mm	d	67945	walnut	b
67702	1–2 mm	d	67946	walnut	a
67703	2–4 mm	c	67947	walnut	a
67704	4–10 mm	a	67948	walnut	a
67705	walnut	a	67960	soil	c
67706	walnut	a	68001	core	a
67707	walnut	a	68002	core	a
67708	walnut	a	68115	rock	d
67710	rake soil	a	68120	soil	a
67711	<1 mm	c	68121	<1 mm	c
67712	1–2 mm	b	68122	1–2 mm	b
67713	2–4 mm	a	68123	2–4 mm	b
67714	4–10 mm	a	68124	4–10 mm	b
67715	walnut	a	68415	rock	d
67716	walnut	a	68416	rock	d
67717	walnut	a	68500	rake soil	a
67718	walnut	a	68501	<1 mm	d
67719	walnut	a	68502	1–2 mm	c
67725	walnut	a	68503	2–4 mm	c
67726	walnut	a	68504	4–10 mm	d
67727	walnut	a	68505	walnut	a
67728	walnut	a	68510	walnut	a
67729	walnut	a	68515	rock	a
67735	walnut	a	68516	walnut	b
67736	walnut	a	68517	walnut	a
67737	walnut	a	68518	walnut	a
67738	walnut	a	68519	walnut	b
67739	walnut	a	68525	walnut	a
67745	walnut	a	68526	walnut	a
57746	walnut	a	68527	walnut	a
67747	walnut	a	68528	walnut	a
67748	walnut	a	68529	walnut	a
67749	walnut	a	68535	walnut	a
67755	walnut	a	68536	walnut	a
67756	walnut	a	68537	walnut	a

Number	Type	Group	Number	Type	Group
68815	rock	d	69942	1–2 mm	c
69001	core	a	69943	2–4 mm	a
69003	CSSD	a	69944	4–10 mm	a
69004	CSSD	c	69945	walnut	c
69920	soil	c	69960	soil	a
69921	<1 mm	a	69961	<1 mm	d
69922	1–2 mm	a	69962	1–2 mm	b
69923	2–4 mm	a	69963	2–4 mm	a
69924	4–10 mm	a	69964	4–10 mm	a
69940	soil	a	69965	walnut	a
69941	<1 mm	d			

Papike, J.J. and Merrill, R.B., eds.
Proc. Conf. Lunar Highlands Crust (1980), p. 51-70
Printed in the United States of America

Recommended classification and nomenclature of lunar highland rocks—a committee report

D. Stöffler[1], H.-D. Knöll[2], U. B. Marvin[3], C. H. Simonds[4], and P. H. Warren[5,6]

[1,2] Institute of Mineralogy, University of Münster, D-4400 Münster, Germany
[3] Smithsonian Astrophysical Observatory, Cambridge, Massachusetts 02138
[4] Northrop Services Incorporated, Houston, Texas 77034
[5] Institute of Geophysics and Planetary Physics, University of California, Los Angeles, California 90024
[6] Present address: Institute of Meteoritics, Department of Geology, University of New Mexico, Albuquerque, New Mexico 87131

Abstract—As a result of the work of the LAPST (Lunar and Planetary Sample Team) Nomenclature Committee the following recommendations for the lunar highland rock classification are made. The rocks should be classified into Groups, Subgroups, Classes, and Subclasses according to a series of textural and compositional properties and to the degree of shock. Root names should be given to the rock classes which are to be subclassified by means of modifying adjectives or prepositions. The proposed system of Groups, Subgroups, and Classes is as follows:

1 *Igneous Rocks.*
   1.1 Volcanic rocks: 1.1.1 Basalts; 1.2 Plutonic rocks: 1.2.1 Anorthosites (Anorthosite, Noritic anorthosite, Gabbroic anorthosite, Troctolitic anorthosite); 1.2.2 Norites (Anorthositic norite, Norite, Gabbroic norite, Olivine norite); 1.2.3 Gabbros (Anorthositic gabbro, Gabbro, Noritic gabbro, Olivine gabbro); 1.2.4 Troctolites (Anorthositic troctolite, Troctolite); 1.2.5 Ultramafics (Pyroxenite, Peridotite, Dunite).
2 *Metamorphic (recrystallized) rocks.*
3 *Breccias.*
   3.1 Monomict breccias: 3.1.1 Cataclastic rocks; 3.1.2 Metamorphic (recrystallized) cataclastic rocks; 3.2 Dimict breccias; 3.3 Polymict breccias: 3.3.1 Regolith (soil) breccias; 3.3.2 Fragmental breccias; 3.3.3 Crystalline melt breccias (impact melt breccias); 3.3.4 Glassy melt breccias ((impact) glass); 3.3.5 Granulitic breccias.

The classes of igneous rocks which are based on the modal or normative composition follow a system similar to that proposed by Prinz and Keil (1977).

1. INTRODUCTION

A number of different classification and nomenclature systems for lunar highland rocks are currently in use in lunar science. This greatly impedes an easy and unequivocal scientific communication on highland rock materials and their genetic interpretation. The reasons for this confusing situation and the needs for an improvement and unification of rock nomenclature have been discussed repeatedly in recent time (Prinz and Keil, 1977; Stöffler *et al.*, 1979 a,b). In view of this, a Nomenclature Committee was established in 1979 on the recommendation of the Lunar and Planetary Sample Team (LAPST) in order to define a new and unified classification and nomenclature system. This paper presents the results of the work of this committee.

2. CLASSIFICATION PRINCIPLES

2.1 First order classification criteria (texture, structure)

The lunar highland rock samples are fragments of impact displaced and metamorphosed lithic units which belong to texturally and compositionally complex, polygenetic breccia formations of the upper part of the lunar crust. Both magmatic and impact processes were important in the evolution of the crustal rocks. Because of their complex genesis such rocks cannot be classified on a single parameter such as mineralogical (or chemical) composition. Several independent parameters such as texture, modal and chemical composition, and grain size are required for an appropriate classification. In view of the dominance of impact processes in the formation of the sampled crustal rocks, we find it necessary to emphasize texture in the classification (see Stöffler *et al.*, 1979 a). Additional properties, such as composition which reflects the nature of the target and which inevitably affects the details of the texture, are treated as modifying terms. With regard to the importance of impact processes it appears reasonable to use the information on rock textures observed at terrestrial impact craters in crystalline bedrock for a first-order *textural classification* of the highland rocks. The present state of knowledge about terrestrial impact formations and their critical implications for the highland rock classification (and genesis) has been reviewed in detail recently (James, 1977; Stöffler *et al.*, 1979 a).

As discussed in some detail by Stöffler *et al*. (1979 a) three basic textural types of highland rocks are distinguished: *igneous rocks, metamorphic rocks,* and *breccias*. Breccias are either monomict, polymict or dimict. The *monomict breccias* are cataclastic rocks formed by in-situ brecciation of a single lithology *(monolithologic)*. *Polymict* or *polylithologic breccias* consist of two main textural components which are termed *matrix* and *clasts*. Such breccias result from the mixing of different lithologies formed under different conditions at different selenological locations. Polymict breccias may have either a *clastic matrix,* a *"melt" matrix* which is crystallized or glassy, or a *metamorphic matrix*. The *clastic matrix* consists of mineral clasts and, in some cases, of additional glassy particles. These clasts form the host for larger rock clasts. The grain size of all clasts is more or less seriate. We define the matrix arbitrarily as the grain size fraction which is smaller than 20–25μm. The *"melt" matrix* is characterized by a variety of textures ranging from holocrystalline (crystalline matrix) to glassy (glassy matrix).

Some crystalline and semi-crystalline textures appear to result from devitrification of glass. The *metamorphic matrix* is typified by a recrystallization texture which is granoblastic to poikiloblastic.

The breccia *clasts* are 1) mineral fragments and fragments of rocks with igneous, metamorphic, and breccia textures, and 2) glassy or partially recrystallized melt bodies or fragments thereof. Many of the clasts are themselves breccias, giving the rock a breccia-in-breccia texture.

Dimict or *dilithologic* breccias are characterized by an unusual structural feature in which a crystalline matrix is combined with a monomict cataclastic breccia texture in an *intrusive-* or *vein-like* relationship.

On the basis of the textural properties outlined above, lunar highland rocks are classified into textural groups, subgroups, and classes in Table 1. This classification has a number of characteristics which are considered to be important for its practical use, in that it

(1) allows the assignment of an unknown sample by standard microscopic examination of thin sections in conjunction with macroscopic observations,
(2) uses equivalent and consistent criteria for the whole classification system,
(3) uses textural and structural properties of a sample as the prime criteria for the classification,
(4) involves a minimum of genetic interpretation, and
(5) is compatible with the classification of terrestrial impact breccias.

The classification system presented here is based on the clast-matrix relationships of the sample as a whole. Thus it is necessary to identify the last formed matrix for each sample considered. This may be difficult or impossible for small samples. Hence, as discussed by Stöffler *et al.* (1979 a), the determination of the parent macroscopic impact formation of any given rock sample may require a very large sample or knowledge of field relationships. Therefore, it is frequently impossible to trace the genesis of a breccia to any particular type of impact process or to a specific parent crater. A detailed discussion of the different textural highland rock classes in terms of their potential selenological provenance (type of parent crater deposit and type and size of parent crater) is given in a previous paper (Stöffler *et al.*, 1979 a).

2.2 Second order classification criteria (composition, texture, grain size, degree of shock)

The wide range of chemical and mineralogical compositions and textures observed in highland breccias make it desirable to supplement the main textural rock classes given in Table 1 by additional characteristics. In the case of breccias, it is most important to link the textural classification with a compositional classification. It is proposed that the same compositional classes as defined for igneous rocks (Table 1, Figs. 1 and 2) be used for the breccias and for the individual breccia components (matrix and clasts).

Table 1. Classification and nomenclature of lunar highland rocks.

Groups	Subgroups	Classes (root names)	Main textural characteristics
Igneous rocks (magmatic rocks)	Volcanic rocks	Basalt	Fine-grained to medium-grained, ophitic-subophitic-intersertal-porphyritic crystallization texture
	Plutonic rocks	Anorthosite Noritic anorthosite Gabbroic anorthosite Troctolitic anorthosite Anorthositic norite Norite Gabbroic norite Olivine norite Anorthositic gabbro Gabbro Noritic gabbro Olivine gabbro Anorthositic troctolite Troctolite Pyroxenite Peridotite Dunite	medium-grained to coarse-grained granular (mosaic) crystallization texture or cumulate texture
Metamorphic rocks (recrystallized rocks)		Granulitic rock[1]	granoblastic to poikiloblastic recrystallization texture
Breccias	Monomict or monolithologic	Cataclastic rock[1]	intergranular in-situ brecciation of a single lithology
		Metamorphic (recrystallized) Cataclastic rock[1]	intergranular in-situ brecciation of a single lithology and partial recrystallization
	Dimict or dilithologic	Dimict breccia	intrusive-like, veined texture of very fine-grained crystallized melt breccia within coarse-grained plutonic or metamorphic rock types

Table 1. *(Continued)*

	Polymict or polylithologic	Regolith breccia or soil breccia	clastic regolith constituents including glass spherules with brown vesiculated matrix glass
		Fragmental breccia	rock clasts in a porous clastic matrix of fine-grained rock debris (mineral clasts)
		(Crystalline) melt breccia or impact melt breccia	rock and mineral clasts in an igneous-textured matrix (granular, ophitic, subophitic, porphyritic, poikilitic, dendritic, fibrous, sheaf-like etc.)
		(Impact) glass or glassy melt breccia	rock and mineral clasts in a coherent glassy or partially devitrified matrix
		Granulitic breccia	rock and mineral clasts in a granoblastic to poikiloblastic matrix

(1) Use plutonic rock name or other rock name (from Figs. 1 and 2), if applicable, for subclassification (e.g., granulitic norite, cataclastic norite, etc.)

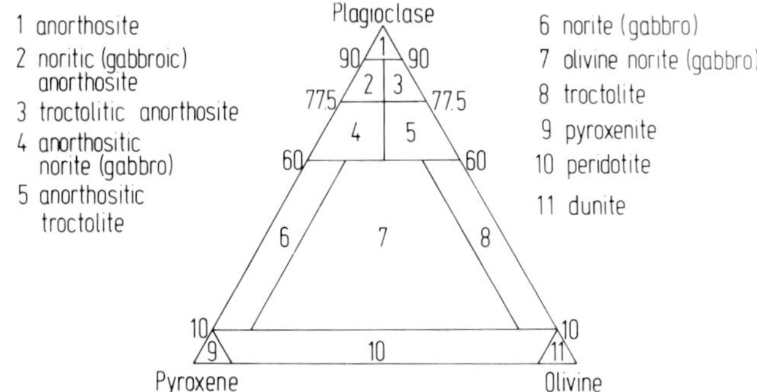

Fig. 1. Classification and names of plutonic lunar highland rocks based on the proposal of Prinz and Keil (1977). Composition in vol. %. The spinel-bearing troctolites are free of pyroxene. Note that the compositional fields are different from the IUGS-system (Streckeisen, 1974).

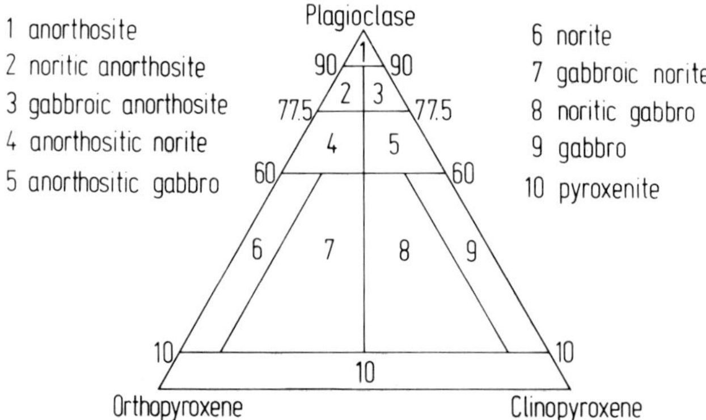

Fig. 2. Composition of noritic and gabbroic lunar highland rocks. Note that the compositional fields are different from the IUGS-system (Streckeisen, 1974).

In general, the following rules should be applied for the subclassification of highland rocks (see Tables 1, 2, 3, and 4):

Igneous rocks

The subclassification of plutonic highland rocks, according to their modal or normative composition, is based on the proposal of Prinz and Keil (1977). The mineralogical composition of the proposed plutonic rock classes in Table 1 is given in Figs. 1 and 2. This classification is preferred over all others (e.g., the IUGS-classification of terrestrial rocks (Streckeisen, 1974) or the purely normative classification of lunar igneous rocks by Engelhardt and Stengelin (1978)) because it allows transitional rock classes between anorthosite and norite (gabbro) or anorthosite and troctolite. Moreover, Engelhardt and Stengelin (1978) make no distinction between igneous rocks and breccias with crystalline matrix.

Additional characterization of igneous rocks may be made on the basis of the crystallization texture

Table 2. Subclassification of breccias by use of a modifier to the root names of Table 1.

breccia class	clast-content	matrix texture	mesostasis-content of matrix	matrix grain-size
fragmental breccias	with melt-particles[1] without melt-particles (cogenetic)[1]	—	—	—
(crystalline) melt breccias or impact melt breccias	clast-poor clast-bearing clast-rich	finely granular ophitic subophitic porphyritic poikilitic dendritic fibrous sheaf-like spherulitic	mesostasis-bearing mesostasis-rich	very fine fine medium coarse
(impact) glasses or glassy melt breccias	clast-poor clast-bearing	vitric partially devitrified	—	—
granulitic breccias	clast-poor clast-bearing	granoblastic poikiloblastic	—	very fine fine medium coarse

very fine grained: <0.03 mm
fine grained: 0.03 mm–0.3 mm
medium grained: 0.3 mm–3.0 mm
coarse grained: >3.0 mm

clast-poor: <10 vol.% clasts
clast-bearing: 10–25 vol.% clasts
clast-rich: >25 vol.% clasts

mesostasis-bearing: <10 vol.% mesostasis
mesostasis-rich: >10 vol.% mesostasis

Table 3. Subclassification of polymict breccias by use of a modifier to the root names of Table 1.

breccia class	composition of matrix	composition of clasts
fragmental breccias	—	
crystalline melt breccias (impact melt breccias) (impact) glasses (glassy melt breccias) granulitic breccias	conventional nomenclature according to modal and/or normative composition, e.g. noritic anorthositic dunitic (see Table 1, Figs. 1 and 2)	volume proportion of clast lithologies (using highland rock classification)

Table 4. Classification of highland rocks according to their degree of solid state shock metamorphism.

Pressure data according to Stöffler, 1971; Kieffer, 1971, 1975; Kieffer *et al.*, 1976; Schaal *et al.*, 1979; Schaal and Hörz, 1977; 1 GPa = 10 kbar

Rock type	degree of shock	estimated peak pressure (GPa)	textural characteristics
Non-porous rocks (igneous and metamorphic rocks, monomict breccias, dimict breccias, crystalline melt breccias, impact glass, granulitic breccias)	unshocked	5	primary rock texture and intragranular texture of mineral grains unchanged
	weakly shocked	5–30	intragranular fracturing and mosaicism in all mineral constituents; planar deformation structures in feldspar, pyroxene and olivine; primary rock texture unchanged
	strongly shocked	30–45	mosaicism and planar deformation structures in mafic minerals; solid state isotropization of plagioclase (maskelynite, diaplectic or thetomorphic glass); primary rock texture unchanged
Porous rocks (regolith breccias, fragmental breccias)	unshocked (friable)	0– 3	highly porous aggregation of breccia constituents
	shocked (coherent)	>3, mostly >10	moderate, small or lacking porosity in the matrix; mosaicism and fracturing of minerals; intergranular melt

Note: Rocks with glass-coating are to be designated "glass-coated" and the grain size. Four grain size classes are proposed for designating the average grain size of an igenous rock:

<0.03 mm (very fine grained)
0.03–0.3 mm (fine grained)
0.3 –3 mm (medium grained)
>3 mm (coarse grained)

A classification is not presented for subgroups of volcanic rocks which are predominantly basalts, because this paper is mainly concerned with the problem of classifying breccias. However, a unified subclassification of basalts should be considered in the future.

Monomict breccias and metamorphic (recrystallized) rocks

They are typically subdivided on the basis of modal mineralogy and mode of origin of the rock prior to brecciation (see Table 1 and Figs. 1 and 2).

Dimict breccias

A more specific characterization of the two lithological components is to be made according to the pertinent properties of each lithology (Tables 1, 2, 3, and 4).

Polymict breccias

The breccia *matrix* should be characterized according to the following properties (Tables 2 and 3):
1) mineralogical or normative composition (Table 3)
2) texture (Table 2)
3) grain size (Table 2)
4) mesostasis content (Table 2)

The *clasts* should be designated according to:
1) the compositional and textural type (Tables 1–4)
2) the volumetric proportion (Table 3)

For the *whole breccia* the clast-content should be given according to the scheme in Table 2.

All highland rocks may be subclassified according to the degree of *shock metamorphism* caused by the latest shock event which predominantly resulted in solid state deformation or transformation (intergranular fracturing, mosaicism, planar deformation structures in the mineral constituents, and solid state isotropization or incipient selective melting of feldspar; Table 4).

3. NOMENCLATURE PRINCIPLES

The basic principle of the nomenclature proposed for the textural classes of lunar highland rocks (Table 1) is to assign short *root names* to the individual rock classes. For a complete and exact naming of a particular sample this root name should be combined with one or more *modifiers* which describe second-order properties of the sample. It is most convenient if the modifiers are adjectives or prepositions (Table 2, 3, and 4; Figs. 1 and 2). We attempted to use only non-genetic, textural terms for the root names and modifiers and to avoid root names which consist of more than two words. However, as the history of lunar rock terminology shows, it is extremely difficult to fulfill these two requirements and to keep a completely uniform scheme of terms without ending in an odd or meaningless nomenclature. Consequently, the proposed names in Table 1 represent a compromise between purely non-genetic and well known, commonly used genetic terms. In some of the latter cases, alternative, non-genetic terms have been suggested.

In the group of *Igneous Rocks,* the naming is based on the proposal of Prinz and Keil (1977). The names of the plutonic rock classes are given in Table 1 and in Figs. 1 and 2. Textural terms commonly used in terrestrial igneous petrology may be applied for the characterization of the crystallization texture of igneous highland rocks (e.g., granular, ophitic, subophitic, intersertal, porphyritic etc.). In addition, igneous rocks may be termed very fine grained, fine grained, medium grained, or coarse grained (see section 2.2 for the proposed grain size ranges).

In the group of *Metamorphic Rocks,* the term granulitic is proposed in conjunction with a compositional name of the kind suggested for plutonic rocks, e.g., *granulitic norite*. Granulitic is meant in a more restricted, textural sense compared to the usage in terrestrial petrography. For lunar highland samples it indicates a granoblastic or poikiloblastic recrystallization texture.

In the group of *Breccias,* the term "breccia", which is part of the root name in most rock classes (Table 1), can be combined with the term "impact" if this appears necessary in a particular case. In the case of monomict breccias, it is suggested to use the compositional plutonic rock names as given in Table 1 in the composite root name, e.g., *cataclastic anorthosite*. For dimict and polymict breccias, the compositional characterization of the breccia components, e.g., the matrix, is made by a modifier, e.g., *anorthositic granulitic breccia* (Table 3). In the breccia class "crystalline melt breccia" ("impact melt breccia"), which comprises a large variety of matrix textures and compositions, the general adjective "crystalline" can be replaced by a term for the matrix texture in any particular case (Table 2), e.g., *poikilitic melt breccia*. The term "impact" could be deleted in such a case. This holds also for the breccia class "impact glass". If a lunar glass body is clast-free, the term "glass" is most appropriate if an impact origin is not certain. For the breccia class "granulitic breccia", there are two options for naming a particular sample, e.g., "granoblastic (or poikiloblastic) granulitic breccia" or just "granoblastic (or poikiloblastic) breccia", in combination with a compositional term, e.g., *noritic granoblastic* breccia (Table 3).

All other terms proposed in Tables 1, 2, 3, and 4 are self-explanatory. A list of previously used rock names or designations, excluding igneous rocks, is given in Table 5 for each newly proposed rock class. The old names are either synonymous with the new name or they designate only a subclass of the rock class in question. Synonyms are italicized in Table 5.

4. DESCRIPTION AND DETERMINATION OF HIGHLAND ROCK CLASSES

A key chart for the classification of a lunar highland rock sample is presented in Fig. 3 which is adapted from Fig. 15 of Stöffler *et al.* (1979 a). The various rock classes listed in Table 1 and Fig. 3 have the general petrographic characteristics outlined in the following sections. These sections are adapted in part from Stöffler

Table 5. List of previous names and designations for non-igneous lunar highland rocks (breccias and metamorphic rocks). Previous names which are fully synonymous with the proposed new name are italicized, the other ones designate only a subclass of the new rock class in question.

A. METAMORPHIC ROCKS
Proposed name: *granulitic rock*[1)]
[1)] plutonic rock names

Previous names or designations	Reference
hornfelsed noritic microbreccia	Chao *et al.*, 1972
granulitic norite breccia	Taylor *et al.*, 1972
granulitic or poikilitic crystalline rock	Simonds *et al.*, 1974
clast-free granulitic impactite	Warner *et al.*, 1977
meta-"rock"	Stöffler *et al.*, 1979a

B. BRECCIAS
1. Monomict breccias
Proposed name: *cataclastic rock*[1)]
[1)] plutonic rock names

Previous names or designations	Reference
brecciated anorthositic rock	Chao, 1973
cataclastic anorthosite	LSPET, 1973a
	Warner *et al.*, 1976
brecciated gabbroic rocks	LSPET, 1973b
Apollo 16 B_1-group (in part)	Wilshire *et al.*, 1973
Apollo 16 C_1-group (in part)	Wilshire *et al.*, 1973
class 4-rock (cataclastic breccia)	Stöffler *et al.*, 1974
cataclasite	James, 1977
cataclastic rock	Phinney *et al.*, 1977
	Stöffler *et al.*, 1979a

Proposed name: *recrystallized cataclastic rock*[1)]
[1)] plutonic rock names

Previous names or designations	Reference
brecciated and recrystallized anorthositic rocks	Chao, 1973
cataclastic breccia with recrystallization	Stöffler *et al.*, 1974
recrystallized anorthosite	Vaniman *et al.*, 1976
metacataclastic rock	Stöffler *et al.*, 1979a

2. Dimict breccias
Proposed name: *dimict breccia*

Previous names or designations	Reference
partially molten breccia	LSPET, 1973a
Apollo 16 B_2-group (in part)	Wilshire *et al.*, 1973
black and white rock	Warner *et al.*, 1973
	James, 1977
	McGee *et al.*, 1979

Table 5. *(Continued)*

Previous names or designations	Reference
white portion: cataclastic rock	Phinney *et al.*, 1977
black portion: basaltic matrix breccia	Phinney *et al.*, 1977
dike breccia	Stöffler *et al.*, 1979a

3. Polymict breccias
Proposed name: *regolith breccia (or soil breccia)*

Previous names or designations	Reference
regolith microbreccia	Chao *et al.*, 1971
	Chao, 1973
(glass-rich) *regolith breccia*	Engelhardt *et al.*, 1972
regolith breccia	Quaide and Wrigley, 1972
	James, 1977
	Stöffler *et al.*, 1979a
grade 1 metamorphic breccia	Warner, 1972
Apollo 14 F_1*-group*	Wilshire and Jackson, 1972
dark matrix breccia (DMB)	LSPET, 1973b
	Delano *et al.*, 1973
	Vaniman *et al.*, 1976
	McGee *et al.*, 1979
glassy breccia	Warner *et al.*, 1973
class 3 - rock (regolith breccia)	Stöffler *et al.*, 1974
vitric matrix breccia	Phinney *et al.*, 1977
soil breccia	Warner *et al.*, 1978

Proposed name: *fragmental breccia*

Previous names or designations	Reference
feldspathic breccia	Chao *et al.*, 1972
glass-poor breccia with fragmental matrix	Engelhardt *et al.*, 1972
grade 3 metamorphic breccia	Warner, 1972
Apollo 14 F_3-group	Wilshire and Jackson, 1972
Apollo 16 B_2-group (in part)	Wilshire *et al.*, 1973
Apollo 16 B_3-group (in part)	Wilshire *et al.*, 1973
polymict breccia	LSPET, 1973a
light matrix breccia	Warner *et al.*, 1973
	Phinney *et al.*, 1977
	McGee *et al.*, 1979
fragmental breccia	Simonds *et al.*, 1974
	McGee *et al.*, 1979
glass-poor feldspathic breccia	James, 1977
clastic breccia	Stöffler *et al.*, 1979a

Table 5. *(Continued)*

Proposed name: *(crystalline) melt breccia or impact melt breccia*

Previous names or designations	Reference
annealed Fra Mauro breccia	Chao *et al.*, 1972
glass-poor fragmental rocks with crystalline matrix	Engelhardt *et al.*, 1972
annealed breccia	Quaide and Wrigley, 1972
Apollo 14 grades 4–8 metamorphic breccias	Warner, 1972
Apollo 14 group F_4	Wilshire and Jackson, 1972
pyroxene poikiloblastic breccia (POIK)	Delano *et al.*, 1973
light matrix breccia (LMB)	Delano *et al.*, 1973
feldspathic intersertal igneous rocks (FIIR)	Delano *et al.*, 1973
Apollo 16 high grade metamorphic rocks	LSPET, 1973a
type 1 and 2 of light gray breccia	LSPET, 1973b; James, 1977
blue gray breccia	LSPET, 1973b; James, 1977
green gray breccia	LSPET, 1973b; James, 1977
metamorphosed breccia	Warner *et al.*, 1973
melted matrix breccia	Warner *et al.*, 1973
mesostasis-rich rocks (breccias)	Warner *et al.*, 1973, 1976
basalts (in part)	Warner *et al.*, 1973
poikilitic rock	Warner *et al.*, 1973, 1976
Apollo 16 groups C_2, B_4, B_5	Wilshire *et al.*, 1973
crystalline breccias with poikilitic, micropoikilitic, subophitic, granular and clast-rich ophitic matrix	Simonds *et al.*, 1974
class 6 rock (melt rock)	Stöffler *et al.*, 1974
recrystallized/remelted noritic breccia	Vaniman *et al.*, 1976
pyroxene poikilitic rock	Vaniman *et al.*, 1976
subophitic-granular-micropoikilitic breccias	Warner *et al.*, 1976
thermally metamorphosed breccia	James, 1977
crystalline matrix breccia	Phinney *et al.*, 1977 Stöffler *et al.*, 1979a
basaltic matrix breccia	Phinney *et al.*, 1977
poikilitic matrix breccia	Phinney *et al.*, 1977
high grade breccia	Phinney *et al.*, 1977
low grade breccia	Phinney *et al.*, 1977
poikilitic breccia	Phinney *et al.*, 1977
impact melt rock	McGee *et al.*, 1979

Proposed name: *(impact) glass or glassy melt breccia*

Previous names or designations	Reference
Apollo 17 glass-bonded agglutinate	LSPET, 1973b
glass and devitrified glass	Warner *et al.*, 1973
Apollo 16 group G	Wilshire *et al.*, 1973
class 1 and 2 rock (glassy agglutinate and glass)	Stöffler *et al.*, 1974
agglutinate	Vaniman *et al.*, 1976
light matrix breccia (?)	Vaniman *et al.*, 1976
glassy breccia	Warner *et al.*, 1976
glassy matrix breccia	Phinney *et al.*, 1977
vitric matrix breccia	Stöffler *et al.*, 1979a

Table 5. *(Continued)*

Proposed name: *granulitic breccia*

Previous names and designations	Reference
Apollo 16 group C_2 (in part)	Wilshire *et al.*, 1973
class 5 rock (polymict metamorphic breccia)	Stöffler *et al.*, 1974
recrystallized noritic breccia (?)	Vaniman *et al.*, 1976
light matrix breccia (in part)	Vaniman *et al.*, 1976
granulitic impactite	Warner *et al.*, 1977
recrystallized ANT-rocks	Warner *et al.*, 1978

et al. (1979 a). Other parts are supplemented and updated with respect to the nomenclature proposed here. Typical microphotographs may be found in Stöffler *et al.* (1979 a).

4.1 Igneous (magmatic) rocks

Rocks of this class display either igneous crystallization or cumulate textures. *Basalts* have various crystallization textures ranging from ophitic, subophitic, intersertal, to porphyritic. Plagioclase is predominant over clinopyroxene, olivine, and ilmenite. Most basalts are fine-grained (<0.3 mm). *Plutonic rocks* display a wide compositional range (Table 1 and Fig. 1) and mostly have granular mosaic textures of plagioclase with interstitial anhedral or intragranular subhedral to euhedral pyroxene and/or olivine. Another plutonic type displays cumulate texture. Igneous rocks commonly have grain sizes of 1–5 mm, i.e. are medium to coarse grained. Except for some dunitic rocks, plagioclase (anorthite) predominates (see reviews of Prinz and Keil, 1977; Meyer, 1977; Warren and Wasson, 1977, 1978; Norman and Ryder, 1979).

4.2 Metamorphic (recrystallized) rocks

The texture of these "granulitic rocks" is typically granoblastic to poikiloblastic (Warner *et al.*, 1977; Bickel and Warner, 1978) and is formed obviously by solid state recrystallization induced by thermal metamorphism of variable source rock types. Usually granulitic rocks are plagioclase-rich (anorthositic, anorthositic-gabbroic, anorthositic-noritic). They display a wide range of grain sizes, very fine-grained to medium-grained (<3 mm). Some seem to be transitional to plutonic rocks. Compared to granulitic breccias, samples of this class are rather rare in the Apollo and Luna rock collections (Bickel and Warner, 1978).

4.3 Breccias

Depending on the number of different lithologic components the breccias are either *monomict, dimict,* or *polymict*. Genetically, these breccias were formed in impact craters by different processes which have been discussed in detail by James (1977), Simonds *et al.* (1977), Phinney *et al.* (1977), McGee *et al.* (1979), and Stöffler *et al.* (1979 a).

4.3.1 Cataclastic rocks

The texture of these rocks is characterized by an intragranular cataclasis and intergranular brecciation of either igneous rocks, granulitic rocks, crystalline melt breccias or granulitic breccias. Small mineral fragments are typically displaced along former grain boundaries of the precursor rock, while large relic grains remain in situ. Such monomict breccias are best developed in coarse-grained source rocks with homogeneous texture (igneous and metamorphic).

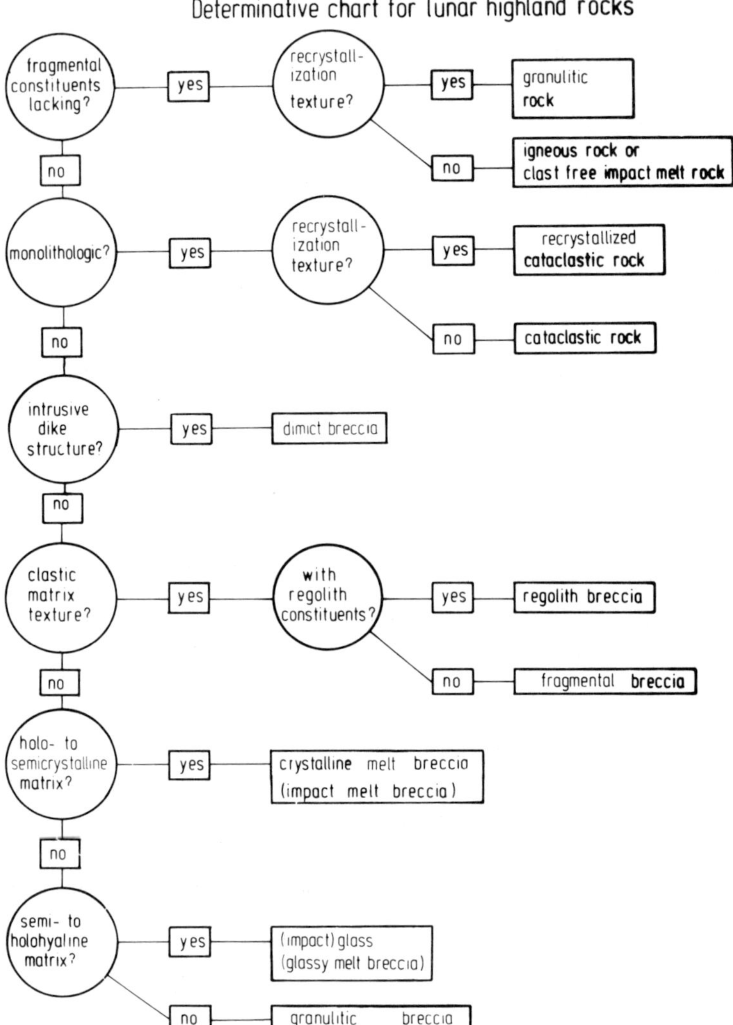

Fig. 3. Determinative chart for the classes of lunar highland rocks adapted from Fig. 15 of Stöffler *et al.* (1979 a). Further subclassification is to be made according to Tables 1, 2, 3 and 4. The term "recrystallization texture" is preferred here over "metamorphic texture" in order to facilitate the recognition of the metamorphic texture.

4.3.2 Recrystallized cataclastic rocks

In some monomict breccias, the fine-grained intergranular highly brecciated regions are affected by a subsequent thermally-induced recrystallization. Thereby a granoblastic texture is formed in those regions and the cataclastic texture appears as a relic in a partially recrystallized breccia. All other characteristics are identical to those of the cataclastic rocks described above. The genetic interpretation of cataclastic rocks and their relation to parent impact formations have been discussed recently in detail (Stöffler *et al.*, 1979 a).

4.3.3 Dimict breccias

Samples of this class are composed of two different types of lithologies with distinctly different textures. One textural unit appears intrusive with respect to the other unit, resulting in a vein- or dike-like structure. The intrusive component commonly contains clasts and has a very fine-grained aphanitic matrix texture; it can be classified per se as a crystalline melt breccia. The clastic inclusions are mostly derived from the adjacent lithology, which belongs to the class of cataclastic rocks in most cases. Texturally these rocks are comparable to terrestrial "pseudotachylites" (Wilshire, 1971; James, 1977; McGee *et al.*, 1979; Stöffler *et al.*, 1979 a).

4.3.4 Regolith (soil) breccias

Samples of this type consist of the fine-grained particulate constituents of the regolith: rock, mineral, and glass fragments; glass spherules; glassy agglutinates; and minor meteoritic fragments. These were agglomerated to a more or less coherent rock by shock lithification or sintering of hot glass (Kieffer, 1975; Simonds, 1973; Simonds *et al.*, 1977). Thereby glass is formed along grain boundaries in the matrix; this glass causes the coherence of the clastic particles. The grain size of the largest clasts in the regolith breccias rarely exceeds 1 centimeter. Most significant for the recognition of regolith breccias is the presence of intact or broken glass spherules, glassy agglutinates, and finely dispersed matrix glass. All other types of highland breccias are free of these constituents. The compositional and textural variation of lithic clasts is larger than in all other breccia types due to the intense impact gardening of the regolith material.

4.3.5 Fragmental breccias

Fragmental breccias are composed of clastic rock debris derived from different lithologies of variable composition, texture, and degree of shock. The grain size of the clasts ranges from several centimeters down to the micron scale. Because of the clastic matrix, these breccias are weakly coherent and porous. According to Table 2, two subclasses may be distinguished. One type contains vitric or devitrified melt particles which have the same chemical composition and appear to be cogenetic, i.e. that the melt was formed in the same cratering event from which the admixed clastic rock material was derived in analogy to the terrestrial suevite breccia. Examples of this type most probably are the Apollo 14 so-called "white rocks" 14063 and 14082 (James, 1977; Stöffler *et al.*, 1979 a). The other type is free of such cogenetic melt particles and may be equivalent to the clastic breccia layers of the bulk ejecta blanket or fallback formations of large terrestrial craters (compare Stöffler *et al.*, 1979a). Possible examples are the "white breccia boulders" (67455, 67475) of North Ray crater (Apollo 16). The compositional and textural variations in the lithic clast population of fragmental breccias is very large, similiar to that of regolith breccias, i.e. distinctly larger than in all other types of polymict breccias.

4.3.6 Crystalline melt breccias (impact melt breccias)

Breccias of this class are the most frequent in the lunar highland rock collection. The most common type is composed of rock and mineral clasts embedded in an *igneous-textured* crystalline *matrix*. The clasts are homogeneously distributed or are concentrated in schlieren-like bodies. In the latter case, the overall rock texture is rather heterogeneous. In some rocks the texture of the crystalline matrix itself may be heterogeneous on the macro- and micro-scale. Typical matrix textures are very fine-grained and range from granular, ophitic, subophitic, poikilitic, to porphyritic. These rocks may contain small amounts of mesostasis. From the chemical and petrographic characteristics of the matrix and of the clast population, it was concluded that these breccias result from the crystallization of a fragment-laden, coherent impact melt (e.g., Simonds *et al.*, 1976; Onorato *et al.*, 1976) or from clast-rich melt agglomerates similar to the very melt-rich suevitic breccias observed at terrestrial impact craters (Stöffler *et al.*, 1978). The most abundant minerals of the igneous-textured matrix types are plagioclase, pyroxene and/or olivine, and minor ilmenite. In the mineral clast population, plagioclase generally dominates over pyroxene and olivine. Accessory minerals generally are ilmenite, spinel, Ni-iron metal, various sulfides, zircon, baddeleyite, and phosphate minerals. Rock fragments in these breccias include variable types of igneous and metamorphic rocks, cataclastic rocks, and polymict breccias with crystalline matrix. Breccias with coarser grained matrix (e.g., pokilitic, intersertal) tend to be poorer in clasts than those with very fine-grained matrix.

A second type of melt breccia is characterized by a *matrix with devitrification texture* and appears to be transitional to the "glassy breccias" ("impact glasses"). The textures range from spherulitic to dendritic, fibrous, and sheaf-like and are very fine-grained or even cryptocrystalline. The most common crystallization products are plagioclase, pyroxene, olivine, and ilmenite. The rock and mineral clasts are similar to those observed in igneous textured melt breccias, but are generally smaller in grain size and less abundant. Breccias of this type are considered to result from a subsolidus crystallization of quenched fragment-laden impact melt.

Vesicles are abundant in some samples of both types of crystalline melt breccia, in others they are lacking. The occurrence and selenological provenance of melt breccias is discussed by James (1977), Phinney *et al.* (1977), Simonds *et al.* (1977), and Stöffler *et al.*, (1979 a).

4.3.7 Impact glass; glassy melt breccias

Samples of this type are typified by a vitric matrix in which fragmental minerals and rocks are embedded. Some glasses are free of clasts. The matrix may be partially devitrified. Schlieren and vesicles are common. Some vesiculated glasses have extremely irregular shapes (agglutinates) and are related to glass-rich regolith breccias. Impact glasses form relatively small particles (0.005–30 mm) of various shapes (spheres or other bodies of revolution, slaggy shapes, and fragments) and are the most characteristic constituents of the regolith. Glasses also occur as clasts in certain types of fragmental breccias or as coatings on larger rocks. Many glasses contain small Ni-bearing iron-spherules and troilite. In general, glasses originate from quenched, ballistically transported fragment-laden impact melts. Some glasses may be of volcanic origin.

4.3.8 Granulitic breccias

In contrast to clast-free granulitic rocks described in section 4.2, these rocks contain mineral and rock clasts or relics of clasts which obviously survived the recrystallization of the breccia matrix. This matrix is granoblastic to poikiloblastic and commonly fine-grained. The clasts may also display a recrystallized texture. The modal composition of the matrix is dominated by plagioclase, forming a mosaic of grains meeting at 120° triple junctions. Small euhedral pyroxene and olivine are embedded within and in between plagioclase (granoblastic breccias). In poikiloblastic textures, plagioclase forms minute euhedral crystals disseminated in larger pyroxene and/or olivine crystals. Ilmenite is rare. Granulitic breccias are interpreted to be formed by thermally induced subsolidus recrystallization of variable breccia precursors (Warner *et al.*, 1977; Bickel and Warner, 1978). The heat may have been delivered by a global crustal metamorphism (Warner *et al.*, 1977; Stewart, 1975) or by a surrounding hot impact melt (Stöffler *et al.*, 1979 b). The latter is commonly observed in shocked clasts of crystalline rocks within terrestrial impact melts.

Acknowledgments—We thank Dr. O. B. James for very helpful comments and suggestions. Thanks are also due to Dr. A. Basu and Dr. K. Keil for their constructive criticism of the manuscript. The first author is indebted to the Deutsche Forschungsgemeinschaft (German Science Foundation) for financial support.

REFERENCES

Bickel C. E. and Warner J. L. (1978) Survey of lunar plutonic and granulitic lithic fragments. *Proc. Lunar Planet. Sci. Conf. 9th.*, p. 629–652.

Chao E. C. T. (1973) Preliminary genetic classification of Apollo 16 breccias (abstract). In *Lunar Science IV*, p. 129. The Lunar Science Institute, Houston.

Chao E. C. T., Boreman J. A., and Desborough G. A. (1971) The petrology of unshocked and shocked Apollo 11 and Apollo 12 microbreccias. *Proc. Apollo 11 Lunar Sci. Conf.*, p. 797–816.

Chao E. C. T., Minkin J. A., and Best J. B. (1972) Apollo 14 breccias: General characteristics and classification. *Proc. Lunar Sci. Conf. 3rd*, p. 645–659.

Delano J. W., Bence A. E., Papike J. J., and Cameron K. L. (1973) Petrology of the 2–4 mm soil fraction from the Descartes region of the moon and stratigraphic implications. *Proc. Lunar Sci. Conf. 4th*, p. 537–552.

Engelhardt W. v., Arndt J., Stöffler D., and Schneider H. (1972) Apollo 14 regolith and fragmental rocks, their compositions and origin by impact. *Proc. Lunar Sci. Conf. 3rd*, p. 753–770.

Engelhardt W. v. and Stengelin R. (1978) Normative composition and classification of lunar igneous rocks and glasses, I. Lunar igneous rocks. *Earth Planet. Sci. Lett.* **42,** 213–222.

James O. B. (1977) Lunar highlands breccias generated by major impacts. In *The Soviet-American Conference on Cosmochemistry of the Moon and Planets*, p. 637–658. NASA SP-370. Washington, D.C.

Kieffer S. W. (1971) Shock metamorphism of the Coconino Sandstone at Meteor Crater, Arizona. *J. Geophys. Res.* **76,** 5449-5473.

Kieffer S. W. (1975) From regolith to rock by shock. *The Moon* **13,** 301–320.

Kieffer S. W., Schaal R. B., Gibbons R. V., Hörz F., Milton D., and Duba A. (1976) Shocked basalt from Lonar Impact Crater (India) and experimental analogues. *Proc. Lunar Sci. Conf. 7th*, p. 1391–1412.

LSPET (Lunar Sample Preliminary Examination Team) (1973a) The Apollo 16 lunar samples: Petrographic and chemical description. *Science* **179,** 23–34.

LSPET (1973 b) Apollo 17 lunar samples: Chemical and petrographic description. *Science* **182,** 659–672.

McGee P. E., Simonds C. H., Warner J. L., and Phinney W. C. (1979) Introduction to the Apollo collections Part II: Lunar breccias. Curator's Office NASA Johnson Space Center, Houston.

Meyer C. E. (1977) Petrology, mineralogy, and chemistry of KREEP basalt. *Phys. Chem. Earth* **10,** 506–508.

Norman M. D. and Ryder G. (1979) A summary of the petrology and geochemistry of pristine highlands rocks. *Proc. Lunar Planet. Sci. Conf. 10th*, p. 531–559.

Onorato P. I. K., Uhlmann D. R., and Simonds C. H. (1976) Heat flow in impact melts: Apollo 17 Station 6 Boulder and some application to other breccias and xenolith laden melts. *Proc. Lunar Sci. Conf. 7th*, p. 2449–2467.

Phinney W. C., Warner J. L., and Simonds C. H. (1977) Lunar Highland rock types: Their implications for impact-induced fractionation. In *The Soviet-American Conference on Cosmochemistry of the Moon and Planets*, p. 91–126. NASA SP-370. Washington, D.C.

Prinz M. and Keil K. (1977) Mineralogy, petrology and chemistry of ANT-suite rocks from the lunar highlands. *Phys. Chem. Earth* **10,** 215–237.

Quaide W. and Wrigley R. (1972) Mineralogy and origin of Fra Mauro fines and breccias. *Proc. Lunar Sci. Conf. 3rd*, p. 771–784.

Schaal R. B. and Hörz F. (1977) Shock metamorphism of lunar and terrestrial basalts. *Proc. Lunar Sci. Conf. 8th*, p. 1697–1729.

Schaal R. B., Hörz F., Thompson T. D., and Bauer J. F. (1979) Shock metamorphism of granulated lunar basalt. *Proc. Lunar Planet. Sci. Conf. 10th*, p. 2547–2571.

Simonds C. H. (1973) Sintering and hot pressing of Fra Mauro composition glass and the lithification of lunar breccias. *Amer. J. Sci.* **273,** 428–439.

Simonds C. H., Phinney W. C., and Warner J. L. (1974) Petrography and classification of Apollo 17 non-mare rocks with emphasis on samples from the Station 6 boulder. *Proc. Lunar Sci. Conf. 5th*, p. 3337–3353.

Simonds C. H., Phinney W. C., Warner J. L., McGee P. E., Gaeslin J., Brown R. W., and Rhodes J. M. (1977) Apollo 14 revisited, or breccias aren't so bad after all. *Proc. Lunar Sci. Conf. 8th,* p. 1869–1893.

Simonds C. H., Warner J. L., Phinney W. C., and McGee P. E. (1976) Thermal model for impact breccia lithification: Manicouagan and the moon. *Proc. Lunar Sci. Conf. 7th,* p. 2509–2528.

Stöffler D. (1971) Progressive metamorphism and classification of shocked and brecciated crystalline rocks at impact craters. *J. Geophys. Res.* **76,** 5541–5551.

Stöffler D., Dence M. R., Graup G., and Abadian M. (1974) Interpretation of ejecta formations at the Apollo 14 and 16 sites by a comparative analysis of experimental, terrestrial, and lunar craters. *Proc. Lunar Sci. Conf. 5th,* p. 137–150.

Stöffler D., Knöll H.-D., and Maerz U. (1979a) Terrestrial and lunar impact breccias and the classification of lunar highland rocks. *Proc. Lunar Planet. Sci. Conf. 10th,* p. 339–375.

Stöffler D., Knöll H.-D., and Maerz U. (1979b) Genetic classification and nomenclature of lunar highland rocks based on the texture and geological setting of terrestrial impact breccias (abstract). In *Lunar and Planetary Science X,* p. 1177–1179. Lunar and Planetary Institute, Houston.

Stöffler D., Knöll H.-D., Stähle V., and Ottemann J. (1978) Textural variations of the crystalline matrix of Fra Mauro breccias and a model of breccia formation (abstract). In *Lunar and Planetary Science IX,* p. 1116–1118. Lunar and Planetary Institute, Houston.

Stewart D. B. (1975) Apollonian metamorphic rocks (abstract). In *Lunar Science VI,* p. 774–776. The Lunar Science Institute, Houston.

Streckeisen A. (1974) Classification and nomenclature of plutonic rocks. *Geol. Rundsch.* **63,** 773–786.

Taylor G. J., Marvin U. B., Reid J. B., and Wood J. A. (1972) Noritic fragments in the Apollo 14 and 12 soils and the origin of Oceanus Procellarum. *Proc. Lunar Sci. Conf. 3rd,* p. 995–1014.

Vaniman D. T., Lellis S. F., Papike J. J., and Cameron K. L. (1976) The Apollo 16 drill core: Modal petrology and characterization of the mineral and lithic component. *Proc. Lunar Sci. Conf. 7th,* p. 199–239.

Warner J. L. (1972) Metamorphism of Apollo 14 breccias. *Proc. Lunar Sci. Conf. 3rd,* p. 623–643.

Warner J. L., Phinney W. C., Bickel C. E., and Simonds C. H. (1977) Feldspathic granulitic impactites and pre-final bombardment lunar evolution. *Proc. Lunar Sci. Conf. 8th,* p. 2051–2066.

Warner J. L., Simonds C. H., and Phinney W. C. (1973) Apollo 16 rocks: Classification and petrogenetic model. *Proc. Lunar Sci. Conf. 4th,* p. 481–504.

Warner R. D., Dowty E., Prinz M., Conrad G. H., Nehru C. E., and Keil K. (1976) *Catalogue of Apollo 16 rake samples from the LM area and Station 5.* Inst. Meteoritics Spec. Publ. No. 13, Univ. New Mexico, Albuquerque. 87 pp.

Warner R. D., Keil K., Nehru C. E., and Taylor G. J. (1978) *Catalogue of Apollo 17 rake samples from Stations 1A, 2, 7 and 8.* Inst. Meteoritics Spec. Publ. No. 18. Univ. New Mexico, Albuquerque, 88 pp.

Warren P. H. and Wasson J. T. (1977) Pristine non-mare rocks and the nature of the lunar crust. *Proc. Lunar Sci. Conf. 8th,* p. 2215–2235.

Warren P. H. and Wasson J. T. (1978) Compositional-petrographic investigation of pristine nonmare rocks. *Proc. Lunar Planet. Sci. Conf. 9th,* p. 185–217.

Wilshire H. G. (1971) Pseudotachylite from the Vredefort Ring, South Africa. *J. Geol.* **79,** 195–206.

Wilshire H. G. and Jackson E. D. (1972) Petrology and stratigraphy of the Fra Mauro formation at the Apollo 14 site. *U.S. Geol. Survey Prof. Paper 785.* 26 pp.

Wilshire H. G., Stuart-Alexander D. E., and Jackson E. D. (1973) Apollo 16 rocks: Petrology and classification. *J. Geophys. Res.* **78,** 2379–2392.

Papike, J.J. and Merrill, R.B., eds.
Proc. Conf. Lunar Highlands Crust (1980), p. 71-79
Printed in the United States of America

On the origins of lunar pristine crustal rocks

Alan B. Binder

Institut für Mineralogie, Universität Münster, 4400 Münster, West Germany

Abstract—The compositional characteristics of the pristine ferroan anorthosites are accounted for by previous models of the formation of the original lunar crust; thus, the suggestion that these rocks are relics of the original cumulate crust seems likely. However, these crustal origin models are based on a mean crustal composition of anorthositic norite and such rocks are seldom found among the pristine rocks. The preservation of only the ferroan anorthosites from the original crust is attributed to their having been formed at the top of the crust. Thus, these rocks were not altered by a phase of metamorphism, partial anatexis, and KREEP injection which occurred in the middle and lower crust according to thermal- and general lunar development models.

A second suite of pristine rocks (i.e., the magnesian troctolites, dunites, and norites) most probably are not relics of the original crust. These rocks are proposed to have formed in plutons which developed during the above mentioned phase of metamorphism, anatexis, and KREEP injection. Such plutons would have as their source materials the original anorthositic norites and norites of the lower crust. Early differentiates from such plutons could form a suite of rocks identical to the pristine magnesian rocks.

INTRODUCTION

The selenochemical and petrological properties of the original crust of the moon provide several of the most important constraints for models of the origin and evolution of the moon. However, identification of these properties is frustrated by the fact that almost all highland samples brought to earth are polymict breccias and melt rocks of impact origin. Thus, there is little, if any, original crustal material which has survived, unaltered, the intense early meteoritic bombardment of the moon. As such, models of the original crust have been developed assuming that, statistically, the major oxide compositions of the current crustal rocks reflect, both reasonably and accurately, those of the original crust (e.g., Wood, 1975; Binder, 1975a, 1976).

However, Warner *et al.* (1976) called attention to the fact that plutonic clasts in rock 76255 and other highland rocks fall into two unrelated groups. The first group consists of magnesian rocks similar to the previously known troctolites, spinel troctolites, and dunites, while the second consists of ferroan anorthosites of the type identified by Dowty *et al.* (1974). Warren and Wasson (e.g., 1977, 1978) have made a concentrated effort to search for such plutonic rocks or clasts. As a result, they have identified a small suite of so called pristine crustal rocks, some of which they consider to be relics of the original crust. These pristine rocks are mainly ferroan anorthosites and magnesian troctolites, dunites, and norites. Warren and Wasson point out that the most abundant crustal rock type, i.e. anorthositic norite (Taylor, 1975) is generally absent from their suite of pristine rocks. They have argued convincingly (Warren and Wasson, 1978) against the proposal (Delano and Ringwood, 1978) that the pristine rocks were formed by differentiation processes in impact melt pools. On the basis of these observations and arguments and some modeling considerations, Warren and Wasson (1979) and Warren (1979) propose that the original crust consisted of early cumulate ferroan anorthosites underlain by later magnesian mafic intrusions of somewhat unspecified origin.

Because of a compositional gap between the ferroan and magnesian suites of rocks and other characteristics, Warren and Wasson (1979) conclude that the two suites of pristine rocks could not have formed from a common magma system. As Warren (1979) discusses, it appears that the ferroan anorthosites were formed during the initial differentiation of the moon or magma ocean, but the magnesian suite of rocks had a different origin. As discussed in the following, earlier models for the formation of the original crust (e.g. Binder, 1975a, 1976) can be used to support Warren's (1979) suggestion regarding the origin of the ferroan anorthosites, while it is proposed that the magnesian rocks are true plutonic rocks formed in differentiating batholiths, stocks, etc. Such plutons are suggested to have formed as the lower crust partially remelted and produced magmas which intruded into the upper crust early in lunar history.

FERROAN ANORTHOSITES: TRUE RELICS OF THE ORIGINAL CRUST

Figure 1 shows the An vs Mg′ (100Mg/(Mg+Fe)) data for the pristine rocks and the calculated compositional limits (along with the mean compositional trend of non-pristine crustal rocks) of the crustal origin model proposed earlier by the author (Binder, 1975a, 1976) and revised by Binder and Lange (1979). On the basis of reviewer comments, it seems necessary to note here that the author's model for the origin of the original crust, used as a basis for this paper, predates the recognition and analysis of the pristine crustal rocks and hence is based on the compositional data of non-pristine rocks. As briefly stated in the introduction, one of the basic assumptions behind the model is that, despite impact modification, metamorphism, and/or contamination by KREEP-rich materials, the average (non-pristine) crustal rock has retained its major oxide, chemical identity since the formation of the original crust. If this basic assumption is correct and if the model is valid or at least partially valid, then true pristine relics of the original crust would lie within the model compositional limits and in the same parts of the An vs Mg′ diagram as their non-pristine counterparts. As can be seen in Fig. 1, the majority of the pristine ferroan anorthosites do fall within the model's boundaries and, as discussed by Binder (1975a, 1976), in that part of the

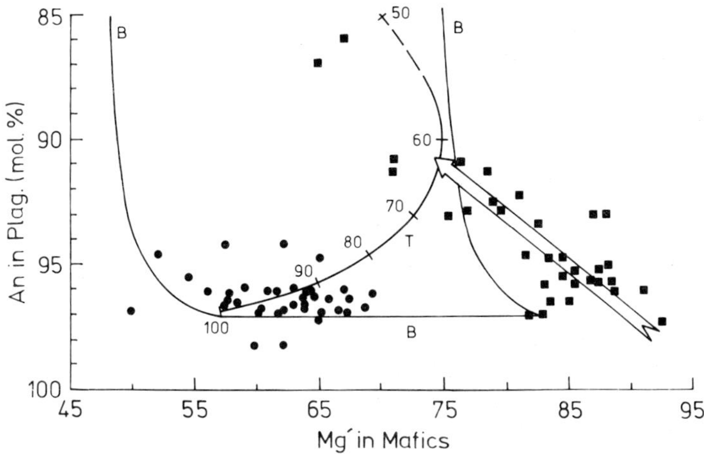

Fig. 1. An content of the plagioclase vs Mg′ of the mafic minerals in pristine rocks. The rock data are taken from Fig. 1 of Warren (1979). The circles represent ferroan anorthosites. The squares represent magnesian rocks. The curves labled B are computed compositional boundaries for original crustal rocks and are taken from Fig. 4 of Binder (1976a, note that in Fig. 4, and other similar figures in this reference, An is given in atomic % and not in mol. % as labled). Curve T is the main sequence of the current crustal rocks; the percentage of plagioclase in the rocks along this trend is indicated. This curve and the plagioclase content data are also taken from Fig. 4 of Binder (1976a). The arrow is the initial part of the differentiation trend of a totally melted anorthositic norite (63% plagioclase) model source rock whose composition plots at the head of the arrow. The earliest differentiates, see text and Fig. 2, of such a melt plot at the tail of the arrow. Note that all the magnesian rocks lie near this trend or its extension.

diagram where the majority of the early anorthosites fall. Also, it is necessary to note here that the anorthosites formed at the top of the original crust. Because of the agreement of the pristine anorthosite data and the model and the apparent old age of the pristine ferroan anorthosites (Warren and Wasson, 1978; Warren, 1979), there appears to be no considerations which contradict the proposal that the pristine ferroan anorthosites are relics of the original crust.

However, if the above is the case, then why are there so few pristine rocks of noritic anorthosite, anorthositic norite, etc. compositions in the Warren and Wasson collection? The author's model and those of Wood and co-workers have an average composition of anorthositic norite for the early crust, with anorthosites being in the minority (Binder, 1976). The answer may be (as discussed in more detail in the next section) that the middle and lower crust, where such more mafic-rich rocks are proposed to have originated, were very hot during the first few hundred million years of lunar history. As such, during this period of time these regions were quite susceptible to metamorphism, partial anatexis, and KREEP injection from the KREEP layer at the crust-mantle boundary (Binder, 1974, 1976). Hence, the average type crustal rocks, i.e. those missing from the Warren and Wasson suite of rocks, most probably were modified by metamorphism, anatexis, and especially injection of KREEP so that they do not now have pristine characteristics [especially the KREEP criteria used by Warren and Wasson, (1977, 1978)]. On the basis of this possibility, it is assumed that the earlier crustal models of Wood or Binder reasonably well define the character of the original crust and that the pristine ferroan anorthosites were members of this original crust and are the only true relics of it.

TRUE PLUTONIC ORIGIN FOR THE MAGNESIAN PRISTINE ROCKS

It is obvious that the temperatures in at least the middle and lower crust were, on a global scale, only slightly below the solidus at the end of the formation of the original crust. Also, thermal models (e.g., Toksöz and Solomon, 1973; Binder and Lange, 1977, 1980) show that temperatures in these parts of the crust remained very high for the first few hundred million years of lunar history. However, such models do not take into account the local or regional heating of the crust caused by 1) impact basin formation, as modeled by Arkani-Hamed (1974), 2) the local or regional concentration of U, Th, and K by impact excavation (Binder, 1975b) and by the intrusion of U, Th and K-rich KREEP materials into the lower crust (Binder, 1974), or 3) possible frictional heating during endogenetic tectonic activity. Hence, it is most plausible to assume that supersolidus temperatures were frequently reached in the middle to lower crust at various selenographic locations during the first 3 to 5×10^8 years of lunar history. Such partial-remelting could have led to large scale plutonic activity in the early crust. It is suggested that the magnesian pristine rocks of Warren and Wasson were formed by the differentiation of such plutons.

Assuming that the author's original crustal model (Binder, 1975a, 1976) is correct, then the distribution of rock types in the crust varied with a continual gradation from anorthosites (An_{97}, Mg′ ~ 60) at the top to norites (An_{85}, Mg′ ~ 70, the main sequence in Fig. 1) at the bottom of the crust. If so, then the original rocks which were most likely to have undergone partial anatexis are the norites and anorthositic norites of the lower crust. Such rocks plot near or along the An-Ol cotectic in the pseudo-ternary diagram for Ol-An-Qtz.

Consider, as an example, a typical anorthositic norite source rock consisting of 63% An_{91}, 28% Py, and 9% Ol, with Mg′ = 75. The composition of such an anorthositic norite plots at the head of the arrow in Fig. 1 and at the filled circle in the An volume near the Ol-An cotectic in the pseudo-ternary phase diagram in Fig. 2. Based on the temperature data given in the pseudo-ternary phase diagram for Fo_{100}-An-Qtz (Anderson, 1915) and from melt experiments on lunar feldspathic rocks (Walker *et al.* 1973), the temperature increase from the Ol-Py-An peritectic to the end of the Ol-An cotectic is only 50–70°C. Hence, it would not have been difficult for total or nearly total melting of the source materials to have occurred. It is proposed that such melts, of cotectic composition, intruded into the cooler, higher levels of the crust and began to solidify and differentiate.

As stated in Fig. 2, the density of the liquid, calculated using the data and

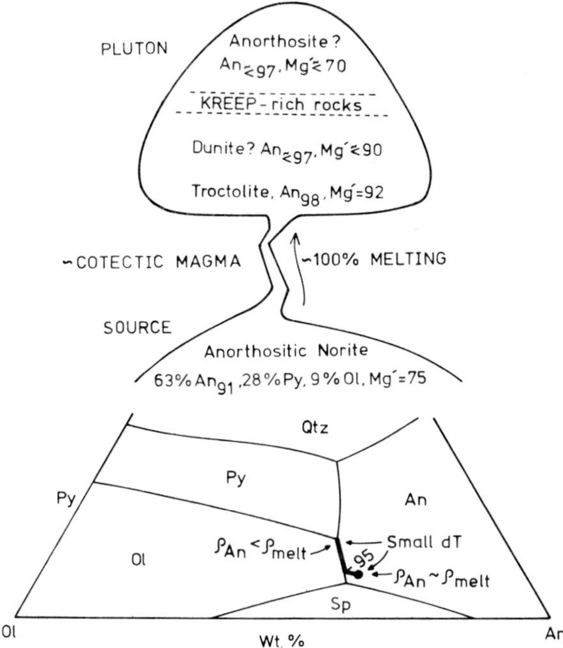

Fig. 2. Melting path in the pseudo-ternary phase diagram of the model anorthositic norite discussed in the text. The compositional point of this rock is the filled circle in the anorthite field to the right of the Ol-An cotectic. The melting path is the heavy line. Such a rock would be about 95% melted before the magma composition would leave the cotectic and move along an anorthite control line. This melt path is also the initial part of the differentiation path as the melt solidifies. If fractional crystallization dominates, then the melt composition would move beyond the Ol-Py-An peritectic to the Py-An-Qtz eutectic. The schematic diagram above the ternary phase diagram shows the sequence of events which could produce the magnesian pristine rocks.

method of Bottinga and Weill (1970), is essentially equal to that of anorthite for total melting and is slightly greater than that of anorthite for melt compositions at or near the Ol-Py-An peritectic. Hence, floatation of plagioclase could have occurred only towards the end of the differentiation sequence. Thus, the first rocks formed at the bottom of the chamber would be a cotectic mixture of Ol and An (i.e., troctolite) with An_{98} and Fo_{92}. Such rocks plot at the tail of the arrow in Fig. 1 and are schematically shown in Fig. 2. Continued differentiation would have produced rocks lying along the magnesian suite trend as shown in Fig. 1. Also, as differentiation continued and the residual melt became more dense, it is possible (but not necessarily the case) that the olivine and anorthite crystals tended to separate giving rise to dunites ($An_{\gtrsim 97}$, $Fo_{\gtrsim 90}$) lying above the early troctolites and anorthosites ($An_{\gtrsim 97}$, $Mg'_{\gtrsim 70}$, anorthite crystals plus some trapped melt) at the top of the pluton (Fig. 2). Clearly, the early troctolites would be incompatible element-poor and the later formed rocks would be incompatible element-rich. Finally, as is indicated in Fig. 2, residual liquids rich in incompatible elements or KREEP-like components should have crystallized giving rise to Fe, Na, and K-rich rocks similar, at least in major oxide composition, to the pristine quartz monzodiorite 15405c (Irving, 1977; Warren and Wasson, 1978). However, mass balance calculations based on differentiating An-Ab and Fo-Fa systems (e.g., see the curves given in Figs. 2 and 13 of Binder, 1975a) show that for Ca and Mg-rich melts of the type considered, the greatest majority of the rocks produced are richer in Ca and Mg than the initial parental magma. This is easily seen since only a small amount of Na and Fe-rich material, such as the quartz monzodiorite (normative plagioclase $Or_{25}Ab_{19}An_{57}$, $Mg' = 35$), is needed to be added to a much larger amount of typical pristine magnesian rock (An_{96}, $Mg' = 85$) to average out to the model source rock considered (An_{91}, $Mg' = 75$). Hence, most of the plutonic rocks produced would lie to the right and below their magma source rock in Fig. 1—and this is the case.

The formation of magnesian dunites, spinel troctolites, or norites as the major rock type in any of the differentiating plutons, would have depended on the composition of the source region before melting. For example, the melt from any original norite or troctolite whose composition plots in the spinel volume of the ternary phase diagram, not too far from the Ol-An-Sp peritectic, would have produced spinel troctolites as its earliest differentiate. Thus, the entire suite of pristine magnesian rocks could have been formed as early differentiates of plutons whose melts were derived by the anatexis of the deep lying norite and anorthositic norite source rocks with $Mg' \sim 75$. The general compositional characteristics of the magnesian rocks (i.e., their being troctolites, dunites, and norites) are a direct consequence of the fact that their parental magmas were Ol-An cotectic in composition. The preservation of the magnesian rocks is attributed to their having been formed at shallow enough depths so that, like the ferroan anorthosites, they were not affected by further metamorphism, anatexis, and KREEP injection.

SUMMARY AND CONCLUSIONS

The compositional characteristics of the pristine ferroan anorthosites are essentially identical to those predicted for anorthosites in crustal models in which there are continuous compositional gradients from ferroan anorthosites (An_{97}, $Mg' \sim 60$) at the top to norites (An_{85}, $Mg' \sim 70$) at the bottom of the crust. The preservation of only the ferroan anorthosites of this original crust is attributed to their having been at the top of the crustal column. Hence, these rocks were too cold to have undergone any metamorphism, partial anatexis, and KREEP injection, all of which must have occurred in the middle and lower crust according to thermal history and general lunar development models. Therefore, even though the original crust probably contained members of the entire suite of ANT rocks and had a mean composition of anorthositic norite, only the anorthosites are now pristine.

The pristine magnesian suite of rocks is proposed to have formed as partial anatexis of the anorthositic norites, and norites (An_{90}, Mg′ ~ 75) of the lower crust gave rise to plutons which intruded into the cooler upper crust. As these plutons cooled and differentiated, they produced the suite of pristine magnesian rocks via normal magmatic differentiation. Such magnesian rocks were then able to retain their pristine character due to their being more favorably placed in the cooler, upper crust after intrusion.

The above discussed model for the formation of the pristine crustal rocks is schematically shown in Fig. 3.

Finally, as is well documented, the crust of the moon underwent considerable modification via meteorite bombardment in the period between 4.6 and 4.0×10^9 years ago. As is discussed in this paper, thermal models show that the middle and lower crust were, in part, hot enough to have undergone thermal metamorphism, anatexis, and KREEP injection from the KREEP layer just below the crust for the first several hundred million years of lunar history. On the basis of these

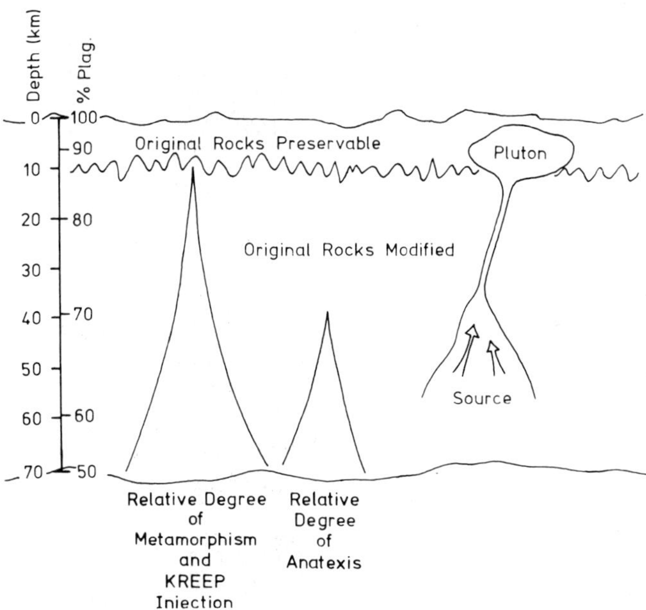

Fig. 3. Schematic representation of the original crust (Binder, 1975a, 1976a) and the processes which acted upon it. Because of metamorphism, KREEP injection from the KREEP layer at the crust-mantle boundary, and anatexis, most or all of the rocks below the first several km of the crust would have lost their pristine characteristics early in lunar history. Only those ferroan anorthosites which also survived the early meteoritic bombardment of the moon are now recognized as pristine rocks. The magnesian suite of rocks are proposed to have formed in plutons which intruded into the upper crust. Such plutonic rocks which also survived the meteoritic bombardment have pristine characteristics—but are *not* relics of the original cumulate crust.

considerations, it seems very likely that considerable *endogenetic* development and modification of the early lunar crust occurred during the first half of a billion years of lunar history.

Acknowledgments—I would like to thank Drs. I. S. McCallum and P. H. Warren for their most helpful official and unofficial, respectively, comments on the paper. This work was supported by the Deutsche Froschungsgemeinschaft.

REFERENCES

Anderson O. (1915) The system anorthite-forsterite-silica. *Amer. J. Sci.* **39,** 407–454.

Arkani-Hamed J. (1974) Effect of a giant impact on the thermal evolution of the moon. *The Moon* **9,** 183–209.

Binder A. B. (1974) On the origin of the moon by rotational fission. *The Moon* **11,** 53–76.

Binder A. B. (1975a) On the petrology and structure of a gravitationally differentiated moon of fission origin. *The Moon* **13,** 431–473.

Binder A. B. (1975b) On the heat flow of a gravitationally differentiated moon of fission origin. *The Moon* **14,** 237–245.

Binder A. B. (1976) On the petrology and early development of the crust of a moon of fission origin. *The Moon* **15,** 275–314.

Binder A. B. and Lange M. (1977) On the thermal history of a moon of fission origin. *The Moon* **17,** 29–45.

Binder A. B. and Lange M. A. (1979) Comparison and synthesis of the Wood and Binder models for the formation of the crust in a convecting magma system (abstract). In *Papers Presented to the Conference on the Lunar Highlands Crust,* p. 6–8. Lunar and Planetary Institute, Houston.

Binder A. B. and Lange M. (1980) On the thermal history, thermal state, and related tectonism of a moon of fission origin, *J. Geophys. Res.* In press.

Bottinga Y. and Weil D. F. (1970) Density of liquid silicate systems calculated from partial molar volumes of oxide components. *Amer. J. Sci.* **269,** 169–182.

Delano J. W. and Ringwood A. E. (1978) Siderophile elements in the lunar highlands: Nature of the indigenous component and implications for the origin of the moon. *Proc. Lunar Planet. Sci. Conf. 9th,* p. 111–159.

Dowty E., Prinz M., and Keil K. (1974) Ferroan anorthosite: a wide spread and distinctive lunar rock type. *Earth Planet. Sci. Lett.* **24,** 15–25.

Irving A. J. (1977) Chemical variation and fractionation of KREEP basalt magmas. *Proc. Lunar Sci. Conf. 8th,* p. 2433–2448.

Taylor S. R. (1975) *Lunar Science: A Post-Apollo View.* 372 pp. Pergamon, N.Y.

Toksöz, M. N. and Solomon S. C. (1973) The thermal history and evolution of the moon. *The Moon* **7,** 251–278.

Walker D., Longhi J., Grove T. L., Stolper E., and Hays J. F. (1973) Experimental petrology and origin of rocks from the Decartes Highlands. *Proc. Lunar Sci. Conf. 4th,* p. 1013–1032.

Warner J. L., Simonds C. H., and Phinney W. C. (1976) Genetic distinction between anorthosites and Mg-rich plutonic rocks: New data from 76255 (abstract). In *Lunar Science VII,* p. 915–917. The Lunar Science Institute, Houston.

Warren P. H. (1979) *Certain* pristine nonmare rocks formed as cumulates from the magma ocean (but others did *not*) (abstract). In *Papers Presented to the Conference on the Lunar Highlands Crust,* p. 192–194. Lunar and Planetary Institute, Houston.

Warren P. H. and Wasson J. T. (1977) Pristine nonmare rocks and the nature of the crust. *Proc. Lunar Sci. Conf. 8th,* p. 2215–2235.

Warren P. H. and Wasson J. T. (1978) Compositional-petrographic investigations of pristine nonmare rocks. *Proc. Lunar Planet. Sci. Conf. 9th,* p. 185–217.

Warren P. H. and Wasson J. T. (1979) Effects of pressure on the crystallization of a "chondritic" magma ocean and implications for the bulk composition of the moon. *Proc. Lunar Sci. Conf. 10th.* In press.

Wood J. A. (1975) Lunar petrogenesis in a well-stirred magma ocean. *Proc. Lunar Sci. Conf. 5th,* p. 1087–1102.

Papike, J.J. and Merrill, R.B., eds.
Proc. Conf. Lunar Highlands Crust (1980), p. 81-99
Printed in the United States of America

Early lunar petrogenesis, oceanic and extraoceanic

Paul H. Warren*‡ and John T. Wasson*†

Institute of Geophysics and Planetary Physics, University of California, Los Angeles, California

Abstract—An attempt has been made to ascertain which (if any) pristine nonmare rocks, other than KREEPy ones, are *not* cumulates from the magma ocean. The only pristine rocks that have bulk densities low enough to have formed by floating above the magma ocean are the ferroan anorthosites, which are easily recognizable as a discrete subset of pristine rocks in general, on the basis of mineral composition relationships.

The other category of pristine nonmare rocks, the Mg-rich rocks, did not form from the same magma that produced the ferroan anorthosites. We suggest that they were formed in layered noritic-troctolitic plutons. These plutons were apparently intruded at, or slightly above, the boundary between the floated ferroan anorthosite crust and the underlying complementary mafic cumulates (i.e., the "mantle"). The parental magmas of the plutons may have arisen by partial melting of either: (a) deep mafic cumulates from the magma ocean, or (b) a still deeper, undifferentiated primordial layer that was not molten during the magma ocean period.

INTRODUCTION

The notion of plagioclase flotation over a primordial moon-wide "ocean" of magma (Wood *et al.,* 1970; Smith *et al.,* 1970) appears to have gained general acceptance as the most likely explanation for the extreme plagioclase enrichment in the nonmare crust. Considering how venerable the magma ocean concept is, it seems at least mildly surprising that, prior to our effort (Warren and Wasson, 1979a), practically no attempt was made to ascertain whether some type(s) of non-extrusive pristine nonmare rocks did *not* accumulate from the magma ocean.

The term "pristine" will be employed repeatedly throughout this article. It means "compositionally endogeneous." Our construction of "pristine" is elaborated in Warren and Wasson (1977, 1978, 1979b)—these papers also contain lists of known (or suspected) pristine nonmare rocks, and inferences about early lunar petrogenesis based on pristine nonmare rock systematics. Synonyms for "pristine" found in the literature include the following: "meteorite-free" (e.g., Anders, 1978), "endogenous" (e.g., Haskin *et al.,* 1979; James, 1979), "indigenous" (e.g.,

---

* Also Department of Earth and Space Sciences

†Also Department of Chemistry

‡ Present address: Institute of Meteoritics, University of New Mexico, Albuquerque, New Mexico 87131

Ryder *et al.*, 1980), "primitive" (e.g., Longhi, 1979). "Plutonic," as used by Warner and co-workers (e.g., Warner *et al.*, 1976), is virtually synonymous but limited to pristine rocks of deep origin. The vast majority of highlands rocks are not pristine but polymict breccias, i.e., mixtures of debris created by meteorite impacts. Pristine rocks are given great emphasis because only they accurately preserve the compositional record of the early, unmixed lunar crust.

A DICHOTOMY AMONG PRISTINE NONMARE ROCKS

As was first suggested by Roedder and Weiblen (1974), and first convincingly documented by Warner *et al.* (1976), on a plot of mean An (molar Ca/(Ca+Na+K)) in plagioclase vs. mean *mg* (molar Mg/(Mg+Fe)) in mafic minerals, pristine rocks are clearly resolved into two distinct groups. Figure 1 is an updated version of this diagram. Following Dowty *et al.* (1974), we refer to the rocks belonging to the lower right cluster as ferroan anorthosites, and following Warner *et al.* (1976), we refer to the ones belonging to the diagonal trend as Mg-rich rocks. The fact that the trends of the two groups have distinctly different slopes (Warner *et al.*, 1976), and the persistence (Bickel and Warner, 1978; Warren and Wasson, 1977, 1979b, and Fig. 1) of a wide gap between the most "Mg-rich" ferroan anorthosite and the most "ferroan" Mg-rich rock, indicates that the two groups are not cogenetic. The gap is probably much more significant than the difference in slopes, however, because Raedeke and McCallum (1979) have found a similar distribution into "normal" (diagonal) and near-vertical trends among cumulates from the Stillwater complex. However, the Stillwater may not be ideal for this sort of comparison, because it probably crystallized as an "open" system. As noted by McCallum *et al.* (1980), no systematic variation exists in olivine composition as a function of stratigraphic height. Olivine is probably preponderantly cumulus wherever it occurs in the Stillwater, so closed-system fractional crystallization should have led to a marked decrease in Fo content with increase in stratigraphic height. Moreover, olivine disappears and then reappears as a cumulus phase (without much change in composition). Lambert and Simmons (1980) arrived at the same conclusion, viz., that the Stillwater is "not the crystallization product of a single magma," based on trace element analysis. Thus, different portions of the Stillwater are not entirely cogenetic. Incidentally, if the Stillwater is a product of open-system fractionation, it is not unlike many other layered intrusions, e.g., the Rhum (Wager and Brown, 1967), the Salt Lick Creek (Wilkinson *et al.*, 1975), the Fiskenaesset (Myers and Platt, 1977), the Kap Edvard Holm (Wager and Brown, 1967), and the Kapalagulu (Wager and Brown, 1967).

If the Mg-rich rocks formed out of the magma ocean, the process obviously seldom involved flotation: Figure 2 is largely self-explanatory. None of the Mg-rich rocks has sufficient plagioclase to be likely to float over moderately evolved "basaltic" magma in which $CaAl_2Si_2O_8$ is a major component. Incidentally, the least dense Mg-rich rock, 76335 (Warren and Wasson, 1978) has recently been

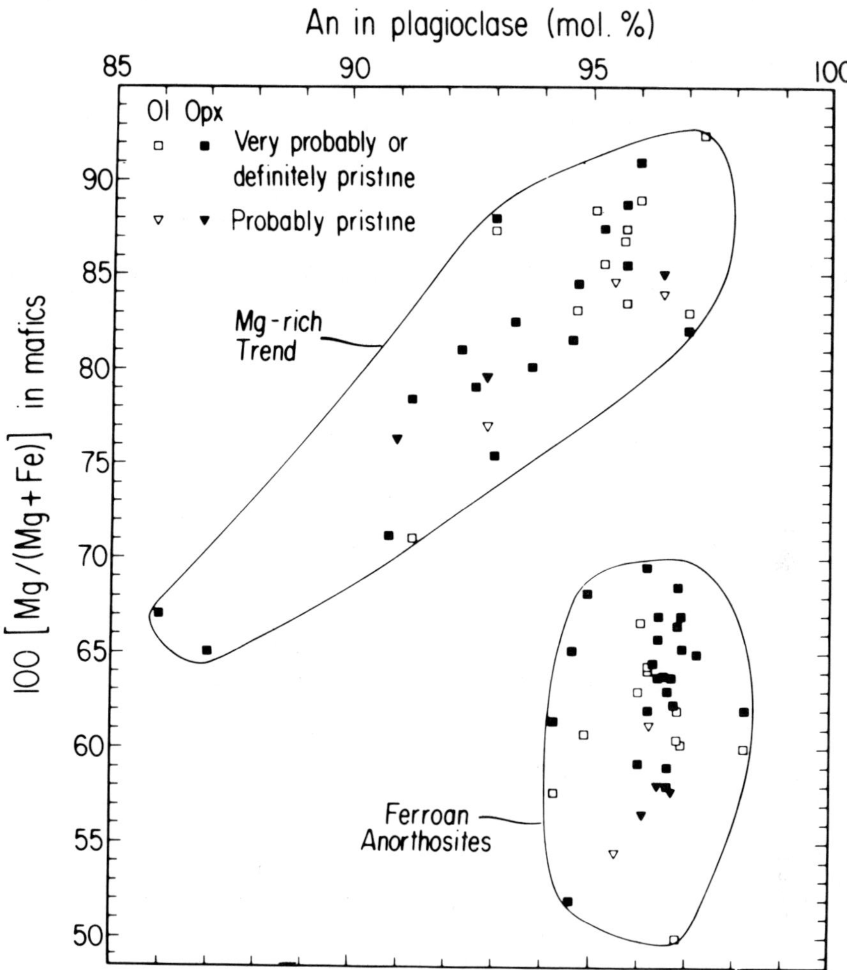

Fig. 1. Mean Mg/(Mg + Fe) in mafics vs. mean An content of plagioclase in pristine nonmare rocks. Note the differing slopes of and the distinct hiatus between the two trends.

studied petrographically by O. James (pers. comm., 1979), and in her thin section 76335 looks very similar to 76535, which was also picked up at Station 6, and which contains only about 60% plagioclase (*Apollo 17 Lunar Sample Information Catalog,* 1973). Considering that all of the samples shown in Figs. 1 and 2 are probably cumulates (which in terrestrial intrusions are known for their tendency to feature "rhythmic" modal banding), the lack of overlap between the two groups, particularly on Fig. 2, is quite remarkable.

On the other hand, Fig. 2 shows that at least the vast majority of the ferroan anorthosites could float over basaltic magma. If any of the Apollo samples represent floated cumulates, the ferroan anorthosites surely must be among them.

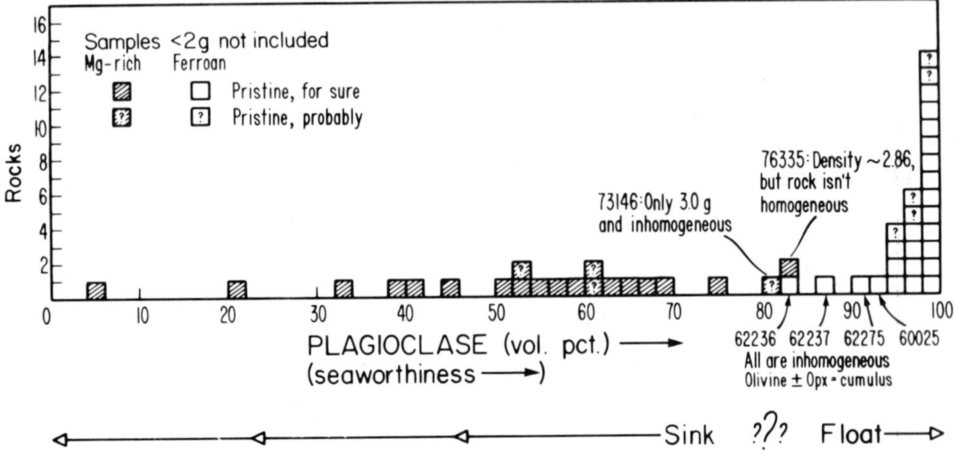

Fig. 2. Plagioclase contents of pristine nonmare rocks. Samples under 2 g in mass were excluded because their modes are too poorly defined. Note the almost complete separation between ferroan anorthosites and Mg-rich rocks.

The evidence for a dichotomy among pristine nonmare rocks has recently been reinforced by Norman and Ryder (1980), who showed that Mg-rich rocks generally (albeit not always) feature Ti/Sm and Sc/Sm ratios much lower than those of ferroan anorthosites.

It has been suggested (Longhi and Boudreau, 1979) that the two groups on Fig. 1 could form from one magma if the Mg-rich rocks consist of cumulus mafics and cumulus plagioclase, while the anorthosites have cumulus plagioclase, but mafics formed from trapped (ferroan) intercumulus liquid. In principle, this could serve to explain why ferroan rocks are generally anorthositic, but it is obviously *ad hoc* as an explanation for the *gap*. The sunken mafic-rich rocks might all be adcumulates (cumulates with negligible trapped liquid contents). Indeed, the highest published estimate of trapped liquid content by far seems to be 16%, for troctolite 76535 (Haskin *et al.*, 1974); there are theoretical reasons to believe that cumulates forming at the bottom of a deep magma ocean would not "trap" much liquid (Ringwood, 1976; Walker *et al.*, 1978); and most terrestrial cumulates are adcumulates (Wager and Brown, 1967, p. 554). However, there is no reason to expect *all* anorthosites to be orthocumulates (cumulates rich in trapped liquid). The gap should be filled with mesocumulates (cumulates with moderate trapped liquid contents). In fact, the low Sm contents of the anorthosites (Fig. 3) strongly indicate that at least most (on a mass basis) of their mafics are primary cumulus crystals. None of the most mafic-rich ferroan rocks (15437, 62236 and 62237) have the high levels of incompatible element concentrations expected from high contents of reasonably Sm-rich trapped liquid, even though some that are intermediate in mafic content (as ferroan anorthosites go), such as the 67455 clasts with extremely Fe-rich "interstitial" pigeonite and olivine (Lindstrom *et al.*, 1977), do have high incompatible levels. Another problem with the trapped liquid hypothesis is the observation (Norman and Ryder, 1979) that no single type of liquid would engender the range of mafic phases found in pristine ferroan anorthosites (calcic pyroxene in some, olivine in others, and so on).

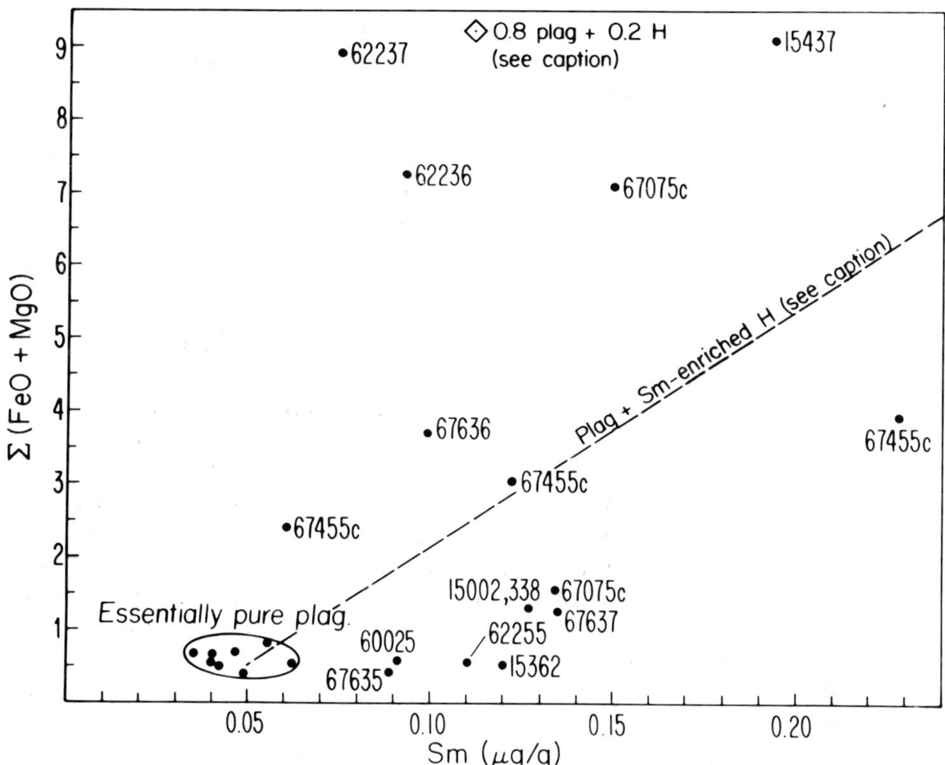

Fig. 3. Bulk rock Sm vs. bulk rock $\sum$(MgO + FeO) (an index of mafic mineral content) in pristine anorthosites. Note how rich in Fe and Mg some of them (e.g., 62236 and 62237) are, *without* being much richer in Sm than the truly monomineralic ones (i.e., pure adcumulates). Based on a plagioclase/liquid distribution coefficient of 0.036 for Sm (McKay and Weill, 1976, 1977), the pure plagioclase samples with ~0.05 μg/g Sm crystallized from liquids with ~1.4 μg/g Sm, i.e., about 6X chondrites. The diamond between 62237 and 15437 shows the composition an anorthosite would have if it were a mixture of 80% essentially pure plagioclase (Sm = 0.05, $\sum$(FeO + MgO) = 0.5) plus 20% trapped liquid of *primitive H chondrite* (silicate portion) composition ($\sum$(MgO + FeO) = 44.4; see Table 1 in Warren and Wasson, 1979a). The dashed line is a mixing line for combinations of essentially pure plagioclase with a liquid with 1.4 μg/g Sm, and an otherwise *primitive H chondrite* composition. For comparison, the compositions used by Walker *et al.* (1973) in locating the olivine-plagioclase cotectic on their Ol-An-SiO_2 pseudoternary diagram have $\sum$(MgO + FeO) = 21.3 − 26.4. Thus, a realistic mixing line would have a slope approximately half as great as even the one shown. Data sources include: Warren and Wasson (1978) (15437, 62236, 62237, etc.); Haskin *et al.* (1973) (67075c); Wänke *et al.* (1975) (67075c); Lindstrom *et al.* (1977) (67455c).

Two different types of rhythmic cumulate layers?

One conceivable way in which adcumulate Mg-rich rocks and cogenetic but orthocumulate anorthosites could be related to yield the relationships shown in Figs. 1 and 2 is the possibility that they come from different rhythmic cumulate layers of a single series of rocks. However, this does not avoid any of the previously cited reasons for believing that the two groups are not cogenetic.

There is also relevant evidence from studies of terrestrial cumulates. For example, the incompatible element P is enriched in leucocratic layers relative to adjacent melanocratic layers in some terrestrial cumulates (Wager and Brown, 1967; Liebenberg, 1960; Henderson, 1970), implying that the leucocratic layers are richer in trapped liquid. However, this trend is absent, if not actually reversed, in at least one layered intrusion (de Waard and Wheeler, 1971). Moreover, pristine lunar anorthosites are as a rule extremely incompatible-poor, even compared to pristine troctolites and norites (see, e.g., Norman and Ryder, 1979).

Himmelberg and Ford (1976) studied the relationship between pyroxene *mg* and modal mafic content among cumulates in the immense Dufek layered intrusion, Antarctica. They found essentially no difference between pyroxenes in anorthosite layers and pyroxenes in melanocratic layers from roughly equal heights in the intrusion. Morse (1979, p. 587) conducted the analogous test among layered cumulates from the Kiglapait intrusion, Labrador. He reported one "very surprising" plagioclase composition "jump" from An_{57} to An_{62}, but observed *no* concomitant change in olivine *mg* in the same sample suite. McBirney and Noyes (1979, p. 521–522) conducted the same test for a single layer of the Skaergaard intrusion. They found that an abrupt change in modal plagioclase from 65 to 15 percent was accompanied by only small changes in olivine *mg* (from 36 to 39) and in plagioclase An content (from 45 to 44). The modal opaques content also abruptly changed from ~7% to ~52%, however, so the constancy in olivine composition is subject to various interpretations. Nevertheless, unless analogy to terrestrial cumulates is badly misleading the two suites of pristine nonmare rocks do not come from interbedded rhythmic layers.

Mg-rich rocks—sunken magma ocean cumulates?

Longhi and Boudreau (1979) suggested that the ferroan anorthosites and Mg-rich rocks could both have formed from the same magma ocean if adcumulate Mg-rich rocks accumulated along the base of the ocean at the same time orthocumulate anorthosites accumulated above the ocean, as originally suggested by Wood (1975). Of course, this model is perfectly consistent with the main message of Fig. 2. Another point in its favor is that Naslund (1976) reported that, for a given An content in plagioclase, olivines in cumulates from the Skaergaard intrusion's floated upper border group have several percent less Fo than olivines from the underlying layered series. The Longhi-Boudreau model might even be consistent with Fig. 1, since the gap between the two groups is conceivably an artifact due to limited sampling and there is some (remote) possibility that (Fig. 4 and the other evidence discussed above notwithstanding) all the magma ocean's floated cumulates are almost devoid of mafics other than the trapped liquid variety.

Several other huge obstacles stand in the way of this variation of the single magma hypothesis. First, if the Mg-rich rocks accumulated at the bottom of the ocean, it is difficult to imagine how any of the ones with extremely basic characteristics (high *mg*, high plagioclase An, etc.; examples are the 67435c spinel troctolite, the troctolites 76255c, 76335 and 76535, and the dunite 72415) could have accumulated at shallow enough depths to ever be subsequently excavated. With little possibility for repeated injections of "fresh" magma from lower depths, as probably frequently occurs in terrestrial layered intrusions (Wager and Brown, 1967), there should be a very regular pattern of "cryptic" layering in the basal cumulates of the magma ocean, i.e., quantities like *mg* and An should *increase monotonically with depth*. So if relatively evolved Mg-rich rocks (e.g., the 76255c norite; Warner *et al.*, 1976) accumulated below the base of the flotation crust, the most extremely basic Mg-rich rocks probably accumulated hundreds of km beneath the surface, according to this model. There is very little agreement about the depth to which basin-forming impacts may have sampled the lunar interior—"limits" range from 27 km (Head *et al.*, 1975) to 330 km (O'Keefe and Ahrens, 1978). By any reckoning, however, samples of the lowest reaches of the former magma ocean should be exceedingly uncommon. Another problem is that, based on mineralogical geobarometry (Herzberg, 1979) at least two of the most basic Mg-rich rocks (72415 and 76535) equilibrated at depths of less than 30 km, surely above the base of the magma ocean during all but its most evolved stages.

Even if we disregard the above-mentioned shortcomings, another very major flaw exists in the variation of the single magma hypothesis that says Fig. 1 is explicable if orthocumulate anorthosites floated at the same time adcumulate Mg-rich rocks were forming along the ocean's base. If both the Mg-rich rocks and the anorthosites contain mainly cumulus plagioclase, then a series of anorthosites should exist with the same An in plagioclase (if not lower, because of their trapped liquid components) as the most evolved members of the Mg-rich group [e.g., the plagioclase in the 76255c gabbro averages An_{86} (Warner *et al.*, 1976)], but with much more ferroan mafics. The curve labeled "trapped liquid" on Fig. 2 in Longhi and Boudreau (1979) shows precisely this prediction, based on their model. Needless to say, few or no such anorthosites have so far been discovered. It might be argued that at some point in its crystallization history the magma's density became so low that plagioclase flotation ceased, before any anorthosites with relatively Na-rich plagioclase had formed, so that thenceforth only Mg-rich sunken cumulates formed. Such a density change is implausible, however, because by far the most important variable affecting the density is the iron content, which increases monotonically in practically any model of a magma's crystallization, at least until extremely late in the crystallization history.

PETROGENESIS OF FERROAN ANORTHOSITES

As discussed above, if we accept that only one of the two groups of pristine nonmare rocks resolved in Fig. 1 formed by floating over the magma ocean, Fig. 2 clearly shows that it must have been the ferroan anorthosites. This conclusion receives puissant support from Rb-Sr isotopic systematics. The (weighted) mean initial $^{87}Sr/^{86}Sr$ ratio of the five pristine ferroan anorthosites (15415, 60015, 60025, 61016c and 64423,13,1) analyzed (a total of 20 times) to date, after correction for inter-laboratory bias (Nyquist, 1977) is 0.698925, with σ_μ (68% confidence error estimate) equal to 0.000005. The analogous values for the only five Mg-rich plutonic rocks for which data exist (15455c, 72255c, 72417, 76535 and 77215) are 0.699013 and 0.000010 (Nyquist, 1977; Compston and Gray, 1974; Nyquist *et al.*, 1979). In the error estimate calculations, no allowance for interlaboratory biases has been made; however, since both types of rocks were analyzed by several labs, biases not precisely compensated should not affect the final results significantly. The anorthosites contain too little radiogenic strontium to be dated precisely. Nevertheless, the initial ratio data strongly indicate that if both the anorthosites and the Mg-rich rocks formed out of one magma, the anorthosites are *generally older*. This is clearly inconsistent with the Mg-rich rocks being derived from the magma ocean, because the fractional crystallization process could not have first produced the low-*mg* magmas parental to the ferroan anorthosites, then the high-*mg* magmas parental to the Mg-rich rocks. In fact, even if they did not form from the same magma, for the ferroan anorthosites to be as young or younger than the Mg-rich rocks, the anorthosites would have to have been produced from magma which formed in a part of the moon that had been poor in Rb since very early in lunar history. But such a part of the moon would probably have been similarly depleted in K, Th and U, which would make it unlikely to melt after the magma ocean heating episode (which probably resulted from accretion).

It is possible to demonstrate (Warren and Wasson, 1979a) that ferroan anorthosites are precisely the sort of rocks that would probably form the bulk of the "upper border group" (using terrestrial nomenclature) of the lunar magma ocean, provided the initial ocean composition was reasonably close to the composition of the earth's mantle [e.g., "pyrolite" (Ringwood and Kesson, 1977)], or to a composition (nearly equivalent to pyrolite) intermediate between the silicate parts of H and IAB group chondrites. We will not repeat all of the arguments behind this conclusion here, as they are rather complex. Briefly, we believe that the reason the anorthosites are ferroan is because they did not accumulate (in any large volume) over the magma ocean until at least 65% of the ocean had crystallized as sunken, magnesian cumulates (decreasing the ocean's depth and at the same time increasing the ocean's concentrations of Ca and Al—prerequisites for the formation of a thick, anorthositic crust), by which time the ocean's *mg* had been greatly reduced.

A crucial phenomenon generally overlooked by modelists (e.g., Wood, 1975; Longhi, 1977) of the lunar magma ocean (the only exceptions we know of are Binder, 1976; and Warren and Wasson, 1979a), has recently been detected among terrestrial layered intrusions (Campbell, 1978): Evidence from several intrusions indicates that crystal nucleation occurs at, or close to, the *bottom* of the magma zone, as long as the magma zone is of large vertical extent. Thus, Campbell's evidence supports theoretical predictions by, e.g., Thomson (1864), Adams (1924), Jackson (1961), and Elsdon (1970). Elsdon specifically concluded that "nucleation near the top of an intrusion only occurs when the vertical extent of the magma column is less than 15 km (i.e., load pressure is approximately 5 kb [equivalent lunar depth: 100 km]). At greater depths . . . nucleation may occur from the base only . . . For a certain range of depths, nucleation can occur at both sites simultaneously." This is so simply because adaibats are much less sensitive to pressure than silicate melting curves (see, e.g., Fig. 2 in Warren and Wasson, 1979a).

Prior to about 1976 (specifically, Warner *et al.*, 1976; Warren and Wasson, 1977), it was widely held that the ferroan anorthosites were only part of a trend that extended on *mg*-An diagrams (such as Fig. 1) upwards and slightly to the left, towards roughly *mg*=90, An=93 (e.g., Steele and Smith, 1971, 1973; Roedder and Weiblen, 1974). This now seems unlikely, due to the gap between the Mg-rich trend and the anorthosite trend, and because the slope of the anorthosite trend is now "normal" (positive) or possibly vertical (Warren and Wasson, 1977, 1979b). Nevertheless, the reason the anorthosites all have such Na-poor plagioclase remains a largely unresolved problem. Most of the explanations previously proposed sought to explain the spurious negative slope, as much as to explain the high Ca/Na ratio (e.g., Steele and Smith, 1973; Roedder and Weiblen, 1974; Drake 1975; Wood, 1975; Longhi, 1977), so they are at least partially inconsistent with the "normal" slope. Likewise, the model of Smith and Steele (1974) was aimed not so much at explaining the anorthosites' high Ca/Na as their low Mg/Fe, by invoking exsolution of Fe out of plagioclase and into mafics. This now seems clearly inadequate, since several anorthosites have been found that contain large cumulus mafics (Dymek *et al.*, 1975; Warren and Wasson, 1978), and yet seem typically "ferroan". The fact that many of those same mafics are pyroxene, not olivine, also militates against Drake's (1975) suggestion that a very low SiO_2 content in the parental magma(s) was responsible for the low Ab content in the plagioclase.

Perhaps the most plausible explanation is basically the same straightforward one originally proposed by Steele and Smith (1973). That is, Na is depleted in the anorthosites because they formed from a magma (i.e., the magma ocean) that had lost Na via volatilization at the moon's surface. In fact Drake (1975), who advocated an alternative explanation, also conceded this model to be plausible *unless* "the lunar magma system rapidly developed a crust analogous to a terrestrial 'chilled margin'." We agree that once a stable crust forms, however thin, Na loss would effectively cease. The loss rate of Na from a liquid silicate system is undoubtedly very high, compared to the probable lifetime of the magma ocean (e.g., Storey, 1973); but as pointed out by Gibson and Hubbard (1972), below 10^{-3} cm depth in the magma, the lithostatic pressure exceeds the vapor pressure. Therefore, unless the surface is a fluid, continually equilibrating with the main magma via some "stirring" mechanism much more efficient than simple diffusion, the effects of volatization will never extend more than "skin deep". It seems likely, however, that before the ferroan anorthosites themselves began to accumulate atop the magma ocean—after at least 65% of its crystallization sequence (see above)—no reasonably *stable* crust existed. Some prior "chilled margin" crystallization undoubtedly did occur, but never enough to produce a stable crust that did not founder (Warren and Wasson, 1979a, p. 2060). Convection probably stirred the main mass of magma extremely thoroughly; but much more significant as an enhancing factor for Na loss, was the early heavy meteoritic bombardment that probably sporadically but persistently stirred, liquified, and sprayed around (indeed, volatilized) the chilled margin crust along with portions of the subjacent liquid ocean (cf. Wood, 1975).

In vacuo studies of volatilization from liquids of mare basaltic (e.g., Gooding and Muenow, 1976) and ordinary chondritic (Gooding and Muenow, 1977) composition show that Na volatilizes as monatomic Na gas. Based on simple Maxwellian mechanics, a Na atmosphere would escape from the lunar gravitational field only at a rate with a half-life of $\sim 10^3$ years. However, strong solar wind and/or ionization effects could easily reduce the time span for total dispersal to 10^2 days, particularly if the overall lunar atmosphere was highly rarefied (Kopal, 1969). It seems probable that, provided the lunar atmospheric temperature was hot enough and/or the pressure was low enough to allow a Na mean free path of $\gtrsim 10$ km, much of the Na could have escaped from the moon without recondensing at the surface.

On the other hand, it seems likely that early in the history of the magma ocean (probably long before the anorthositic crust formed), large quantities of volatiles, probably H_2O or CO (Chou *et al.*, 1975; Wasson *et al.*, 1976), were outgassing, and that Na was a subordinate component. A plausible model is that these gases periodically erupted, perhaps at points where impacts caused pervasive fracturing of the chilled crust, reaching gas pockets just below the base of the crust. Two conditions must be met if the erupted cloud of gas is to escape the moon and entrain the Na (and other minor volatiles including K and Rb) with it: At the moment it reaches the surface its pressure must significantly exceed the ambient pressure of the lunar atmosphere; and the mean velocity of the dominant carrier gas must exceed the lunar escape velocity. The pressure of the gas when it first started ascending towards the surface would depend on the overburden pressure above the chamber (e.g., 5 atm at a depth of 100 m). Of course, some of this pressure would dissipate via expansion during the gas pocket's ascent to the surface, but even at the surface vent it might have been several times greater than the mean pressure of the ambient atmosphere, particularly if the gas was suddenly released due to an impact breaching the crust. The velocity of the gas is difficult to ascertain. The thermal velocity of CO or H_2O would probably be considerably less than the lunar escape velocity. However, the upward velocity of the parcel of gas could have exceeded the escape velocity if the eruptive fissure were relatively narrow. It seems probable that many such eruptions did result in the loss of volatiles from the Moon, with many minor components including Na entrained together with the CO and/or H_2O carrier.

Clearly, the precise mechanisms by which Na was probably (a) volatilized from the magma ocean into the early lunar atmosphere, and (b) dispersed from the atmosphere into deep space, are matters for some speculation. Nevertheless, it seems equally clear that plausible mechanisms do exist.

Steele and Smith (1973) argued that a possible consequence of a Na volatilization model is that "the age of crystallization is progressively later as the An-content of the plagioclase increases. . . ," the opposite of the sequence expected from igneous crystallization. This demands, however, that Na volatilization continue at a significant pace after anorthosites begin to form. It seems far more likely that Na volatilization was essentially complete about the time anorthosites began forming, because: (a) The anorthosites probably didn't form until late in the ocean's crystallization sequence (Warren and Wasson, 1979a)—by which time the rate of meteorite impacting had probably greatly diminished, leaving no mechanism for frequently and violently breaching the crust, as required to permit volatilization. (b) The anorthosites (almost by definition) accumulated as a floating crust on top of the ocean; and the thicker the crust became, the less frequently it would be breached, even if the meteorite flux remained constant. It therefore seems more likely that the An content of the plagioclase decreased during crystallization, and thus the positive slope of the anorthosite trend on Fig. 1 is not at all inconsistent with Na loss via volatilization.

PETROGENESIS OF Mg-RICH ROCKS

Reasons have been reviewed above for concluding that the Mg-rich rocks did not form out of the same magma, even at a separate time or place, that produced the ferroan anorthosites. Therefore, if it is true that the ferroan rocks formed as flotation cumulates over the ocean, this implies that the Mg-rich rocks did *not* form from the ocean, and that the Mg-rich rocks are *younger* than the ferroan rocks. Isotopic data consistent with the model have just been discussed.

On the basis of Fig. 2 it appears that at least most of the Mg-rich rocks formed, as almost all terrestrial cumulates do, along the floors of their magma chambers. The trend of the Mg-rich rocks on Fig. 1 is likewise conventional, inasmuch as it has a positive slope of approximately 2.0. The simplest explanation is that they formed in layered troctolitic-noritic intrusions similar to those on the Earth.

One of the phenomena that this model can account for is the preponderance of Mg-rich rocks among pristine rocks from Apollo 15 and Apollo 17, as opposed to Apollo 16, where ferroan anorthosites are preponderant. This relationship implies that the upper part of the original crust was ferroan anorthosite, while the deeper levels plumbed only by basin-forming impacts are rich in Mg-rich rocks (Warren and Wasson, 1977). Apparently the parental magmas of the intrusions were denser than the ferroan anorthosites, just as the magma ocean had been, so they tended to be emplaced below the older, ferroan anorthositic crust (but above the mantle, which in any magma ocean model consists of dense, sunken cumulates).

If there was much assimilation by the magmas of these hypothetical intrusions, there are two important materials that probably were involved. First, if the parental magmas were initially of noritic-troctolitic composition, they probably assimilated some Ca and Al from the ferroan anorthosite "country" rock. This might have affected Eu anomalies and $^{87}Sr/^{86}Sr$ ratios much more than major element compositions. Second, it would be surprising if the intruding magmas did not frequently assimilate pockets of *urKREEP,* defined by Warren and Wasson (1979c) as the incompatible element rich residuum of the magma ocean, thought to have collected between the crust and the mantle. Assimilation of urKREEP probably gave the parental magmas of the Mg-rich rocks their KREEP-like patterns, i.e., enrichments of light rare earths relative to heavy (Haskin *et al.,* 1974; Winzer *et al.,* 1975; Laul and Schmitt, 1975; McKay *et al.,* 1979), which otherwise would seem somewhat at odds with their overall primitive nature (the intrusions could probably never assimilate enough urKREEP to have much effect on their major element compositions, however).

The time at which the major increase in Rb/Sr of KREEP occurred appears on the basis of Rb/Sr model ages to be about 4.46 G.y. ago (Palme, 1977; Warren and Wasson, 1979c). Although this value is distinctly lower than the remarkably high Rb-Sr age of 4.61±0.07 G.y. reported by Papanastassiou and Wasserburg (1976) for troctolite 76535, we do not believe this is a problem for our model. The Sm-Nd age is only 4.26 ± 0.06 G.y., an unresolved discrepancy with the Rb-Sr age, and the uncertainty in the time of emplacement of urKREEP is fairly large (in fact, Papanastassiou and Wasserburg consider the model ages "only suggestive"); thus the age data seem compatible with emplacement of the Mg-rich plutons after the formation of urKREEP.

The "frozen crust" and early "rockbergs" as sources of Mg-rich rocks

Taylor and Bence (1975) suggested that several of the Mg-rich rocks formed as part of a surficial "frozen crust" over the magma ocean. However most of the Mg-rich rocks, including the ones Taylor and Bence specifically discussed, are almost certainly *cumulates,* and (based on Fig. 2) certainly *not* floatable over basaltic magma, so it seems inconceivable that they are representative of any surficial crust. The slim possibility that they come from atypically mafic-rich layers within a more feldspathic (and hence, more floatable) crustal zone has been discussed above.

Herbert *et al.* (1977) proposed that "rockbergs" formed a considerable crust over the magma ocean, rich enough in plagioclase to be stable against foundering, *early* in the ocean's history, before plagioclase would normally have crystallized. If this model is correct, much of the upper crust (and hence, many of the pristine nonmare rocks) might be Mg-rich rocks, especially magnesian anorthosites, formed at an early stage of crustal growth. Such magnesian anorthosites are unknown, and it is doubtful if they would float over a magnesian ocean. Also, as pointed out by Warren and Wasson (1979a), the theoretical *modus operandi* for rockberg growth (repeated foundering of the "frozen" crust, followed by rapid sinking of dense mafic crystals, leaving the magma locally rich in Ca and Al) can be viable only so long as the crust continues to founder, so it seems logically inconsistent to suppose that a significant portion of the present crust formed by the rockberg mechanism. The evidence (see above) that most Mg-rich rocks come from deeper levels in the crust than most ferroan anorthosites clearly indicates that most Mg-rich rocks were formed by some other mechanism. In addition, the arguments cited above against "frozen crust" models are almost equally inconsistent with early-forming "rockberg" models.

Mixed magmas as parents of Mg-rich rocks

Longhi and Boudreau (1979) proposed that within the upper parts of the magma ocean, magmas with primitive compositions often mixed with magmas with evolved compositions, analogous to O'Hara's (1977) postulate that a residium from fractional crystallization within a magma chamber beneath a volcano sometimes mixes with a fresh batch of parental magma from depth. Longhi and Boudreau envisage a pocket of residuum crystallizing in isolation to reach the eutectic in the pseudoternary system olivine-anorthite-SiO_2 (Walker *et al.*, 1973), mixing with some "primitive" magma, and then *repeating* this cycle as many as 7 times, all the while retaining its separate identity as a discrete parcel of matter within the magma ocean. They propose that "turbulent convection" in the ocean was responsible for maintaining these discrete parcels grossly out of equilibrium with the main magma, in lieu of the many tens of km of solid rock separating the magma chamber from the parental magma source region in O'Hara's scheme. It is unfortunate that Longhi and Boudreau did not give more details about their mechanism for isolating parcels of magma, because it seems to call for systematized *turbulence*. The objection raised by Wager and Brown (1967, p. 340) to Jackson's (1961) model, which was equivalent to a single cycle version of Longhi and Boudreau's process, ought to have its cogency multiplied several times over when applied to Longhi and Boudreau's model: "Chemical and temperature changes would not easily be maintained in an isolated part of the magma . . . diffusion and convection [and *turbulence*] . . . would be continuous processes tending to eliminate steep compositional and temperature gradients."

Possible origins of magmas parental to Mg-rich layered intrusions

Since we have postulated that magmas of roughly noritic-troctolitic composition once formed layered intrusions below the ferroan anorthosite zone, we should at least speculate on the origins of those magmas. The magmas are thought to have intruded in the vicinity of some of the last rocks to crystallize from the magma ocean. Obviously, they must have originated by partial melting of some deeper part of the moon. There are basically two types of potential source regions for the partial melts: (a) remelted Mg-rich cumulates formed near the base of the magma ocean; or (b) an undifferentiated primordial interior layer that was below the original ocean, and hence retained most or all of its complement of heat-producing elements. Experts on mare basalt petrogenesis will recognize the former as the type of source most authors advocate for mare basalt melts, and the latter as the type of source advocated by Ringwood (e.g., 1976). Unfortunately, the precise compositions of the parent noritic-troctolitic magmas cannot be determined directly, and the importance of assimilation in altering their compositions while the plutons were being emplaced is also undeterminable, so it may be even harder to determine the correct source than it has been for the mare basalts.

One thing to be noted is that there is no reason to conclude, just because the oldest (4.0 G.y.) mare basalts we have found are much younger than the oldest Mg-rich rocks (approximately 4.5 G.y.; Papanastassiou and Wasserburg, 1976, p. 2049), that the source region for the mare basalts must have been completely unrelated to the source region for the Mg-rich rock plutons. As was pointed out by Ryder and Taylor (1976), the age of the "oldest" mare basalt samples appears to coincide with the age of the late heavy meteoritic bombardment. It is entirely possible that the early intense bombardment and subsequent volcanism obliterated or buried many very ancient volcanic rocks, some of which may have had quite normal mare basaltic compositions, and others of which may have been much more noritic-to-troctolitic. It is interesting to note that the two oldest mare basalts are both unusually Al-rich (Papanastassiou and Wasserburg, 1971).

Of course, there are great differences in composition between the mare basalts and the Mg-rich rocks, most notably in Mg/Fe ratios and Al and Ti contents. While this does suggest that the petrogenetic link between these two groups is not too tight, it should be borne in mind that a range of magmas may be produced from a single source by varying such factors as extent of melting, pressure, and assimilation subsequent to the melting. Assimilation processes are bound to be more important in intrusive magmas than in rapidly cooled extrusive ones, and also are bound to be more important on the anhydrous moon as compared to the "wet" earth (because anhydrous silicate melting curves have positive dP/dT, but "wet" melting curves are negatively sloped; hence anhydrous melts become "superheated" upon ascent, but "wet" melts do not). The composition of the terrestrial source region of peridotitic komatiites (Bickel *et al.*, 1976) appears remarkably similar to the composition of the source region of oceanic tholeiites (Ringwood, 1975), even though komatiites have vastly higher Mg/Fe, lower Al and lower Ti than tholeiites. Furthermore, in cumulate remelting models there is no reason to assume a uniform source region, because the cumulates presumably vary in composition, at least vertically.

The fact that the Mg-rich rocks as a group have lower Ca/Na than the ferroan anorthosites (Fig. 1) is most simply explained by a model in which the Mg-rich parental magmas formed by partial melting of a deep layer of primitive whole moon material, endowed with a full lunar complement of Na (in contrast to the anorthosites which probably formed from the magma ocean after it had lost Na via volatilization—see above). However, the Ca/Na data are also compatible with the cumulate remelting model, if we make the reasonable assumption that the amount of Na that had been lost when these early cumulates formed was much less than the loss that had occurred by the time the anorthosites began to form.

An examination of potential heat sources might eventually make possible a resolution between the two partial melting models. However, most of the same arguments involved in discussions of mare basalt petrogenesis are equally relevant here, and thus far no solid consensus has been reached on mare basalts, despite a decade of debate.

One reasonable criticism of any deep partial melting model for the parental Mg-rich magmas (Alan Binder, pers. comm., 1980) is that thermal models (e.g., Toksöz and Solomon, 1973; Hubbard and Minear, 1976; Binder and Lange, 1977) generally indicate that partial melting of the deep interior does not commence until several hundred million years after formation of the moon. However, Solomon and Longhi's (1977) "petrologically most complete" thermal history model (i.e., the "evolving magma ocean" model, summarized in their Fig. 8) indicates that deep partial melting commences even before the magma ocean has quite finished solidifying, only ~50 m.y. after formation of the moon.

Ringwood (1976) has criticized the cumulate remelting model (for mare basalt petrogenesis) on the grounds that no effective heat source is available, without invoking implausibly high contents of "trapped" liquid (rich in heat-producing elements) in the proposed source rocks. This is probably a valid argument, but there are several alternatives to trapped liquid as a heat source. For instance, as Dowty (1975) pointed out, magma ocean crystallization would tend to produce a mantle with dense Fe-rich mafics overlying lighter Fe-poor mafics. This system would be highly unstable with respect to convection, particularly if just a trace of partial melting (perhaps triggered by large meteorite impacts) lubricated the mass of crystals. Alternatively, movement might have started entirely in the solid state, triggered solely by a gradual build-up of radiogenic heat. In either case, considerable melting would take place within the Mg-rich materials, due to the resultant unloading of pressure.

In summary, none of the available evidence clearly favors one or the other of the two types of deep partial melting models for the origin of the Mg-rich parental magmas. It may even be that some combination of the two source types was tapped. Indeed, if tidal friction played a significant role in contributing heat for the melting, it would have been localized in regions of density contrast (Wones and Shaw, 1975), and one such place was probably the interface between the first (lower-most) sunken magma ocean cumulates and the top of the "core" of undifferentiated whole moon material.

Acknowledgments—This work was mainly supported by NASA grant NGL 05-007-367 (J. Wasson) and partly by NASA grant NGL 32-004-063 (K. Keil). We thank I. S. McCallum, G. A. McKay, and G. J. Taylor for constructive criticism, and P. L. Tu for expert typing.

REFERENCES

Adams L. H. (1924) Temperatures at moderate depths within the earth. *Wash. Acad. Sci. J.* **14,** 459–472.

Anders E. (1978) Procrustean science: Indigenous siderophiles in the lunar highlands, according to Delano and Ringwood. *Proc. Lunar Planet. Sci. Conf. 9th,* p. 161–184.

Apollo 17 Lunar Sample Information Catalog (1973) MSC-03211, NASA Johnson Space Center, Houston. 447 pp.

Bickel C. E. and Warner J. L. (1978) Survey of lunar plutonic and granulitic lithic fragments. *Proc. Lunar Planet. Sci. Conf. 9th,* p. 629–652.

Bickel M. J., Hawkesworth C. J., Martin A., Nisbet E. G., and O'Nions R. J., (1976) Mantle composition derived from the chemistry of ultramafic lavas. *Nature* **263,** 577–580.

Binder A. B. (1976) On the implications of an olivine dominated upper mantle on the development of a moon of fission origin. *The Moon* **16,** 159–173.

Binder A. B. and Lange M. (1977) On the thermal history of a moon of fission origin. *The Moon* **17,** 29–45.

Campbell I. H. (1978) Some problems with the cumulus theory. *Lithos* **11,** 311–323.

Chou C.-L., Boynton W. V., Sundberg L. L., and Wasson J. T. (1975) Volatiles on the surface of Apollo-15 green glass and trace-element distributions among Apollo-15 soils. *Proc. Lunar Sci. Conf. 6th,* p. 1701–1727.

Compston W. and Gray C. M. (1974) Rb-Sr age of the civet cat clast, 72255, 41. In *Interdisciplinary Studies of Samples from Boulder 1, Station 2, Apollo 17,* Vol. 1 (J. A. Wood, ed.). LSI Contr. No. 210D, 139–143. Smithsonian Astrophysical Observatory, Cambridge.

de Waard D. and Wheeler E. P. 2nd (1971) Chemical and petrologic trends in anorthositic rocks of the Nain massif, Labrador. *Lithos* **4,** 367–380.

Dowty E. (1975) Shallow origin of mare basalts (abstract). In *Papers Presented to the Conference on Origins of Mare Basalts and Their Implication for Lunar Evolution,* p. 35–39. The Lunar Science Institute, Houston.

Dowty E., Prinz M., and Keil K. (1974) Ferroan anorthosite: A widespread and distinctive lunar rock type. *Earth Planet. Sci. Lett.* **24,** 15–25.

Drake M. J. (1975) Lunar anorthosite paradox: An alternative explanation. *Proc. Lunar Sci. Conf. 6th,* p. 293–299.

Dymek R. F., Albee A. L., and Chodos A. A. (1975) Comparative petrology of lunar cumulate rocks of possible primary origin: Dunite 72415, troctolite 76535, norite 78235, and anorthosite 62237. *Proc. Lunar Sci. Conf. 6th,* p. 301–341.

Elsdon R. (1970) Crystallization behavior of mafic intrusives. *Earth Planet. Sci. Lett.* **9,** 313–316.

Gibson E. K. and Hubbard N. J. (1972) Thermal volatilization studies on lunar samples. *Proc. Lunar Sci. Conf. 3rd,* p. 2003–2014.

Gooding J. L. and Muenow D. W. (1976) Activated release of alkalis during the vesiculation of molten basalts under high vacuum: Implications for lunar volcanism. *Geochim. Cosmochim. Acta* **41,** 435–438.

Gooding J. L. and Meunow D. W. (1977) Experimental vaporization of the Holbrook chondrite. *Meteoritics* **12,** 401–408.

Haskin L. A., Helmke P. A., Blanchard D. P., Jacobs J. W., and Telander R. (1973) Major and trace element abundances in samples from the lunar highlands. *Proc. Lunar Sci. Conf. 4th,* p. 1275–1296.

Haskin L. A., Korotev R. L., and Lindstrom M. M. (1979) Are norite and KREEP only skin deep? What are the highlands made of (abstract)? In *Papers Presented to the Conference on the Lunar Highlands Crust,* p. 47–49. Lunar and Planetary Institute, Houston.

Haskin L. A., Shih C. -Y., Bansal B. M., Rhodes J. M., Wiesmann H., and Nyquist L. E. (1974) Chemical evidence for the origin of 76535 as a cumulate. *Proc. Lunar Sci. Conf. 5th,* p. 1213–1225.

Head J. W., Settle M., and Stein R. S. (1975) Volume of material excavated from major lunar basins and implications for the depth of excavation of lunar samples. *Proc. Lunar Sci. Conf. 6th,* p. 2805–2829.

Henderson P. (1970) The significance of the mesostasis of basic layered igneous rocks. *J. Petrol.* **11,** 463–473.

Herbert F., Drake M. J., Sonett C. P., and Wiskerchen (1977) Some constraints on the thermal history of the lunar magma ocean. *Proc. Lunar Sci. Conf. 8th,* p. 573–582.

Herzberg C. T. (1979) Identification of pristine lunar highland rocks: criteria based on mineral chemistry and stability (abstract) In *Lunar and Planetary Science X,* p. 536–539. Lunar and Planetary Institute, Houston.

Himmelberg X. and Ford A. B. (1976) The Dufek intrusion of Antarctica. *J. Petrol.* **17,** 219–234.

Hubbard N. J. and Minear J. W. (1976) Petrogenesis in a modestly endowed moon. *Proc. Lunar Sci. Conf. 7th,* p. 3421–3435.

Irving A. J. (1978) A review of experimental studies of crystal/liquid trace element partitioning. *Geochim. Cosmochim. Acta* **42,** 743–770.

Jackson E. D. (1961) Primary textures and mineral associations in the ultramafic zone of the Stillwater Complex, Montana. *U.S. Geol. Survey Prof. Paper 358,* 106 pp.

James O. B. (1979) Petrology of the major rock types of the lunar highlands (abstract). In *Papers Presented to the Conference on the Lunar Highlands Crust,* p. 92–94. Lunar and Planetary Institute, Houston.

Kopal Z. (1969) *The Moon.* Reidel, Dordrecht. 525 pp.

Lambert D. D. and Simmons E. C. (1980) Cumulate minerals as liquid analogues, a new approach to modelling trace elements in mafic layered intrusions: Application to the Stillwater Complex, Montana. *EOS (Trans. Amer. Geophys. Union)* **61,** 410.

Laul J. C. and Schmitt R. A. (1975) Dunite 72417: A chemical study and interpretation. *Proc. Lunar Sci. Conf. 6th,* p. 1231–1254.

Liebenberg C. J. (1960) The trace elements of the rocks of the Bushveld Igenous Complex. *Publ. Univ. Pretoria,* No. 12, p. 1–79.

Lindstrom M. M., Nava D. F., Lindstrom D. J., Winzer S. R., Lum R. K. L., Schuhmann P. J., Schuhmann S., and Philpotts J. A. (1977) Geochemical studies of the White Breccia Boulders at North Ray Crater, Descartes region of the lunar highlands. *Proc. Lunar Sci. Conf. 8th,* p. 2137–2151.

Longhi J. (1977) Magma oceanography: 2. Chemical evolution. *Proc. Lunar Sci. Conf. 8th,* p. 529–594.

Longhi J. (1979) Magma oceanography revisited: formation of the primitive lunar crustal rocks (abstract). In *Papers Presented to the Conference on the Lunar Highlands Crust,* p. 105–107. Lunar and Planetary Institute, Houston.

Longhi J. and A. E. Boudreau (1979) Complex igneous processes and the formation of the primitive lunar crustal rocks. *Proc. Lunar Planet. Sci. Conf. 10th,* p. 2085–2105.

McBirney A. R. and Noyes R. M. (1979) Crystallization and layering of the Skaergaard intrusion. *J. Petrol.* **20,** 487–554.

McCallum I. S., Raedeke L. D. and Mathez E. A. (1980) Investigations of the Stillwater Complex: Part I. Stratigraphy and structure of the banded zone. *Amer. J. Sci.* **280.** In press.

McKay G. A. and Weill D. F. (1976) Petrogenesis of KREEP. *Proc. Lunar Sci. Conf. 7th,* p. 2427–2447.

McKay G. A. and Weill D. F. (1977) KREEP (abstract). In *Lunar Science VIII,* p. 649–651. The Lunar Science Institute, Houston.

McKay G., Wiesmann H., and B. Bansal (1979) The KREEP-magma ocean connection (abstract). In *Lunar and Planetary Science X,* p. 804–806. Lunar and Planetary Institute, Houston.

Morse S. A. (1979) Kiglapait geochemistry I: Systematics, sampling and density. *J. Petrol.* **20,** 555–590.

Myers J. S. and Platt R. G. (1977) Mineral chemistry of layered Archaean anorthosite at Majorqap qava, near Fiskenaesset, Southwest Greenland. *Lithos* **10,** 59–72.

Naslund H. R. (1976) Mineralogical variations in the upper part of the Skaergaard intrusion, East Greenland. *Carnegie Inst. Wash. Yearb.* **75,** 640–644.

Norman M. D. and G. Ryder (1979) A summary of the petrology and geochemistry of pristine highlands rocks. *Proc. Lunar Planet. Sci. Conf. 10th,* 531–559.

Norman M. D. and Ryder G. (1980) Geochemical evidence for the role of ilmenite and clinopyroxene in the early lunar differentiation (abstract). In *Lunar and Planetary Science XI*, p. 821–823. The Lunar and Planetary Institute, Houston.

Nyquist L. E. (1977) Lunar Rb-Sr chronology. *Phys. Chem. Earth* **10**, 103–142.

Nyquist L. E., Wiesmann H., Wooden J., and C.-Y. Shih (1979) Age and REE abundances of anorthositic norite from 15455 (abstract). In *Papers Presented to The Conference on the Lunar Highlands Crust*, p. 122–124. Lunar and Planetary Institute, Houston.

O'Hara M. J. (1977) Geochemical evolution during fractional crystallization of a periodically refilled magma chamber, *Nature* **266**, 503–507.

O'Keefe J. D. and T. J. Ahrens (1978) Impact flows and crater scaling on the moon. *Phys. Earth Planet. Inter.* **16**, 341–351.

Papanastassiou D. A. and Wasserburg G. J. (1971) Rb-Sr ages of igneous rocks from the Apollo 14 mission and the age of the Fra Mauro formation. *Earth Planet. Sci. Lett.* **12**, 36–48.

Papanastassiou D. A. and Wasserburg G. J. (1976) Rb-Sr age of troctolite 76535. *Proc. Lunar Sci. Conf. 7th*, pp. 2035–2054.

Raedeke L. D. and McCallum I. S. (1979) Fractionation in the Stillwater Complex: An analog for the lunar crust and upper mantle (abstract). In *Papers Presented to the Conference on the Lunar Highlands Crust*, p. 125–127. Lunar and Planetary Institute, Houston.

Ringwood A. E. (1975) *Composition and Petrology of the Earth's Mantle*. McGraw-Hill, N.Y. 618 pp.

Ringwood A. E. (1976) Limits on the bulk composition of the moon. *Icarus* **28**, 325–349.

Ringwood A. E. and Kesson S. E. (1977) Composition and origin of the moon. *Proc. Lunar Sci. Conf. 8th*, p. 371–398.

Roedder E. and P. W. Weiblen (1974) Petrology of clasts in lunar breccia 67915. *Proc. Lunar Sci. Conf. 5th*, p. 303–318.

Ryder G. and G. J. Taylor (1976) Did mare-type volcanism commence early in lunar history? *Proc. Lunar Sci. Conf. 7th*, p. 1741–1755.

Ryder G., Norman M. D., and Score R. A. (1980) Ni, Co content of metal grains for the identification of indigenous rocks (abstract). In *Lunar and Planetary Science XI*, p. 968–970. Lunar and Planetary Institute, Houston.

Smith J. V., Anderson A. T., Newton R. C., Olsen E. J., Wyllie P. J., Crewe A. V., Isaacson M. S., and Johnson D. (1970) Petrologic history of the moon inferred from petrography, mineralogy, and petrogenesis of Apollo 11 rocks. *Proc. Apollo 11 Lunar Sci. Conf.*, p. 897–925.

Smith J. V. and Steele I. M. (1974) Intergrowths in lunar and terrestrial anorthosites with implications for lunar differentiation. *Amer. Mineral.* **59**, 673–680.

Solomon S. C. and Longhi J. (1977) Magma oceanography: 1. Thermal evolution. *Proc. Lunar Sci. Conf. 8th*, p. 583–599.

Steele I. M. and Smith J. V. (1971) Mineralogy of Apollo 15415 "genesis rock": Source of anorthosite on Moon. *Nature* **234**, 138–140.

Steele I. M. and Smith J. V. (1973) Mineralogy and petrology of some Apollo 16 rocks and fines: General petrologic model of Moon. *Proc. Lunar Sci. Conf. 4th*, 519–536.

Storey W. C. (1973) Volatilization studies on a terrestrial basalt and their applicability to volatilization from the lunar surface. *Nature* **241**, 154–157.

Taylor S. R. (1975) *Lunar Science: A Post-Apollo View*. Pergamon, N.Y. 372 pp.

Taylor S. R. and Bence A. E. (1975) Evolution of the lunar highland crust. *Proc. Lunar Sci. Conf. 6th*, p. 1121–1141.

Thomson W. (1864) On the secular cooling of the earth. *Trans. Roy. Soc. Edinburgh* **23**, 157–169.

Toksöz M. N. and Solomon S. C. (1973) Thermal history and evolution of the moon. *The Moon* **7**, 251–278.

Wager L. R. and Brown G. M. (1967) *Layered Igneous Rocks*. W. H. Freeman and Co. 588 pp.

Wänke H., Palme H., Baddenhausen H., Dreibus G., Jagoutz E., Kruse H., Palme C., Spettel B., Teschke F., and Thacker R. (1975) New data on the chemistry of lunar samples: Primary matter in the lunar highlands and the bulk composition of the moon. *Proc. Lunar Sci. Conf. 6th*, p. 1313–1340.

Walker D., Grove T. L., Longhi J., Stolper E. M., and Hays J. F. (1973) Origin of lunar feldspathic rocks. *Earth Planet. Sci. Lett.* **20,** 325–336.

Walker D., Stolper E. M., and J. F. Hays (1978) A numerical treatment of melt/solid segregation size of the eucrite parent body and stability of the terrestrial low-velocity zone. *J. Geophys. Res.* **83,** 6005–6013.

Warner J. L., Simonds C. H., and Phinney W. C. (1976) Genetic distinction between anorthosites and Mg-rich plutonic rocks: New data from 76255 (abstract). In *Lunar Science VII,* p. 915–917. The Lunar Science Institute, Houston.

Warren P. H. and Wasson J. T. (1977) Pristine nonmare rocks and the nature of the lunar crust. *Proc. Lunar Sci. Conf. 8th,* p. 2215–2235.

Warren P. H. and Wasson J. T. (1978) Compositional petrographic investigation of pristine nonmare rocks. *Proc. Lunar Planet. Sci. Conf. 9th,* p. 185–217.

Warren P. H. and Wasson J. T. (1979a) Effects of pressure on the crystallization of a "chondritic" magma ocean and implications for the bulk composition of the moon. *Proc. Lunar Planet. Sci. Conf. 10th,* p. 2051–2083.

Warren P. H. and Wasson J. T. (1979b) The compositional-petrographic search for pristine non-mare rocks: Third foray. *Proc. Lunar Planet. Sci. Conf. 10th,* p. 583–610.

Warren P. H. and Wasson J. T. (1979c) The origin of KREEP. *Rev. Geophys. Space Phys.* **17,** 73–88.

Warren P. H. and Wasson J. T. (1980) Further foraging for pristine nonmare rocks (abstract). In *Lunar and Planetary Science XI,* p. 1217–1219. Lunar and Planetary Institute, Houston.

Wasson J. T., Boynton W. V., Kallemeyn G. W. Sundberg L. L., and Wai C. M. (1976) Volatile compounds released during lunar lava fountaining. *Proc. Lunar Sci. Conf. 7th,* p. 1583–1595.

Wilkinson J. F. G., Duggan M. B., Herbert H. K., and Kalocsai G. I. Z. (1975) The Salt Lick Creek layered intrusion, East Kimberley region, Western Australia. *Contrib. Mineral. Petrol.* **50,** 1–23.

Winzer S. R., Nava D. F., Lum R. K. L., Schuhmann S., Schumann P., and Philpotts J. A. (1975) Origin of 78235, a lunar norite cumulate. *Proc. Lunar Sci. Conf. 6th,* p. 1219–1229.

Wones D. R. and Shaw H. R. (1975) Tidal dissipation: A possible heat source for mare basalt magmas (abstract). In *Lunar Science VI,* p. 878–880. The Lunar Science Institute, Houston.

Wood J. A. (1975) Lunar petrogenesis in a well-stirred magma ocean. *Proc. Lunar Sci. Conf. 6th,* p. 1087–1102.

Wood J. A., Dickey J. S., Marvin U. B., and Powell B. N. (1970) Lunar anorthosites and a geophysical model of the moon. *Proc. Apollo 11 Lunar Sci. Conf.,* p. 965–988.

Papike, J.J. and Merrill, R.B., eds.
Proc. Conf. Lunar Highlands Crust (1980), p. 101-111
Printed in the United States of America

Candidate samples for the earliest lunar crust*

S. Jovanovic and G. W. Reed, Jr.

Chemistry Division, Argonne National Laboratory, Argonne, Illinois 60439

Abstract—A group of non-mare samples has a Cl/P_2O_5 ratio that is much lower than in other lunar samples. Two clusters of samples with order(s) of magnitude differences in minor and trace element contents make up the group. They were all returned from non-mare sites; no mare basalt or Apollo 16 anorthositic samples are included. It is proposed that these samples could be relics from an original lunar crustal layer which evolved more or less independently but concurrently with the differentiation of an underlying deep magma ocean from which all other samples were eventually derived.

INTRODUCTION

We will explore the possibility that a group of lunar samples identified by their Cl/P_2O_5 ratio may be relics of the primitive crust of the moon. Lunar sample Cl/P_2O_5 ratios exhibit discrete values in contrast to most trace element ratios. The majority of lunar samples fall into one of three Cl/P_2O_5 ratio groups and each group includes both basaltic and anorthositic samples (Jovanovic and Reed, 1975, 1976, 1977, 1978). The basalts may be derived from cumulate source material complementary to the sources of the more aluminous samples. Thus both cumulate sources as well as intermediate differentiates could have originated from a single, well mixed magma with a fixed $Cl-P_2O_5$ content.

There are a few samples that do not fall into one of the main Cl/P_2O_5 ratio groups. A group of these has a ratio that is significantly different. These samples are unique in that there is no basaltic member. This absence of a mare basalt, e.g., a view from below, as well as other relevant factors to be noted later provide the rationale for the discussion in this paper.

Since $Cl-P_2O_5$ systematics are the bases for the point of view to be explored, two aspects of the early moon that must be taken into consideration because of their relation to these systematics are the presence of large-scale lateral heterogeneities and the presence of convection cells in the early lunar magma ocean. The status of both will be briefly reviewed along with a summary of the $Cl-P_2O_5$ systematics.

---

* Work supported by the National Science Foundation grant number EAR-7908357.

With this background, the special group of samples will be examined and a model suggested to explain them. The model will identify these samples with a crust that covered the original magma ocean. This crust was subjected to rupturing as a result of convection processes in the underlying ocean and by impacting objects. The fragments drifted over the ocean, collided and were uplifted or were subducted or foundered and possibly remelted. Two distinct crustal regimes are postulated: the quenched magma layer at the surface and the underlying plagioclase-rich layer which was accumulated from the convecting magma(s) below the solidified crust. The evolution of the top-most crust must be considered from the perspective of the evolution of the subcrust.

The need for a relatively undifferentiated layer to have existed at the surface has been recognized by others, for instance Taylor and Jakeš (1974) and Brown (1977), to explain the magnesian olivines and pyroxenes in the near surface regions and by Walker *et al.* (1975) to explain the presence of anorthositic norites.

BACKGROUND

Cl-P_2O_5 systematics

The three major Cl/P_2O_5 ratio groups of lunar samples are interpreted as being derived from discrete volumes of differentiating magma. Cl and P_2O_5 are considered to be incompatible elements and were restricted to late residual liquids during the solidification of the magma(s). It was necessary to have well stirred volumes of liquid (convection cells) in order for the ratio to be constant at both the top and bottom of the magma ocean where presumably anorthositic and mafic cumulates existed along with intercumulus liquid and in intermediate zones where KREEP-rich residual liquids concentrated. The discrete ratios could have been the result of regional accretional inhomogeniety or chemical or physical processes which occurred in the respective volumes of magma (Jovanovic and Reed, 1978).

Large convection cells were postulated (Jovanovic and Reed, 1975). A well homogenized moon should exhibit a constant Cl/P_2O_5 ratio throughout or random ratios if mixing over the whole ocean was not complete. There is evidence of neither in the Cl/P_2O_5 ratios. We have not identified other explanations for the data. Mineral control cannot play a role because of the wide variations in the mineralogy of the samples in any given ratio group. The results from measurements on separated minerals from lunar basalts support this (Jovanovic and Reed, 1980). Whatever the physical and chemical processes that may have caused loss or gain of trace element bearing phases, they apparently did not cause partition between Cl and P_2O_5 since concentrations of both range over two orders of magnitude in a given ratio group. Table 1 is the Appendix table in Jovanovic and Reed (1978) modified to emphasize the constancy of the Cl/P_2O_5 ratios irrespective of the variety of material constituting the different ratio groups.

Convection

Geochemical consequences of convection in the lunar magma ocean have been discussed from three perspectives. Wood (1975) and Walker *et al.* (1975) and Dymek *et al.* (1975) invoked convection, in one case to cause mixing of mafic and anorthositic components and, in the other case to cause crystallization from the bottom of the magma chamber in order to account for the inefficient separation of these components. Jovanovic and Reed (1975) proposed convection cells as a means for both mixing large volumes in the magma ocean and isolating them from one another. Such a mechanism could cause and/or conserve chemical differences within these volumes as noted above. Herbert *et al.* (1978) proposed that convection-related accumulation of solidified anorthositic blocks, "rock-bergs", at the surface might have caused the residual magma to have developed a fractionated REE pattern which became imprinted on the sinking mafic cumulates.

Lateral heterogeneity

Geophysical evidence which could support lateral heterogeneity cited in 1975 included electrical conductivity anomalies (Sonett *et al.,* 1974) and remanent magnetism localization (Murthy and Banerjee, 1973). Further support may be implied by different crustal thicknesses between at least two areas of the near side of the moon. These areas are separated roughly by a SW-NE line between mare Humorum and mare Serenitatis (Solomon and Head, 1979). To the east, the crust is significantly thicker than to the west. This is attributed to a heterogeneity in crustal temperature that existed 3.6–3.8 b.y. ago. Possible causes of the heterogeneity could "include concentrations of radioactive heat sources, lateral variations in the thermal conductivity and variation in heat flow at the base of the lithosphere. . ." (Solomon and Head, 1979). It is of interest that this division coincides with one of the projected boundaries between convection cells (Jovanovic and Reed, 1976).

In addition to the interpretation of radioactive heat sources based on orbital data by Hubbard (1979), other geochemical data have also been interpreted as implying lateral heterogeneity. Irving (1975) suggests that lateral (and/or radial) variations in Al and other elements might explain "the apparent tendency for some concentration of the aluminous mare basalt magmas at certain landing sites." He also suggests lateral heterogeneity in K and other LIL elements as a plausible explanation for the variation of these elements in Apollo 11 and 15 mare basalts.

DISCUSSION

Lateral heterogeneity and convective mixing are moon-wide features. The special samples will be examined in the light of these hypotheses and it will be concluded

Table 1. Three Cl/P_2O_5 ratio groups subdivided on the basis of the K_2O contents of the samples (Jovanovic and Reed, 1976). All three groups contain anorthositic and basaltic end members a-b, g-h, ff′-i; the 0.009 group includes samples with intermediate compositions.

$Cl_r/P_2O_5 = 0.021 \pm 0.005$

	Cl_r/P_2O_5*	Cl_r	P_2O_5
	0.025 ± 0.006 Basaltic		
a			
70002[1]	0.028	21	*0.076*
70006[2]	0.033	13	*0.039*
70019[a]	0.019	15	*0.084*
70181	0.03	19	0.07
74001[3]	0.017	3.8	*0.023*
74001[4]	0.025	0.66	*0.003*
74241	0.030	27	*0.091*
75075	0.02	12	0.05
75080	0.02	12	0.07
78501	0.02	12	0.05
	0.021 ± 0.005		
b			
60016	0.021	17	*0.080*
60618	0.03	2.7	*0.009*
61221	0.020	22	0.11
65501	0.021	23	0.11
67115	0.017	5.9	*0.034*
67460	0.030	12	*0.040*
67941	0.017	17	0.10
68415	0.020	11	*0.056*
68815	0.016	28	0.18
66095	0.022	53	0.24
L-20	0.02	12	0.06

$Cl_r/P_2O_5 = 0.009 \pm 0.002$

	Cl_r/P_2O_5	Cl_r	P_2O_5
	0.010 ± 0.002		
g			
61241	0.011	14	0.13
63321	0.013	9.9	*0.079*
63501	0.009	9.4	0.10
64501	0.011	17	0.14
64801	0.011	14	0.13
65785	0.007	8.6	*0.12*
	0.008 ± 0.002 Basaltic		
h			
10044	0.007	2.6	0.04
10047	0.009	9.6	0.11
10061	0.005	6.8	0.14
10084	0.009	9.9	0.12
L-16	0.011	14	0.12
10017	0.007	11	0.16
10072	<0.008	<14	0.17
70005[5]	0.013	14	*0.11*
70017	0.009	0.26	*0.003*
70135	0.006	2.2	0.04
71501	0.009	5.7	*0.067*
75035	0.008	2.4	0.03
78526	0.006	0.94	*0.015*
79115	0.010	2	*0.020*
24077	0.010	3.2	*0.033*
24174	0.006	3.6	*0.058*
24182	0.008	3.8	*0.049*

	Cl_r/P_2O_5	Cl_r	P_2O_5
	0.009 ± 0.001		
c			
14064	0.010	17	*0.17*
15081	0.010	16	*0.16*
15091	0.007	11	0.16
15211	0.008	14	*0.20*
15301	0.009	14	*0.16*
15427	0.011	3.9	*0.036*
15459	0.011	17	0.15
15501	0.008	14	0.17
15505	0.008	14	*0.18*
15515	0.008	15	*0.19*
72215†	0.007	8.9	*0.12*
72701	0.008	12	0.15
73141	0.008	10	0.12
76281	0.012	11	0.09

* Chlorine and italicized P_2O_5 data are from our laboratory. We are reporting here the residual chlorine (Cl_r) that is left in the sample after 10 min hot water (≤100°C) leaching. Other P_2O_5 data are from Lunar Sample Analyses Data Base. XRF, COL. GRAV. data are favored over EMP. [1] = 256 cm, [2] = 94 cm, [3] = 4 cm, [4] = 11 cm, [5] = 135 cm. [a] Breccia. † Gabbroic anorthosite.

Table 1. *(Continued)*

$Cl_r/P_2O_5 = 0.009 \pm 0.002$			
	Cl_r/P_2O_5	Cl_r	P_2O_5
	0.009 ± 0.001		
d			
12070	0.008	26	0.31
14064 cl	0.009	20	*0.22*
15041	0.010	20	0.20
15251	0.009	21	0.24
15265	0.008	19	0.25
15431	0.009	17	*0.18*
15445	0.007	12	*0.17*
72215	0.010	57	0.57
72275	0.008	28	0.36
73235	0.009	17	0.19
	0.009 ± 0.002		
e			
12034	0.009	46	0.53
14163	0.008	42	0.51
14259	0.008	37	0.48
15205	0.010	58	0.56
15415	0.009	52	*0.57*
14321	0.012	46	0.38

$Cl_r/P_2O_5 = 0.004 \pm 0.001$			
	Cl_r/P_2O_5	Cl_r	P_2O_5
	0.004 ± 0.001 Basaltic		
f'			
15065	0.0049	3.7	*0.076*
15085	0.0037	2.5	*0.068*
15535	0.0037	2.2	*0.060*
15557	0.0046	3.2	*0.070*
	0.003 ± 0.001 Basaltic		
f			
12021	0.004	3.3	0.09
12022	0.0020	2.6	0.13
12040	0.0032	1.4	0.044
12052	0.0024	1.9	0.08
12053	0.0031	4.4	0.14
70019[b]	0.0033	7.8	*0.24*
71055	0.0047	2	*0.043*
74275	0.004	2.8	0.07
75055	0.0049	3.4	*0.070*
	0.004 ± 0.001		
i			
14312	0.0028	18	*0.65*
72275††	0.0047	28	*0.60*
72255	0.0030	7.4	*0.25*
72395	0.0026	7.4	0.29
73275	0.0042	11	0.26
76315	0.0047	15	0.32
77035	0.0038	9.6	0.25
14305	0.0032	20	0.62

†† Clast #2.

[b] Glass coating.

that two very different regimes were necessary in the early lunar magma ocean.

The group of exceptional samples is listed in Table 2. The trace and minor element concentrations in these samples fall into two clusters with an order of magnitude or greater difference between them. Thus, they appear to have sampled material representing both extreme depletion and moderate enrichment in at least P_2O_5, U and K_2O (and also F and REE) relative to most lunar samples. The depleted-sample cluster includes the cumulates 15415 and 76535, the enriched samples have MKFM composition.

Each highland site is represented. This dispersion of samples over the surface suggests that they may be ubiquitous. The group can be characterized as:

a. All are highland or pre-mare rocks (>17% Al_2O_3); no basaltic complement has been identified.
b. There is no single "rock" type: 14310 is a KREEPy basalt, 60315, 65015, 73215 and 15455-dk are noritic anorthosites and 76535 is a troctolite.
c. There is no "typical" Apollo 16 anorthosite such as 60015, 65785, 67115 or 67075.
d. The areal distribution is random.

Two sources for this group of samples were suggested (Jovanovic and Reed, 1976). Orientale ejecta might account for the lack of a mare basalt component; this ejecta would be expected to be widely distributed. The Orientale basin is outside the projected boundaries of convection cells with which other samples have been associated (Jovanovic and Reed, 1976). The 0.001 Cl/P_2O_5 ratio for the special group could be the signature for this region. An attractive aspect of an Orientale source is the possibility for sampling deep in the lunar crust. Some

Table 2. Samples with Cl_r/P_2O_5 ratio of 0.0011 ± 0.0003.*

Sample	Cl_r ppm	P_2O_5 %	U ppm	K_2O %
60315,103[a]	9.7[a]	0.45	2.2	0.38
14310,124	4.2	(0.37)	3.5	0.52
65015,32	3.1	0.26	1.9	0.33
73215,81 bl. mtx.	3.3	0.33	1	0.30
15455,3 dark	1.1	0.13	0.14	0.17
15415,42	0.12	(0.01)	0.007	<0.01
15418,30-80	0.44	(0.03)	0.016	0.02
76535,23	0.22	(0.03)	0.03	0.05

* Residual Cl(Cl_r), U, and some P_2O_5 are from our laboratory; P_2O_5 data in parentheses and K_2O data are from Lunar Sample Analyses Data Base. Cl_r is residual Cl after hot water leach of samples.

[a] Wänke *et al.* (1976), data for total Cl.

authors have invoked the necessity of an origin at depth to account for the equilibration in the minerals in troctolite 76535, Gooley *et al.* (1974), Dymek *et al.* (1975) Stewart (1975) and others.

An original early lunar crust could also be the source. In this case, all the samples could be "local", i.e., not ejecta from afar. The only way this may have been possible would be through the formation of a thin laterally well-mixed crust which was isolated from the underlying convection cells with their respective Cl/ P_2O_5 signatures. The rationale and model for the evolution of such a crust might be the following:

a. An original crust was necessary to retain heat long enough for the magma ocean to differentiate (Walker *et al.,* 1975).
b. This crust could have been a boundary layer through which heat was conducted and below which heat was convected.
c. The convecting magma, along with impacts, caused disruption of the crust and migration (drifting) of solidified sheets over the surface until they collided, were uplifted or subducted or until they foundered. Terrestrial plate tectonics or, perhaps more appropriately, lava lake crusts (Duffield, 1972; Wright *et al.,* 1977) are models.
d. Fractionation and differentiation processes occurring in this topmost layer would not be greatly different from those below where the main lunar differentiation was occurring since magmas with similar compositions were involved. It may have been somewhat less differentiated and, therefore, remained more magnesian.

   A crust derived from magma that was not highly differentiated is suitable for this model. A plagioclase crust could have been invaded by an undifferentiated magma that could have crystallized out mafic phases (Brown, 1977). Such a crust could also arise from: crystallization of an anorthositic norite of density ~2.9 *vs* liquid density ~3.0 (Walker *et al.,* 1975), a low density contrast between plagioclase and a 14310 type KREEP liquid (Ford *et al.,* 1977) or entrapment of intercumulus liquid (Hess *et al.,* 1977; Brown, 1977). If volatiles and gases were being released at the surface, a highly vesicular chilled outer layer of the crust might result. This seems not unreasonable for very early in lunar history. Even though the moon is supposed to have been depleted in volatiles, there is evidence that low temperature volatiles have persisted until today (Reed *et al.,* 1971). Ford *et al.,* (1977) suggest that wet igneous processes, *vis a vis* the usual assumption of a dry moon, can also "account for all the observed feldspathic rocks and leave a residue capable of producing mare basalts". A volatile-rich magma could have other important consequences: (1) its lower density might permit it to remain relatively isolated from underlying, possibly denser magmas; (2) if the magma was hydrous, equilibration reactions could occur more rapidly and at lower temperatures (Ford *et al.,* 1977); (3) near-surface solidified material could be porous, due to gases, causing higher density, more mafic rocks to remain at or near the surface—in some cases continued exposure to high temperatures could cause collapse of voids and hence eventual sinking into near-surface magma chambers (Brown, 1977).

e. If chambers of primitive mafic magmas developed within the hot crust, fractionation could have occurred leading to plutonic rock types (Brown, 1977).

f. Melt-rock textures and relic clast inclusions could develop as readily in crust penetrating melts as in molten impact ejecta.

SAMPLES

Samples of the special groups are distributed randomly at all non-mare sites. They tend to fall within the K_2O-Al_2O_3 compositional fields (Irving, 1975, Fig. 2) of non-mare melt rocks. We have already noted that Apollo 16 anorthositic samples are not included in this group; the anorthositic basalt field is not populated. No samples were available to us to check the Apollo 15 and 17 KREEP basalt fields. In the Sm-Al_2O_3 compositional diagram, Irving (1975) Fig. 1, the special samples with the low trace element contents plot in the ANT field and those with higher concentrations in the Apollo 16 and 17 poikilitic field which includes the VHA basalt and Apollo 14 type KREEP fields. The aluminous mare basalt and anorthositic basalt fields are not populated. A difference between the K-Al_2O_3 and Sm-Al_2O_3 patterns for the special group of samples may be reasonable since K is less subject to mineral control than Sm and thus because of its persistence in intercumulus liquids will be more generally dispersed in the solidifying magma. It appears that a more restricted differentiation sequence or trace element partitioning applied to the special group than to the bulk of the non-mare samples. We may thus be dealing with a less evolved system than that from which the majority of the non-mare melt rocks were derived.

The samples are of two general types: cumulates, 15415 and 76535, and melt rocks. The latter category is perhaps equivocal since 14310 is a crystalline rock and may have been magmatic in origin. The anorthosite 15415 would presumably form at the top of a magma body by flotation. Troctolite 76535 has been relegated to a source at depth (>7 Km, Stewart, 1975) in the lunar crust. The most general interpretation of the melt rocks is that they are impact melts.

Samples 65015 and 60315 are poikilitic but not the typical Apollo 16 poikilitic rock based on Simonds *et al.*'s (1973) statement that exclusively LKFM composition is associated with poikilitic texture. These rocks are MKFM (Taylor and Bence, 1975) based on their K_2O and P_2O_5 contents. Albee *et al.* (1973) have described 65015. Two of their observations are germane to the point of view developed here. The only phosphate mineral found was whitlockite whereas most lunar rocks contain both whitlockite and apatite. They suggest that F and Cl were lost by decomposition of apatite. The low Cl/P_2O_5 ratio is consistent with their observations; a near surface origin of the sample would be consistent with F and Cl loss. A fluid phase, probably containing H_2O, might have been the agent for displacing Cl from apatite or from the residual liquid from which phosphates crystallized. The P_2O_5 in the residual liquid may have controlled how much Cl could be lost. A second observation was the presence of "corroded" Fe globules and balls which apparently predated the metamorphism to which the rock was subjected. The corrosion could have been the result of fluids which are no longer present.

15455-dark is a magnesian (Ryder and Bower, 1977), 100% LKFM rock (Taylor *et al.*, 1973). Ryder and Bower (1977), however, note that the whole rock contains clasts of deep-seated origin and further that LKFM is not the composition of the upper crust. Thus, they require a large scale impact to excavate the sample. We suggest that this LKFM type material could have formed in the differentiation of early upper-crustal magma and incorporated clasts from depth during the violent turbulence and mixing that must have occurred during the early epoch. The presence of meteorite contamination (Hertogen *et al.*, 1976) could be due to the fact that early differentiating magma was itself accretive material.

76535 is especially enigmatic. The problem with this cumulate is how to maintain it at high temperatures (~1000°C) for ~600 m.y. during which time it was completely equilibrated or annealed chemically, mineralogically and structurally (Gooley *et al.*, 1974; Dymek *et al.*, 1975; Smyth, 1975; Nord, 1976; and others) and then excavate it with little or no shock or reheating. The evolutionary history must also explain isotope ages of 4.61 b.y. via Rb-Sr and 4.25–4.0 b.y. via Sm-Nd, Pb-Pb and K-Ar as determined and/or summarized and discussed by Papanastassiou and Wasserburg (1976).

If hydrous fluids were present, lower melt temperatures would require shorter cooling times. Chemical and isotopic exchange would have been facilitated until the fluids were eventually lost or degassed, possibly at about 4.2 b.y. ago. Lower temperatures would have required a less deep burial. If this was the case, a major impact may not have been needed for excavation.

Only Rb-Sr gives a very old 4.61 b.y. internal isochron age which is based primarily on olivine. Papanastassiou and Wasserburg (1976) attribute this to the presence of inclusions trapped during the initial crystallization of olivine which then effectively isolated the inclusions from interaction and exchange with the outside. The presence of a hydrous fluid offers the possibility that grain growth could be more rapid, permitting trapping of material which might have been excluded during a 0.4–0.6 b.y. cooling time.

The positive Eu anomaly in 76535 is attributed by Haskin *et al.* (1974) to the presence of cumulus plagioclase. This suggests that the rock was not involved in the major lunar differentiation but accumulated its olivine and plagioclase in a primary fractionation of the primitive "crustal" magma.

SUMMARY

It seems possible that the samples identified here as being unique on the basis of their Cl/P_2O_5 ratio may also be categorized as special in other respects and may represent material which evolved in a different manner than the bulk of the returned lunar samples. Two early differentiations, occurring simultaneously, could explain the different groups of samples. One was in a strictly near-surface crustal layer; the other in a convectively mixed subcrustal deep magma ocean. Little or no mixing occurred between the two regions. Lateral homogenization of the crustal layer resulted from tectonic processes similar to continental drift or lava lake chilled crust behavior. This could account for the presence of samples of this special group at all non-mare sites. The two clusters of samples with order of magnitude or greater differences in trace and minor elements and the very different petrogeneses of the rocks in the two clusters suggest that the controlling factor was probably the relative amounts of a late residual liquid incorporated in the various source regions. Very small amounts of late residual liquid became associated with the cumulates and larger amounts in the differentiates forming later.

REFERENCES

Albee A. L., Gancarz A. J., and Chodos A. A. (1973) Metamorphism of Apollo 16 and 17 and Luna 20 metaclastic rocks at about 3.95 AE: Samples 61156, 64423,14-2, 65015, 67483,15-2, 76055, 22006, and 22007, *Proc. Lunar Sci. Conf. 4th,* p. 569–595.

Brown G. M. (1977) Extraterrestrial mineralogy—Two major igneous events in the evolution of the moon. *Phil. Trans. Roy. Soc. London* **A286,** 439–451.

Duffield W. A. (1972) A naturally occurring model of global plate tectonics. *J. Geophy. Res.* **77,** 2543–2555.

Dymek R. F., Albee A. L., and Chodos A. A. (1975) Comparative petrology of lunar cumulate rocks of possible primary origin: Dunite 72415, troctolite 76535, norite 78235 and anorthosite 62237. *Proc. Lunar Sci. Conf. 6th,* p. 301–341.

Ford C. E., O'Hara M. J., and Spenser P. N. (1977) The origin of lunar feldspathic liquids. *Phil. Trans. Roy. Soc. Lond.* **A285,** 193–197.

Gooley R., Brett R., Warner J., and Smyth J. R. (1974) A lunar rock of the deep crustal origin: Sample 76535. *Geochim. Cosmochim. Acta* **38,** 1329–1339.

Haskin L. A., Shih, C.-Y., Bansal B. M., Rhodes J. M., Wiesmann H., and Nyquist L. E. (1974) Chemical evidence for the origin of 76535 as a cumulate. *Proc. Lunar Sci. Conf. 5th,* p. 1213–1225.

Hertogen J., Janssens M.-J., Takahashi H., Palme H., and Anders E. (1977) Lunar basins and craters: Evidence for systematic compositional change of bombarding population. *Proc. Lunar Sci. Conf. 8th,* p. 17–45.

Herbert F., Drake M. J., Sonett C. P., and Wiskerchen M. J. (1977) Some constraints on the thermal history of the lunar magma ocean. *Proc. Lunar Sci. Conf. 8th,* p. 573–582.

Hess P. C., Rutherford M. J., and Campbell H. W. (1977) Origin and evolution of LKFM basalt. *Proc. Lunar Sci. Conf. 8th,* p. 2357–2373.

Hubbard N. (1979) Regional chemical variations in the lunar crust (abstract). In *papers presented to the Conference on the Lunar Highlands Crust,* p. 81–83. Lunar and Planetary Institute, Houston.

Irving A. J. (1975) Chemical, mineralogical and textural systematics of non-mare melt rocks: Implications for lunar impact and volcanic processes. *Proc. Lunar Sci. Conf. 6th,* p. 363–394.

Jovanovic S. and Reed G. W. Jr. (1975) Cl and P_2O_5 systematics: Clues to early lunar magmas. *Proc. Lunar Sci. Conf. 6th,* p. 1737–1751.

Jovanovic S. and Reed G. W. Jr. (1976) Convection cells in the early lunar magma ocean: Trace-element evidence. *Proc. Lunar Sci. Conf. 7th,* p. 3447–3459.

Jovanovic S. and Reed G. W. Jr. (1977) Trace element geochemistry and early lunar differentiation. *Proc. Lunar Sci. Conf. 8th,* p. 623–632.

Jovanovic S. and Reed G. W. Jr. (1978) Trace element evidence for a laterally inhomogeneous moon. *Proc. Lunar Planet. Sci. Conf. 9th,* p. 59–80.

Jovanovic S. and Reed G. W. Jr. (1980) Cl, P_2O_5, Br and U partioning among mineral separates from mare basalt 75055 (abstract). In *Lunar and Planetary Science XI* p. 517–519. Lunar and Planetary Institute, Houston.

Lunar Sample Analysis Data Base (1975) Curator's Office, NASA Johnson Space Center, Houston.

Murthy V. R. and Banerjee S. K. (1973) Lunar evolution: How well do we know it now? *The Moon* **7,** 149–171.

Nord G. L. Jr. (1976) 76535: Thermal history deduced from pyroxene precipitation in anorthite. *Proc. Lunar Sci. Conf. 7th,* p. 1875–1888.

Papanastassiou D. A. and Wasserburg G. J. (1976) Rb-Sr age of troctolite 76535. *Proc. Lunar Sci. Conf. 7th,* p. 2035–2054.

Reed G. W. Jr., Goleb J. A., and Jovanovic S. (1971) Surface related Hg in lunar samples. *Science* **172,** 258–261.

Ryder G. and Bower J. F. (1977) Petrology of Apollo 15 black-and-white rocks 15445 and 15455—Fragments of the Imbrium impact melt sheet? *Proc. Lunar Sci. Conf. 8th,* p. 1895–1923.

Simonds C. H., Warner J. L., and Phinney W. C. (1973) Petrology of Apollo 16 poikilitic rocks. *Proc. Lunar Sci. Conf. 4th,* p. 613–632.

Smyth J. R. (1975) Intracrystalline cation order in a lunar crustal troctolite. *Proc. Lunar Sci. Conf. 6th,* p. 821–832.

Solomon S. C. and Head J. W. (1979) Lunar mascon basins: Lava filling, tectonics, and evolution of the lithosphere. *Rev. Geophys. Space Phys.* In press.

Sonett C. P., Smith B. F., Schubert G., Colburn D. S., and Schwartz K. (1974) Polarized electromagnetic response of the moon. *Proc. Lunar Sci. Conf. 5th,* p. 3073–3089.

Stewart D. B. (1975) Apollonian metamorphic rocks—The products of prolonged subsolidus equilibration (abstract). *Lunar Science VI,* p. 774–776. The Lunar Science Institute, Houston.

Taylor S. R. and Bence A. E. (1975) Evolution of the lunar highland crust. *Proc. Lunar Sci. Conf. 6th,* p. 1121–1141.

Taylor S. R. and Jakeš (1974) The geochemical evolution of the moon. *Proc. Lunar Sci. Conf. 5th,* p. 1287–1305.

Taylor S. R., Gorton M. P., Muir P., Nance W., Rudowski R., and Ware N. (1973) Lunar highlands composition: Apennine front. *Proc. Lunar Sci. Conf. 4th,* p. 1445–1459.

Walker D., Longhi J., and Hayes J. F. (1975) Differentiation of a very thick magma body and implications for the source regions of mare basalts. *Proc. Lunar Sci. Conf. 6th,* p. 1103–1120.

Wänke H., Palme H., Kruse H., Baddenhausen H., Cendales M., Dreibus G., Hofmeister H., Jagoutz E., Palme C., Spettel B., and Thacker R. (1976) Chemistry of lunar highland rocks: A refined evaluation of the composition of the primary matter. *Proc. Lunar Sci. Conf. 7th,* p. 3479–3499.

Wood J. A. (1975) Lunar petrogenesis in a well-stirred magma ocean. *Proc. Lunar Sci. Conf. 6th,* p. 1087–1102.

Wright T. L., Peck D. L., and Shaw H. R. (1976) Kilawea lava lakes: Natural laboratories for study of cooling, crystallization, and differentiation of basaltic magma. *Amer. Geophys. Union Monograph 19,* p. 375–390.

Papike, J.J. and Merrill, R.B., eds.
Proc. Conf. Lunar Highlands Crust (1980), p. 113-132
Printed in the United States of America

The cordierite- to spinel-cataclasite transition: Structure of the lunar crust

Claude T. Herzberg and Michael B. Baker

Lunar and Planetary Institute, 3303 NASA Road 1, Houston, Texas 77058

Abstract—The moonwide magmatic processes which were responsible for the formation of the lunar crust and mantle cannot be understood without stratigraphic constraints. Some meteorite-excavated lunar highlands rocks contain this information in their mineral chemistry and stability relations, as do nodules in some basaltic rocks and kimberlites on the earth. Therefore, from an understanding of the physical-chemical conditions which are required to form certain nonmare mineral assemblages and their mineral chemistry, inferences can be drawn about their preexcavation stratigraphic location.

The cordierite- and spinel-cataclasites (olivine + plagioclase + high alumina orthopyroxene ± spinel ± cordierite) exhibit phase changes and considerable variations in their mineral chemistry at depths appropriate to the lunar crust. These have been calibrated by experiment in the system MgO-Al_2O_3-SiO_2 and extended to the natural lunar chemical system by solution thermodynamics. The results show that spinel cataclasites from Apollo 15 and 17 were located at depths greater than or equal to about 12 to 32 kilometers prior to excavation. An ultramafic spinel cataclasite clast in 15445,177 may have been located in the uppermost mantle of the moon.

From geochemical relationships between spinel cataclasites and the other members of the Mg-rich and anorthositic series given in plots of Mg/(Mg+Fe) in mafic minerals vs. An content of plagioclase, and from relative abundances of these rocks, a two layer crust of the moon can be modelled. In this model the uppermost stratigraphic unit consists of members of the anorthositic series, and is probably about 12 to 20 kilometers thick. The lower Mg-rich unit may constitute the greatest volume of the crust, and consist of rocks with roughly cotectic mineralogical proportions (i.e., norites, troctolites, spinel troctolites and their buffered low-K Fra Mauro and KREEP compositions).

INTRODUCTION

Lunar highlands plutonic and volcanic rocks contain information about the moonwide process of crust and mantle formation. Yet after nearly 10 years of research, no durable and comprehensive model of this process exists. In part this is because there are few observational constraints which can be used to test models of lunar crust formation, the vertical and lateral petrologic structure of the crust being largely unknown.

Perhaps the many problems encountered by such models can best be appreciated by considering the progress made in understanding the formation of layered intrusions of the Earth such as Skaergaard and Stillwater. The papers contained in the American Journal of Science Jackson Memorial Volume (1980) report some of the geochemical, physical, and thermal processes which no single unifying model has yet succeeded in integrating. That is, no model is currently able to predict the stratigraphy of rocks in layered intrusions. The problem then becomes compounded when similar attempts are extended to the lunar crust because of the absence of stratigraphic constraints which are routinely considered by geologists working on layered intrusions. The rocks which composed the lunar crust column are now meteorite-excavated surficial debris which may or may not have had cogenetic igneous histories.

Most models of lunar crust and mantle formation have been articulations of the magma ocean paradigm in which the entire thickness of a plagioclase-rich crust was a floated crystallization product (e.g., Wood *et al.*, 1970; Smith *et al.*, 1970; Wood, 1975; Drake, 1976; Solomon and Longhi, 1977; Longhi and Boudreau, 1979; Hubbard and Minear, 1975; Minear and Fletcher, 1978). The textures, mineralogy, and chemistry of this primordial crust were in most cases extensively modified by meteorite impact, thus obscuring the earliest record of the moon's history. However, some primordial material is believed to have survived to varying degrees the chemical and physical effects of meteorite bombardment. These are called pristine lunar highlands rocks (Warren and Wasson, 1977; Warner and Bickel, 1978). Such rocks, therefore, are likely to yield the most information about the nature and formation of the crust and mantle.

Pristine rocks define two distinct geochemical trends (Roedder and Weiblen, 1974; Warner *et al.*, 1976; Warren and Wasson, 1977; see Fig. 2): an anorthositic trend of varying Mg/(Mg+Fe) in the mafic minerals at essentially constant An content of plagioclase, and an Mg-rich trend (i.e., spinel-cataclasites, troctolites, dunite, norites) of complementary variations in these two parameters. The formation of these two suites of highland rocks can be modelled from the crystallization of a global magma ocean; the most detailed model has the ferroan anorthosites stratigraphically above the Mg-rich rocks (Longhi and Boudreau, 1979).

Some recent alternative models have tended to explain both pristine rock suites by formation from two separated magmas (Warner *et al.*, 1976; Warren, 1979; James, 1979; Binder, 1979); the ferroan anorthosites were formed as floated crystallization products of a lunar magma ocean whereas the Mg-rich rocks formed later as intruding plutons from below. Although a possible weakness in the "two magmas" hypothesis has been illustrated by analogy with similar mineralogical trends in the Stillwater Complex which crystallized from one magma (Raedeke and McCallum, 1979; DePaolo and Wasserburg, 1979), the model may account for a layered lunar crust similar in character to that proposed by Longhi and Boudreau (1979).

On the basis of melt-rock samples and their clasts from the Apollo 15 and 17 landing sites, Ryder and Wood (1977) constructed a layered model of the crust with features similar to those above. The uppermost stratigraphic horizons were inferred to consist of anorthosite and noritic anorthosite which grade downwards to more ferromagnesian-rich rocks with higher Mg/(Mg+Fe). This model is consistent with the observation that the ferroan anorthosites are preponderant among pristine rocks from the Apollo 16 landing site which is the furthest removed from large impact basins (Warren and Wasson, 1977). Conversely, the Mg-rich rocks are preponderant among pristine rocks from the Apollo 15 and 17 landing sites which, being near the rims of the Imbrium and Serenitatis impact basins, are the most likely to contain excavated material from the lower crust. A consensus, therefore, appears to be developing in which the anorthositic and Mg-rich suites of highland rocks are excavated samples of a layered crust.

Any layering of the lunar crust can have far-reaching consequences on the models themselves. For example, in the absence of any accurate estimate of the bulk composition of the crust, the compositional layering may range from dominantly anorthosite with minor variations in the plagioclase/mafic mineral ratio and Mg/(Mg+Fe), to layering in which anorthosite is volumetrically overwhelmed by norite-troctolite, or low-K Fra Mauro material (Ryder and Wood, 1977; Charette *et al.,* 1977). In the latter case, "the incentive for calling on a lunar magma ocean disappears as no anorthositic layer of global proportions need be accounted for" (Walker, 1979). Indeed, the latter crustal profile could be modelled with or without a magma ocean. The interaction of convecting magma of a global ocean with the underside of anorthositic rockbergs, entrapment of mafic crystals in the advancing plagioclase flotation crust, lack of efficient gravitational separation caused by convection are all possibilities in the magma ocean concept (Minear and Fletcher, 1978; Longhi and Boudreau, 1979; Herzberg, 1978a). Alternatively, an Mg-rich layered crust with a profile similar to sections of the Stillwater Complex (Raedeke and McCallum, 1979) may also be the product of a lunar interior which underwent extensive partial melting. Indeed, the distinction between the two models may be somewhat misleading, the magma ocean requiring something to be partially or totally melted. If the products of a magma ocean differ from those of a partially melted lunar interior, these may be manifest in the kind of crust each may produce. The stratigraphy of each may differ and, accordingly, become identified as constraints on the competing hypotheses.

Assuming that the products of the two models will become understood, it is clear that the formation of the lunar crust will remain in a shroud of mystery unless its petrological and geochemical structure becomes known. We have directed our efforts at supplying such stratigraphic constraints by developing thermodynamic criteria which permit depth determinations to be made on various mineral assemblages. Of the highlands lithologies which, given proper calibration of the relevant equilibria, exhibit a range of mineral compositions and phase changes appropriate to the depths of the lunar crust, the spinel cataclasites are the best candidates (Herzberg, 1978a, 1979). Olivine-bearing norites, and olivine-two pyroxene anorthosites (e.g., 62275,4, Prinz *et al.,* 1973; Herzberg, 1979) are also subject to depth estimates upon calibration of the relevant equilibrium discussed in Herzberg (1979).

Depth estimates for spinel cataclasite clasts (olivine + plagioclase + high alumina orthopyroxene + spinel ± cordierite) in 72435, 73263, and 15445 were previously made (Herzberg, 1978a) based on published mineral analyses and the best available thermochemical data at that time. It was noted, however, that such estimates were subject to a possible error of about ±30 kilometers, based on uncertainties in the thermochemical data. Accordingly, we have experimentally calibrated the equilibrium which bounds the T-P stability fields of spinel cataclasites and their low-pressure cordierite breakdown products. The results of these experiments are reported here and their application to the depth of origin of lunar spinel cataclasites discussed. Details of the experimental methods and thermodynamic analysis of the results are rather lengthy, and will be reported elsewhere (Herzberg, in preparation). Finally, we suggest some possible relationships between spinel cataclasites and the other members of the anorthositic and Mg-rich suites, and offer a stratigraphic model of the lunar crust.

THE CORDIERITE- TO SPINEL-CATACLASITE BOUNDARY IN THE SYSTEM MgO-Al_2O_3-SiO_2

Ideally, the best way to experimentally establish the depth of origin of natural spinel cataclasite assemblages would be to "cook" the assemblage at the temperatures at which they equilibrated and the appropriate range of pressures (i.e., 1 to 3000 bars), and observe any breakdown products or changes in the mineral chemistry. However, spinel cataclasites being plutonic rocks which presumably cooled from magmatic temperatures, there are good reasons to expect temperatures of equilibration in the range of 600°–1000°C. These temperatures, however, render direct experimental determination an impossible task due to the exceptionally low reaction rates in anhydrous silicate systems. An alternative method is to perform the experiments at high temperatures (e.g., ≥1200°C) and extrapolate. For natural chemical systems this approach is inadequate because the many chemical variables and equilibria (e.g., 7–8 important oxide components) preclude a thermodynamic basis for any extrapolation at this time. Linear T-P extrapolations of a limited set of high temperature data are invalid because changing configurational entropy terms due to order-disorder and solid solution effects (O'Hara *et al.,* 1971; Herzberg, 1978b) invariably result in pyroxene-bearing facies boundaries which are curves in T-P space.

The method chosen here involves experimental determination of the cordierite- to spinel-cataclasite boundary in the simple system MgO-Al_2O_3-SiO_2 which represents about 80–90 wt % of the relevant components in natural complex spinel cataclasites. The other important components are FeO, Cr_2O_3, TiO_2 and CaO. With certain testable assumptions, the effects of these extra components on the temperature and depth estimates are considered through solution thermodynamics.

In the system MgO-Al_2O_3-SiO_2 the cordierite- to spinel-cataclasite boundary is represented by the univariant equilibrium (Herzberg, 1978a):

$$\underset{\text{forsterite}}{5Mg_2SiO_4} + \underset{\text{cordierite}}{Mg_2Al_4Si_5O_{18}} = \underset{\text{enstatite}}{5Mg_2Si_2O_6} + \underset{\text{spinel}}{2MgAl_2O_4}. \quad (1)$$

This is a simple end-member reaction which does not obviously show the aluminous nature of orthopyroxene. Orthopyroxene compositions buffered by forsterite and spinel vary with temperature and pressure according to the reaction (Dankwerth and Newton, 1978; Herzberg, 1978a):

$$\underset{\text{enstatite}}{Mg_2Si_2O_6} + \underset{\text{spinel}}{MgAl_2O_4} = \underset{\substack{\text{magnesium}\\ \text{tschermak's}\\ \text{pyroxene}}}{MgAl_2SiO_6} + \underset{\text{forsterite}}{Mg_2SiO_4}. \tag{2}$$

A mass balance representation of the reaction forsterite + cordierite to aluminous orthopyroxene + spinel is a combination of reactions (1) and (2), and takes the form:

$$\begin{aligned} &\underset{\text{forsterite}}{5(1 - x)Mg_2SiO_4} + \underset{\text{cordierite}}{Mg_2Al_4Si_5O_{18}} \\ &\quad = \underset{\text{aluminous orthopyroxene}}{5[(1 - x)Mg_2Si_2O_6 + xMgAl_2SiO_6]} + \underset{\text{spinel}}{(2 - 5x)MgAl_2O_4} \end{aligned} \tag{3}$$

where x is the mole fraction of $MgAl_2SiO_6$ in aluminous orthopyroxene [equivalent to $(X^{M1}_{Al})_{opx}$, that is the mole fraction of Al in the Ml site of orthopyroxene].

The T-P locations of reaction (1) and the alumina content of orthopyroxene can be quantified in part by consideration of the free energy relation:

$$\begin{aligned} (\Delta G^\circ)_{T,P,1,2} &= (\Delta H^\circ - T\Delta S^\circ + P\Delta V^\circ)_{T,P,1,2} \\ &= -RT\ \ln K_{1,2} \end{aligned} \tag{4}$$

where the equilibrium constants:

$$K_1 = \frac{(a^{opx}_{Mg_2Si_2O_6})^5\ (a^{sp}_{MgAl_2O_4})^2}{(a^{ol}_{Mg_2SiO_4})^5\ (a^{cord}_{Mg_2Al_4Si_5O_{18}})} \tag{5}$$

and

$$K_2 = \frac{(a^{opx}_{MgAl_2SiO_6})\ (a^{ol}_{Mg_2SiO_4})}{(a^{opx}_{Mg_2Si_2O_6})\ (a^{sp}_{MgAl_2O_4})} \tag{6}$$

become for multisite mixing of cations:

$$K_1 = \frac{(X^{M2}_{Mg})^5_{opx}\ (X^{M1}_{Mg})^5_{opx}\ (\gamma_{Mg_2Si_2O_6})^5_{opx}\ (X^{TET}_{Mg})^2_{sp}\ (X^{OCT}_{Al})^4_{sp}\ (\gamma_{MgAl_2O_4})^2_{sp}}{(X^{OCT}_{Mg})^{10}_{ol}\ (\gamma_{Mg_2SiO_4})^5_{ol}\ (X^{OCT}_{Mg})^2_{cord}\ (\gamma_{Mg_2Al_4Si_5O_{18}})_{cord}} \tag{7}$$

and

$$K_2 = \frac{(X^{M1}_{Al})_{opx}\ (\gamma_{MgAl_2SiO_6})_{opx}\ (X^{OCT}_{Mg})^2_{ol}\ (\gamma_{Mg_2SiO_4})_{ol}}{(X^{M1}_{Mg})_{opx}\ (\gamma_{Mg_2Si_2O_6})_{opx}\ (X^{TET}_{Mg})_{sp}\ (X^{OCT}_{Al})^2_{sp}\ (\gamma_{MgAl_2O_4})_{sp}} \tag{8}$$

In Eqs. (7) and (8), $(X^b_a)^d_c$ refers to the mole fraction of cation a in site b of phase c to the power d, and $(\gamma_e)^g_f$ refers to the activity coefficient of component e in phase f to the power g. The activity coefficients expressed in this way combine interactions of cations within and between energetically distinct sites (cf., Sack, 1979; Wood and Nicholls, 1978). It is a macroscopic formulation which does not allow the inter- and intrasite cationic interactions to be considered separately.

With the standard state taken to be the pure phases at the temperature and pressure of interest, all activity coefficients in the simple system are unity except those for the two pyroxene components. For the latter, evidence favouring a close approach to ideality (i.e., $\gamma \rightarrow 1$) has been suggested by Dankwerth and Newton (1978) on the basis of experimental and calorimetric data.

In this simple system, Eqs. (4) and (8) can be rearranged and quantified:

$$T = \frac{-P[0.18(X_{Al}^{M1})_{opx} - 0.005] - 4522}{\ln K_2 - 1.811}, \tag{9}$$

where

$$K_2 = \frac{(X_{Al}^{M1})_{opx}}{(1 - X_{Al}^{M1})_{opx}}. \tag{10}$$

In Eq. (9) P is in bars and T is in °K. This equation was derived from Dankwerth and Newton (1978), and includes the entropy, enthalpy, and volume changes of reaction (2) and the effects of thermal expansion of the phases. Thus, for any value of $(X_{Al}^{M1})_{opx}$ in this simple system at a designated pressure, the temperature can be calculated. Figure 1 shows the T-P distribution of these isopleths of alumina from Eqs. (9) and (10).

Figure 1 shows also the experimentally determined temperatures and pressures at which reaction (1) occurs [or more descriptively reaction (3); Herzberg, in preparation]. These are the conditions at which the cordierite cataclasite assemblage (i.e., forsterite + cordierite + orthopyroxene ± anorthite) reacts to the spinel cataclasite assemblage (i.e., forsterite + spinel + orthopyroxene ± anorthite) in this simple system (note that anorthite is an extra phase which does not participate in any of the reactions, except in contributing small amounts of Ca to orthopyroxene). Combining Eqs. (4) and (8), the experimental data can be fitted to the equation:

$$P = \frac{RT}{-1.43}\left[\frac{401.6}{T} - 1.839 - \ln K_1\right] \tag{11}$$

where

$$K_1 = [1 - (X_{Al}^{M1})_{opx}]^5. \tag{12}$$

R is the gas constant, and the value −1.43 is the volume change of equilibrium (1) in cal bar^{-1}. At any temperature along the equilibrium boundary, K_1 can be evaluated from the alumina isopleths of Fig. 1 or Eqs. (9) and (10).

Equation (11) permits an extrapolation of the cordierite- to spinel-cataclasite boundary to all temperatures as shown in Fig. 1. The accuracy of this extrapolation is tested by comparison with the experimental results of Seifert (1974) and Fawcett and Yoder (1966) on the same reaction in the water-saturated system by means of the hydration model of Newton and Wood (1979). Using this model, which is based on H_2O solubility data in Mg-cordierite (Mirwald and Schreyer, 1977), the location of the boundary in both the wet and dry systems satisfy the experimental data to within ±0.20 kbar over the temperature range 780°–1300°C.

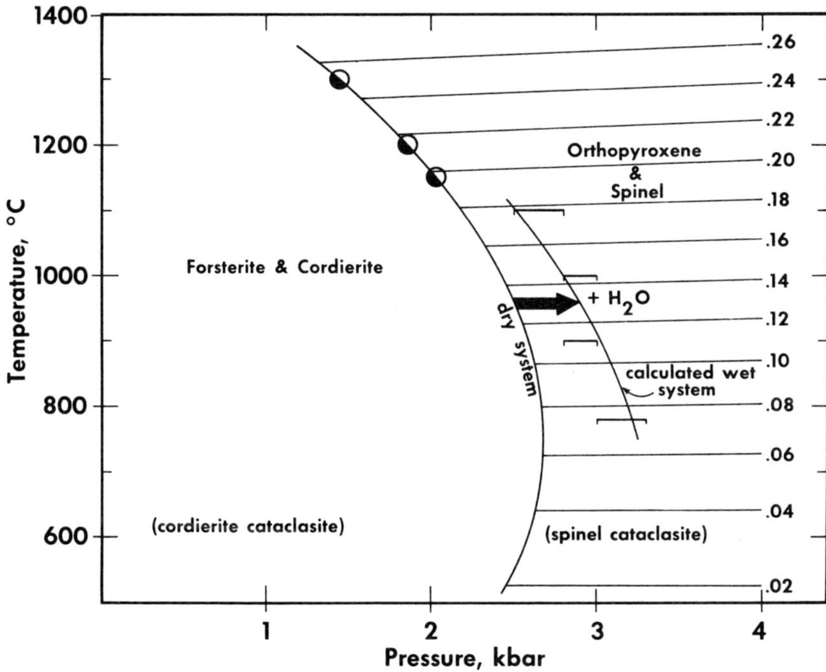

Fig. 1. The cordierite- to spinel-cataclasite boundary in the system MgO-Al_2O_3-SiO_2 ($\pm H_2O$). Half-filled circles are experimental results from Herzberg (in preparation), and extrapolated curve is from equation (11) in the text. Brackets are experimental data from Seifert (1974) and reversal data from Fawcett and Yoder (1966) in the wet system. Curve passing through these brackets is calculated from the dry reference state using the cordierite hydration model of Newton and Wood (1979). Numbered lines are isopleths of alumina in orthopyroxene [$(X_{Al}^{M1})_{opx}$] from Dankwerth and Newton (1978) using Eq. (9) in the text.

APPLICATION OF THE CORDIERITE- TO SPINEL-CATACLASITE BOUNDARY IN THE SYSTEM MgO-Al_2O_3-SiO_2 TO LUNAR SPINEL CATACLASITES

If a lunar spinel cataclasite assemblage was confined to the system MgO-Al_2O_3-SiO_2 (+ CaO to form anorthite), it would be stable at some pressure greater than or equal to those shown in Fig. 1 (i.e., dry system) at any specified temperature. However, as reviewed by Ryder and Bower (1977), small but significant amounts of Cr_2O_3 and FeO are present in the mafic phases. Because these two components partition preferentially into spinel, they can significantly expand the spinel stability field at the expense of the cordierite field (Herzberg, 1978a). Therefore, depth estimates must consider these extra components. To do so requires that the equilibrium constant K_1 be calculated from the mineral analyses of the spinel cataclasite assemblage, and the pressure solved from Eq. (11). In this natural system, K_1 must be evaluated from Eq. (7). However, because the activity coefficients in Eq. (7) are not known, a first-order approximation of ideality can be made and tested (i.e., γ's remain equal to unity). Thus:

$$K_1 = \frac{(X_{Mg}^{M2})_{opx}^5 (X_{Mg}^{M1})_{opx}^5 (X_{Mg}^{TET})_{sp}^2 (X_{Al}^{OCT})_{sp}^4}{(X_{Mg}^{OCT})_{ol}^{10} (X_{Mg}^{OCT})_{cord}^2} \tag{13}$$

It is clear from inspection of Eq. (13) that the site occupancies for each mineral from their chemical analysis must be estimated. For orthopyroxene this involves construction of a structural formula as outlined previously (Herzberg, 1978a). It is known, however, that for orthopyroxenes in the system $Mg_2Si_2O_6$-$Fe_2Si_2O_6$, Fe^{2+} partitions preferentially into the larger M2 site (Virgo and Hafner, 1969; Saxena and Ghose, 1971). Not only is it safe to assume that this partitioning will be maintained for natural aluminous orthopyroxenes, but because of the lattice contraction effect of Al in the M1 site, it is possible that it will be enhanced somewhat relative to the simple binary system. To simplify the calculations, however, values for the site occupancies can be made by the approximation that Mg/(Mg+Fe) in M1 and M2 are equal to the whole pyroxene value. Although this approximation will result in slight underestimations in the calculated pressure (e.g., for 15445,177; 1.42 kbar instead of 1.43 kbar using the partitioning data in Virgo and Hafner (1969) and Saxena and Ghose (1971)), these are not significant.

Spinel site occupancies are simply calculated from the cation proportions:

$$(X_{Mg}^{TET})_{sp} = Mg/(Mg + Fe^{2+} + Mn); \tag{14}$$

$$(X_{Al}^{OCT})_{sp} = Al/(Al + Cr + Si + Ti). \tag{15}$$

These occupancies are those for a fully ordered normal spinel. Although some disordering is likely to occur and be similar for $MgAl_2O_4$-$FeAl_2O_4$ spinel solutions (Navrotsky and Kleppa, 1967) the boundary shown in Fig. 1 and evaluated from equation (11) considers this (Herzberg, in preparation).

Mg site occupancy for olivine consisting of two energetically similar sites is simply:

$$(X_{Mg}^{OCT})_{ol} = Mg/(Mg + Fe + Ca + Mn). \tag{16}$$

For spinel cataclasites which contain no cordierite, the Mg octahedral site occupancy [i.e., Mg/(Mg + Fe + Mn)] must be calculated for a hypothetical cordierite that would be in equilibrium with the other mafic phases at the temperatures and pressures of reaction (1). This was discussed previously (Herzberg, 1978a), and can be calculated from the relation:

$$\frac{(X_{Mg}^{OCT})_{cord}}{(1 - X_{Mg}^{OCT})_{cord}} = \frac{(X_{Mg}^{TET})_{sp}}{0.15(X_{Fe}^{TET})_{sp}} \tag{17}$$

where the value 0.15 is the Fe-Mg partition coefficient of the cordierite-spinel exchange reaction. From terrestrial occurrences and the lunar cordierite-spinel pair (Dymek *et al.*, 1976; Baker and Herzberg, in preparation) this value varies little for Mg/(Mg+Fe) of spinels greater than about 0.20. That is, this partition is affected little by changes in temperature and pressure. Large errors in this value of 0.15 result in insignificant errors in the Mg site occupancy of cordierite. For example, using the average spinel analysis for 15445,177 in Table 1A, $(X_{Mg}^{OCT})_{cord}$ = 0.96, 0.97, and 0.98 for the partition coefficients 0.20, 0.15 and 0.10 respectively.

Inspection of Eq. (11) shows also that the temperature of formation of spinel cataclasites must be known in order to solve for the pressure. The available selenothermometers which permit such temperatures to be estimated are the Fe^{2+}-Mg exchange between olivine and spinel [Evans and Frost (1975) minus about 50°C to correct for $Fe^{3+} = 0$ from Medaris (1975)], olivine and orthopyroxene (Sack, 1980), and reaction (2) based on the alumina content of orthopyroxene buffered by olivine and spinel. The latter can be calculated from Eq. (9) by assuming ideal solid solutions and:

$$K_2 = \frac{(X_{Al}^{M1})_{opx}\,(X_{Mg}^{OCT})_{ol}^2}{(X_{Mg}^{M1})_{opx}\,(X_{Mg}^{TET})_{sp}\,(X_{Al}^{OCT})_{sp}^2}\,. \tag{18}$$

It has been suggested that the experimental data of Fujii (1976), which results in a free energy equation similar to Eq. (9), can result in temperature underestimations of as much as 390°C (Grove *et al.,* 1979) based on experimental results on complex systems similar to spinel cataclasites. The experimental results of Mori (1977) also indicate that temperature underestimates will be made from experimental data in the system MgO-Al_2O_3-SiO_2. It was shown, however, that experiments on natural complex systems yielded results similar to those in the system CaO-MgO-Al_2O_3-SiO_2 (Mori, 1977). Comparing the data of Mori (1977) to Eq. (9) indicates that such underestimates may be closer to about 180°C rather than 390°C. Regardless of the error, it appears that small amounts of Ca in orthopyroxene can substantially reduce the amount of alumina dissolved in orthopyroxene. This will require that $(\gamma_{MgAl_2SiO_6})_{opx}$ of Eq. (8) be greater than unity. Little systematic data is available to quantify these activity coefficients as functions of composition, temperature, and pressure. Consequently, the temperature underestimates generated from Eq. (9) should be born in mind. This, however, does not create serious difficulties for estimating the depths of spinel cataclasites because the cordierite- to spinel-cataclasite boundary is largely a function of pressure (see Table 1).

Temperature and pressure estimates have been made from mineral chemistry data on 73263,1,11 (Bence *et al.,* 1974) and our data on 15445,177, 72435,30, and the cordierite-bearing spinel cataclasite in 72435,8. The latter has been briefly described by Dymek *et al.* (1976). Detailed petrographic work and extensive microprobe surveys for all three will be reported in Lunar and Planetary Science XI (Baker and Herzberg, in prep.). An example calculation is given in the Appendix for our representative data on 15445,177 listed in Table 1A. It is emphasized that the representative analyses must be truly representative of a statistically significant number of mineral chemistry data points. Only by such exhaustive surveys (e.g., 44 olivine, 24 pyroxene, and 58 spinel analyses of 15445,177) can foreign polymict material be detected (i.e., by Fe-Mg partition criteria; Baker and Herzberg, in prep.), inhomogeneities due to endogenic process (e.g., retrograde metamorphic effects) be assessed, and the original precataclasis mineral assemblage be reconstructed. These temperature and pressure estimates are given in Table 1.

Although in most cases the composition of olivine is highly uniform and homogeneous (e.g., $Fo_{92\pm1}$ from 44 analyses), there is generally a small but significant range in composition of the orthopyroxenes and spinels. These are generally variations in Al/(Al + Cr) in spinel and $(X_{Al}^{M1})_{opx}$ which could be due to a number of endogenic processes. These include crystal-liquid interactions of the igneous precursors, kinetic effects of cooling from magmatic temperatures to those of the subsolidus, and interaction of the clasts with impact melt. Accordingly, there are a range of possible temperatures and pressures at which these rocks could have reached partial or complete equilibrium prior to excavation by meteorite impact. However, part of the T-P variation is due to insufficient information on the activity-composition relations of orthopyroxene and spinel. The latter, however, can be assessed on a very crude quantitative level as discussed below; that is,

Table 1. Temperature and pressure estimates of various spinel cataclasites

Selenothermometer 1: Fe-Mg/olivine-spinel (Evans and Frost, 1975; Medaris, 1975)
Selenothermometer 2: Fe-Mg/olivine-orthopyroxene (Sack, 1980)
Selenothermometer 3: Equilibrium 2 in text and Appendix
Selenothermometer 4: Equilibrium 2 from Grove *et al.* (1979)

15445,177

Selenothermometer	T(°C)	Depth (km) I	Depth (km) II	Depth (km) III
1	700–800	≥26	≥26	≥12
2	700	≥26	≥26	≥12
3	810–1020	≥28	≥28	≥12
4	1230–1310	≥34	≥34	≥14

I: from lowest Al_2O_3 in opx and highest Mg/(Mg + Fe) and Al/(Al + Cr) in spinel
II: from average Al_2O_3 in opx and average spinel composition
III: from highest Al_2O_3 in opx and lowest Mg/(Mg + Fe) and Al/(Al + Cr) in spinel
Depth is calculated from the lowest T if a range is indicated.

73263,1,11 (Bence *et al.*, 1974)

Selenothermometer	T(°C)	Depth (km) I	Depth (km) II
1	750–1000	≥28	≥20
2	750– 900	≥28	≥20
3	860– 950	≥30	≥22
4	1250–1290	≥36	≥24

I: from opx 2 and spinel 11 (Table 5 of Bence *et al.*, 1974)
II: from opx 1 and spinel 10 (Table 5 of Bence *et al.*, 1974)
Depth is calculated from the lowest T where a range is indicated.

Table 1. *(Continued)*

72435,30

Olivine (Fo_{73}), Orthopyroxene [Mg/(Mg + Fe) = 0.78, Al_2O_3 = 3.83 wt %, Ti/Al = .14], Spinel [Mg/(Mg + Fe) = 0.57 − 0.64, Al/(Al + Cr) = .87 − .94], Plagioclase (An_{96})

Selenothermometer	T(°C)	Depth (km) I	Depth (km) II
1	1000–1200	≥32	≥12
2	800–1200	≥28	≥12
3	680– 810	≥26	≥12
4	1170–1230	≥32	≥12

I: from highest Al/(Al + Cr) and Mg/(Mg + Fe) in spinel
II: from lowest Al/(Al + Cr) and Mg/(Mg + Fe) in spinel

72435,8

Olivine (Fo_{73}), Orthopyroxene [Mg/(Mg + Fe) = .75, Al_2O_3 ≃ 4 wt %, Ti/Al ≃ 0.11], Cordierite [Mg/(Mg + Fe) = .84], Spinel [Mg/(Mg + Fe) = .45, Al/(Al + Cr) = 0.81], Plagioclase (An_{97}).

Selenothermometer	T(°C)	Depth (km)* I		Depth (km)* II
1	700	−6		−12
2	no solution		?	
3	950–1020	−10		−18
4	1290–1310	−16		−24

* 72435,8 is a univariant mineral assemblage. In principle, a specific T and P can be determined.
I: from lowest Al_2O_3 in opx
II: from highest Al_2O_3 in opx

inferences can be made on geological grounds as to whether neglect of the activity coefficients in the calculations [i.e., Eqs. (7) and (8)] will result in pressure under- or overestimates.

Inspection of the results on 15445,177, 73263,1,11 and 72435,30 in Table 1 shows that all the clasts resided at depth in the lunar crust prior to excavation. Generally, a ±100°C uncertainty in the temperature of equilibration generates an uncertainty of about ±2 kilometers in the minimum depth of burial. In most cases, temperatures estimated from the Fe-Mg exchange equilibria are lower than those calculated from the aluminous orthopyroxene-olivine-spinel buffer reaction. This may be due in part to uncertainties in the calibration of these equilibria and the relevant activity coefficients. However, they may also represent real differences in the closure temperatures due to kinetic differences associated with reaction, diffusion, and nucleation. On equilibrium cooling from magmatic temperatures to say 700°C, in order for the aluminous orthopyroxene to continuously lower its alumina content (see Fig. 1) it must first be in contact with an olivine grain, diffusion must operate, and finally spinel (the recipient of alumina from orthopyroxene) must nucleate and grow. This is a complex process compared to simple Fe-Mg diffusion between olivine/orthopyroxene and olivine/spinel coexisting pairs.

It should also be noted that the lowest pressure estimates result from use of the most iron- and chrome-rich spinels of any one clast. For the cordierite-bearing sample in 72435,8, which departs furthest from the $MgO-Al_2O_3-SiO_2$ system [i.e., lowest Mg/(Mg+Fe) in mafic phases] *negative* pressures are calculated. On geological grounds this is an absurd solution, which indicates that important activity coefficient terms arise for mineral compositions which depart to such an extent from the system $MgO-Al_2O_3-SiO_2$. From Eq. (7) it can be shown that:

$$\frac{(\gamma_{Mg_2Si_2O_6})^5_{opx}\,(\gamma_{MgAl_2O_4})^2_{sp}}{(\gamma_{Mg_2SiO_4})^5_{ol}\,(\gamma_{Mg_2Al_4Si_5O_{18}})_{cord}} > 1 \quad (19)$$

and thus $(\gamma_{Mg_2Si_2O_6})^5_{opx}\,(\gamma_{MgAl_2O_4})^2_{sp} > (\gamma_{Mg_2SiO_4})^5_{ol}\,(\gamma_{Mg_2Al_4Si_5O_{18}})_{cord}$. Indeed, a ratio of about 1.4 to 1.8 of Eq. (19) would result in lunar surficial pressures. If 72435,8 was in fact a deep-seated rock, the above activity coefficient ratio would have to be higher still. This result is consistent with expectations drawn from activity coefficients arising from reciprocal lattice effects in solid solutions containing two or more energetically nonequivalent sites (i.e., orthopyroxenes and spinels; Wood and Nicholls, 1978; Sack, 1980). That is, activity coefficients for orthopyroxene and spinel are likely to be significantly different from those for olivine and cordierite in which the cations Fe-Mg mix in two sites which are energetically similar.

The results for 72435,8 thus provide important information for assessing the error in depth estimates for the other spinel cataclasites. Because the mineral compositions in 15445,177, 73263,1,11 and 72435,30 approximate more closely the simple system $MgO-Al_2O_3-SiO_2$ than 72435,8, their ratios of activity coefficients in equation (19) should be less than 1.4 to 1.8, but remain greater than unity. This indicates that the minimum pressure estimates given in Table 1 are also underestimated somewhat, perhaps by 0 to 10 kilometers. We suggest, therefore, that the depths of ≥12 kilometers for 15445,177 determined from the most iron- and chrome-rich spinels are much too low, and the values ≥26 kilometers from the most aluminous spinels are more realistic. This is consistent with the most aluminous spinels being volumetrically the most abundant in 15445,177 (Baker and Herzberg, in prep.). Unfortunately, an error propagation analysis is precluded because sufficient information is not available on the activity-composition relations. However, with a suggested error of ±0.2 kbar in the T-P location of reaction (1), the minimum pressure estimates in Table 1 may be accurate to about ±4 km + (0 to 10 km).

THE BEARING OF SPINEL CATACLASITES ON THE STRUCTURE OF THE LUNAR CRUST AND MANTLE

With the possible exception of the cordierite-bearing clast in 72435,8, the remainder of the samples considered appear to have had a deep-seated origin in the crust and/or mantle. Because only minimum pressure estimates can be made, it is not possible to be more specific about their preexcavation stratigraphic location. That is, they may have resided in the middle or the deepest parts of the crustal column.

Although an uppermost mantle origin for some cannot be established from these equilibria alone, it also cannot be ruled out. Clasts in 15445 consist of olivines with forsterite contents averaging 92 (this work and review in Ryder and Bower, 1977) and modal proportions ranging from 40 to 61%. Section 15445,177 consists of the mode: 15% plagioclase, 44% olivine, 6–7% spinel, 2–3% orthopyroxene and 32% matrix material. Judging from the heterogeneities observed in other clasts from 15445 (Ryder and Bower, 1977), this mode is probably not representative of the entire clast. Regardless of this variation, these are some of the most ultramafic highland rocks yet reported.

Apollo 17 spinel cataclasites are generally more plagioclase-rich and lower in Mg/(Mg+Fe). For 72435,8,30, the mineralogical proportions are: 83–89% plagioclase, 2–7% olivine (Fo_{73-75}), 2–11% spinel, and 2–6% orthopyroxene. Bence and McGee (1976) have reported a mode of 57–67% plagioclase, 25–27% olivine (Fo_{90}), 2–8% orthopyroxene, and 5–7% spinel with trace amounts of ilmenite and metal for two sections of 73263.

On the plot of Mg/(Mg+Fe) in mafic minerals and anorthite content of plagioclase shown in Fig. 2, the spinel- and cordierite-cataclasites which have been reported are located at the Mg-rich extremities of the Mg-rich and anorthositic suites. It is possible that their restricted distributions are a simple artifact of the limited number of samples that have been reported; future work on other clasts may show a broad distribution of spinel cataclasites in each suite. It should also be noted that the gap between the two trends is maintained.

Placing the spinel cataclasites in a stratigraphic context can be fraught with difficulties because it is not known how representative they are at depth in the crust. Indeed, they may be interleaved with other highland rocks of greater abundance, including pyroxene-free spinel troctolites. Their lateral distribution is also unknown. They may be of global extent, or they may be confined to the crustal columns below the Imbrium and Serenitatis Basins. It is also not yet possible to establish if there is any relationship between depth estimates for spinel cataclasites and the other members of the Mg-rich and anorthositic suites of Fig. 2. However, with all these uncertainties in mind, we propose the stratigraphic model shown in Fig. 3.

The fundamental assumption in the construction of this model is that the members of the two trends of Fig. 2 represent samples from two distinct lithologic units (e.g., Warren and Wasson, 1977), and their depths within each are reflected in their values of Mg/(Mg+Fe). Figure 3 is similar in overall features to models which have been proposed previously (Ryder and Wood, 1977; Warren and Wasson, 1977) in that an anorthositic layer overlies a layer consisting of members of the Mg-rich trend. We would suggest that the anorthositic layer is between 12 and 20 kilometers thick, based on the depth values given in Table 1. That is, if spinel cataclasite in 72435,30 is a member of the anorthositic suite as Fig. 2 suggests, it must have been located at a depth greater than or equal to about 12 kilometers. However, spinel cataclasite in 73263,1,11, being a member of the

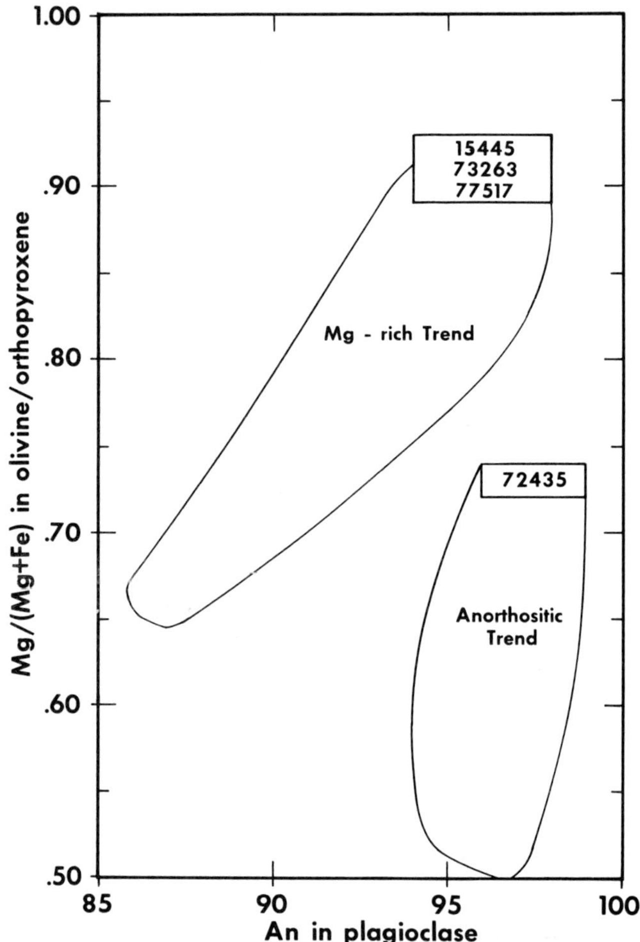

Fig. 2. Mg/(Mg+Fe) in mafic minerals vs. An content of plagioclase for pristine nonmare lunar rocks, after Warren (1979). Analyses for spinel cataclasites in 15445 and 72435 are from Baker and Herzberg (in preparation). Data for 73263 and 77517 are from Bence *et al*. (1974) and Warner *et al*. (1978).

lower Mg-rich series, was located at a depth greater than or equal to about 20 kilometers. Since the equilibria from which these estimates are made provide only minimum possible depths in the crust, the 12 to 20 kilometer range must also be considered a *minimum* thickness range for the anorthositic layer. Indeed, the seismic discontinuities at about 20 kilometers depth modelled by Goins *et al*. (1979) could reflect the interface of this two layer crust. That this interface is sharply defined rather than gradual is implied in Fig. 2. Accordingly, the boundary between these two layers given in Fig. 3 shows abrupt changes in the mineral chemistries and mineralogical proportions.

If the anorthositic layer exhibits stratigraphic variations in its mineralogical abundances, we would suggest that 72435,30, being the most primitive member of the anorthositic series (Fig. 2), and having about 83–89% plagioclase, may constrain these variations to those shown in Fig. 3. That is, there may be a strong increase in the plagioclase content upward from the base of the anorthositic layer. This would be accompanied by the iron enrichment trend as given in Fig. 2. An iron-rich spinel horizon may or may not be present at the base of this layer, depending on how representative 72435 is. As with the Middle Banded Zone of the Stillwater Complex (Raedeke and McCallum, 1979), the anorthositic layer may consist of a number of lithologies, including norites and troctolites. Thus the actual volume of anorthosite sensu stricto may be less than that of the anorthositic layer itself. Heterogeneities and exceptions to the simplicity of Fig. 3 must be expected. Indeed, it is possible that the dunite 72415 was contained in the anorthositic layer, based on other equilibria discussed elsewhere (Herzberg, 1979).

The Mg-rich layer may constitute the largest volume fraction of the lunar crust, and consist of many lithologies whose minerals may be roughly in cotectic proportions. Plagioclase may occupy about 40–60% of this layer, with its volume fraction decreasing (perhaps highly irregularly) toward the crust-mantle boundary. Troctolites and norites, with perhaps their major element buffered low-K Fra Mauro and KREEP compositions (Walker *et al.*, 1979), may be the dominant lithologies of this lower layer, and hence the lunar crust as a whole. If the depth of excavation is reflected inversely in the abundance of a lithology and plagioclase content (a crude criterion because of polymict processes), the pleonaste spinel rocks, being relatively rare and having some of the lowest plagioclase contents, would be located at the deepest stratigraphic levels (e.g., 15445,177). Values for

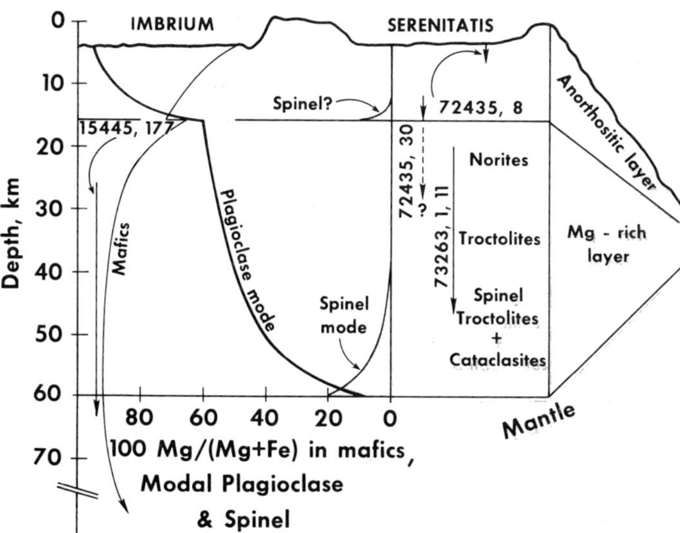

Fig. 3. Some possible stratigraphic features of the lunar crust and mantle. Vertical bars with spinel cataclasite clast numbers indicate their possible stratigraphic locations prior to excavation by meteorite impact.

Mg/(Mg+ Fe) could range from 0.90–0.92 over a considerable thickness of the lower crust. As pointed out elsewhere (Walker *et al.,* 1975; Herzberg, 1978a), Mg/(Mg+ Fe) would then have to decrease from the crust-mantle boundary to the depths of the mantle in order for there to have been a suitable source region for mare basalts.

CONCLUSIONS

Nodules in some basaltic rocks and kimberlites represent a stratigraphy of the Earth's crust and upper mantle. Similarly, pristine highlands clasts of a plutonic nature represent stratigraphic sections of the lunar crust, and possibly the upper mantle. Obviously, the mode of transport of these deep-seated materials to the surfaces of each planet were different, being volcanic in the case of the Earth, and excavation by meteorite impact for the Moon.

Of the plutonic pristine highlands rocks, which are commonly the norites, troctolites (±spinel), and anorthosites, only the spinel- and cordierite-cataclasites (olivine + plagioclase + high alumina orthopyroxene ± spinel ± cordierite) exhibit phase changes and considerable variations in their mineral chemistry at depths appropriate to the lunar stratigraphic column. These have been calibrated by experiment in the system $MgO\text{-}Al_2O_3\text{-}SiO_2$ and applied to the natural lunar spinel- and cordierite-cataclasites by solution thermodynamics.

The results show that the cordierite + spinel cataclasite in 72435,8 was probably of near-surface origin, whereas spinel cataclasites in 72435,30, 73263,1,11 and 15445,177 were located at depths greater than or equal to about 12 to 32 kilometers. For the cordierite-free types, only *minimum* depths of burial prior to excavation by meteorite impact can be determined. This precludes the possibility of unequivocally identifying lunar upper mantle rocks from the equilibria of concern here; however, the ultramafic spinel cataclasite clast in 15445,177, which resided at depths greater than or equal to about 26 kilometers, is a logical candidate.

The highly magnesian spinel cataclasites in 15445 and 73263 are associated with troctolites and norites of the Mg-rich series of rocks in a plot of Mg/(Mg+ Fe) in mafic minerals vs. An content of plagioclase. However, the Fe-rich spinel- and cordierite-cataclasites in 72435 are associated with the anorthositic members of this plot. All plot at the most Mg-rich extremities of both series. If it is assumed that the members of the two series represent samples from two distinct lithologic units of the lunar crust, and their depths within each are reflected in their values of Mg/(Mg+ Fe), a two layer model of the crust can be modelled. The uppermost stratigraphic unit consists of the members of the anorthositic series, and is probably about 12 to 20 kilometers thick. The lower Mg-rich unit may constitute the greatest volume of the crust, and consist of lithologies with roughly cotectic mineralogical proportions (i.e., norites, troctolites, spinel troctolites, and their buffered low-K Fra Mauro and KREEP compositions). The boundary separating the two stratigraphic units may be sharp, rather than gradational.

Acknowledgments—Colin M. Graham at Edinburgh University and the Natural Environment Research Council of Britain are thanked for generously providing the experimental facilities. Part of this research was financed by NASA grant NGL 09-015-150 to J. A. Wood, and the remainder by the Lunar and Planetary Institute which is operated by the Universities Space Research Association under Contract No. NSR 09-051-001 with the National Aeronautics and Space Administration. Thanks go to Graham Ryder, Lewis Ashwal, Charles Bickel, and Jeffrey Warner for their reviews of the manuscript. Bob Newton from the University of Chicago is also thanked for his critical review of another manuscript (Herzberg, in preparation) within which the experimental data and thermodynamic analysis are discussed in detail. All support is gratefully acknowledged. This paper is Lunar and Planetary Institute Contribution No. 402.

REFERENCES

Bence A. E., Delano J. W., Papike J. J., and Cameron K. L. (1974) Petrology of the highlands massifs at Taurus-Littrow: An analysis of the 2–4 mm soil fraction. *Proc. Lunar Sci. Conf. 5th,* p. 785–827.

Bence A. E. and McGee J. (1976) Significance of the assemblage anorthite-aluminous enstatite-forsterite-aluminous spinel in the lunar highlands. *Abstracts with Programs,* Geol. Soc. Amer. **8,** 772.

Binder A. B. (1979) Pristine primary crustal rocks or secondary plutonic rocks (abstract)? In *Papers Presented to the Conference on the Lunar Highlands Crust,* p. 3–5. Lunar and Planetary Institute, Houston.

Charette M. P., Taylor S. R., Adams J. B., and McCord T. B. (1977) The detection of soils of Fra Mauro basalt and anorthositic gabbro composition in the lunar highlands by remote spectral reflectance techniques. *Proc. Lunar Sci. Conf. 8th,* p. 1049–1061.

Danckwerth P. A. and Newton R. C. (1978) Experimental determination of the spinel peridotite to garnet peridotite reaction in the system MgO-Al_2O_3-SiO_2 in the range 900°–1100°C and Al_2O_3 isopleths of enstatite in the spinel field. *Contrib. Mineral. Petrol.* **66,** 189–201.

DePaolo D. J. and Wasserburg G. J. (1979) Sm-Nd age of the Stillwater complex and the mantle evolution curve for neodymium. *Geochim. Cosmochim. Acta* **43,** 999–1008.

Drake M. J. (1976) Evolution of major mineral compositions and trace element abundances during fractional crystallization of a model lunar composition. *Geochim. Cosmochim. Acta* **40,** 401–411.

Dymek R. F., Albee A. L., and Chodos A. A. (1976) Petrology and origin of boulders #2 and #3. Apollo 17 Station 2. *Proc. Lunar Sci. Conf. 7th,* p. 2335–2378.

Evans B. W. and Frost B. R. (1975) Chrome-spinel in progressive metamorphism—a preliminary analysis. *Geochim. Cosmochim. Acta* **39,** 959–972.

Fawcett J. J. and Yoder H. S., Jr. (1966) Phase relationships of chlorites in the system MgO-Al_2O_3-SiO_2-H_2O. *Amer. Mineral.* **54,** 1645–1653.

Fujii T. (1976) Solubility of Al_2O_3 in enstatite coexisting with forsterite and spinel. *Carnegie Inst. Wash. Yearb.* **75,** 566–571.

Goins N. R., Dainty A. M., and Toksöz M. N. (1979) Lunar crustal structure beneath the Apollo seismic stations (abstract). In *Papers Presented to the Conference on the Lunar Highlands Crust,* p. 24–26. Lunar and Planetary Institute, Houston.

Grove T. L., Bence A. E., and Lindsley D. H. (1979) Pressure—temperature estimates of spinel cataclasites: Comparison of experimental results in complex and simple systems (abstract). In *Papers Presented to the Conference on the Lunar Highlands Crust,* p. 30–32. Lunar and Planetary Institute, Houston.

Herzberg C. T. (1978a) The bearing of spinel cataclasites on the crust-mantle structure of the moon. *Proc. Lunar Planet. Sci. Conf. 9th,* p. 319–336.

Herzberg C. T. (1978b) Pyroxene geothermometry and geobarometry: Experimental and thermodynamic evaluation of some subsolidus phase relations involving pyroxenes in the system CaO-MgO-Al_2O_3-SiO_2. *Geochim. Cosmochim. Acta* **42,** 945–957.

Herzberg C. T. (1979) Identification of pristine lunar highland rocks: Criteria based on mineral chemistry and stability (abstract). In *Lunar and Planetary Science X,* p. 537–539. Lunar and Planetary Institute, Houston.

Hubbard N. J. and Minear J. W. (1975) A physical and chemical model of early lunar history. *Proc. Lunar Sci. Conf. 6th,* p. 1057–1085.

James O. B. (1979) Petrology of the major rock types of the lunar highlands (abstract). In *Papers Presented to the Conference on the Lunar Highlands Crust,* p. 92–94. Lunar and Planetary Institute, Houston.

Longhi J. and Boudreau A. R. (1979) Complex igneous processes and the formation of the primitive lunar crustal rocks. *Proc. Lunar Planet. Sci. Conf. 10th,* p. 2085–2105.

Medaris L. G. Jr. (1975) Coexisting spinel and silicates in alpine peridotites of the granulite facies. *Geochim. Cosmochim. Acta* **39,** 947–958.

Minear J. W. and Fletcher C. R. (1978) Crystallization of a lunar magma ocean. *Proc. Lunar Planet. Sci. Conf. 9th,* p. 263–283.

Mirwald P. W. and Schreyer W. (1977) Die stabile und metastabile Abbareaktion von Mg-cordierit in Talk, Disthen and Quartz und ihre Abhängigkeit vom Gleichgewichtswassergehalt des Cordierits. *Fortschr. Mineral.* **55,** 95–97.

Mori T. (1977) Geothermometry of spinel lherzolites. *Contrib. Mineral. Petrol.* **59,** 261–279.

Navrotsky A. and Kleppa O. J. (1967) The thermodynamics of cation distributions in simple spinels. *J. Inorg. Nucl. Chem.* **29,** 2701–2714.

Newton R. C. and Wood B. J. (1979) Thermodynamics of water in cordierite and some petrologic consequences of cordierite as a hydrous phase. *Contrib. Mineral. Petrol.* **68,** 391–405.

O'Hara M. J., Richardson S. W., and Wilson G. (1971) Garnet-peridotite stability and occurrence in crust and mantle. *Contrib. Mineral. Petrol.* **32,** 48–68.

Prinz M., Dowty E., Keil K., and Bunch T. E. (1973) Spinel troctolite and anorthosite in Apollo 16 samples. *Science* **179,** 74–76.

Raedeke L. D. and McCallum I. S. (1979) Fractionation in the Stillwater Complex: An analog for the lunar crust and upper mantle (abstract). In *Papers Presented to the Conference on the Lunar Highlands Crust,* p. 125–127. Lunar and Planetary Institute, Houston.

Roedder E. and Weiblen P. W. (1974) Petrology of clasts in lunar breccia 67915. *Proc. Lunar Sci. Conf. 5th,* p. 303–318.

Ryder G. and Bower J. F. (1977) Petrology of Apollo 15 black-and-white rocks 15445 and 15455—fragments of the Imbrium impact melt sheet? *Proc. Lunar Sci. Conf. 8th,* p. 1895–1923.

Ryder G. and Wood J. A. (1977) Serenitatis and Imbrium impact melts: Implications for large-scale layering in the lunar crust. *Proc. Lunar Sci. Conf. 8th,* p. 655–668..

Sack R. O. (1980) Some constraints on the thermodynamic mixing properties of Fe-Mg orthopyroxenes and olivines. *Contrib. Mineral. Petrol.* In press.

Saxena S. K. and Ghose S. (1971) Mg^{2+}-Fe^{2+} order-disorder and the thermodynamics of the orthopyroxene crystalline solution. *Amer. Mineral.* **56,** 532–559.

Seifert F. (1974) Stability of sapphirine: A study of the aluminous part of the system MgO-Al_2O_3-SiO_2-H_2O. *J. Geol.* **82,** 173–204.

Smith J. V., Anderson A. T., Newton R. C., Olsen E. J., and Wyllie P. J. (1970) A petrologic model for the moon based on petrogenesis, experimental petrology, and physical properties. *J. Geol.* **78,** 381–405.

Solomon S. C. and Longhi J. (1977) Magma oceanography: I. Thermal evolution. *Proc. Lunar Sci. Conf. 8th,* p. 583–599.

Virgo D. and Hafner S. S. (1969) Fe^{2+}, Mg order-disorder in heated orthopyroxenes. *Mineral. Soc. Amer. Spec. Pap.* **2,** 67–81.

Walker D. (1979) The lunar magma ocean revisited (abstract). In *Papers Presented to the Conference on the Lunar Highlands Crust,* p. 180–182. Lunar and Planetary Institute, Houston.

Walker D., Longhi J., and Hays J. F. (1975) Differentiation of a very thick magma body and implications for the source regions of mare basalts. *Proc. Lunar Sci. Conf. 6th,* p. 1103–1120.

Walker D., Stolper E. M., and Hays J. F. (1979) Basaltic volcanism: The importance of planet size. *Proc. Lunar Planet. Sci. Conf. 10th,* p. 1995–2015.

Warner J. L. and Bickel C. E. (1978) Lunar plutonic rocks: A suite of materials depleted in trace siderophile elements. *Amer. Mineral.* **63,** 1010–1015.

Warner J. L., Simonds C. H., and Phinney W. C. (1976) Genetic distinction between anorthosites and Mg-rich plutonic rocks: New data from 76255 (abstract). In *Lunar Science VII,* p. 915–917. The Lunar Science Institute, Houston.

Warner R. D., Taylor G. J., Mansker W. L., and Keil K. (1978) Clast assemblages of possible deep-seated (77517) and immiscible-melt (77538) origins in Apollo 17 breccias. *Proc. Lunar Planet. Sci. Conf. 9th,* p. 941–958.

Warren P. H. (1979) Certain pristine nonmare rocks formed as cumulates from the magma ocean (but many others did not) (abstract). In *Papers Presented to the Conference on the Lunar Highlands Crust,* p. 192–194. Lunar and Planetary Institute, Houston.

Warren P. H. and Wasson J. T. (1977) Pristine nonmare rocks and the nature of the lunar crust. *Proc. Lunar Sci. Conf. 8th,* p. 2215–2235.

Wood B. J. and Nicholls J. (1978) The thermodynamic properties of reciprocal solid solutions. *Contrib. Mineral. Petrol.* **66,** 389–400.

Wood J. A. (1975) Lunar petrogenesis in a well-stirred magma ocean. *Proc. Lunar Sci. Conf. 6th,* p. 1087–1102.

Wood J. A., Dickey J. S., Jr., Marvin U. B., and Powell B. N. (1970) Lunar anorthosites and a geophysical model of the moon. *Proc. Apollo 11 Lunar Sci. Conf.,* p. 965–988.

Appendix

Representative mineral analyses in spinel cataclasite 15445,177 are given in Table 1A. The average orthopyroxene has the following structural formula:

M2	M1	TET
0.899 Mg	0.828 Mg	1.883 Si
0.077 Fe^{2+}	0.071 Fe^{2+}	0.115 Al
0.020 Ca	0.080 Al	
0.004 Mn	0.012 Ti	
	0.012 Cr	

Therefore, $(X_{Mg}^{M2})_{opx} = 0.899$, $(X_{Mg}^{M1})_{opx} = 0.828$, and $(X_{Al}^{M1})_{opx} = 0.080$.

Determining the temperature at which this orthopyroxene was in equilibrium with olivine and the average spinel requires calculating K_2 from Eq. (18). Thus

$$K_2 = \frac{(X_{Al}^{M1})_{opx}(X_{Mg}^{OCT})_{ol}^2}{(X_{Mg}^{M1})_{opx}(X_{Mg}^{TET})_{sp}(X_{Al}^{OCT})_{sp}^2}$$

$$= \frac{(0.080)(0.924)^2}{(0.828)(0.833)(0.920)^2}$$

$$= 0.117.$$

Substituting now into Eq. (9), T becomes 1145°K (872°C) at 1000 bars, and 1148°K (875°C) at 2000 bars. Using this temperature estimate, a pressure can now be determined by calculating first K_1 of Eq. (13), and substituting into Eq. (11).

In order to calculate K_1, the hypothetical cordierite composition must be known. From Eq. (17) and the average spinel analysis:

$$(X_{Mg}^{OCT})_{cord} = 0.97$$

and thus

$$Kl = \frac{(X^{M2}_{Mg})^5_{opx}(X^{M1}_{Mg})^5_{opx}(X^{TET}_{Mg})^2_{sp}(X^{OCT}_{Al})^4_{sp}}{(X^{OCT}_{Mg})^{10}_{ol}(X^{OCT}_{Mg})^2_{cord}}$$

$$= \frac{(0.899)^5(0.828)^5(0.833)^2(0.920)^4}{(0.924)^{10}(0.97)^2}$$

$$= 0.27$$

$$\ln K_1 = -1.3.$$

Equation (11) now appears numerically:

$$P \geq \frac{(1.987)(1148)}{-1.43}(-0.350 - 1.839 + 1.3)$$

$$\geq 1400 \text{ bars}$$

$$\geq 28 \text{ kilometers (i.e., where 1000 bars = 20 kilometers).}$$

Table 1A. 15445,177: Representative mineral analyses[1]

			Orthopyroxenes			Spinel		
	Plagioclase	Olivine	Lowest-Al	Average-Al	Highest-Al	Lowest Cr/Cr+Al	Average Cr/Cr+Al	Highest Cr/Cr+Al
SiO_2	44.22	41.00	55.38	54.54	54.30	0.07	0.09	0.25
Al_2O_3	35.95	0.02	4.09	4.80	6.03	62.90	61.86	57.91
TiO_2	0.09	0.07	0.45	0.51	0.46	0.05	0.05	0.01
FeO	0.13	7.40	5.11	5.13	5.05	7.58	7.74	8.70
MnO	0.00	0.10	0.07	0.14	0.08	0.13	0.11	0.20
MgO	0.05	51.42	33.68	33.59	33.14	22.05	22.34	21.20
CaO	19.15	0.02	0.54	0.52	0.57	0.04	0.04	0.09
K_2O	0.03	n.a.	n.a.	n.a.	n.a.	n.a.	n.a.	n.a.
Na_2O	0.38	n.a.	n.a.	n.a.	n.a.	n.a.	n.a.	n.a.
Cr_2O_3	n.a.	0.00	0.37	0.40	0.52	7.32	7.89	11.97
Total	100.00	100.03	99.68	99.63	100.15	100.14	100.10	100.33
Si	2.043	.992	1.906	1.883	1.863	.004	.004	.008
Al	1.957	.000	.168	.195	.242	1.859	1.836	1.746
Ti	.004	.000	.012	.012	.012	.000	.000	.000
Fe	.004	.148	.148	.148	.145	.160	.164	.188
Mn	.000	.004	.004	.004	.004	.004	.004	.004
Mg	.004	1.855	1.730	1.727	1.695	.824	.840	.809
Ca	.949	.000	.020	.020	.020	.000	.000	.004
K	.000	—	—	—	—	—	—	—
Na	.035	—	—	—	—	—	—	—
Cr	—	.000	.012	.012	.016	.145	.156	.242
Total	4.996	3.000	4.000	4.000	3.996	2.996	3.004	3.000

[1] Oxides in wt. %; formula units based on eight oxygens for plagioclase, six for pyroxene, and four for olivine and spinel.

n.a. = not analyzed.

Papike, J.J. and Merrill, R.B., eds.
Proc. Conf. Lunar Highlands Crust (1980), p. 133-153
Printed in the United States of America

A comparison of fractionation trends in the lunar crust and the Stillwater Complex

L. D. Raedeke and I. S. McCallum

Department of Geological Sciences, University of Washington, Seattle, Washington 98195

Abstract—On a plot of molar Mg/(Mg+Fe) in mafic phases versus molar An in coexisting plagioclase, data from the banded zone of the Stillwater Complex show two fractionation trends remarkably similar to those of lunar highlands samples. Significant differences include (1) the higher Na content of the Stillwater plagioclases reflecting higher alkali contents in the magma, (2) the absence of a gap between the two trends for the Stillwater samples and (3) a shallower slope for the normal fractionation trend in the Stillwater rocks. The normal fractionation trend for the Stillwater rocks is defined exclusively by samples from the Lower and Upper Banded zones where the crystallization sequence, modal abundances and cryptic variation are consistent with those predicted by phase equilibria for a basaltic system undergoing near-perfect fractional crystallization accompanied by accumulation on the floor of the magma chamber. A vertical fractionation trend is defined exclusively by the anorthositic rocks of the Middle Banded zone and can be explained by a process of equilibrium crystallization of "trapped" intercumulus liquid in a plagioclase-rich crystal mush. The abundance of plagioclase relative to intercumulus liquid in effect buffers the plagioclase composition at a near constant value. Mafic phases crystallizing from the trapped liquid are relatively iron-rich. The vertical trend can be duplicated exactly by a simple model in which the proportions of cumulus phases and intercumulus liquid are varied within narrow limits.

A similar model is successful in explaining the fractionation trend exhibited by the lunar ferroan anorthosite suite. However, in contrast to the Stillwater case, it can be shown that the rocks of the ferroan anorthosite suite *cannot* be generated from the same magma that formed the rocks of the Mg-rich plutonic suite. This conclusion is consistent with trace element data. The lunar anorthosites are probably remnants of the primitive lunar crust that, at a slightly later time, was intruded by a set of plutons that gave rise to the Mg-rich cumulates.

INTRODUCTION

The concept that the lunar crust and upper mantle formed by a process of crystal/liquid fractionation and accumulation in a global magma ocean in a manner analogous to terrestrial layered intrusions was first proposed by Wood *et al.* (1970). While increasingly complex magma ocean models have been required to explain the growing body of petrologic, geochemical and geophysical data, the concept is still widely embraced by selenologists. The model does provide a framework for understanding such large scale lunar features as the contrast between crust and mantle compositions and the formation of distinctive source reservoirs for mare basalts and KREEP basalts. Even if the magma ocean hypothesis in its present form should not prove equal to the task of explaining all the observational data, it seems highly probable that the concept of crystal/liquid fractionation and segregation will endure. We accept as a basic premise for this paper that these processes were important during the early evolution of the moon and will attempt to explain the observed fractionation trends in lunar highland rocks within this framework.

A major problem in the interpretation of geochemical and petrologic data on lunar highlands rocks is the absence of stratigraphic constraints. In effect, the stratigraphic relationships are a "free parameter" in existing models of the evolution of the lunar crust and upper mantle. To add to this problem, primary igneous textures in almost all highland samples have been either destroyed or severely modified. Since it is generally agreed that the rocks of the lunar crust and upper mantle formed originally by igneous cumulate processes, it is our contention that detailed study of an appropriate terrestrial analog, in which the stratigraphy is well known and the original textures preserved, will help us understand the evolution of the lunar crust and upper mantle. We believe that the Stillwater Complex provides an excellent analog. The lithologies observed in the Stillwater cumulates, i.e., dunite, harzburgite, orthopyroxenite, norite, gabbro, gabbronorite, troctolite, and anorthosite, provide an excellent match with the lithologies observed in lunar highlands samples and those inferred to exist in the lunar mantle. In addition, the original igneous textures and structures are preserved and we now have a detailed knowledge of the stratigraphic section (Raedeke, 1979; McCallum *et al.,* 1980).

LUNAR FRACTIONATION TRENDS

On a plot of mole % An in plagioclase vs. mole % Mg/(Mg+Fe) in coexisting mafic phases, pristine rocks of the lunar highlands define two distinct trends (Fig. 1). The "vertical" trend on Fig. 1 is defined by anorthosites, noritic anorthosites, and troctolitic anorthosites of the ferroan anorthosite suite from the Apollo 15 and 16 landing sites. An oblique trend is defined by rocks relatively richer in mafic phases, dominantly norites and troctolites from the Apollo 15 and 17 sites. This is the so-called "normal" trend predicted for a magmatic system undergoing simple fractional crystallization with all minerals accumulating together. There is an obvious gap between the two trends.

The two distinct groups of primitive lunar crustal rocks have puzzled petrologists since their discovery. The first indication of the vertical fractionation trend was provided by Steele and Smith (1973) for anorthositic fragments from Apollo 11, 14, 15 and 16. They interpreted the trend, which they thought to have a steep negative slope, as resulting from preferential loss of sodium from the lunar surface as the ANT rocks develop. Roedder and Weiblen (1974) noted the existence of two trends, i.e., the near vertical trend observed in anorthositic samples and a

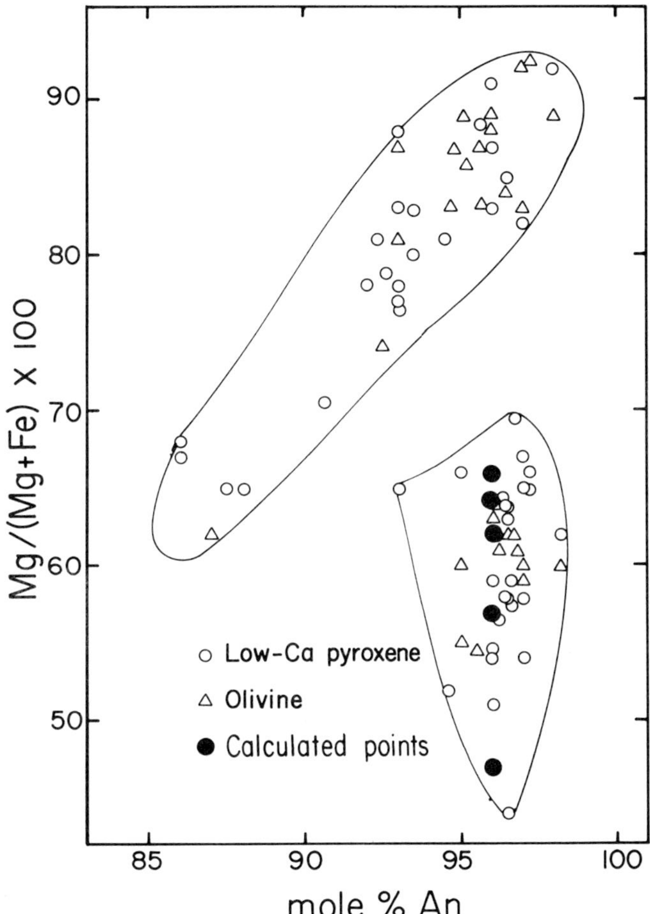

Fig. 1. Plot of mole % An in plagioclase versus mole % Mg/(Mg+Fe) in coexisting mafic minerals for pristine lunar highlands samples. Data are taken from Warren and Wasson (1979a) and Warner, *et al.* (1976). Calculated points plotted as large filled circles are from Table 4 (Data set 11, 25% intercumulus liquid).

trend with a positive slope comparable to that observed for terrestrial layered instrusions. According to Roedder and Weiblen, ". . . the negative slope (of the near vertical trend) . . . requires either disequilibrium or a process like Na volatilization or, if the trend is vertical, fractional crystallization of olivine and equilibrium crystallization of plagioclase, an obscure process that has been proposed but which we do not fully understand." Warner *et al.* (1976) expanded the data base and confirmed the two distinct trends, each with a different slope. The "normal" trend was defined by Mg-rich plutonic rocks (dunite, norite, troctolite and gabbro) and the "vertical" trend by anorthosites. Warner *et al.* concluded that the two suites are not cogenetic, having formed from different magmas, but that ". . . they almost certainly are plutonic and probably of cumulate origin."

Warren and Wasson (1977, 1978, 1979a) made a comprehensive survey of pristine non-mare rocks and showed that the two trends are even more clearly defined when only definite pristine rocks are considered. Their data show that a negative correlation for the anorthosite trend is not justified (see Fig. 1). Warren and Wasson made the important observation that anorthositic rocks with *cumulus* olivine, plotted on the vertical trend. They also pointed out the distinct compositional gap between the two trends and interpreted this to mean that the two suites must have crystallized from different magmas.

Bickel and Warner (1978) surveyed a large number of lunar plutonic and granulitic lithic fragments and showed that the coarse-grained plutonic fragments plotted in one of the two trends discussed above, whereas the fine-grained granulitic impactites have mineral compositions that, in general, fall in the gap between the two groups. Bickel and Warner suggested that most granulitic impactites are recrystallized mixtures of igneous precursors belonging to both the primary trends.

Several attempts have been made to explain the vertical anorthosite trend. Drake (1975) proposed, on the basis of experimental data on olivine/liquid and plagioclase/liquid equilibria, that the observed trend is a predictable consequence of crystallization from a lunar bulk composition that is poorer in silica than the bulk compositions of terrestrial layered intrusions.

The most comprehensive attempts at an explanation of the two distinct series are those of Longhi (1977, 1978, 1979) and Longhi and Boudreau (1979). Longhi maintained that the displacement of the two series of rocks is such that the Mg-rich cumulates and the anorthosites could not have formed from the same magmas. In addition, the presence of a distinct gap in mineral compositions between the two series attests to separate magmatic processes. Longhi and Boudreau suggested that, ". . . these relations seem consistent with a similar initial parent for the two rock series but different evolutionary paths." Following a suggestion by Herbert *et al*. (1977, 1978), they have developed a complex model involving the formation of anorthositic "rockbergs" over regions of convective downwelling. The observed fractionation trends can be approximately modeled if both equilibrium and fractional crystallization processes are involved along with contamination of the magma by foundering crust, and magma mixing.

STILLWATER COMPLEX FRACTIONATION TRENDS

As a result of ~12 man-months of field work in the Stillwater Complex and the laboratory examination of over 1000 well-documented samples, we have determined the detailed modal stratigraphy for the complex which serves as the fundamental constraint in petrogenetic models. This stratigraphy is described in detail in Raedeke (1979) and McCallum *et al*. (1980). In addition, rocks of the

Stillwater Complex are generally unaltered, with original igneous textures well-preserved.

Electron microprobe analysis of ~60 Stillwater samples has resulted in an interesting discovery. The two distinct fractionation trends observed in lunar samples plotted on a Mg# vs. An diagram (Fig. 1) are closely matched by rocks from the Stillwater Complex (Fig. 2). While the similarity between Figs. 1 and

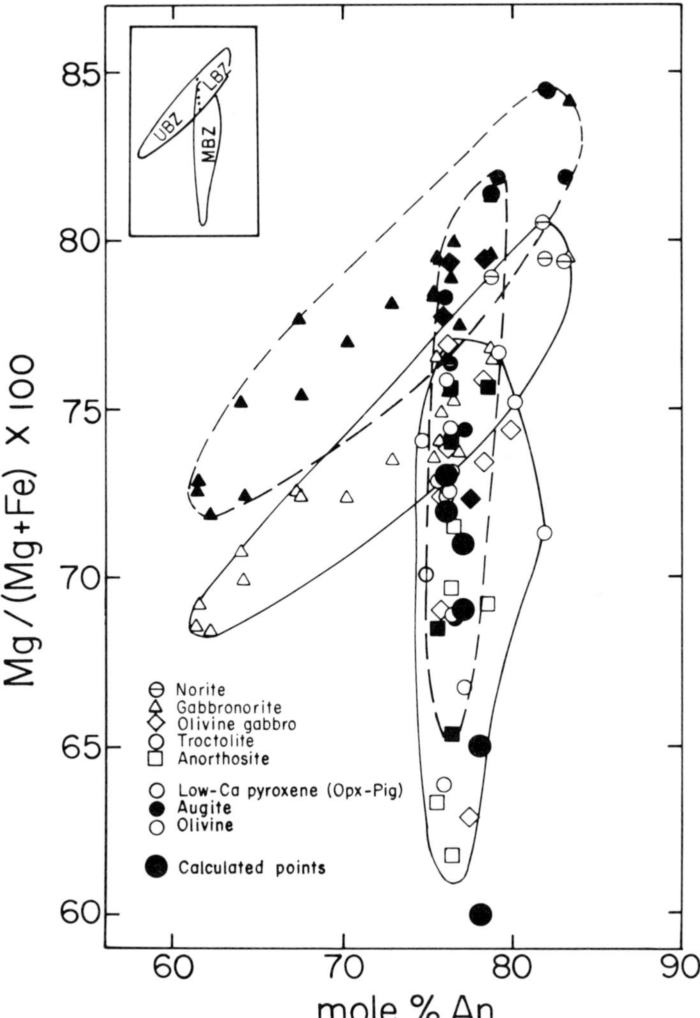

Fig. 2. Plot of mole % An in plagioclase versus mole % Mg/(Mg+Fe) in coexisting mafic minerals for samples from the Banded zone of the Stillwater Complex. Dashed line encloses augite/plag data and solid line encloses ol-opx/plag data. Calculated points are listed in Table 2 (Data set 3, 20% intercumulus liquid). The most iron-rich calculated point is not plotted.

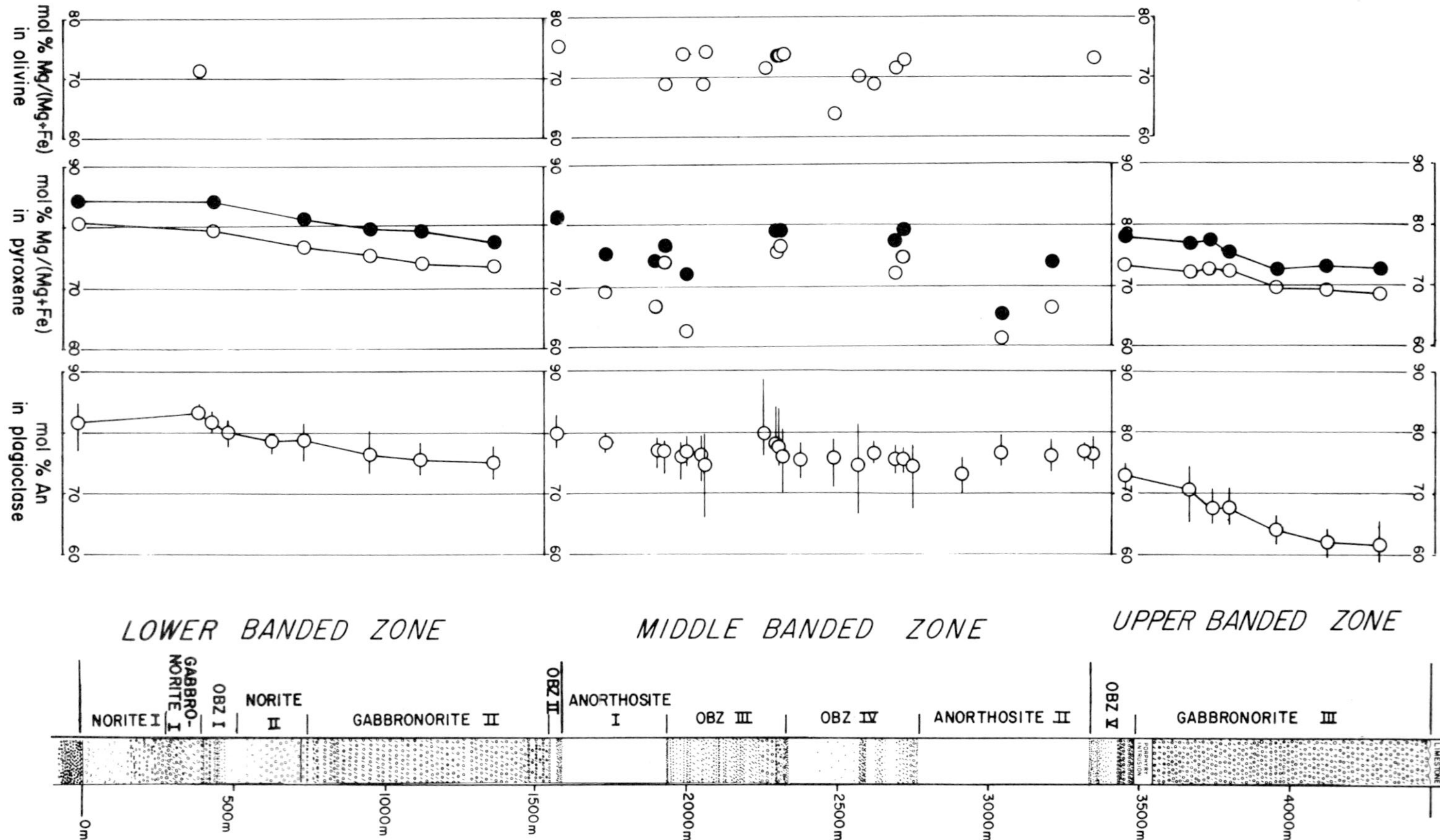

mol % Mg/(Mg+Fe) in olivine
mol % Mg/(Mg+Fe) in pyroxene
mol % An in plagioclase
LOWER BANDED ZONE
MIDDLE BANDED ZONE
UPPER BANDED ZONE
NORITE I
GABBRO-NORITE I
OBZ I
NORITE II
GABBRONORITE II
OBZ II
ANORTHOSITE I
OBZ III
OBZ IV
ANORTHOSITE II
OBZ V
GABBRONORITE III
PORPHYRY INTRUSION
LIMESTONE
0m
500m
1000m
1500m
2000m
2500m
3000m
3500m
4000m

2 is obvious, it is important to point out the following differences: (1) The Stillwater plagioclases are more albitic reflecting the higher alkali content of the Stillwater magma. (2) Augite and low-Ca pyroxene (orthopyroxene and inverted pigeonite) are present in approximately equal amounts in the Stillwater rocks in contrast to the low abundance of augite in the lunar rocks. In Fig. 2, augite defines a second set of trends parallel to those defined by olivine and orthopyroxene but plotting at higher Mg values reflecting the fact that $(Fe/Mg)^{opx} \sim 1.3$ $(Fe/Mg)^{cpx}$ at magmatic temperatures (McCallum, 1968). (3) The Stillwater data do not show a gap between the two trends. (4) When plotted on the same diagram, the Stillwater data define a shallower slope for the normal fractionation trend than do the lunar data. We believe that the Stillwater trend is modified by the process of mixing fractionated magma with fresh magma batches. (5) Data from ultramafic cumulates from the Stillwater (containing *postcumulus* plagioclase) plot on a *near horizontal* trend on the An versus Mg# diagram. Comparable data are, of course, not available for lunar rocks. (6) In the case of the Stillwater Complex, field observations clearly indicate that a single magma, or a set of closely related magmas from a homogeneous source, produced the entire observed rock suite. This conclusion is reinforced by isotopic data. DePaolo and Wasserburg (1979) measured Sm and Nd isotopic abundances on a suite of samples representing all lithologies and from a wide range of stratigraphic and geographic locations. All samples plot on a single Sm-Nd isochron (2.7 AE) indicating a uniform initial $^{143}Nd/^{144}Nd$ value.

Figure 3 shows a greatly simplified representation of the stratigraphic section of the Banded zone of the complex. The rocks of the Banded zone, all of which contain cumulus plagioclase, have been broadly subdivided into the Lower, Middle and Upper Banded zones according to lithology (Fig. 3). The rocks of the

---

Fig. 3. Simplified stratigraphic section of the Banded zone of the Stillwater Complex. Pyroxene compositional data include both low-Ca pyroxene data (open circles) and augite data (filled circles). The plagioclase compositional data show the mean as an open circle and the range of compositions within a single probe mount as a horizontal line. The range in olivine and pyroxene compositions in any one sample is smaller than the size of the symbol. Detailed descriptions of the subzones are as follows:

Gabbronorite III: Throughout this uniform subzone, plagioclase, augite and low-Ca pyroxene occur in approximately cotectic proportions. While planar lamination is almost always present, there is no preferred mineral orientation within the plane of layering. Below 3765 m orthopyroxene is clearly a cumulus mineral. Between 3765 and 3975 m, orthopyroxene occurs as poikilitic crystals containing abundant "inclusions" of small rounded augites. Above 3975 m, orthopyroxene occurs as poikilitic crystals containing numerous sets of oriented "001" augite exsolution lamellae. These sets of lamellae outline domains corresponding to original cumulus pigeonite crystals. Insofar as there is no significant change in mineral proportions associated with these textural changes, we conclude that low-Ca pyroxene formed as a cumulus mineral throughout this subzone, the poikilitic texture being the result of postcumulus recrystallization accompanying the inversion reaction.

Olivine-bearing subzone V (OBZ V): The basal member of this unit is a well-banded troctolite containing varied amounts of plagioclase, normally in excess of cotectic pro-

portions. Modally graded layering, cross bedding and cut-and-fill structures are locally present indicating strong current action during the deposition of this troctolite.
Anorthosite II: This uniform unit is the thickest anorthosite in the Banded zone. Post-cumulus augite and inverted pigeonite make up 10–12% of the rocks and disseminated sulfides are concentrated in two narrow layers near the base and top of this unit. The average grain size of plagioclase is approximately twice that of plagioclase in two- and three-phase cumulates.
Olivine-bearing subzone IV (OBZ IV): The lower boundary of this subzone is placed at the base of a well banded troctolite and the upper boundary at the base of the second *thick* anorthosite. The sequence troctolite-anorthosite-anorthositic troctolite-olivine gabbro (± gabbronorite) is repeated three times. The contact between, troctolite and underlying olivine gabbro ranges from gradational to sharp, sinuous and discordant, whereas the upper contact between troctolites and overlying anorthosites is invariably sharp, planar and concordant. The uppermost troctolite is structureless, highly discordant, contains ameboidal olivine aggregates, and is virtually identical to the troctolite at the top of OBZ II.
Olivine-bearing subzone III (OBZ III): The predominant rock types in this 400 m of section are troctolite, anorthositic troctolite, anorthositic gabbro, olivine gabbro, and olivine gabbronorite. The various members are complexly interlayered, often on a centimeter scale. Olivine is present as a cumulus mineral through 80% of the subzone, but, except for eight narrow troctolite layers in which it comprises 40 modal percent, olivine abundances are low, making up between 1 and 10% of the cumulus mineral assemblage. The major units can be traced laterally but there are substantial lateral variations in the thickness, modes and textures of the thinner members. Two distinct cyclic sequences are repeated several times in the lower 275 m of this subzone: troctolite-anorthosite-anorthositic gabbro-olivine gabbro, and troctolite-anorthositic troctolite-olivine gabbro. Olivine gabbros and gabbronorites are usually isomodal with a well defined planar lamination. A distinctive feature of these rocks is the occurrence of polycrystalline aggregates of plagioclase (~2X2X1 cm), the long axes of which are aligned parallel to the lamination. By contrast, the anorthositic gabbros commonly show an erratic layering defined by alternating mafic and felsic layers. The upper 125 m of this subzone are characterized by a four-phase cumulate. In some samples, small embayed olivines are present in the cores of orthopyroxenes, clearly indicating a reaction relationship. In other samples, however, coexisting orthopyroxene and olivine do not show the reaction relationship and both appear to be cumulus minerals.
Anorthosite I: This subzone is composed entirely of a uniform plagioclase cumulate with postcumulus augite and inverted pigeonite. The average grain size of the plagioclase is coarser and more uniform than in the two- and three-phase cumulates above and below. Disseminated sulfides occur in the upper 8 m of this unit and reach a maximum concentration in the upper 2 meters.
Olivine-bearing subzone II (OBZ II): The upper 8 m of this subzone is a remarkable association of gabbro, troctolite, and gabbroic pegmatite. The contact between gabbro and overlying troctolite is sinuous and discordant, in contrast to the planar and conformable contact between the troctolite and overlying anorthosite. Irregular patches of gabbro are enclosed within troctolite and vice versa. Olivine in the troctolite occurs as large (>10 mm) "ameboidal" grains with inclusions of plagioclase set in a matrix of relatively small (1-2 mm) plagioclase crystals.
Gabbronorite II: Variable proportions of cumulus augite, orthopyroxene and plagioclase are seen in the lower section with four interlayered anorthosite members, two of which contain sulfides. In the central part of this subzone, mineral proportions are near cotectic and planar lamination is well developed. The upper 50 m are composed of five well developed but laterally discontinuous cyclic units showing modally graded layering defined by an upward increase in plagioclase/pyroxene ratio.

Norite II: Rhythmic inch-scale layering is well developed in the basal anorthositic norites. Upward in the section, modal proportions become more pyroxene-rich, approaching cotectic values, and layering becomes correspondingly more uniform (isomodal). Toward the upper part of the norite the habit of the cumulus orthopyroxene changes from subrounded to highly elongate. A distinctive, laterally extensive, mafic layer (2 m thick) and a complementary anorthosite layer occur ~20 m below the top of the subzone.
Olivine-bearing subzone I (OBZ I): The basal contact of this complex subzone is marked by the reappearance of cumulus olivine. The sequence troctolite-anorthosite-norite-gabbronorite is repeated five times with minor variations. The fourth troctolite contains abundant sulfides. The noritic members in the lower part of the subzone are characterized by a distinctive wispy banding; higher in the subzone, layering in the norites becomes progressively more regular. This zone can be traced along strike for at least 20 km, but individual members are laterally variable, both in thickness and mode. Modal proportions are highly variable and generally non-cotectic.
Gabbronorite I: The lower contact of this unit is placed at the first appearance of cumulus augite. Proportions of augite, low-Ca pyroxene and plagioclase are largely non-cotectic with plagioclase increasing up section. Layering is poorly defined or isomodal. Pyroxenite inclusions are common in the upper 30 meters.
Norite I: Orthopyroxene and plagioclase are in approximate cotectic proportions in the lower uniform part of the unit, but plagioclase increases in abundance in the upper 100 m and layering becomes more pronounced. Layering is defined by alternating pyroxene-rich and plagioclase-rich layers with modally graded layering, scour-and-fill and slumping structures common near the top of the unit.

Lower Banded zone (LBZ) are predominantly norites and gabbronorites with minor anorthosite and troctolite members. Throughout most of the LBZ, cumulus minerals of the norites and gabbronorites occur in the cotectic proportions predicted by phase equilibria in the dry basaltic system at low pressures. Mole percent Mg/(Mg+Fe) in mafic minerals and mole percent An in plagioclase decrease concurrently with increasing stratigraphic height in the LBZ, i.e., the "normal" fractionation trend (Fig. 3).

Anorthosites and anorthositic rocks are dominant in the 1750 meter thick Middle Banded zone (MBZ). Plagioclase comprises 82% (by volume) of this zone. Average grain size of plagioclase in the MBZ is about twice that found in the LBZ and UBZ. The MBZ is composed of two thick anorthosites, one at the base and one at the top of the zone, with two complex olivine-bearing subzones sandwiched between. The latter are composed of troctolites, olivine-bearing anorthosites, gabbros and olivine-gabbros. Modes of cumulus minerals in the olivine-bearing subzones are varied and nearly always non-cotectic. Mineral compositions of the MBZ define the "vertical" fractionation trend of Fig. 2.

The Upper Banded zone (UBZ) is dominantly a uniform cotectic gabbronorite. As in the LBZ, Mg/(Mg+Fe) in mafic minerals and mole % An in plagioclase decrease with increasing stratigraphic height, i.e., the "normal" fractionation trend of Fig. 2. The UBZ is unconformably overlain by Paleozoic sediments that obscure an unknown thickness at the top of the intrusion.

There is a simple correlation between the stratigraphy of the Banded zone and the Mg# vs. An plot as seen in the inset of Fig. 2 (cf. Fig. 3). The envelope of the oblique trend is defined by samples of the LBZ and UBZ exclusively, while the vertical trend is defined by rocks of the MBZ. The continuous Na-enrichment in the plagioclase and Fe-enrichment in the mafic minerals of the LBZ (Fig. 3) is interrupted by the Middle Banded zone in which the plagioclase composition remains virtually constant, while the mafic minerals show a relatively wide range of Mg#. The normal fractionation trend resumes in the UBZ.

In the LBZ and UBZ the crystallization sequence and cryptic variation of mineral compositions are consistent with those predicted by phase equilibria where all crystallizing minerals accumulated on the *floor* of the magma chamber (see McCallum *et al.*, 1980). Transportation by magmatic density currents was probably the dominant mechanism and not simple crystal settling. It is noteworthy that the "mafic cumulate trend" defined by the norites and gabbronorites of the Lower and Upper Banded zones forms a smoothly varying continuous sequence even though the 1750 meter thick MBZ is sandwiched between the LBZ and UBZ.

DISCUSSION

The vertical trend

The Stillwater Complex data

The vertical trend in Fig. 2 is defined largely by data from anorthosites but also includes points from troctolites and olivine gabbros of the MBZ. The latter lithologies contain cumulus augite and/or olivine in addition to cumulus plagioclase. There is no systematic correlation between mafic mineral compositions and stratigraphic position in the MBZ. Mass balance calculations indicate that the "excess" plagioclase was that which failed to accumulate on the floor of the magma chamber during formation of the Ultramafic zone and the LBZ. Trace elements provide indirect support for this conclusion. Nearly monomineralic pyroxenites (>95% orthopyroxene) from the upper part of the Ultramafic zone show small negative Eu anomalies indicating crystallization and removal of plagioclase. The coarse grain size and uniform composition of plagioclase in the MBZ indicates an extended period of growth and continuous equilibration in a relatively homogeneous, well mixed liquid.

The very complex processes of crystallization and accumulation in a convecting magma of the dimensions of the Stillwater Complex will not be discussed in this paper (see McCallum *et al.*, 1980). Suffice it to say that during the formation of the Banded zone, plagioclase had a density approximately equal to or slightly less than that of the melt. Such plagioclase would not have accumulated on the floor of the magma chamber by a simple settling mechanism. Since plagioclase did accumulate along with mafic minerals during the formation of the LBZ and UBZ it is probable that it was transported and deposited via a density current mechanism. Irvine (1980) has shown that such a mechanism is capable of depositing plagioclase on the floor of a magma chamber even when it is less dense than the melt. However, a significant fraction of the plagioclase will be carried as a dilute suspension in the general convective circulation. The slight flotation tendency of plagioclase will be enhanced in regions of low fluid velocity near the upper parts of the chamber.

For ultimate segregation and accumulation of the plagioclase we envision a situation somewhat analogous to the "rockberg" model proposed by Herbert *et al*. (1977, 1978) for the lunar case. In this model, the suspended plagioclase would begin to accumulate in laterally accreting "rockbergs" while gabbronorite of the LBZ accumulated below. The rockbergs would remain suspended either at the top of or at some intermediate level in the magma chamber so long as their density was approximately that of the magma. The initiation of crystallization of intercumulus mafic minerals would increase the bulk density of the rockbergs, and at the point where $\rho_{rockberg} > \rho_{magma}$, they would sink. The rockbergs were very likely large, conceivably with maximum dimensions of meters or even hundreds of meters (the thickness of the uppermost anorthosite alone is 600 meters). A consideration of simple Stokes Law settling where the settling rate is directly proportional to the radius squared, shows that when the density contrast was greater than zero, the rockbergs would sink catastrophically. There is indeed field evidence that huge blocks of anorthosite did sink into the top of the LBZ, trapping liquid and causing reaction with the newly accumulated gabbronorites.

The segregation of the plagioclase into anorthosites and anorthositic rocks of the MBZ would not immediately trap intercumulus liquid. Rather, much liquid would escape into the main magma body by processes of compaction, diffusion and metasomatic infiltration (Irvine, 1979). Throughout this compaction process, postcumulus crystallization would continue with plagioclase plus one or more mafic phases crystallizing in predictable cotectic proportions. At some point, however, the crystal/liquid ratio would increase to the point where the liquid was effectively "trapped." At that point crystallization would continue by an equilibrium process.

From the plagioclase composition data (Figs. 2 and 3), it appears that the suspended plagioclase which later formed the Middle Banded zone was in equilibrium with the liquid crystallizing to form the cumulates at the top of the LBZ. Assuming cotectic equilibrium crystallization of the intercumulus liquid and using $K_d^{pl/L}$ (Drake, 1976) and $K_d^{opx/L}$ and $K_d^{cpx/L}$ (Nielsen and Drake, 1979) we can calculate backwards to the crystal and liquid compositions in mutual equilibrium at the time of accumulation of the uppermost gabbronorite of the LBZ. Note that the ultimate mineral compositions plotted in Fig. 2 are actually a product of 1) crystallization of cumulus minerals in equilibrium with the main magma body, and 2) postcumulus crystallization of the intercumulus liquid. In order to proceed with this calculation then, we require some assumption of the ratio of crystals to intercumulus liquid at the time when the postcumulus liquid is effectively "trapped" and equilibrium crystallization begins.

We know from the observation that total modes of the thick anorthosites of the Stillwater Complex are on the average 90% cumulus plagioclase, 5% postcumulus pigeonite and 5% postcumulus augite. If postcumulus crystallization forms minerals in the cotectic proportions of 60% plagioclase, 20% pigeonite, 20% augite as discussed by McCallum *et al.* (1980), then the final mode requires crystallization of 25% intercumulus liquid in the presence of 75% crystalline plagioclase (i.e., total plagioclase (90) = cumulus plagioclase (75) + postcumulus plagioclase (.6 × 25), with similar expressions for pyroxenes). A reasonable value for the proportion of intercumulus liquid in the crystal/liquid mush is therefore ~25%. This value will be higher or lower depending on the effectiveness of processes such as compaction, deformation and metasomatic infiltration occurring in the crystal mush. A range of values from 15% to 30% intercumulus liquid is therefore considered.

The results of the calculations outlined above are listed in Table 1. Using appropriate values of K_d at 1150°C, compositions of crystals (A) and liquid (B) at mutual equilibrium were calculated. The four data sets represent varied proportions of intercumulus liquid in the original cumulate gabbronorite. The calculated assemblages will crystallize by an equilibrium process to yield a gabbronorite of the composition observed at the top of the LBZ, i.e., An 76, $En^{opx}73$, $En^{cpx}77$.

As an example, consider the case of data set 3 (Table 1). Liquid B would be saturated with plagioclase (An 84) + orthopyroxene ($En^{opx}79$) + clinopyroxene ($En^{cpx}82$). Liquid B is the composition of the main magma body at the time of crystallization of the upper LBZ (assuming 20% intercumulus liquid in the gabbronorite), as well as the initial composition of the "trapped" intercumulus liquid.

Table 1. Compositions of crystals (A) and intercumulus liquid (B) in equilibrium which will ultimately yield a rock of the composition of the uppermost gabbronorite of the LBZ (An 76, $En^{opx}73$, $En^{cpx}77$). Four sets of values are tabulated corresponding to various proportions of intercumulus liquid in the gabbronorite. K_d values at 1150°C were used in the calculations. See text for discussion.

Data set	% intercumulus liquid		A (Crystal)	B (Liquid)
1	30	pl	An 88	An 49
		opx	En 82	En 56
		cpx	En 85	En 56
2	25	pl	An 86	An 46
		opx	En 80	En 55
		cpx	En 84	En 55
3	20	pl	An 84	An 42
		opx	En 79	En 53
		cpx	En 82	En 53
4	15	pl	An 83	An 39
		opx	En 78	En 51
		cpx	En 81	En 51

It is expressed in Table 1 in terms of anorthite component (An 42) and enstatite component (En 53) calculated using the appropriate equations in Drake (1976) and Nielson and Drake (1979). If this assemblage of 80% crystals (A), and 20% liquid (B) were allowed to crystallize by an equilibrium process, it would yield a gabbronorite of the required composition (An 76, $En^{opx}73$, $En^{cpx}77$).

If we now consider a pure plagioclase mush such as that presumed to have been present immediately after settling of the "rockbergs", then plagioclase alone will comprise 100% of the crystal matrix. Again, the proportion of intercumulus liquid in the crystal/liquid mush may range from 15% to 30%. Let us assume, for illustration, that the intercumulus liquid associated with this pure plagioclase segregation comprises 25% of the system with the plagioclase comprising the remaining 75%. Since no cumulus pyroxene crystals were initially present, equilibrium cotectic crystallization of the 25% intercumulus liquid of the composition of data set 3 will produce orthopyroxene and clinopyroxene at mutual equilibrium as determined by $K_d^{opx/cpx}$. The resulting pyroxene compositions are $En^{opx}50$ and $En^{cpx}55$ (Table 2). By contrast, the proportion of plagioclase formed by cumulus processes of fractional crystallization (An 84) to plagioclase formed by equilibrium crystallization of the intercumulus liquid (An 42) is 75:15. Therefore, the ultimate plagioclase composition resulting from these two processes is An 77 (Table 2). If a small amount of cumulus pyroxene were present at the time of segregation and "trapping" of the liquid, the resulting pyroxene compositions would be more magnesian but the plagioclase composition would be only slightly changed (Table 2). It is obvious that the relative abundances of cumulus crystals is important in controlling ultimate compositions and we use the term "buffering capacity of cumulates" to describe this process. Irvine (1974) was first to point out the possible dependence of Mg/Fe in mafic minerals on cumulus crystal/intercumulus liquid ratio in the mush (see Irvine, 1974, Fig. 36). Figure 4 presents a graphical illustration of the process as described in this paper (see figure caption for details).

It is apparent from inspection of Table 2 that the vertical trend of Fig. 2 can be completely explained by a model of equilibrium crystallization of ~20% trapped liquid, using data set 3 (i.e., 20% intercumulus liquid of Table 1), by simply varying the proportions of cumulus plagioclase to mafics. These values are plotted in Fig. 2. Plagioclase must, however, comprise ≥70% of the rock, because, as is evident from Table 2, when cumulus pyroxene exceeds 30%, the points fall in the oblique trend of Fig. 2 rather than on the vertical trend. Therefore, the following conditions are required by this model: 1) high modal percent plagioclase, 2) increasing modal percent plagioclase with increasing Fe/Mg in the vertical trend, and 3) values of ~20% intercumulus liquid. These conditions are consistent with observations since, 1) plagioclase comprises ~82% of the entire MBZ and ~73% of the olivine-bearing subzones of the MBZ; 2) as a general rule, points with highest modal % plagioclase plot lowest on the vertical trend of Fig. 2, and those with less plagioclase plot higher, and 3) values of 20–25% for the intercumulus liquid are within the range of estimation, based on field observations of actual modes. It is of interest to note that within each vertical data set

Table 2. Calculated ultimate plagioclase, orthopyroxene and clinopyroxene for Stillwater data in Table 1. Percentage cumulus pyroxene is varied from 0 to 30 and percentage intercumulus liquid is varied from 15 to 30. Cumulus pyroxene mode is assumed to be 50% orthopyroxene and 50% clinopyroxene (cotectic proportions). Cumulus plagioclase mode is the complement to the sum of the total pyroxene and intercumulus liquid modes. Data sets 1–4 refer to those of Table 1. Percentage intercumulus liquid refers to that affecting the rock in question, i.e., independent of the values listed in Table 1 which refer uniquely to the gabbronorite at the top of the LBZ. See text for further discussion.

Modal % cumulus px		DATA SET 1 % Intercumulus liquid		DATA SET 2 % Intercumulus liquid			DATA SET 3 % Intercumulus liquid				DATA SET 4 % Intercumulus liquid			
		30	25	30	25	20	30	25	20	15	30	25	20	15
0	pl	An 80	An 81	An 78	An 79	An 81	An 75	An 77	An 79	An 80	An 74	An 76	An 77	An 79
	opx	En 53	En 53	En 52	En 52	En 52	En 50	En 50	En 50	En 50	En 48	En 48	En 48	En 48
	cpx	En 58	En 58	En 57	En 57	En 57	En 55	En 55	En 55	En 55	En 53	En 53	En 53	En 53
5	pl	An 80	An 81	An 77	An 79	An 80	An 75	An 76	An 78	An 80	An 73	An 75	An 77	An 79
	opx	En 61	En 62	En 60	En 61	En 62	En 58	En 59	En 60	En 63	En 56	En 57	En 59	En 61
	cpx	En 66	En 67	En 65	En 66	En 67	En 63	En 64	En 65	En 68	En 61	En 62	En 64	En 66
10	pl	An 79	An 81	An 77	An 78	An 80	An 74	An 76	An 78	An 79	An 73	An 75	An 76	An 78
	opx	En 66	En 67	En 64	En 66	En 67	En 63	En 64	En 65	En 67	En 61	En 62	En 64	En 66
	cpx	En 71	En 72	En 69	En 71	En 72	En 68	En 69	En 70	En 72	En 66	En 67	En 69	En 71
15	pl	An 78	An 80	An 76	An 78	An 80	An 74	An 76	An 77	An 79	An 72	An 74	An 76	An 78
	opx	En 69	En 71	En 67	En 69	En 70	En 65	En 67	En 69	En 71	En 64	En 65	En 66	En 70
	cpx	En 74	En 75	En 72	En 74	En 74	En 70	En 72	En 74	En 75	En 69	En 70	En 71	En 74
20	pl	An 78	An 80	An 75	An 77	An 79	An 73	An 75	An 77	An 79	An 71	An 74	An 76	An 78
	opx	En 71	En 72	En 70	En 70	En 72	En 67	En 69	En 71	En 72	En 66	En 66	En 69	En 71
	cpx	En 75	En 76	En 74	En 74	En 76	En 72	En 74	En 75	En 76	En 71	En 71	En 74	En 75
25	pl	An 77	An 79	An 75	An 77	An 79	An 72	An 74	An 76	An 79	An 70	An 73	An 75	An 77
	opx	En 72	En 74	En 70	En 72	En 73	En 69	En 71	En 72	En 73	En 66	En 69	En 71	En 72
	cpx	En 76	En 78	En 74	En 76	En 77	En 74	En 75	En 76	En 77	En 71	En 74	En 74	En 76
30	pl	An 76	An 78	An 74	An 76	An 78	An 71	An 74	An 76	An 78	An 70	An 72	An 75	An 77
	opx	En 74	En 75	En 72	En 73	En 74	En 71	En 72	En 73	En 74	En 69	En 70	En 72	En 73
	cpx	En 78	En 79	En 76	En 77	En 78	En 75	En 76	En 77	En 78	En 74	En 74	En 76	En 77

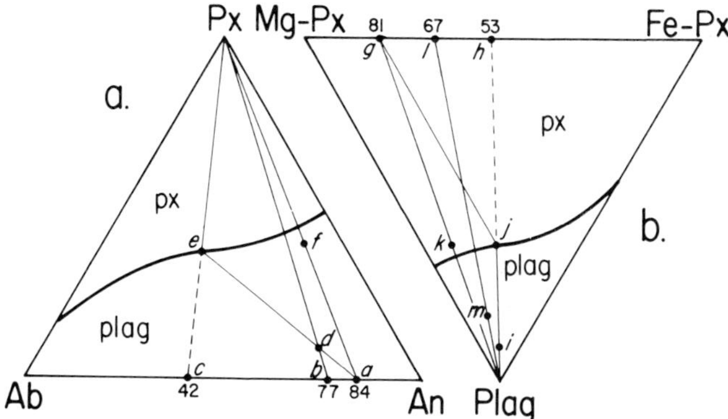

Fig. 4. Schematic sections through the pseudo-quaternary system Ab-An-MgPx-FePx (a) at $X_{Mg} \sim 0.5$ and (b) $X_{An} \sim 0.75$. The field boundaries are based on those determined by Bowen (1915) in the Di-Ab-An system. While the field boundaries are schematic, correct geometrical relationships are retained. The data plotted are from data set 3 with 25% intercumulus liquid (see Table 2).

(a) The bulk composition of the system (*d*) contains 75% cumulus plagioclase (*a*—An 84) and 25% intercumulus liquid (*e*—extrapolates to *c* An 42). The field boundary is drawn such that the tangent to the cotectic at *e* intersects the pyroxene-plagioclase tie line at *f* (60 plagioclase : 40 pyroxene). With equilibrium crystallization, plagioclase composition changes to *b* (An 77). The bulk system (*d*) eventually crystallizes to a mixture of 90 plagioclase (*b*—An 77) plus 10 pyroxene (En 53). The pyroxene will be 5% cpx ($En^{cpx}55$) and 5% opx ($En^{opx}50$), as determined by $K_d^{opx/cpx}$.

(b) Consider the simple case where only plag + opx crystallize from the intercumulus liquid in 60 plagioclase : 40 orthopyroxene cotectic proportions. The bulk composition (*i*) contains 75% cumulus plagioclase (An 84) and 25% intercumulus liquid (*j*—extrapolates to En 53). Point *g* (En 81) is the composition of orthopyroxene in equilibrium with liquid *j*. The tangent to the cotectic at *j* intersects the pyroxene-plagioclase tie line at *k* (60 plagioclase : 40 orthopyroxene). With equilibrium crystallization, plagioclase composition changes to An 77 while pyroxene of composition *h* (En 53) crystallizes.

Point *m* represents a bulk composition containing 65% cumulus plagioclase (An 84) plus 10% cumulus orthopyroxene (En 81) plus 25% intercumulus liquid (*j*). With equilibrium crystallization, plagioclase changes composition to An 76 and orthopyroxene changes composition to En 67 (*l*).

of Table 2, i.e., where all parameters are held constant except modal proportions of cumulus pyroxene/plagioclase, the trend produced will have a steep negative slope when plotted on Fig. 2. The original negative slope of Steele and Smith (1973) for lunar anorthosites, therefore, may well be real.

It must be re-emphasized that other processes were also operating, so such a simple model as that outlined above is perhaps a bit naive. For example, there is abundant field evidence for the migration of differentiated liquid from one portion of the mush to another as a result of deformation during compaction. This would leave, on the one hand, more magnesian mafic minerals where the liquid was removed, and more iron-rich mafics where the liquid was added. Indeed, by such a process one might expect iron-enrichment well beyond the predicted limits listed in Table 2. However, as a first estimate, it appears that the modal proportions of minerals actually accumulating strongly affect the ultimate mineral compositions in the rock. Also, proportion of intercumulus liquid has a measurable effect on ultimate mineral compositions as do the relative proportions of mafic minerals crystallizing. These various effects and their interdependencies will be discussed at greater length in a subsequent paper.

The lunar data

The concept of a lunar magma ocean in which plagioclase floated and was somehow segregated into anorthosites to form the lunar crust, is well-entrenched in current models of the moon (e.g., Herbert *et al.*, 1977, 1978; Longhi, 1977, 1978, 1979; Longhi and Boudreau, 1979). However, there is much controversy as to how the vertical trend of Fig. 1 defined by these same anorthosites, can be explained. Although it is imprudent to apply directly models developed for terrestrial layered intrusions to the moon, such analogs are very valuable in testing their plausibility. Considering the success of the model outlined above in explaining the vertical trend for the Stillwater data, it seems reasonable to test its applicability to the moon. In order to consider these same arguments with reference to the moon, however, we need some estimation of the magma composition with which the suspended plagioclase was in equilibrium. To do this we must again have a known rock composition from which we can calculate backwards to the crystal and liquid compositions which ultimately form the rock. By analogy with Fig. 2 we could extrapolate the vertical trend of Fig. 1 to the intersection with the oblique trend (An 96, En 80) and consider the liquid and crystal assemblage that would give that composition. This assumes that rocks defining the vertical trend were at one time in equilibrium with the same liquid which crystallized rocks of the oblique trend (columns IA and B of Table 3). Alternatively, we could

Table 3. Compositions of crystals (A) and intercumulus liquid (B) in equilibrium which will ultimately yield: (I) a rock of composition An 96, En 80, plotting at the extrapolated intersection of the vertical trend of Fig. 1 with the oblique trend, and (II) a rock of composition An 96, En 70, plotting at the top of the vertical trend of Fig. 1. Four sets of data are tabulated corresponding to various proportions of intercumulus liquid. K_d values at 1200°C were used in the calculations.

% intercumulus liquid		Data set	I A (Crystals)	I B (Liquid)	Data set	II A (Crystals)	II B (Liquid)
30	pl	5	An 98	An 91	9	An 98	An 91
	opx		En 87	En 65		En 78	En 50
25	pl	6	An 98	An 90	10	An 98	An 90
	opx		En 86	En 64		En 77	En 49
20	pl	7	An 98	An 88	11	An 98	An 88
	opx		En 85	En 62		En 76	En 47
15	pl	8	An 98	An 86	12	An 98	An 86
	opx		En 84	En 60		En 74	En 45

consider the upper extent of the vertical trend as plotted in Fig. 1 (An 96, En 70) as the ultimate composition from which we calculate a liquid composition for the rocks exhibiting this trend (columns IIA and B in Table 3). The results of calculations using both assumptions are listed in Table 3.

Inspection of Table 3 shows that the assumption that rocks defining the vertical trend were at one time in equilibrium with the liquid complementary to the rock composition An 96, En 80, at the extrapolated intersection with the oblique trend, cannot be true if our present model is reasonable. Even if one considers the extreme case where data set 8B is used as the intercumulus liquid composition (i.e., the most iron-rich), and we allow it to be trapped in a pure plagioclase matrix, the ultimate composition of the resulting intercumulus orthopyroxene will be only En 60. This is far above the actually observed value of En 44 (Fig. 1). This simple exercise confirms the likelihood that the gap between the suites in Fig. 1 is real and that the two suites of rocks indeed do represent two distinct lunar magmas.

By contrast, the data from columns IIA and B in Table 3 do yield compositions of the "ferroan anorthosite" as seen in Table 4. The entire suite can be generated simply by varying the proportions of intercumulus liquid and cumulus mafics relative to plagioclase. The best fit of the calculated values to the actual observed trend is seen where the rock has 25% intercumulus liquid of the composition of data set 11 (i.e., 20% intercumulus liquid of Table 3). These calculated values are plotted in bold face in Fig. 1.

As is apparent from these manipulations, the compositions generated by a coupling of cumulus processes of fractional crystallization with those of equilibrium processes of postcumulus crystallization, are affected principally by: 1) the composition of the initial liquid (columns IB and IIB of Table 3), 2) the proportions of cumulus mafics to plagioclase, and 3) the proportion of intercumulus liquid relative to crystals. Also recall the necessary assumptions of this model are that 1) plagioclase is the dominant cumulus phase, 2) the intercumulus liquid at some stage becomes "trapped" and crystallizes by a process of equilibrium crystallization, and 3) crystallization of the intercumulus liquid occurs in predictable cotectic proportions.

Trace element data support the conclusion that the Mg-rich plutonic suite and the ferroan anorthosites formed from different magmas. Norman and Ryder (1980) have compiled available data on all known members of both suites and show that Ti/Sm and Sc/Sm depletions are features common to all pristine norites and troctolites, whereas anorthosites have Ti/Sm near the chondritic ratio. The single dunite sample is an anomaly; it shows Ti/Sm and Sc/Sm values characteristic of the anorthosite suite, yet plots on the Mg-rich plutonic trend in the Mg# versus An diagram. It is of interest to note that calculated parent liquids for the dunite and anorthosite have remarkably similar absolute and relative REE abundances (Laul and Schmitt, 1975; and McKay *et al.*, 1979). Calculated parent liquids for troctolites and norites have distinctly higher REE abundances and show patterns similar to KREEP (Haskin *et al.*, 1974; McCallum and Mathez, 1975; Blanchard and Budahn, 1979). The KREEP-norite/troctolite connection

Table 4. Calculated ultimate plagioclase and orthopyroxene compositions for lunar highlands data in columns II of Table 3. Notations are analagous to those for Table 2. Since augite is not an abundant phase in the lunar rocks, the norite cotectic (60% plagioclase, 40% orthopyroxene) was used for modeling the crystallization of the intercumulus liquid. See text for discussion.

Modal % cumulus opx		DATA SET 9 % Intercumulus liquid		DATA SET 10 % Intercumulus liquid			DATA SET 11 % Intercumulus liquid				DATA SET 12 % Intercumulus liquid			
		30	25	30	25	20	30	25	20	15	30	25	20	15
0	pl	An 97	An 97	An 96	An 97	An 97	An 96	An 96	An 97	An 97	An 96	An 96	An 96	An 97
	opx	En 50	En 50	En 49	En 49	En 49	En 47	En 47	En 47	En 47	En 45	En 45	En 45	En 45
5	pl	An 96	An 97	An 96	An 97	An 97	An 96	An 96	An 97	An 97	An 95	An 96	An 96	An 97
	opx	En 58	En 59	En 57	En 58	En 60	En 55	En 57	En 58	En 60	En 53	En 55	En 56	En 58
10	pl	An 96	An 97	An 96	An 96	An 97	An 96	An 96	An 97	An 97	An 95	An 96	An 96	An 97
	opx	En 63	En 64	En 61	En 63	En 65	En 60	En 62	En 63	En 65	En 58	En 60	En 61	En 63
15	pl	An 96	An 97	An 96	An 96	An 97	An 96	An 96	An 96	An 97	An 95	An 96	An 96	An 97
	opx	En 66	An 67	An 65	En 66	En 67	En 63	En 64	En 66	En 68	En 61	En 62	En 64	En 66
20	pl	An 96	An 96	An 96	An 96	An 97	An 95	An 96	An 96	An 97	An 95	An 95	An 96	An 97
	opx	En 68	En 69	En 67	En 68	En 69	En 65	En 66	En 68	En 69	En 63	En 64	En 66	En 67

extends to the Ti/Sm and Sc/Sm values (Norman and Ryder, 1980). Anorthosites have initial $^{87}Sr/^{86}Sr$ ratios that are slightly more primitive than those of the Mg-rich suite which can be interpreted to mean that they are slightly older (Warren, 1979). The ferroan anorthosites (and possibly the dunite) may represent the original crust, whereas the Mg-rich suite (and possibly KREEP) may have been derived by fractional crystallization of a set of later intrusions emplaced at a relatively high level within the primitive crust. Support for this idea comes from the non-random distribution of pristine samples that led Ryder and Wood (1977) to postulate a primary zonation of the lunar crust with anorthosites common in the upper parts and the Mg-rich suite forming the deeper layers.

CONCLUSIONS

Data from the Stillwater Complex show trends analogous to those of lunar highlands samples on a plot of Mg/(Mg+Fe) in mafic minerals vs. An content in coexisting plagioclase. The oblique trend observed for the Stillwater rocks on such a plot, is defined exclusively by samples of the Lower and Upper Banded zones where the crystallization sequence and cryptic variation are consistent with those predicted by phase equilibria for the condition that all crystallizing minerals accumulate together on the floor of the magma chamber, i.e., a system approaching perfect fractional crystallization. The vertical trend of Fig. 2 is defined exclusively by plagioclase-rich rocks of the Middle Banded zone and can be explained by a process of equilibrium crystallization of "trapped" intercumulus liquid in a plagioclase-rich crystal mush. If the mafic mineral grains have cumulus cores they will tend to be more magnesian but the compositions of coexisting plagioclase will not be affected.

By applying the model derived for the anorthositic rocks of the Stillwater Complex to the lunar highlands rocks, major element variations in the "ferroan anorthosite" suite can be explained. Furthermore it can be shown that, not even under extreme circumstances, can the rocks of the "ferroan anorthosite" suite be generated from the same liquid that crystallized rocks of the Mg-rich cumulates. In other words, the two suites evidently represent two distinct lunar magmas. The anorthositic suite may represent the original lunar crust, whereas the Mg-rich suite may have been derived from later high level intrusions.

Acknowledgments—This study has been supported by NASA Grant NSG 7313. We are indebted to John Longhi for suggesting that we plot the Stillwater data on the Fo versus An diagram. This manuscript was prepared while I. S. M. was a visiting scientist at the Lunar and Planetary Institute which is operated by the Universities Space Research Association under Contract No. NSR-09-051-001 with the National Aeronautics and Space Administration. The assistance of the LPI staff is greatly appreciated. Lila Mager kindly typed the manuscript. This research was conducted as a portion of the senior author's Ph.D. thesis at the University of Washington. This paper is Lunar and Planetary Institute Contribution No. 404.

REFERENCES

Bickel C. E. and Warner J. L. (1978) Survey of lunar plutonic and granulitic lithic fragments. *Proc. Lunar Planet. Sci. Conf. 9th,* p. 629–652.

Blanchard D. P. and Budahn J. R. (1979) Remnants from the ancient lunar crust: Clasts from consortium breccia 73255. *Proc. Lunar Planet. Sci. Conf. 10th,* p. 803–816.

Bowen N. L. (1915) The crystallization of haplobasaltic, haplodioritic and related magmas. *Am. J. Sci.,* **40,** 161–185.

DePaolo D. J. and Wasserburg G. J. (1979) Sm-Nd age of the Stillwater Complex and the mantle evolution curve for neodymium. *Geochim. Cosmochim. Acta* **43,** 999–1008.

Drake M. J. (1975) Lunar anorthosite paradox: An alternative explanation. *Proc. Lunar Sci. Conf. 6th,* p. 293–299.

Drake M. J. (1976) Plagioclase-melt equilibria. *Geochim. Cosmochim. Acta* **40,** 457–465.

Haskin L. A., Shih C.-Y., Bansal B. M., Rhodes J. M., Wiesmann H., and Nyquist L. E. (1974) Chemical evidence for the origin of 76535 as a cumulate. *Proc. Lunar Sci. Conf. 5th,* p. 1213–1225.

Herbert F., Drake M. J., and Sonett C. P. (1978) Geophysical and geochemical evolution of the lunar magma ocean. *Proc. Lunar Planet. Sci. Conf. 9th,* p. 249–262.

Herbert F., Drake M. J., Sonett C. P., and Wiskerchen M. J. (1977) Some constraints on the thermal history of the lunar magma ocean. *Proc. Lunar Sci. Conf. 8th,* p. 563–582.

Irvine T. N. (1974) Petrology of the Duke Island Ultramafic Complex, southeastern Alaska. *Mem., Geol. Soc. Amer. 138,* 240 pp.

Irvine T. N. (1979) Infiltration metasomatism, adcumulus growth, and secondary differentiation in the Muskox intrusion. *Geophys. Lab. Yearb.* **78,** 743–751.

Irvine T. N. (1980) Magmatic density currents and cumulus processes. *Amer. J. Sci., Jackson vol.* In press.

Laul J. C. and Schmitt R. A. (1975) Dunite 72417: A chemical study and interpretation. *Proc. Lunar Sci. Conf. 6th,* p. 1231–1254.

Longhi J. (1977) Magma oceanography 2: Chemical evolution and crustal formation. *Proc. Lunar Sci. Conf. 8th,* p. 601–621.

Longhi J. (1978) Pyroxene stability and the composition of the lunar magma ocean. *Proc. Lunar Planet. Sci. Conf. 9th,* p. 285–306.

Longhi J. (1979) Magma oceanography revisited: Formation of the primitive lunar crustal rocks (abstract). In *Papers presented to the Conference on Lunar Highlands Crust* p. 105–107. Lunar and Planetary Institute, Houston.

Longhi J. and Boudreau A. E. (1979) Complex igneous processes and the formation of the primitive lunar crustal rocks. *Proc. Lunar Planet. Sci. Conf. 10th,* p. 2085–2106.

McCallum I. S. (1968) Equilibrium relationships among the coexisting minerals in the Stillwater Complex, Montana. *Ph.D. thesis, Univ. Chicago.* 175 pp.

McCallum I. S. and Mathez E. A. (1975) Petrology of noritic cumulates and a partial melting model for the genesis of Fra Mauro basalts. *Proc. Lunar Sci. Conf. 6th,* p. 395–414.

McCallum I. S., Raedeke L. D., and Mathez E. A. (1980) Investigations in the Stillwater complex: Part I, stratigraphy and structure of the Banded zone. *Amer. J. Sci.,* Jackson vol. In press.

McKay G. A., Wiesmann H., and Bansal B. M. (1979) The KREEP-magma ocean connection (abstract). *In Lunar and Planetary Science X,* p. 804–806. Lunar and Planetary Institute, Houston.

Nielson R. L. and Drake M. J. (1979) Pyroxene-melt equilibria. *Geochim. Cosmochim. Acta* **43,** 1259–1272.

Norman M. D. and Ryder G. (1980) Geochemical evidence for the role of ilmenite and clinopyroxene in the early lunar differentiation (abstract). In *Lunar and Planetary Science XI.* p. 821–823. Lunar and Planetary Institute, Houston.

Raedeke L. D. (1979) Stratigraphy and petrology of the Stillwater Complex, Montana. *M.S. thesis, Univ. Washington.* 89 pp.

Roedder E. and Weiblen P. W. (1974) Petrology of clasts in lunar breccia 67915. *Proc. Lunar Sci. Conf. 5th,* p. 303–318.

Ryder G. and Wood J. A. (1977) Serenitatis and Imbrium impact melts: Implications for large scale layering in the lunar crust. *Proc. Lunar Sci. Conf. 8th,* p. 655–668.

Steele I. M. and Smith J. V. (1973) Mineralogy and petrology of some Apollo 16 rocks and fines: General petrologic model of the Moon. *Proc. Lunar Sci. Conf. 4th,* p. 519–536.

Warner J. L., Simonds C. H., and Phinney W. C. (1976) Genetic distinction between anorthosites and Mg-rich plutonic rocks: New data from 76255 (abstract). In *Lunar Science VII,* p. 915–917. The Lunar Science Institute, Houston.

Warren P. H. (1979) Certain pristine nonmare rocks formed as cumulates from the magma ocean (but many others did not) (abstract). In *Papers presented to the Conference on Lunar Highlands Crust* p. 192–194. Lunar and Planetary Institute, Houston.

Warren P. H. and Wasson J. T. (1977) Pristine nonmare rocks and the nature of the lunar crust. *Proc. Lunar Sci. Conf. 8th,* p. 2215–2235.

Warren P. H. and Wasson J. T. (1978) Compositional-petrographic investigation of pristine nonmare rocks. *Proc. Lunar Planet. Sci. Conf. 9th,* p. 185–217.

Warren P. H. and Wasson J. T. (1979a) The compositional-petrographic search for pristine nonmare rocks: Third foray. *Proc. Lunar Planet. Sci. Conf. 10th,* p. 583–610.

Warren P. H. and Wasson J. T. (1979b) Effects of pressure on the crystallization of a "chondritic" magma ocean and implications for the bulk composition of the Moon. *Proc. Lunar Planet. Sci. Conf. 10th,* p. 2051–2084.

Wood J. A., Dickey J. S., Marvin U. B., and Powell B. N. (1970) Lunar anorthosites and a geophysical model of the Moon. *Proc. Apollo 11 Lunar Sci. Conf.,* p. 965–988.

Papike, J.J. and Merrill, R.B., eds.
Proc. Conf. Lunar Highlands Crust (1980), p. 155-171
Printed in the United States of America

Dropping stones in magma oceans: Effects of early lunar cratering

William K. Hartmann

Planetary Science Institute, Tucson, Arizona 85719

Abstract—A new methodology is used to calculate the accumulation rate of megaregolith materials for two models of early lunar cratering, both with and without episodes of late cataclysmic cratering. Results show that the pulverization of early rock layers was an important process competing with the formation of a coherent rock lithosphere at the surface of the hypothetical lunar magma ocean. If a magma ocean existed, then its initial cooling was marked by a period of pre-lithospheric chaos in which impacts punched through the initially thin rocky skin, mixing rock fragments with splashed magma. Furthermore, the results show that intense brecciation and pulverization of rock materials must have occurred to a depth of at least tens of kilometers in the first few hundred years of lunar history regardless of whether a "terminal lunar cataclysm" occurred around 4.0 G.y. ago. The predicted pattern of brecciation and the ages of surviving rock fragments is similar to that actually observed among lunar samples. More reliable dating of basin-forming events and models of rock exhumation and survival are needed in order to understand better the relation between the early intense bombardment of the moon and the samples collected on the moon today.

INTRODUCTION

Recent literature has dealt with thermal, physical and chemical models of lunar crustal evolution (e.g., Herbert *et al.*, 1978; Solomon and Longhi, 1977). These clarify evidence that the lunar lithosphere (outermost coherent rock layer) formed from mineral cumulates solidified in a molten layer, or "lunar magma ocean." Models of crustal evolution have attempted to deduce the lithosphere's chemistry and its increasing thickness as a function of time. For the sake of simplicity, these numerical models consider only endogenic processes, so that the lunar lithosphere has been pictured as accumulating solely through magma ocean cooling, with various assumptions about convection, "rockberg" formation, efficiency of heavy mineral settling, etc.

Earlier petrologic studies introducing the magma ocean concept did discuss more complexities: rapid convection, sinking of dense cumulates, and pressure effects in the lower ocean (Wood, 1975; Walker *et al.*, 1975). Wood's 1975 study, "Lunar petrogenesis in a well-stirred magma ocean" emphasized that stirring could explain a number of chemical trends in lunar rocks, but attributed this stirring to convection, without emphasizing that cratering had a strong stirring effect.

The purpose of this paper is to emphasize that an exogenic process—intense cratering by meteoritic planetesimals—was also an important environmental constraint on the nature of the rocks in the early lunar lithosphere. The problem can be viewed as a competition between rock-forming processes that tend to create coherent igneous layers, and rock-destroying processes that tend to create megaregolith by pulverizing rock. Which process had the upper hand at any given time in lunar history had important consequences for the chemical and physical nature of the lithospheric rock, and controlled the type of primeval rock samples available on the surface of the moon today. Intense impact stirring may have produced many of the effects considered by Wood (1975).

MEGAREGOLITH PRODUCTION MODEL

A key to analyzing the competition between the rock-forming and rock-destroying processes is a model of the effect of the intense early cratering. Suppose a semi-infinite coherent igneous rock layer was created at the time $t = 0$. The question being asked, then, is how fast this was "chewed through" by cratering of a specified intensity. More specifically, for a certain crater production rate, what is the increase in depth of megaregolith as a function of time? Like models of the primeval lunar lithosphere and hypothetical magma ocean, models of megaregolith are in only early stages of development. Housen *et al.* (1979) produced an analytical model of regolith evolution that is perhaps the most sophisticated to date. However, this model assumes a crater production function with a steep slope (size index $b > 2$ in the cumulative size distribution law $N = kD^{-b}$, where N = cumulative number of craters larger than diameter D), appropriate only to the sub-kilometer lunar craters which are widely believed to be secondary impact features. While that model is applicable to the lunar mare regoliths and hypothetical asteroid regoliths, it is inappropriate to the development of deeper megaregolith in the lunar highland, because the multi-kilometer craters involved in megaregoliths have a different size distribution. Strong points of the Housen model are that it analytically keeps track of ejecta accumulated in some regions as craters are formed in other regions, and allows for ejecta blown off of smaller planetary bodies. Nonetheless, it would become considerably more complex analytically if it attempted to allow for the more complex diameter distribution observed among craters from sub-kilometer to multi-kilometer sizes on the moon. To treat the evolution of megaregolith by larger scale craters, I have developed a model which is simpler and which tabulates not volumes of ejecta removed, but rather the percentage area of the moon covered by craters penetrating deeper than a specified depth. Consider any arbitrary production function for lunar craters, such as that actually observed on lunar maria. The observed counts of craters, in the format that I have favored in past publications (e.g., Hartmann, 1967; Hartmann *et al.*, 1981) create a histogram giving the number of craters per square kilometer in a series of logarithmic diameter increments, henceforth called bins. It is thus trivial to calculate the percentage area covered by craters in each diameter bin.

Such a calculation is shown in Table 1, which illustrates the essence of the method. Column 1 gives the crater diameter intervals and Column 2, the observed number densities, n, of craters in the lunar maria. Because the method is non-analytic, we are not constrained to use power laws or other approximations to the diameter distribution, and can use numbers of craters actually observed.

Table 1. Area coverage of craters (See text for explanation)

OBSERVED (LUNAR MARIA)			CALCULATED Area Covered (%) for N × No. of Craters on Lunar Maria							
D Diameter (m & km)	n No. Impacts/km² in √2 diam. increment (1=ave. mare)	A Fractional area covered in √2D increment (%)	N=1/2	N=2	N=4	N=10	N=20	N=40	N=100	
7.8m	[10,000]	68	34							
11.0	[4,400]	60	30							
15.6	[1,800]	49	24							
22.1	[700]	38	19							
31.2	[280]	30	15	61						
44.2	[100]	22	11	43						
62.5	[35]	15	7.5	30	60					
88.5	[12]	10.3	5	21	42					
125	[4.2]	7.3	3.6	15	30	73				
177	[1.4]	4.8	2.4	9.6	19	48				
250	[0.48]	3.3	1.6	6.6	13	33	66			
354	.16	2.2	1.1	4.4	8.8	22	45			
500	5.9 (−2)	1.6	0.8	3.2	6.4	16	33			
707m	1.8 (−2)	1.0		2.0	4.0	10	20	40		
1km	4.0 (−3)	0.44		0.2	1.8	4	8.8	17		
1.41	1.3 (−3)	0.29			1.2	3	5.8	12		
2	3.5 (−4)	0.16			0.6	2	3.1	6.2		
2.83	1.8 (−4)	0.16			0.6	1.6	3.2	6.4		
4	8.8 (−5)	0.16			0.6	2	3.1	6.2		
5.66	5.0 (−5)	0.18			0.7	2	3.6	7.2		
8	2.6 (−5)	0.18			0.7	2	3.7	7.4		
11.3	1.2 (−5)	0.17			0.7	2	3.4	6.8		
16	4.5 (−6)	0.13			0.7	1	2.6	5.2		
22.6	3.6 (−6)	0.20			0.8	2	4.2	8.2		Σ
32	1.6 (−6)	0.18			0.7	2	3.6	7.3		AREA
45.3	1.2 (−6)	0.27			1.1	3	5.5	11		=
64	2.6 (−7)	0.12			0.5	1	2.4	4.7		
90.5	2.6 (−7)	0.24			1.0	2	4.7	9.5	24	200%
128	2.0 (−7)	0.36			1.5	4	7.2	14	36	
181	[6 (−8)]	0.22			0.9	2	4.4	8.8	22	
256	[3 (−8)]	0.22			0.9	2	4.4	9.8	22	
362	2.6 (−8) = 1 per Moon									100%
512	[1.5 (−8)]	EXCLUDED:			0.87	2	4.4	8.8	22	
724	[7 (−9)]	LESS THAN 1 CRATER			0.82	2	4.1	8.2	21	
1024	[3.4 (−9)]	PER WHOLE MOON				2	4.0	7.9	20	50%
1448	[1.5 (−9)]						3.5	7.0	17	
2045	[6 (−10)]								14	

[1] Brackets [ ] indicate estimates.

Crater numbers in brackets are best estimates. Numbers of small craters in Column 1 are extrapolated from observed numbers at small sizes to allow for assumed saturation effects. Numbers of larger craters are derived from highland counts at large sizes, where mare statistics are poor. Column 3 gives the percentage area of the moon covered by craters in each bin (n times area of log-mean-sized crater in bin).

As a reference, define the amount of cratering sustained by the average lunar maria as "1 unit" of cratering (following Hartmann, 1979). Then, based on Column 3 we can tabulate the area covered by craters in the various diameter bins for other, arbitrary N units of cratering. This is done in Columns 4 through 10.

For any specified number of units of cratering, it is easy to tabulate the cumulative area covered by the largest craters, starting with the largest available crater on the moon. For instance, for one unit of cratering (mare densities), the largest diameter bin contains only 0.2 percent of the area of the moon, the largest two bins, 0.4%, and so on. As we cumulate toward smaller craters, a percentage of 50% is reached at crater diameter about 60 m (following 50% dashed line to intercept in Column 1). This condition might be defined as indicating an immature regolith development. That is, 50% of the area of the moon (or a bit less, due to stochastic overlapping) would be covered by craters larger than this size and deeper than this corresponding depth. Note that the depth of regolith is defined not as the observed depth of the floors of the craters, but rather as the underlying depth at which fractures have penetrated and broken the rock. Based on results from terrestrial craters, this is probably about two to three times typical observed floor depths, or about 0.25 to 0.4 of the observed crater diameter (Dence *et al.*, 1977). At diameters above 10 km on the moon, the fractional observed depth of the floor begins to decrease (Pike, 1977, Fig. 8) and the fractional depth of fracturing may also decrease. Such a decrease is highly problematical, since one might expect the fractures to extend throughout a nearly hemispherical shocked volume (depth = 0.5 crater diameter) underneath any crater, even though the observed depth/diameter ratio of the basins decreases with increasing size. Based on discussions at the Lunar Highlands Conference, I have assumed that the fractional depth of the fracturing decreases to 0.10 to 0.25 in the diameter range of 30 to 100 km.

As we cumulate area toward still smaller sizes of craters, a cumulative area of 100% is reached, which might be defined as a moderately mature regolith indicator. Here, 100% of the area (or somewhat less due to stochastic overlapping with the difference being area that is "doubly" cratered) is covered by craters deeper than the indicated dimension. In the lunar maria, the 100% regolith criterion is reached for craters of a diameter of about 30 meters. This has several interesting consequences. First, the corresponding predicted mare regolith is about 7 to 12 m deep, in good agreement with observation. Second, the rolling topography characteristics of a crater-saturated surface is reached only at subtelescopic dimensions. Third, even 10% coverage is reached at crater diameter no more than 300 m, so that the regolith is controlled (as in the Housen model) by sub-kilometer craters with a very steeply sloping size distribution such that the regolith is being created most efficiently by the smallest craters, while the largest craters have negligible effect.

The situation is strikingly different in the highlands, where N $\simeq$32 times the mare crater density, and the 100% criterion is reached at diameters of about 8 kilometers. This means that the "regolith" due to the observable cratering in the uplands is really a "megaregolith" with a depth at least 2 to 3 kilometers as found by Short and Forman (1972) and Hartmann (1973). It is controlled by large, multi-kilometer craters.

Of course, there is no guarantee that the observed amount of cratering in the uplands (N = 32 units) is the maximum amount ever sustained. The observed upland cratering, which is about 32 units (32 times the average mare crater density) may be merely a saturation or equilibrium level of cratering. Hundreds of units of cratering may have accumulated on a hypothetical igneous lunar lithosphere formed at some earlier date. One of the advantages of the methodology outlined above is that any amount of cratering can be specified, even amounts many times the 32 units observed in the lunar uplands, and megaregolith depths can be calculated for these cratering levels.

Figure 1 summarizes results of the method. This dramatically shows an effect very important in giving the highlands their character: as the amount of cratering increases (to the right along the bottom), the depth of regolith (right scale) explosively increases as a crater density around 16 to 64 units is reached. This happens because the craters covering 100% of the surface are no longer on the

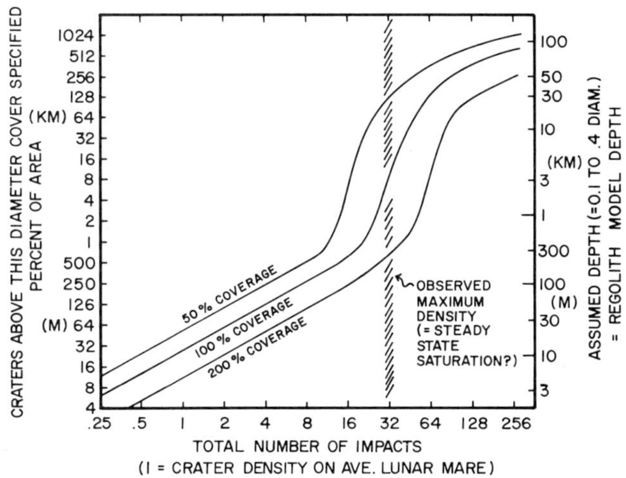

Fig. 1. Growth of regolith depth (right scale) as a function of increasing accumulated crater density (bottom). Left scale gives crater diameter D such that craters larger than D cover specified percentages of lunar area (see text). Explosive growth of megaregolith occurs at crater density around 32 $\times$ values in lunar maria, i.e., at values observed in the lunar highlands.

steeply sloping sub-kilometer part of the distribution, where only small craters add significant area, but are in the multi-kilometer diameter range, where all diameter bins add roughly equal area. When these crater densities are reached, multi-kilometer craters throughout the size spectrum suddenly begin to contribute to regolith production.

The sudden jump in regolith depth at crater densities around N = 32 marks the qualitative difference between ordinary regolith as observed in lunar mare sites, with intact bedrock layers exposed at depths of the order 10 m, and what I (1973) termed the megaregolith. The sudden jump, which corresponds to a very short time interval, also explains the characteristic breakdown of lunar topography into two types of terrain: highlands terrain and (post-highlands) sparsely-cratered mare terrain.

Figure 1 also emphasizes that the highest crater densities observed on the moon (N $\simeq$ 32) lie in the midst of the sudden transition from a thin regolith "skin" to a profound, deep megaregolith. Probably N = 32 is the greatest number of craters that can be recorded on a surface, given the size distribution involved in crater production, due to a geometric saturation. In the highlands, each $\sqrt{2}$ diameter bin contains craters covering about 6% of the moon's area. The idea that N = 32 corresponds to an upper limit produced by a steady state saturation is supported by the fact that the sub-hectometer diameter craters involved in producing mare regoliths (according to our 100% coverage criterion) also lie along the N = 32 line, as do the craters in the densest-cratered regions of several other planets (Hartmann, 1979).

To meet our goal of a more detailed model of highlands evolution, we next need a model of early bombardment as a function of time.

Models of early intense bombardment

The nature of cratering in the first billion years of lunar history has been a matter of some controversy and deserves a brief review. When pre-Apollo studies of the lunar cratering rate showed that the maria must be relatively old, it became apparent that the extensive pre-mare cratering required an "early intense bombardment" (Hartmann, 1966). The pattern of mare basalt ages derived from even the earliest lunar samples clearly showed that pre-mare time was relatively short and that the average levels of cratering during this time interval were very large. As lunar rock statistics accumulated, rock analysts were struck by the paucity of dates older than 4.0 G.y., and Tera *et al.*, (1974) introduced a concept of a "terminal lunar cataclysm" to explain this observation.

This concept has been carried forward in the literature as if empirically established, for example by Papanastassiou and Wasserburg (1976), who write of a 4.61 troctolite sample which was unusually protected so that its age and chemical properties were "minimally effected by terminal lunar cataclysm." Similarly, Dyal *et al.*, (1979) speak of ". . . magnetized material from the intense bombardment epoch 4 b.y. ago." It is usually pictured as a "spike" in a graph of cratering rate versus time, added to a much lower background of "normal" cratering, and large enough to reset radiometric ages and deeply brecciate the lunar highlands. This concept has led some observers, in my opinion, to minimize the importance of the intensive bombardment that was occurring throughout the interval from 4.6 to 4.0 G.y.

As an alternative, I pointed out that although minor "spikes" would appear in the curve of cratering rate versus time, due to meteorite production processes (Hartmann, 1969), even a smooth fall-off of intense bombardment would proceed in such a way as to pulverize most rocks older than 4.0 G.y. without any major "spike" (Hartmann, 1975). Thus, a cataclysm seemed unnecessary to explain the rock observations alone, at least if one assumed resetting of isotopic clocks by shock and pulverization, accompanying production of multi-kilometer craters.

Meanwhile, Wetherill (1975, 1976, 1977) developed dynamical models of planetesimal evolution that indicated a high likelihood that planestesimals of basin-forming energy passed through the inner solar system as late as 4.0 G.y. ago, 600 m.y. after planet formation. One source would be planetesimals stored among the outer planets. Some of these might be perturbed into the inner solar system and directly hit the moon as individuals. Others might have broken apart in the inner solar system, providing a sudden burst of cratering bodies of all sizes—a true "spike." Wetherill proposed that the most likely cause of breakup was a near encounter with the Earth or Venus, causing a tidal disruption and suddenly releasing a host of fragments that would be swept up on a short timescale. Though considering a large cometary body from the outer solar system as a candidate for such a breakup, Wetherill (1977) concluded that "the most promising source for the peak is a tidal disruption by Earth or Venus of a Ceres-sized asteroid initially in a Mars-crossing orbit."

These dynamical studies illuminate the cratering history in the first 1 G.y. of geologic time, but leave several scenarios still plausible. In particular, three cases could be considered. In Case A, a monotonic decline in primeval cratering occurred, and the "spikes" were small and unimportant. In Case B, a *spike* or *spikes* occurred in which one or a few sporadic big basin-forming bodies hit the moon without the rest of the meteorite size spectrum being affected, and at a time when the background cratering rate had declined substantially from its earlier values. This case could correspond to direct hits by planetesimals from the outer solar system, without tidal breakup. In Case C, a *spike* or *spikes* occurred in which the number of meteorites suddenly increased manyfold across the entire size spectrum, due to tidal breakup of a large planetesimal in the inner solar system after the cratering rate had declined substantially from its earliest value. The difference between Cases B and C is important (Hartmann, 1975); Case B would create one or a few young basins and throw young ejecta across much of the moon, but cause minimal change in the underlying megaregolith in regions farthest from the basins. Case C would intensely churn megaregolith all over the moon and possibly reset radio-isotopic clocks in virtually all surface material, affecting even material located at some depth prior to the *cataclysm*.

Though not proven, Case C may be the most realistic of these three scenarios.

Wetherill (1979, pers. comm.) argues that this is so because the dynamics show that large, late planetesimals are likely, and although we don't know the mechanisms of tidal breakup, we know that several near misses are likely before such planetesimals collide directly with a terrestrial planet or satellite. A variant of Case C, favored by Wetherill, is that tidal breakup produces not the ordinary power law size spectrum of fragments, but rather a few large fragments with a relative paucity of smaller fragments. The "spike" in this case would have a different size distribution than the background cratering.

The key to empirical confirmation of a terminal lunar cataclysm is accurate dating of several major lunar basins. Wetherill *et. al.,* (in press) find that the Imbrium impact occurred at about 3.80 G.y. ago and Orientale, after that date but before the earliest mare at about 3.8 G.y. Serenitatis, one of the stratigraphically oldest basins with barely-discernable ring structures, also has a suggested age of about 3.9 ± .05 G.y. based on Apollo 17 samples (Minkin *et al.,* 1978; Jessberger *et al.,* 1978), but identification of this rock age with the Serenitatis basin-forming event is questionable. If this were the Serenitatis age, it would argue for an enormous cratering cataclysm that created numerous basins, then overlayed them with enough craters in 100 m.y. or so to make them look stratigraphically old, and then formed the stratigraphically young basins. All this would have occurred in the interval 3.8 to 4.0 G.y. On the other hand, Carlson and Lugmair (1979) discuss a "large cataclysmic episode" about 4.18 ± 0.04 G.y. ago, recorded in other Apollo 17 highland samples, and this age has also been interpreted as a Serenitatis age by some workers in discussions at the Lunar Highland Conference. Debate at the conference even led to questioning of the "conventional" 3.8 to 3.9 G.y. ago for the Imbrium impact as well as the efficiency and mechanism of hypothetical radio-isotopic clock resetting during impacts. These controversies erode any empirical rock evidence for a terminal cataclysm.

In the light of these debates I represent the cratering history of the moon by two models, as in Fig. 2. The dashed line refers to Case A, where the "spikes" or irregularities are assumed to be negligible. This case provides a reference model that describes cratering effects without any cataclysms. The dotted line represents Case C, where the cratering rate experiences two late cataclysmic "spikes" with 15 m.y. half-lives and the same size distribution as the background. Of course, the precise configuration of "spikes" (if any) is unknown, so the curve is merely an assumed model. Case B is harder to model since it involves effects of single sporadic cataclysmic explosions, rather than increases in cratering at all sizes. However, the effects of Case B might be between Cases A and C.

The curves in Fig. 2 are constructed in four steps as follows. Step 1 fixes the present lunar cratering or accretion rate in the lower right corner. Various measurements of terrestrial accretion rate lie in the range of 10^{11} to 10^{12} g/yr with the lower value favored (Wetherill 1979, pers. comm.). That rate corrected by the relative area of the Moon gives a lunar accretion rate of the order of 4×10^9 g/yr, as read on the left ordinate. (Note correction to values used in my abstract; Hartmann, 1979.) The right ordinant gives corresponding rates relative to this

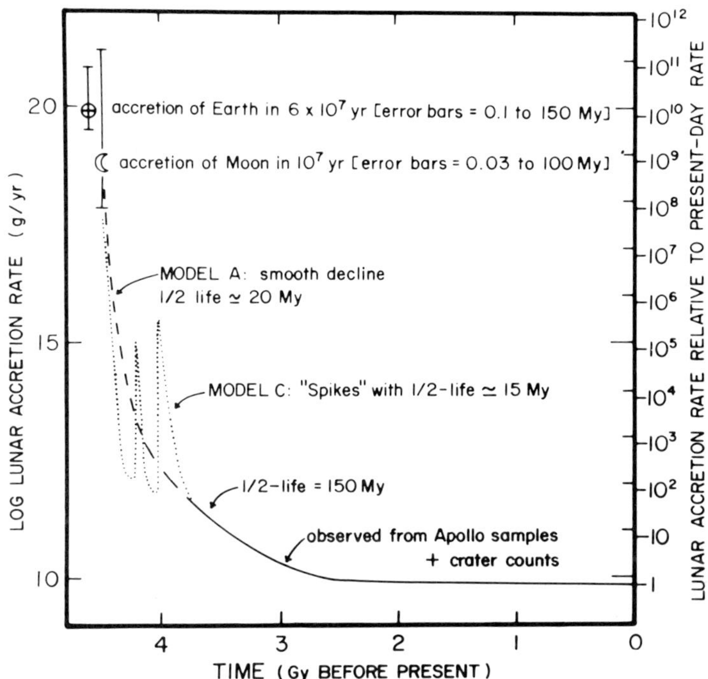

Fig. 2. Lunar cratering rate as a function of time. Models, A and C, for the early intense bombardment, distinguish between a case without "cataclysms" (episodes of unusually heavy bombardment) and a case with large "cataclysms." Construction of the curve is described in the text.

present-day figure. Step 2 gives the relative rates determined as far back as 3.9 G.y. from lunar samples, (as documented by Hartmann, *et. al.*, 1980). Step 3 plots the points in the upper left, giving the minimum bombardment rates necessary to accumulate the entire masses of the Earth and Moon in time intervals equal to their estimated formation intervals, estimated at 0.03 to 150 m.y. from radio isotopic studies, (e.g., Pepin and Phinney, 1975). In other words, the rates plotted are the terrestrial and lunar masses divided by the time interval of formation. Step 4 fills in the curves in the missing interval 4.5 to 3.9 G.y. according to Models A and C discussed above. Curve C assumes one or several 10^{21} G.y. (100 km-diameter) bodies and accompanying fragments spread through the mass spectrum and falling in each of two 50 m.y. bursts, giving two "spikes."

Note that the cratering histories empirically determined in Fig. 2, aside from being based on empirical crater/age data from lunar landing sites, is very consistent with entirely independent dynamical analyses. Wetherill (1977) presents a similar curve calculated for cratering in the earth-moon system by earth-crossing asteroids with initial elements: a = 0.90 A.U., e = 0.27, and i = 6°. That curve declines from an initial high cratering rate with a half-life of 17 m.y., lengthening to 186 m.y. by 4 G.y. ago, and lengthening still further to 900 m.y. by 1.5 G.y. ago. The cratering from these bodies was about 10^2 times the present rate 3.8 G.y. ago, very close to the empirical result in Fig. 2, measured from crater counts and rock ages.

Consequences of megaregolith formation in early lunar history

Figures 3 and 4 show the depth of material penetrated by cratering as a function of time, for rock layers created at different moments (designated "o" along the bottom). Figure 3 shows the results for Case A, where the cratering declines monotonically with time, and rock layers forming later are correspondingly less pulverized. For example, a rock layer 1 km deep formed before 4.25 G.y. ago would be quickly pulverized. If it formed 4.2 G.y. ago, its top would be thoroughly pulverized by now, but its base would be only slightly fractured. If it formed 4.0 G.y. ago or later, less than about 100 meters of its upper layers would be pulverized; scattered deep craters would eject intact rocks from it.

Figure 4 shows the more complex results for Case C. Here, the two bursts of cratering at 4.2 and 4.0 G.y. have devastating effects, just as assumed in the hypothesis of the "terminal lunar cataclysm." Of course, the magnitudes and times of the episodes are assumed and could be varied in other models. Qualitatively, though, the models show that cratering episodes capable of forming

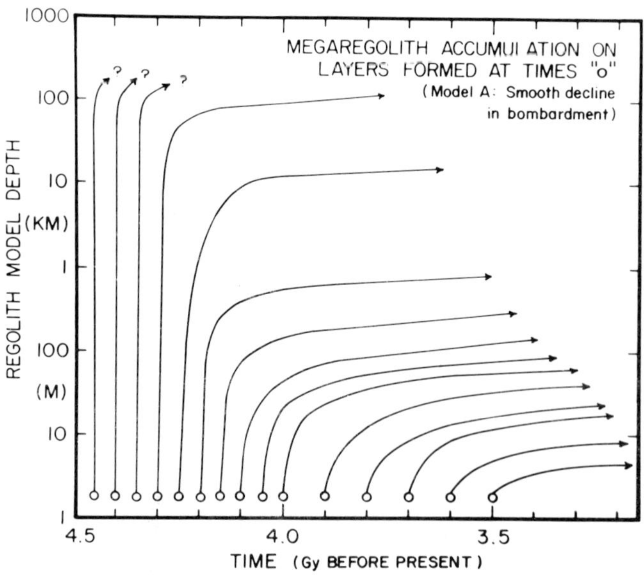

Fig. 3. Increase in megaregolith depth as a function of time in semi-finite coherent rock layers instantaneously formed at times "o." under cratering Model A. The upper kilometer of rock formed prior to about 4.2 G.y. ago is thoroughly pulverized.

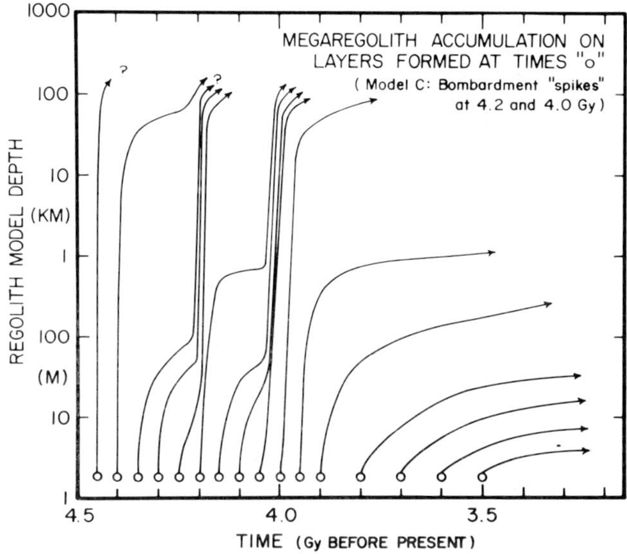

Fig. 4. As in Fig. 3, but calculated for cratering Model C, with a terminal cataclysm at 4.0 G.y.

basins in 50 m.y. intervals, if they also created craters according to the observed crater-size spectrum, are capable of pulverizing pre-existing rock layers to a depth of many tens of kilometers (probably more than 100 km if fractures under basins and large craters penetrate that deep). This intense shocking and pulverization would probably re-set isotopic clocks at 4.0 G.y. in most accessible lunar surface material, particularly since the upper few hundred meters of material would have been impacted enough to pulverize most rocks not just once, but many times over. Judging which of these two results, Fig. 3 or 4, most closely predicts the distribution of ages found on the moon would require a more detailed calculation of the survival probabilities of rocks at different depths during stochastic cratering, and a followup calculation of the probability of their ejection to the surface. Such calculations have not been made in conjunction with the present megaregolith model. However, present results do emphasize that "genesis" rocks would be mostly pulverized by cratering in both models. They would thus be rare, even in the absence of a terminal crater cataclysm 4 G.y. ago.

Competition of rock formation and rock destruction: lithosphere structure

If the early lunar lithosphere formed by solidification of a magma ocean, then its structure must have been greatly affected by the early intense bombardment, which was strongest during the putative solidification period. Published numerical models of magma ocean solidification ignore these effects. Figure 5 illustrates this point. The dashed and dotted curves show the depth of burial required for

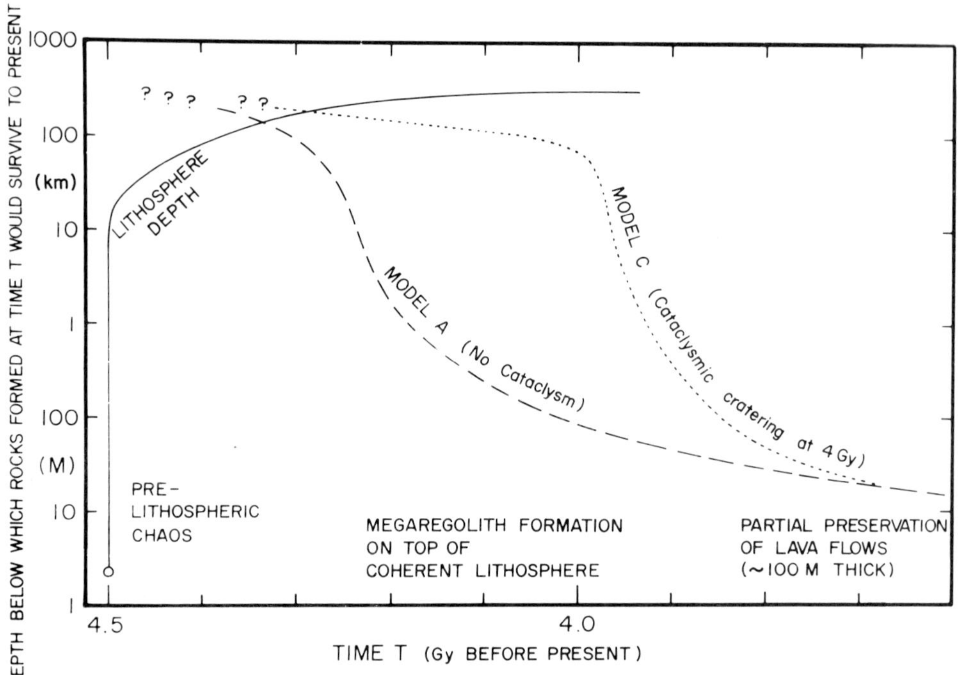

Fig. 5. The competition between formation of coherent lithosphere (magma ocean solidification in absence of cratering, solid curve) and pulverization of lithospheric layers (dotted and dashed curves). Results show that in first 10^8 yr or so (before cross-over of curves) cratering would be so intense that any lithospheric crust would be 100% penetrated on a timescale short compared to the solidification time. Later, solid lithosphere layers could form at the bottom of a megaregolith tens of kilometers (10^2 km?) deep, probably partially welded into coherent breccia. Plausibly-thick lava flows on the surface would avoid complete pulverization only after about 3.8 to 4.2 G.y. ago.

a freshly formed, coherent, igneous rock layer to survive to the present. Any rocks solidified at a depth less than about 100 km around 4.5 G.y. ago would be destroyed by cratering very rapidly. They would have to be formed at a deeper level and be shielded to escape destruction. But rocks were not forming at 100 km depth or deeper until more than 100 m.y. after the magma ocean began cooling, according to current ocean-cooling models (Solomon and Longhi, 1977; Herbert *et al.*, 1978; Minear and Fletcher, 1978). Therefore, contrary to some magma ocean models, no coherent, floating lithosphere of igneous rock could form within about 10^2 km of the surface in the first 100 m.y. or so. This is indicated in more detail by Fig. 5's solid curve which shows the increasing depth of the solid lithosphere calculated from the numerical thermal models mentioned above. The fact that the dashed and dotted "preservation depth" curves exceed the lithosphere depth shows that the first stage would not be coherent lithosphere formation.

Rather, the first stage would be a period of "pre-lithospheric chaos," in which intense cratering would repeatedly penetrate and pulverize any lithospheric skin formed on the ocean. The initial solid material would be a brecciated chaos frequently broken and intimately mixed with splashes of molten magma from below. No coherent igneous global rock layer could form during this period.

The constant and chaotic mixing of early lithospheric and magma would affect the chemistry and cooling rate of the magma ocean as well as the lithosphere itself. The cooling might be faster than the numerical models predict due to stirring and splashing of magma, and also due to creation of thinly capped holes (impact scars) where heat losses would be higher than calculated in magma ocean cooling models. Petrology and differentiation may also have been affected if denser minerals trapped on the rocky matrix were broken loose by bombardment and allowed to sink, while shards of splashed glass from intermediate depth magmas helped weld megaregolith materials into weakly coherent breccias. Therefore, a quantitative determination of the timescales and other characteristics is difficult, and requires revised models of magma ocean cooling.

Warren and Wasson (1979) present a model of magma ocean evolution with different consequences than those of the numerical thermal models discussed above. Warren and Wasson argue that solidification is principally from the bottom upward, and that the ocean is roughly 70% solidified before a "crust" more than a small fraction of the ocean depth can form. (They use the term crust to indicate this solid surface layer, similar to my use of the term lithosphere.) Though they do not present a quantitative timescale, their model implies interesting possibilities when combined with the present megaregolith modelling technique. Their thin "crust" would be continuously pulverized during most of the ocean's evolution, so that the concept of "pre-lithospheric chaos" remains valid. If the floor of the ocean solidified slowly enough to stay below the "preservation depth" that cratering could reach, a coherent solid rock layer could form below the ocean and might reach closer to the surface than the 10^2 km depth mentioned above. One could imagine more complex scenarios in which the ocean floor begins to solidify below the range of the crater pulverization, then reaches a shallow enough depth that impacts splash through the shallow ocean and pulverize the top of the rocky layer. This is especially likely during a Model C cataclysmic cratering episode. This could then be followed by a rapid decline in cratering, leaving the bottom of the ocean to continue solidifying in peace. The result would be a layered structure, with a megaregolith layer (produced during the cataclysm) sandwiched at some depth ($<10^2$ km) between coherent rock layers. In all cases, the surface would be covered with megaregolith at least a few kilometers deep.

The detailed structure of layers in the upper 100 km of the oldest highlands, if revealed by seismology or other techniques, could thus become a test of models of early lunar lithosphere evolution and cratering.

Figure 6 clarifies the events schematically with a cartoon diagram showing cross sections through the lunar outer layers at different times. Note the non-linear depth and time scale. The diagram presents a plausible scenario as a synthesis of the models discussed above. Four periods can be distinguished. During the first period, of "pre-lithospheric chaos," impacts repeatedly penetrate and

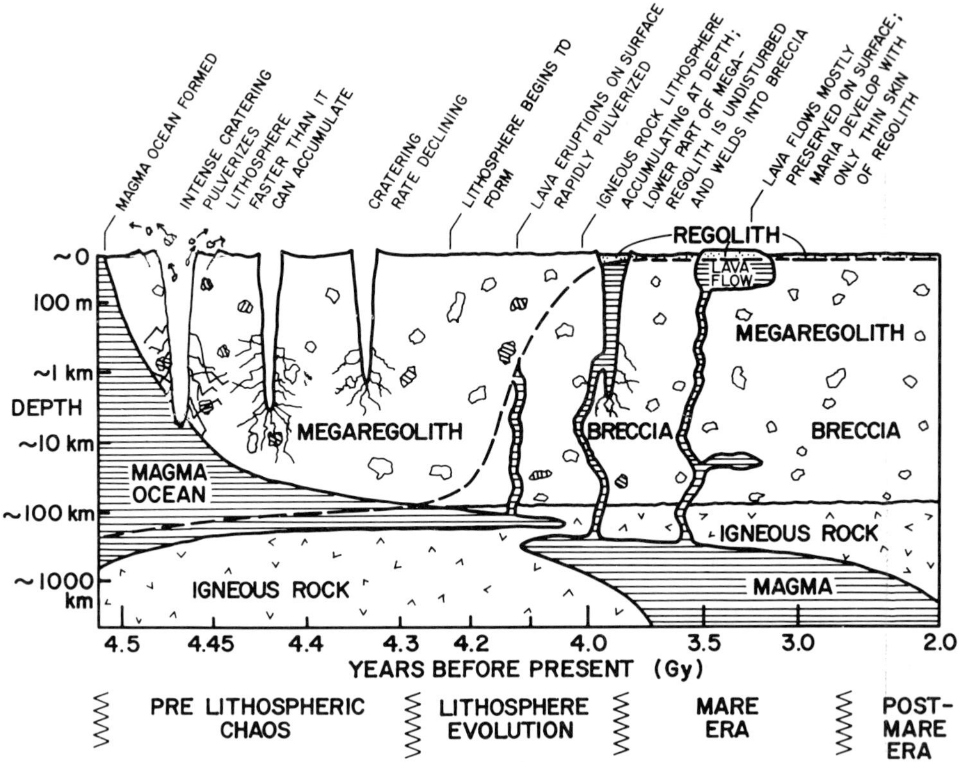

Fig. 6. Schematic early history of lunar lithosphere. Shaded magma areas are based on published thermal models. Dashed curve is depth of megaregolith generated in any given 100 m.y. interval. During period of pre-lithosphere chaos, surface lithosphere is breached as fast as it forms, so that only a heavily cratered megaregolith of fragments and splashed lava can accumulate on the surface. During period of lithosphere evolution, coherent igneous rock layers can accumulate on roof of magma layer, perhaps 10^2 km below surface. Overlying breccia, now protected from cratering, welds into coherent breccia. Surface flows are still quickly obliterated by cratering. During mare era, surface flows partially survive. Post-mare era begins when frequent magma access to surface ceases.

pulverize the thin surface layer of solid material, creating a megaregolith of mixed mineral fragments, dust, and splashed magma (glass?). The dashed line shows the depth that can be pulverized by cratering within 100 m.y. The ocean floor, solidifying upward below this level, creates coherent igneous rock layers. The second period; evolution of a true lithosphere of coherent rock layers overlying magma begins when the dashed line crosses the solidification curve, probably within 200 m.y. of magma ocean formation. New rock solidifying below this level is protected, because the cratering rate falls rapidly, and remains as a layer of coherent igneous rock. Material just above this has already been pulverized, but is under pressure and may weld into coherent breccia. This layer could be a permanent source of old, upland breccias, to be unearthed by later cratering. Material above the dashed line is gardened on a timescale of 100 m.y. and retains its character as less coherent or non-coherent megaregolith. Since surface layers as deep as 100 m or more could be pulverized rapidly during this period, surface lava flows would quickly be obliterated, explaining why premare volcanism is difficult to confirm in the uplands. The megaregolith and breccia layers established during this period include anorthositic fragments from the deep coherent layers, fragments of old and newly-created breccia, glass fragments, and dust from the finely pulverized surface layers.

The third period, or mare era, is characterized by surface eruptions in an environment where flows are at least partially preserved. Only their surfaces are converted to regolith, while their deeper parts remain as sources of basaltic boulders, ejected by moderate-scale craters. The starting time for this period depends on the thickness and rate of accumulation of flows. For isolated flows on the order of 30 to 50 m thick, this stage would begin 3.8 to 3.9 G.y. ago, a date typical of the oldest mare rock samples. This result is the same for Models A and C and thus does not depend on the occurrence of a terminal cataclysm. Such flow thicknesses are consistent with thick flow units observed on Mare Imbrium under low lighting, which include flow fronts 12 to 52 m thick (Schaber, 1973, Fig. 12).

CONCLUSIONS

Thermal and chemical models of the early lithospheric evolution, with or without a lunar magma ocean, should take into account the effects of early intense cratering. These effects include a period of pre-lithospheric chaos in which rock chips and splashed magma were intimately mixed, altering the petrology and cooling rates of the materials. Some of these effects might be preserved and detectable in the rock chemistry of the earliest lunar samples, especially those found in the highlands.

The early intense cratering in the first ten to one hundred m.y., or so, strongly inhibited formation of coherent igneous rock. The lunar highlands thus began not as coherent igneous rock layers, (or true lithosphere), but as a megaregolith of pulverized rocks and dust tens of kilometers deep. This material would have been maintained in a loose form on the surface, but the degree to which it may have welded by magma splashes or metaphosed into a coherent mass at depth is uncertain. Several kilometers of loose rubble on the surface may have been underlain by tens of kilometers of loosely welded breccia. If we assume that a magma ocean was present, then after the period of pre-lithospheric chaos ended, coherent igneous lithospheric rock layers began to accumulate at the bottom of this megaregolith. This lithosphere thickened until the magma ocean entirely solidified. Concurrently, cratering "ate into" this layer, deepening the megaregolith and exhuming various igneous boulders and chips.

Throughout geologic history the igneous layers at the base of the megaregolith served as a source of rock chips, formed perhaps 4.4 to 4.5 G.y. ago. These were incorporated into breccias and scattered into the megaregolith and on the surface.

Any surface lava flows formed prior to 4 G.y. ago were highly fractured or obliterated by concurrent intense bombardment and possibly also by catastrophic episodes of cratering as late as 4.0 to 3.9 G.y. ago. If flows accumulated fast enough on the surface in this period, lower portions of them may have been preserved and might be a source of igneous chips of age about 4.0 to 4.4 G.y. Basin-forming events were either stochastic, large isolated impacts (Models A and B) or associated with episodes of unusually intense bombardment added to the general pulverization (Model C). They would thus show up in the rock records as local (A and B) or moon-wide (C) clusters of ages of metamorphosed rocks. An example might be the age of 4.18 ± 0.04 G.y. "identified as a major metamorphic event by K-Ar and Pb ages for various Apollo 17 highland breccias and clasts" (Carlson and Lugmair, 1979).

Lava flows that filled basins and other depressions after 4.0 G.y. ago had a higher chance of survival. Flows a few tens of meters thick are known from the relief in Mare Imbrium and isolated flows of such thickness had (on the average) only their top portions converted into megaregolith after 3.8 G.y. ago, according to Figs. 2 and 3. Their bottom layers served as sources of boulders and chips, exhumed by the ubiquitous craters as small as 100 meters across. These provided the mare basalt samples.

In summary, these models of the moon's early cratering history show that the intense early bombardment prior to 4.0 G.y. was an important environmental influence in determining the moon's lithospheric evolution and in affecting the rock samples collectible today in the uplands and maria. Furthermore, since rocks older than 4 G.y. would be rare even in the absence of a terminal cataclysm, more detailed models of rock survival statistics and better data on basin ages are needed before we can assess the geologic importance of "spikes" of enhanced cratering rate during early lunar history.

Acknowledgments—I thank George Wetherill, Richard Greenberg, Sean Solomon, Michael Drake, and other colleagues for helpful discussions regarding problems of early cratering and magma oceans, and Paul Warren and Charles Simonds for helpful manuscript reviews and suggestions. This work was partially supported by NASA-contract NASW-2909.

REFERENCES

Carlson R. W. and Lugmair G. W. (1979) Early Lunar History recorded by norite 78236 (abstract). *In Papers Presented to the Conference on the Lunar Highlands Crust,* p. 9–11. Lunar and Planetary Institute, Houston.

Dence M. R., Grieve R., and Robertson P. (1977) Terrestrial impact structures: Principle characteristics and energy considerations. In *Impact and Explosion Cratering* (D. J. Roddy, R. O. Pepin, and R. B. Merrill, eds.), p. 247–275. Pergamon, N.Y.

Dyal P., Daily W., and Vanyan L. (1978) Crustal evolution inferred from Apollo magnetic measurements. *Proc. Lunar Planet. Sci. Conf. 9th,* p. 231–248.

Hartmann W. K. (1966) Early lunar cratering. *Icarus* **5,** 406–418.

Hartmann W. K. (1967) Lunar crater counts II: Three lunar surface-type areas. *Commun. Lunar and Planetary Lab.* **6,** 39–41.

Hartmann W. K. (1969) Preliminary note on lunar cratering rate and absolute time-scales. *Icarus* **12,** 131–133.

Hartmann W. K. (1973) Ancient lunar mega-regolith and sub-surface structure. *Icarus* **18,** 634–636.

Hartmann W. K. (1975) Lunar "cataclysm:" A misconception? *Icarus* **24,** 181–187.

Hartmann W. K. (1979) Formative conditions controlling structure of planetary highlands (abstract). In *Papers Presented to the Conference on the Lunar Highlands Crust,* p. 42–44. Lunar and Planetary Institute, Houston.

Hartmann W. K., Strom R., Weidenschilling S., Blasius K., Woronow A., Dence M., Grieve R., Diaz J., Chapman C., Shoemaker E., and Jones K. (1981) Chronology of planetary volcanism by comparative studies of planetary cratering. In *Basaltic Volcanism on the Terrestrial Planets,* Chap 8. Pergamon, N.Y. In press.

Herbert F., Drake M. J., and Sonnett C. P. (1978) Geophysical and geochemical evolution of the lunar magma ocean. *Proc. Lunar Planet Sci. Conf. 9th,* p. 249–262.

Housen K. R., Wilkening L., Chapman C., and Greenberg R. (1979) Regolith development and evolution or asteroids and the moon. In *Asteroids* T. Gehrels, ed., Univ. Arizona Press, Tucson.

Jessberger E. K., Staudacher T., Dominik B., and Kirstin T. (1978) Argon-argon ages of aphanite samples from consortium breccia 73255. *Proc. Lunar Planet. Sci. Conf. 9th,* p. 841–854.

Minear J. and Fletcher C. (1978) Crystallation of a lunar magma ocean. *Proc. Lunar Planet Sci. Conf. 9th,* p. 263–283.

Minkin J., Thompson C., and Chao E. (1978) The Apollo 17 Station 7 boulder: Summary of Study 6, The International Consortium. *Proc. Lunar Planet Sci. Conf. 9th,* p. 877–903.

Papanastassiou D. and Wasserburg G. (1976) Rb-Sr age of troctolite 76535. *Proc. Lunar Sci. Conf. 7th,* p. 2035–2054.

Pepin R. and Phinney D. (1975) The formation interval of the Earth (abstract). In *Lunar Science VII,* p. 682–684. The Lunar Science Institute, Houston.

Pike R. J. (1977) Size-dependence in the shape of fresh impact craters on the moon. In *Impact and Explosion Cratering* (D. J. Roddy, R. O. Pepin, and R. B. Merrill, eds.), p. 489–509. Pergamon, N.Y.

Schaber G. G. (1973) Lava flows in Mare Imbrium: Geologic Evolution from Apollo orbital photography. *Proc. Lunar Sci. Conf. 4th,* p. 73–92.

Short N. and Forman M. (1972) Thickness of impact crater ejecta on the lunar surface. *Mod. Geol.* **3,** 69.

Solomon S. C. and Longhi J. (1977) Magma oceanography: 1. Thermal evolution. *Proc. Lunar Sci. Conf. 8th,* p. 583–599.

Tera F., Papanastassiou D., and Wasserburg G. (1974) Isotopic evidence for a terminal lunar cataclysm. *Earth Planet. Sci. Lett.* **22,** 1–21.

Walker D., Longhi J., and Hays J. (1975) Differentiation of a very thick magma body and its implications for the source regions of mare basalts. *Proc. Lunar Sci. Conf. 6th,* p. 1103–1120.

Warren P. H. and Wasson J. T. (1979) Effects of pressure on the crystallization of a "chondritic" magma ocean and implications for the bulk composition of the moon. *Proc. Lunar Planet. Sci. Conf. 10th,* p. 2051–2083.

Wetherill G. W. (1975) Late heavy bombardment of the Moon and terrestrial planets. *Proc. Lunar Conf. 6th,* p. 1539–1561.

Wetherill G. W. (1976) The role of large bodies in the formation of the Earth and Moon *Proc. Lunar Sci. Conf. 7th,* p. 3245–3257.

Wetherill G. W. (1977) Pre-mare cratering and early solar system history. In *The Soviet-American Conference on Cosmochemistry of the Moon and Planets,* p. 553–567. NASA SP-370. Washington, D.C.

Wetherill G. W., Allegre C., Brooks C., Eberhardt P., Hart S., Murthy V., Tera F., and Van Schmus R. Radiogenic and stable isotopes, radiometric chronology, and basaltic volcanism. In *Basaltic Volcanism on the Terrestrial Planets.* Chp. 7. Pergamon, N.Y. In Press.

Wood J. A. (1975) Lunar petrogenesis in a well-stirred magma ocean. *Proc. Lunar Sci. Conf. 6th,* p. 1087–1102.

Papike, J.J. and Merrill, R.B., eds.
Proc. Conf. Lunar Highlands Crust (1980), p. 173-196
Printed in the United States of America

Cratering in the lunar highlands: Some problems with the process, record and effects*

Richard A. F. Grieve

Gravity and Geodynamics Division, Earth Physics Branch, Dept. Energy, Mines and Resources, Ottawa, Canada K1A 0Y3

Abstract—There is a general appreciation of the far-reaching effects of impact cratering on the physical and chemical characteristics of the lunar highlands. There are, however, a number of problems of detail and interpretation, some of which are discussed here with reference to data on terrestrial cratering. It is not clear, for example, if the preserved number of highlands craters represents production or saturation. If they were in production, they constrain the severity of impact-induced effects such as melting, metamorphism, reset ages and lateral and vertical mixing of crustal materials. The relatively common occurrence of thermal effects, however, can be interpreted as indicating that, unless they are in part due to internal heat sources, the cratering record was probably one of saturation. Terrestrial data indicate that only impact melt rocks have their ages directly reset by impact. The Fra Mauro formation is most likely a relatively low-temperature clastic ejecta deposit from Imbrium with a high percentage of admixed local material and it is possible, therefore, that the ages of the Apollo 14 breccias do not directly reflect the Imbrium event. Thus the status of the terminal lunar cataclysm at 3.9 ± 0.1 b.y., which depends critically on the interpretation of an age for Imbrium, remains in doubt.

An analysis of cratering mechanics indicates that it may be theoretically possible to differentiate primary from local secondary ejecta on the basis of shock. In practice, this is complicated by complex motions during excavation, by secondary mass transport during deposition and by multiple impact events which have imposed a range of shock features on local material. The depth of excavation and depth of sampling in basin-forming events is a major area of controversy. Terrestrial data favor some form of proportional growth in which the initial cavity has approximately the same parabolic geometry, irrespective of the size of the event. On this basis, experimental data on ejecta distribution suggest maximum depths of sampling of the order of 60 km. Such deep-seated samples would be volumetrically rare and would be most likely to occur at those sites closest to the edges of the major basins, such as Apollo 15 for Imbrium and Apollo 17 for Serenitatis. These depths are compared to 30 km derived from non-proportional growth models, which call for shallower cavities and are based on the relative volumes of material excavated and rim restoration models for cavity modification. A model for the formation of multi-ring structures is presented, in which the modification of relatively deep initial cavities is accounted for by uplift of the cavity floor and slumping of the cavity rim.

The principal chemical effect of impact on the highlands was to produce mixed compositions of more primary rock types. This must be remembered when considering the petrogenesis of such compositional groupings as highland basalt and LKFM basalt, which are found only in impact-produced lithologies. Some of the chemical and textural criteria for pristine highland samples are duplicated in terrestrial impact melt rocks and some caution in their rigid application is therefore advisable.

INTRODUCTION

Numerous photogeologic, geophysical and petrographic observations indicate the moon underwent a period of intense hypervelocity impact cratering early in its

* Contribution from the Earth Physics Branch No. 839.

history. This early bombardment had profound effects on the nature of the upper crust of the lunar highlands. It was directly responsible for features ranging from the second-order topography of the moon to the siderophile-enriched character of the bulk of the returned highlands samples. It is apparent that, with the exception of a few so-called pristine rocks (Warren and Wasson, 1977 and 1978), the highland samples are impact products (Simonds *et al.*, 1976). If the available data base is to be fully utilized to interpret the nature of the primary lunar crust, it is necessary that the cratering record be adequately described and the effects of impact cratering be deciphered and separated from those of internal lunar processes. To this end, major efforts have been made in recent years to understand the cratering process (e.g., Roddy *et al.*, 1977). Nevertheless, some of the details still remain contentious or are poorly understood.

The problem is that cratering is a disruptive process. It disturbs with varying degrees of efficiency the primary physical-chemical relationships of the target materials. Almost all of our present understanding of the cratering process comes from laboratory or high-explosive and nuclear experiments, studies at specific terrestrial impact structures and a limited number of computer studies (see papers in Roddy *et al.*, 1977). In all cases, the data are from relatively small-scale, single events in well-controlled or relatively well-known environments. In applying these data to the lunar cratering record, additional uncertainties arise. The low gravity, lack of an atmosphere and the difference in target materials must be accounted for (Chabai, 1977; Settle, 1979). There is also a problem of scale. Major basin-forming events, such as Imbrium, had impact energies of 10^{25}–10^{26} J (O'Keefe and Ahrens, 1978); energies 15 orders of magnitude greater than some laboratory craters (Stöffler *et al.*, 1975) and 2–3 orders of magnitude greater than the largest recognized terrestrial impact event (Dence *et al.*, 1977). Most importantly, the physical and chemical disordering that accompanies impact is complicated on the moon by the virtual absence of geologic control and the potential for multiple impact events contributing not only to the diversity of samples from a particular landing site but also to the various components within a single sample.

Due to these complications, our detailed understanding of the cratering record and its effects on the lunar highlands is relatively poor and many interpretations are, by necessity, model dependent. This contribution focuses on some of the problems, in the hope that it will provide a starting point for future integrated studies into the relationship between impact cratering and the evolution of the highlands crust. Emphasis is given to those areas of the cratering process and the record in the highlands which are poorly understood and to some of the interpretations of the samples that are at variance with terrestrial experience.

THE CRATERING PROCESS

It is desirable to review our knowledge of the cratering process as it applies to the highlands and the returned samples. The process is commonly discussed under a number of progressive stages: compression, rarefaction, crater excavation and ejection, and crater modification. Although these stages represent successive steps, they do not occur simultaneously everywhere within the evolving crater. For more detailed and quantitative treatments of the various aspects of cratering, the reader is referred to Cooper (1977), Dence *et al.* (1977), Gault *et al.* (1968), Grieve *et al.* (1977), Kieffer and Simonds (1980), Oberbeck (1975) and Shoemaker (1960).

Compression

On impact, a compressional shock wave is propagated radially into the crust and backwards into the projectile. Peak pressures are a function of impact conditions and for stony and iron projectiles impacting crystalline rocks at 15–20 km s^{-1} are the order of several hundred GPa. The shock wave travels initially with speeds similar to the impact velocity and the crustal rocks are compressed and accelerated radially with initial particle velocities of at least half that of the shock wave. The shock wave decays exponentially, with model calculations indicating low rates close to the point of impact and higher rates of approximately $P \alpha R^{-2}$ at several projectile radii, where R is distance normalized to the projectile radius (Ahrens and O'Keefe, 1977). Field observations on shock metamorphic features at terrestrial impact structures suggest rates ranging from R^{-3} to $R^{-5.5}$, where R is normalized to the initial cavity and correction is made for the movement of the "autochthonous" rocks of the cavity floor and wall (Robertson and Grieve, 1977). Considering the current uncertainties involved in restoring the initial cavity to its pre-movement geometry, the correspondence between the observed and calculated rates is relatively good.

As the target material and the projectile are compressed, the projectile penetrates to some depth within the crust. This is also a function of the conditions of impact, with equation of state calculations indicating that, for the impact conditions used earlier, the penetration depth is the order of 2–3 times the projectile radius (Kieffer and Simonds, 1980; Shoemaker, 1960). This provides a minimum estimate for depth of the initial cavity and, with the observations on shock attenuation, has a bearing on the problem of the potential depth of sampling, as discussed below in the treatment of cavity geometry and modification.

Rarefaction

Rarefaction waves generated at free surfaces, such as the original ground surface and the back of the projectile, follow the compressional wave. The rarefaction waves travel with velocities similar to the compressional wave. They, therefore, travel faster than the particle velocities imparted by the compressional wave and overtake the moving target material. The rarefaction wave fronts, however, are not parallel to the compressional wave, except directly below the impacting body, and the resultant stress vectors deflect the original radial particle motions upward and outward excavating a cavity (Gault *et al.*, 1968).

The increase in internal energy, which accompanies compression, is not accomplished isentropically and some energy is trapped irreversibly on pressure release. Most of this excess energy is manifested as heat, melting and/or vaporizing the projectile and that portion of the target shocked to pressures in excess of 60–100 GPa. Rocks shocked to lower pressures, but above their Hugoniot elastic limit, which for highland gabbroic anorthosite is ~3.5 GPa (Ahrens *et al.*, 1973), develop characteristic deformational features such as the conversion of plagioclase to maskelynite and formation of planar features and planar fractures. Many of these features have been reproduced in shock experiments and can be correlated with specific pressure ranges (Reimold and Stöffler, 1978; Stöffler and Hornemann, 1972 and others). Thus, for a given attenuation rate, rocks shocked to particular levels and from a known source crater can be placed in their original pre-impact radial positions with respect to the point of origin of the shock wave.

Excavation and ejection

Material near the point of impact is ejected first. Since it was subjected to the highest pressures, it has the highest particle velocities and travels the greatest range. As cavity growth continues, successive segments containing progressively lower shocked material are excavated. They have lower ejection velocities and impact nearer to the cavity rim, with the deepest material landing closest to the rim of the excavated cavity (Fig. 9 in Oberbeck and Morrison, 1976). Not all the material set in motion is ejected. The particle motions directly beneath the projectile are such that some of the impact-produced melt plus vapor, and some comminuted crystalline debris, remain within the cavity. This material forms a lining to the cavity and takes the final form of a basal melt pool in simple craters or a flat-lying coherent sheet in complex, central-uplift or ring structures (Grieve *et al.*, 1977).

The stratigraphy established by compression is to some degree preserved in the ejecta but in an inverted form, with the highest shocked material from close to the surface being furthest from the cavity (Oberbeck and Morrison, 1976; Stöffler *et al.*, 1975). If an ejection angle is assumed, generally taken from experimental cratering results as 45° or less (Oberbeck, 1975; Stöffler *et al.*, 1975), the ejection velocities required to transport ejecta a particular range from the excavated cavity can be calculated from equations for ballistic transport. As the ejection velocities are related to the initial particle velocities within the cavity, they can be translated into minimum shock pressures, using Hugoniot data for lunar or lunar-like rock types. Thus the minimum shock level required for transport can be estimated and primary ejecta can theoretically be differentiated on the basis of shock grade from secondary admixed local material.

It is difficult, however, to reach unambiguous conclusions when applying these principles to the highlands. The above model is somewhat oversimplistic. In a cratering event mass motions within the cavity are extremely complex (Grieve *et al.*, 1977) and the relationships between shock pressure, initial particle velocities and final ejection velocities, which are derived from flat plate geometries, are unlikely to be directly applicable. Primary ejecta at a given range will be a mixture of shock grades and any remaining regular distribution of shock will be further disrupted during the complex mass motions that accompany ejecta deposition and produce considerable lateral and vertical mixing of primary ejecta (Hörz and Banholzer, 1980).

In the highlands with its complicated cratering history, it is probable that local material will also exhibit a range of shock features produced by prior local impact events. In addition, as there is no atmosphere on the moon, ejecta can land with undiminished velocity. This will result in a second event, with the potential to re-shock the primary ejecta and produce shock features, of approximately the same grade as those in the primary ejecta, in the local material. The effects of a second shock event on already shocked material has yet to be investigated experimentally. It seems intuitively that since the shocked material in the primary ejecta would have a reduced density and therefore a different Hugoniot, relative to its original pre-shock state, the second shock event might produce additional shock features.

Nevertheless, an analysis of the level of shock has been applied to some Apollo 14 samples by Hawke and Head (1977), who conclude that, because of their low shock grade, the high Al mare basalts found as clasts in breccias such as 14321 and 14312 or believed to have been clasts, such as 14072, are candidates for local material, as is much of the material in the White Rocks 14064 and 14082. A similar analysis has been carried out relating the Apollo 17 grey breccias to Serenitatis, and an attempt made to place the various clast types in their pre-excavation position within the crust and to model the matrix composition as a sample of the entire 60 km crustal column (Dence *et al.*, 1976). In this treatment, a critical assumption was that the excavated cavity in large basin-forming events has a geometry similar to that of simple craters. The geometry of the original cavity, however, is one of the more contentious points in cratering mechanics and is discussed below.

Cavity geometry and modification

Most workers agree that for simple craters, $D \lesssim 15$ km on the moon (Pike, 1977), the initial transient cavity would have had parabolic form. From terrestrial and experimental studies, this transient cavity would have had a depth/diameter ratio of 1:3 to 1:4, with up to half its depth due to compression (Dence *et al.*, 1977; Stöffler *et al.*, 1975). Following excavation, the cavity modifies by the inward collapse of the walls (Dence, 1968; Melosh, 1977), producing a crater partially filled by breccias derived from the wall rocks and the melt lining (Grieve, 1978a). On the moon, the final form is that of a bowl with a depth/diameter ratio of ~1:5 (Pike, 1977).

There is less agreement in the case of large transient cavities, which modify to central uplift or ring structures with final depth/diameter ratios as large as 1:50

or greater. Estimates for Imbrium range from 160–190 km for the depth of a transient cavity with radius 285 km (Dence, 1976) to 30 km for the maximum depth of sampling recorded in the ejecta from a cavity with radius 485 km (Head *et al.,* 1975). It should be noted, however, that these definitions of depth are not synonymous. The depth of sampling estimate does not exclude the possibility that the cavity extended deeper and that material excavated from greater depths remained within the transient cavity or even within the final modified cavity. Also the depth of the transient cavity according to Dence (1976) is only partly due to excavation. When this is taken into account and the experimental data for the distribution of ejecta is applied (Stöffler *et al.,* 1975), a cavity of 285 km radius would deposit material from maximum depths of the order of 60 km at a radial distance of 485 km. That is, the disparity in the two estimates may be closer to a factor of two, rather than five to six. Nevertheless, estimates of original cavity depth for multi-ring structures is a major problem.

In the proportional growth hypothesis the transient cavity has a parabolic form, irrespective of the size of the event (Dence, 1976; Dence *et al.,* 1976; Howard *et al.,* 1974). In multi-ring structures, one of the inner rings, such as the Inner Rook Mountains at Orientale, has been interpreted as the trace of the cavity rim and the outer rings, the Outer Rooks and Cordillera, are interpreted as collapse features produced during cavity modification. Some form of proportional growth is favored by observations at terrestrial structures. Stratigraphic and structural data indicate that the amount of uplift in complex craters increases with increasing diameter (Fig. 1) and suggests transient cavities become deeper with increasing diameter. In addition, the level of shock at the base of simple craters and at the center of complex structures is relatively constant at 25–30 GPa (Robertson and Grieve, 1977). Unless there are major differences in shock wave attenuation with size, this also suggests increasing depth with increasing size. However, *strict* proportional growth is not demanded by the data. Small differences in the volumes of melt and in shock attenuation rates between complex and simple craters (Grieve *et al.,* 1977; Robertson and Grieve, 1977) may indicate that large structures are slightly less efficiently excavated.

The shallow cavity or non-proportional growth hypothesis is based on comparisons of estimated volumes of ejecta and excavated volumes for particular cavity geometries (Head *et al.,* 1975) and the failure of restoration models to predict parabolic cavities for lunar craters with $D \geq 50$–70 km (Settle and Head, 1979; Malin and Dzurisin, 1978). The lack of equivalence between the so-derived depth of sampling estimates for major multi-ring basins of ≤ 30 km and possible transient cavity depth is exemplified by the fact that these sampling depths are less than the estimated diameters of the impacting bodies (O'Keefe and Ahrens, 1978). However, the difference between the estimates for the volumes of ejecta based on proportional growth and those used as an argument for non-proportional growth may not be that great. Head *et al.* (1975) compared a 485 km radius cavity for Imbrium with a volume of 3–10×10^6 km^3 for the ejecta exterior to the rim. Although there is considerable uncertainty associated with the prediction of ejecta thicknesses and volumes on the moon (Pike, 1974), this volume is similar to the

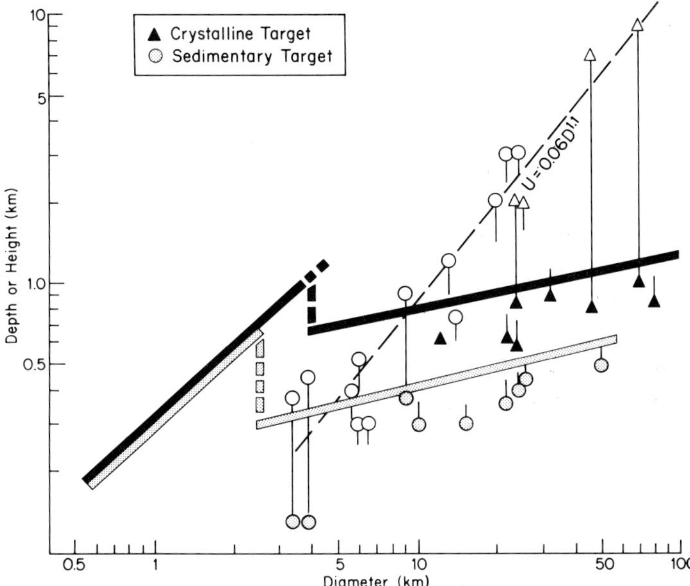

Fig. 1. Depth/diameter relationships (solid line) and structural uplift (broken line) in terrestrial craters. Structures in sedimentary targets develop complex form at smaller diameters than those in crystalline targets, reflecting the physical properties of target. Physical heights of central uplifts or rings (closed symbols) never exceed final crater depth, indicating apparent height (depth) is equilibrium feature. Amount of structural uplift (open symbols) increases with diameter, suggesting initially deep excavated cavities with the amount of uplift a function of impact energy.

7–8.5 × 10^6 km^3 calculated by Dence (1976) for the volume of material excavated from a 285 km radius parabolic cavity and landing beyond approximately one radius from the original rim, i.e., at a distance only slightly greater than 485 km. The correspondence may be even closer if bulking is considered. Croft (1979) has suggested an increase of 40–50% in apparent volume is associated with changes in the bulk density of ejecta over the original material from within the cavity.

A third hypothesis considers the outer ring in multi-ring basins as the limit of excavation and the inner rings as traces of the rims of a series of cavities within the overall transient cavity (Hodges and Wilhelms, 1978). This "nested-crater" model has similarities to that for terrestrial craters such as the Ries, which formed in layered targets (Hörz and Ostertag, 1979). In lunar basins, the rims of the inner cavities would correspond to the base of the low-velocity, fractured crust at 20–25 km and the crust-mantle interface at 55–60 km. The depth/diameter ratio of the overall transient cavity would be 1:10–1:15, which gives calculated transient cavity depths similar to those of the proportional growth model.

Ejecta fall-back, rim slumping and uplift of the cavity floor all contribute to the modification of the transient cavity. The contribution from fall-back is least important. Fall-back of ejecta within the cavity at terrestrial craters is minor, as is the contribution of melt sheets and breccias to the shallowing of complex structures (Grieve *et al.*, 1977). Calculations by Settle (1979) suggest that for 100 km structures fall-back is of the order of 10% of the total ejecta volume. This represents an upper limit for the highlands, as ejection occurs under vacuum conditions and in a weaker gravity field.

Rim slumping can account for the modification of parabolic cavities on the moon up to final diameters of ~40 km and possibly 50–70 km (Settle and Head, 1979). At larger diameters, slumping fails to account for all the modification. The amount of structural uplift in terrestrial craters, normalized to the equivalent impact energy for the moon, suggests that some of the additional modification required can be accounted for by uplift (Fig. 2). The uplift curve (Fig. 2) represents a minimum in the lunar case as no allowance was made for erosion in the terrestrial data. In addition, uplift is a function of the volume of material compressed. This will be approximately the same on the earth and moon for structures produced in crystalline rocks by the same impact energy. On pressure release, the lower gravity of the moon, however, may play a role in reducing the forces acting against uplift.

The driving force behind uplift is not clear. Deep centripetal collapse has been suggested (Dence, 1973; Melosh, 1977), as has elastic rebound (Ullrich *et al.*, 1977). An extreme position is the "tsunami" model of Baldwin (1972), in which the rings are likened to a frozen gravity wave. This suggestion has been modified by Murray (1980), who considers the outer rings to represent the rim of the

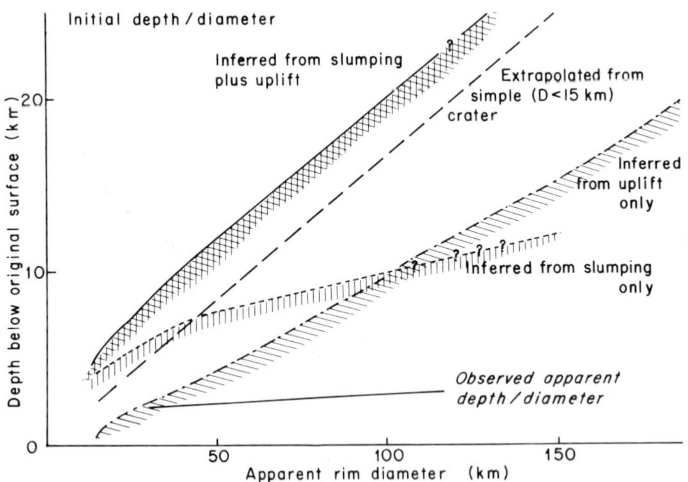

Fig. 2. Depth/diameter relationships in lunar craters. Observed relationships and those extrapolated from simple (D ≲ 15 km) craters from Pike (1977), inferred initial depths from slumping only from Settle and Head (1979) and uplift only from terrestrial data of Dence and Grieve (1979), normalized to equivalent impact energy on the moon. Combined effects of slumping and uplift are consistent with deep initial cavities for large structures. The combined depths are greater than predicted by simple crater relationship, which does not account for breccia and melt fill within the crater.

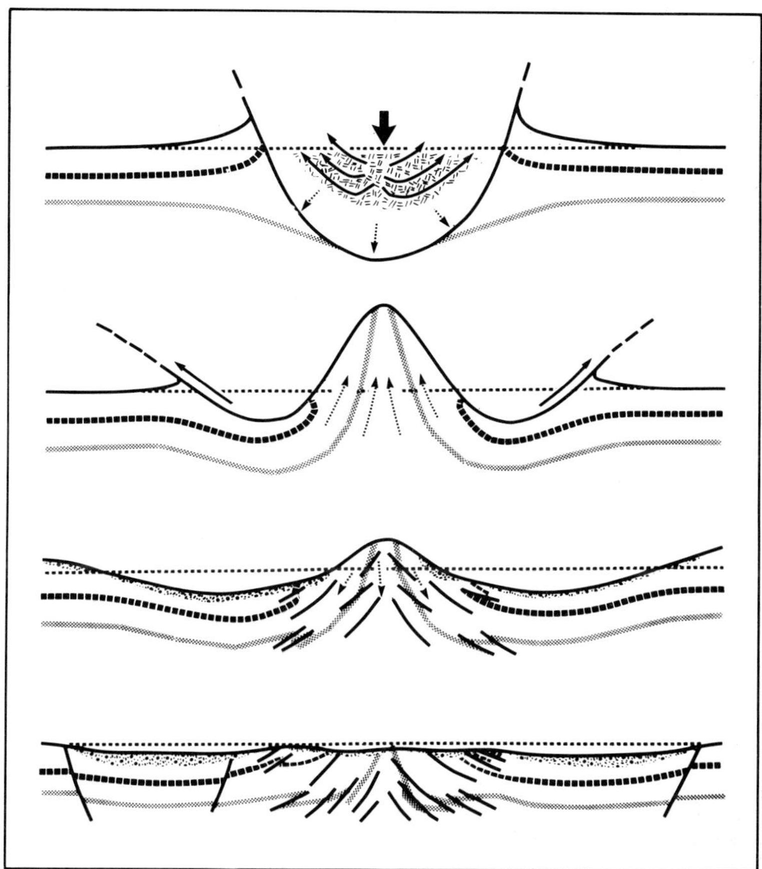

Fig. 3. Model for the formation of ring structures. Top-formation of transient cavity by excavation and compression. Second top-rebound of compressed base of transient cavity above ground surface. Excavation from near surface at periphery is continuing. Second bottom-collapse of uplift peak under its own weight with development of reverse faulting in central area. Major excavation has ceased, although melt and breccias (stipple) are still in motion within cavity. Bottom-final configuration. Rim and uplift have reached equilibrium height controlled by rock strength and gravity. Excess volume in initial uplift, which can not be accommodated by reverse faulting, appears as a ring. Faulting and slumping has occurred on rim. Movement of material within final crater has ceased.

original cavity. Dence and Grieve (1979) have proposed a model which incorporates elements of several hypotheses. They envisage a deep parabolic central cavity, which rebounds following compression and collapses under its own weight to form a series of rings (Fig. 3). During the uplift stage, excavation is still proceeding at the outer edges of the cavity and the net result is a sombrero-shaped cavity (Fig. 3). Finally, slumping occurs around the rim area, further enlarging the final crater. In cross-section, the combined excavated volume is deep in the center and shallow towards the rim, not unlike the nested craters of Hodges and Wilhelms (1978).

It is apparent that there are several hypotheses for the geometry of transient cavities and their modification in large impact events. The choice of a correct model is of paramount importance if highland samples believed excavated during the formation of the multi-ring basins are to provide information on the vertical composition of the lunar crust. Examples of deep-seated crustal materials are most likely to occur within the samples from those sites closest to major basins, that is Apollo 15 for Imbrium and Apollo 17 for Serenitatis (Dence *et al*., 1974 and 1976). Depending on the model for the initial cavity geometry for multi-ring basins, they would represent sampling depths of 30–60 km, with the terrestrial data favoring the greater depth.

THE CRATERING RECORD

Saturation or production

A large body of data is available on the crater size-frequency distribution on various lunar terrains. Crater superposition and degradation (Fig. 4) result in the interpretation of cratering in the highlands being the least precisely constrained. The average size-frequency distribution for the highlands indicates an average cratering density (N/km^2) of ~32 times that of the mare (Fig. 5 and Hartmann, 1979). This reflects crater preservation and it is not certain if it is the result of a production or saturation population. That is, it is not clear whether the observed

Fig. 4. Photograph of heavily cratered southern highlands. Note apparent saturation with superposition of several generations of craters.

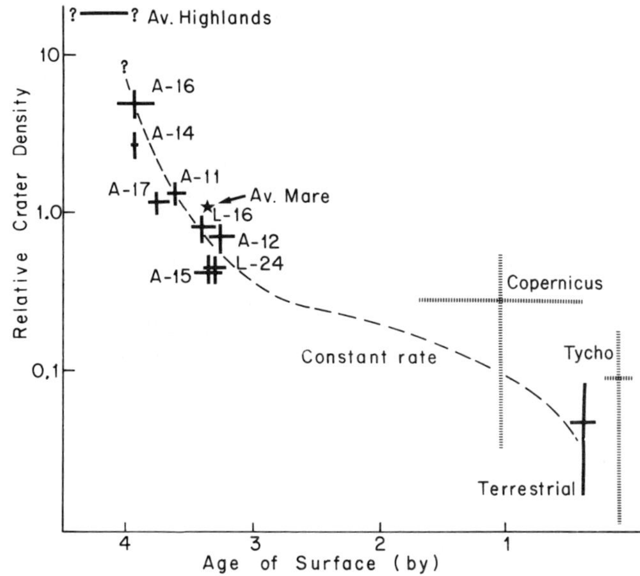

Fig. 5. Variation in preserved crater density with time. Data from Hartmann *et al.* (1980) indicate an exponentially decreasing flux from ~4.0 b.y. till ~3.0 b.y., when a constant flux may be achieved. No definite age relationship is given to preserved rate of ~32 times mare rate recorded in the highlands.

population of craters represents a complete record of the bombardment history of the lunar crust or represents empirical saturation with significant destruction of earlier craters.

The more widely-quoted view is that the cratering record represents production. This is based primarily on detailed analyses of small variations in the size-frequency distributions between different terrains within the highlands (Strom and Whitaker, 1976; Wilhelms *et al.,* 1978 and others). The conclusion that the record is one of saturation or equilibrium finds support in the observation that the crater population generally follows an $N \alpha D^{-2}$ distribution and that this distribution and a crater density equivalent to that of the highlands also occurs on the heavily-cratered surfaces of Mars and Mercury (Hartmann, 1972 and 1979).

The question of whether the observed crater densities represent production or saturation is of great importance. If they are a production population they can be used to model the time-dependency of the cratering rate and derive the relative and approximate absolute ages for surfaces within the highlands (Baldwin, 1971; Neukum *et al.,* 1975). In addition, these model histories can be used to estimate such parameters as the average depth of sampling, average number of impacts per sample etc. (Hörz *et al.,* 1976), and the total volume of impact melt produced within the highlands (Lange and Ahrens, 1979a). A production population, however, also requires that if the thermal effects in the samples, such as metamorphic textures and reset ages, cannot be accounted for by the observed cratering record, with some adjustment for the 30–50% of earlier craters which are destroyed even in a production population (Woronow, 1978), they must be partially ascribed to endogenic processes.

If the record is one of saturation the observed size-frequency distribution can only provide a minimum estimate for models of the bombardment history and the integrated thermal effects of the cratering record will be more severe. There is some evidence, such as the discrepancy between the known rock ages and the amount of impact melt as a function of age for model cratering rates (Lange and Ahrens, 1979b), in favor of the interpretation that all the characteristics of the samples cannot be produced if the observed cratering record is a true estimate of the total bombardment history of the highlands. Although there is general agreement that the cratering rate decreased over several orders of magnitude during the first 1.5 b.y. after the moon's formation, the question of whether the record in the highlands is one of production or saturation still remains largely unsettled.

The "cataclysm"

The clustering of ages in the interval 4.1–3.9 b.y. is interpreted in terms of the intense bombardment of the highlands. It has been called the "terminal cataclysm" by Tera *et al*. (1974), who associate it with the formation of Imbrium and possibly several other major multi-ring basins at 3.9 ± 0.1 b.y. It is not clear from the interpretation of the geochronological data whether the clustering of ages represents an actual period of increased cratering with the formation of a number of large basins within ~0.2 b.y. or is simply the effective termination of an extremely high early flux of impacting bodies (Kirsten and Horn, 1974; Turner, 1977). The same equivocal conclusion is reached from the observed cratering record (Baldwin, 1974; Hartmann, 1975; Wilhelms *et al*., 1978). The reconstruction of the early cratering rate from the observed post ~4.0 b.y. rate, extrapolated to the rate required to accrete the moon within a time frame of ~100 m.y., suggests flux rates in the period 4.5–4.0 b.y. of 10^2–10^5 times that recorded on the mare and does not *require* an anomalous increase in the cratering rate at 3.9 ± 0.1 b.y. (Hartmann, 1979).

On the basis of photogeologic evidence, the multi-ring basins have been placed in sequence from youngest to oldest of: Orientale, Imbrium, Crisium, Humorum, Nectaris, Serenitatis (Stuart-Alexander and Howard, 1970). Thus the ages of the "young" Imbrium and the "old" Serenitatis basins assume prime importance in the cataclysm hypothesis. The Fra Mauro formation at the Apollo 14 site is generally considered as Imbrium ejecta (Sutton *et al*., 1972) and an age of 3.95 b.y., typical of the Apollo 14 breccias, taken as that of Imbrium. The Taurus-Littrow massifs at the Apollo 17 site are believed to contain ejecta from the south Serenitatis impact (Reed and Wolfe, 1975) and an age of 3.98 b.y. for the Apollo 17 melt rocks may define Serenitatis. [The ages quoted are restricted to ^{40}Ar–^{39}Ar results from Turner and Cadogan (1975) to remove bias.] Thus it would appear that there is support for several major basin-forming events occurring at 3.9–4.1 b.y. However, there is no general agreement as to the interpretation of the age data. Schaeffer *et al.* (1976) interpret their data, and that of other workers, as favoring the major basin-forming events occurring over an extended period, with Serenitatis being as old as 4.26 b.y. A third alternative is that there is no strong relationship between the age of the samples and the large multi-ring basins (Kirsten and Horn, 1974).

Central to the interpretation is the assumption that specific samples can be related to basin-forming impacts. This owes its origin to pre-mission rationales, with Apollo 14 designed to sample Imbrium ejecta and Apollo 17 to sample Serenitatis ejecta. The case for the Apollo 17 sample containing a significant melted component of primary Serenitatis ejecta is fairly strong (Dence *et al.*, 1976; James and Blanchard, 1976). The case, however, is considerably weaker for the correlation between the Apollo 14 sample and the Imbrium Basin. Cratering experiments indicate that continuous ejecta deposits contain increasing amounts of admixed local material with increasing radial range (Oberbeck *et al.*, 1974). Although the Fra Mauro formation may have been formed by ejecta from Imbrium, its position ~3 crater radii from the original excavated cavity suggests that the Apollo 14 sample is dominated by re-excavated local material and contains only ~1/3 primary Imbrium ejecta (Hawke and Head, 1977).

The nature of the samples is relevant to this problem. The Apollo 14 breccias represent the result of multiple impact events (Chao *et al.*, 1972; Dence and Plant, 1972; Simonds *et al.*, 1977). There is little apparent annealing associated with their final assembly in the Fra Mauro (Chao *et al.*, 1972; Grieve *et al.*, 1975) and the Fra Mauro formation has been interpreted as a relatively low-temperature clastic ejecta deposit from Imbrium with a high percentage of admixed local material, not unlike the Bunte breccia of the Ries (Dence and Plant, 1972; Engelhardt *et al.*, 1972; Hörz and Banholzer, 1980). As discussed in detail later, such deposits do not have their ages directly reset by the impact event. This being the case, an age of 3.95 b.y. for the Apollo 14 samples would define only an upper limit for the Imbrium event.

From the preceding, it is obvious that the terminal cataclysm is not yet a tenet of the highlands cratering record. It is an attractive hypothesis to explain the peak in highland ages at 3.9 ± 0.1 b.y. It must remain as a hypothesis, however, until geochronologic studies, which include the petrography and chemistry of the samples (Maurer *et al.*, 1978), are fully integrated with the cratering record, observed photogeologically and recorded in the samples, and constrained by what is known about the physics of cratering and ejection processes.

THE EFFECTS OF CRATERING

A listing of some major direct and indirect results of the bombardment of the lunar highlands is given in Table 1. Within the limits of this contribution it is impossible to summarize adequately all the effects, and discussion is restricted to some effects recorded in the sample.

Table 1. Major effects of cratering on highland crust.

GEOPHYSICAL	
Topography —	2nd order topography and resultant dichotomy between highlands and mare.
Gravity —	2nd order gravity field; highs associated with mascons within multi-ring basins, lows associated with brecciation and ejecta.
Magnetics —	local high intensity anomalies may be associated with high temperature impact deposits.
Seismicity —	low seismic velocities in upper 20–25 km of crust due to fracturing.
GEOLOGICAL	
Distribution —	many geologic units related to major ejecta deposits, recovered samples from ejecta and transported from original location.
Physical nature —	90% of samples are impact products, many are physical mixtures of pre-existing lithologies and show effects of shock.
Chemical nature —	bulk chemistry of samples often a mixture of previous rock types, commonly enriched in "meteoritic" siderophile elements and with reset ages.

Physical nature of the sample

The overwhelming influence of cratering on the petrogenetic history of the highlands is reflected in approximately 60% of the returned sample being impact breccias and 30% being impact melt rocks (Simonds *et al.*, 1976). This is a minimum estimate of the relative abundance of impact melt rocks, because there is a significant melt component as clasts and in the matrices of some breccias (Simonds *et al.*, 1977). Studies at terrestrial craters (Grieve *et al.*, 1977) and equation of state calculations (Kieffer and Simonds, 1980) indicate that the amount of melt produced in a single impact event is $\lesssim$10% of the excavated volume. This is a maximum for a lunar event, because for an equivalent impact energy (equivalent volume of melt produced) the volume of more weakly to unshocked material ejected under the lower lunar gravity may be 50% greater than in the terrestrial case. Monte Carlo simulations of the observed cratering record indicate that the average highland sample underwent 1–2 impact events (Hörz *et al.*, 1976). As noted by Hörz *et al.* (1976), the 1–2 impacts per sample is probably an underestimate because the effects of ejecta blankets and secondary cratering were not simulated. Nevertheless, the high percentage of returned melt rocks may indicate that the actual cratering record was more severe than that preserved. There are other possible interpretations. There may be a sampling bias towards more coherent, melt-bearing lithologies which can withstand long periods of impact erosion. An apparent decrease in the relative amounts of other lithologies to melt rocks in the Apollo 16 drill core is explicable in terms of their susceptibility to comminution (Vaniman *et al.*, 1976).

The breccia-in-breccia texture observed in most highland breccias has been used to define the minimum number of impact events recorded in particular samples (Grieve *et al.*, 1975). Conclusions on the number of events, based solely on textural relationships, must be viewed with caution. Secondary impacting can result in breccia-in-breccia textures (Dence *et al.*, 1976) and equivalent textures, produced during the primary excavation process, have been noted from Brent and the Ries (Grieve, 1978a; Hörz *et al.*, 1977).

Metamorphic textures are present in many highland samples. Thermal metamorphism, however, is localized and rare at terrestrial impact structures, with high temperature metamorphism restricted to target rocks occurring as inclusions or in contact with the melt (Grieve, 1975 and 1978a). This could be the provenance of samples such as 15415 and 76535; however, it is more likely that in their case the heat source was endogenic and recrystallization took place within the crust during Apollonian metamorphism (Stewart, 1975). For obvious impact-generated rocks, which have metamorphic textures, the heat source is more problematical. For example, the granulitic impactites from Apollo 16 and 17 are believed to have been held at temperatures of ~1000°C for extended periods of time (Warner *et al.*, 1977). Studies at the Ries, however, indicate that post-shock temperatures in the high-temperature suevitic breccias are only of the order of 500–600°C for geologically short time periods (Miller and Wagner, 1979). Thus it is unlikely that the granulitic impactites owe their texture to auto-metamorphism within their own ejecta blanket. The temperatures determined for the Ries suevite also suggest that the metamorphism in the granulitic impactites is unlikely to be achieved by simply burying still hot breccias beneath additional ejecta deposits.

The likelihood of this scenario is only slightly improved if the protoliths to the granulitic impactites were impact melts. Preservation of a distinct clast-matrix texture indicates an initial period of rapid heat loss (Simonds *et al.*, 1976); however, it is *possible* that these impactites (*sensu stricto*) were buried by suevitic ejecta or another melt deposit while at sub-solidus temperatures of 1000°C or greater. Almost by default, an additional endogenic heat source may be necessary. In particular, burial to some depth beneath a series of ejecta blankets has been suggested but under conditions of high internal heat flow from the crust (Warner *et al.*, 1977). Whether or not an additional heat source is also required to explain the textures in less strongly metamorphosed breccias must await thermal and kinetic equilibration models similar to those of Uhlmann *et al.* (1977), which are constrained by the cratering record of the highlands and data on the cooling history of terrestrial impact deposits (Miller and Wagner, 1979; Onorato *et al.*, 1977).

The age and chemistry of the samples

The impact process is efficient at resetting ages only in melt rocks (Bottomley *et al.*, 1978). Autochthonous rocks shocked to 25–30 GPa show only partial gas loss (Mak *et al.*, 1976; Wolfe, 1971) and shocked lithic and mineral debris in suevite breccias show no apparent age reduction (Jessberger *et al.*, 1978). From studies at the Ries, Hörz and Banholzer (1980) suggest it is erroneous to simply assume that highlands sampling sites conform to a layer cake model of overlapping ejecta blankets and that these blankets were sufficiently hot to reset the argon radiometric clock. For shock pressures less than required for complete melting, it is the pre- or post-shock thermal history that determines the degree of relationship between the age of ≲90% of the material excavated and the impact event. The ineffectiveness of impact in producing significant argon loss has also been noted for recent lunar craters. Ejecta from Camelot, Cone, and North Ray craters have been dated by exposure ages but in no case has significant resetting of the ^{40}Ar–^{39}Ar clock also been established (Turner, 1977).

What then is the significance of the highland ages? If the ages reflect a number of medium-sized local cratering events (Kirsten and Horn, 1974) and resetting is accomplished solely by impact melting, the clustering of ages at 3.9 ± 0.1 b.y. cannot be produced by the preserved cratering record. Lange (1979) has shown that for the observed number of craters with $30 \leq D \leq 500$ km, 75% of the ages would be >4.0 b.y. This is compared to ~70% of the ages for highland samples being in the range 4.1–3.9 b.y. (Schaeffer and Schaeffer, 1977).

Approximately 60% of the returned samples are breccias and their age relationship to specific impact events requires further consideration. As discussed earlier, this is of particular interest for the case of the Apollo 14 breccias and the lunar "cataclysm" at 3.9 ± 0.1 b.y. If these breccias are related to Imbrium, their average shock grade is too low for them to have a direct shock-induced Imbrium age. A somewhat *ad hoc* model has been suggested, in which their age could reflect an excavation age, in much the same manner as has been proposed for the granulitic impactites (Warner *et al.*, 1977), with the argon system being closed by excavation from pre-existing hot ejecta deposits. This appears unlikely in view of the results from the Ries (Hörz and Banholzer, 1980; Jessberger *et al.*, 1978). In such a case, however, the age of the breccias would still reflect the age of Imbrium. If, as is most likely, the Apollo 14 breccias are a mixture of local material and Imbrium ejecta and there was little thermal metamorphism associated with their assembly in the Fra Mauro (Chao *et al.*, 1972; Dence and Plant, 1972; Grieve *et al.*, 1975), it may be the age of the youngest clasts which defines an upper limit to Imbrium.

It is obvious that several interpretations are possible and may even be required to explain the reset ages. In view of the relative inefficiency of impact in resetting the ages of large volumes of ejecta, considerable attention must be paid to the possible pre- and post-impact thermal histories of the dated samples. Whether the reset ages of a sample are produced by impact melting, excavation or burial in a hot ejecta blanket, it would appear that the lack of "old" ages for the highland samples as a whole suggests an impact flux in excess of the preserved record.

The breccias and melt rocks in the highlands sample are potential mixtures of pre-existing lithologies. Intimate mixing of target components in melt rocks is well-documented and arises from the high velocity, turbulent flow of a mixture of vapor, super-heated melt and crystalline debris during crater formation (Grieve *et al.*, 1977; Phinney *et al.*, 1978). Mixing model calculations have been performed on highland melt rocks and breccias to determine their primary components (Dence *et al.*, 1976; Schonfeld, 1974; Wänke *et al.*, 1976). A major problem, however, is the lack of geologic control with respect to the original target composition. As a result, the petrogenetic significance of such ubiquitous highland compositions as Low K Fra Mauro basalt, and to a lesser extent highland basalt, remains uncertain. They occur only as impact products (Bickel and Warner, 1978; Reid *et al.*, 1977) and as such may be mixtures, even though the Low K Fra Mauro basalt composition lies close to the plagioclase-olivine cotectic and is a candidate for a partial melt origin (Hess *et al.*, 1977).

Many breccias and melt rocks show appreciable enrichments in siderophile elements. This is generally accepted as the chemical signature of the bodies which produced the highland craters, although the degree of contamination and the indigenous siderophile content of pristine highland lithologies is a point of considerable debate (Anders, 1978; Delano and Ringwood, 1978, Wänke *et al.*, 1978). Without wishing to add to the debate, some discussion of the converse argument that low-siderophile contents are a feature of pristine highlands samples is in order. Warren and Wasson (1977, 1978) and Warner and Bickel (1978) list several criteria for pristine highlands rocks, principally: siderophile contents $<3 \times 10^{-4}$ C-1 concentrations, negligible KREEP component and coarse igneous texture, >3 mm, with phase homogeneity and no clasts. Some of these features are found in terrestrial impact melts.

The impact melts at Manicouagan, L. St. Martin, W. Clearwater and 9 of 11 samples from Mistastin all have siderophile contents $<3 \times 10^{-4}$ C-1 abundances (Göbel *et al.*, 1980; Palme *et al.*, 1978; Wolf *et al.*, 1980). In some cases this may be a real reflection of the composition of the projectile. However, at the twin Clearwater structures, which are known to be contemporaneous (Reimold, pers. comm) and probably formed from the break-up of a single projectile, the melt rocks at the eastern structure have 5–10% C-1 contamination (Palme *et al.*, 1979). At the western structure, however, there is an apparent absence of a meteoritic signature and this is ascribed to sampling from non-equivalent stratigraphic positions in the respective melt sheets and the inhomogeneous distribution of meteoritic elements within the melt rocks (Grieve, 1978b). Thus impact melt rocks *may* have siderophile contents consistent with those of the target rocks with no detectable siderophile enrichment.

Low KREEP contents are a characteristic of the granulitic impactites (Warner *et al.*, 1977) and it is possible that sizeable bodies of KREEP did not exist until relatively late in the sequence of highland crustal evolution (Irving, 1977). If no sizeable bodies of KREEP were present in the target area and the impact melt did not have an obvious clast-matrix texture, then an igneous-textured rock with no KREEP component could be produced. Fine grain size and abundant lithic and mineral inclusions are characteristic of *recognized* highland melt rocks (Fig. 6a), the coarsest-grained and most clast-poor example being 14310, which has a grain size of $\lesssim 1.0$ mm (James, 1973). These textural features are not necessarily present in all terrestrial impact melt samples. Samples from the uppermost portion of the $\sim$200m thick melt sheet at the 75 km Mincouagan structure, are clast-free, have well-developed subophitic igneous textures (Fig. 6b), with phenocrysts of

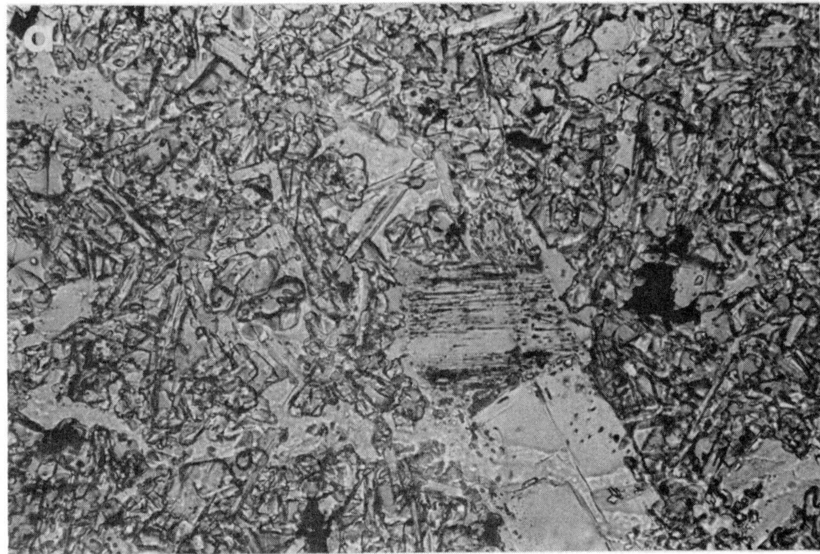

Fig. 6a. Photo-micrograph of typical lunar impact melt clast in breccia 77035. Note obvious plagioclase clast with shock planar features and fine grain-sized igneous texture of matrix. Width of field of view 0.90 mm.

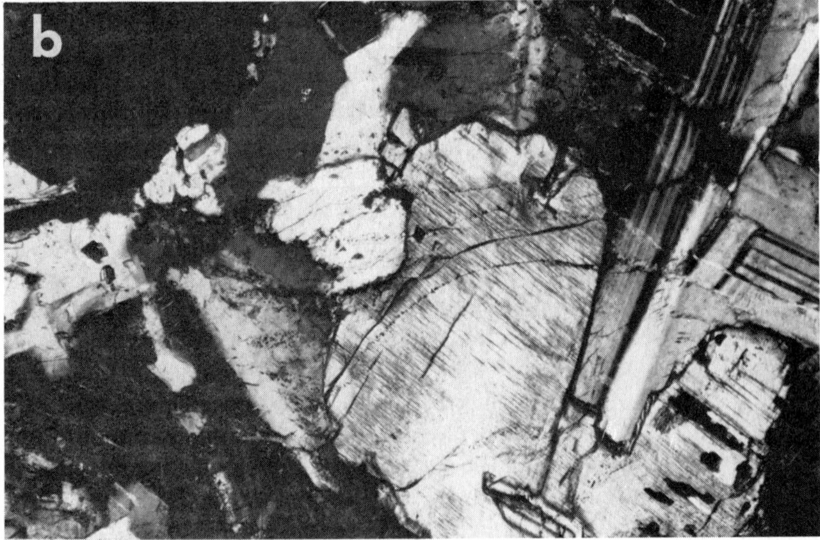

Fig. 6b. Photo-micrograph of medium-grained impact melt from 75 km Manicouagan structure, Quebec. Note lack of obvious clasts and well-developed sub-ophitic texture with exsolution lamellae in pyroxene. Width of field of view 2.28 mm. This melt has no detectable siderophile enrichment and at its coarsest development contains pyroxene phenocrysts up to 1 cm in length.

augitic pyroxene up to 1 cm in length and containing exsolution lamellae of orthopyroxene (Floran *et al.*, 1978). It is not known what the effect of composition will be on relative grain size in terrestrial and lunar impact melt rocks with similar cooling histories. However, the lowering effect of water on the viscosity of terrestrial melts may be compensated for by the absence of alkalies in the lunar case (Simonds, pers. comm.).

The above statements are not necessarily an endorsement of the arguments of Delano and Ringwood (1978). They are advanced to emphasize the potential complications and advise against the indiscriminate use of rigid criteria for differentiating pristine highland samples. Obviously, if the lunar crust has some form of compositional layering, it is virtually impossible to envisage how sizeable bodies of ultramafic impact melt, without an additional less mafic component, could be produced. On the other hand, it would be possible to produce siderophile-free impact melts of anorthositic composition with a clast-free, "coarse grained" igneous texture. They would in some ways correspond by definition to "pristine" (Warren and Wasson, 1977) or "plutonic" (Warner and Bickel, 1978), although they would not be indigenously igneous.

CONCLUDING REMARKS

There is a general comprehension of the profound consequences of cratering on the crustal and surface evolution of the highlands. This contribution, however, has stressed many of the problems and uncertainties regarding the details. Some questions which are still unanswered or in doubt are:

1. Is the cratering history of the highlands one of production or saturation? In other words, are the relative amounts of melt rocks, the breccia-in-breccia and metamorphic textures and abundance of reset ages consistent with the preserved cratering record?
2. What is the geometry of transient cavities in multi-ring structures and what are the consequences for the depth of sampling?
3. Are the cratering history, petrography and ages consistent with a basin or local derivation for particular samples and lithologies?
4. Can the definition of "pristine" categorically differentiate between endogenic igneous rocks and melt rocks from early, large impact structures?

It is not the intention to leave a negative impression and suggest that these problems are insolvable. Because answers to these problems are fundamental to understanding the early evolution of the lunar crust, they are prime goals of the Lunar Highlands Initiative. Solutions or a consensus will require considerable multi-disciplinary co-operation between those in the field of impact cratering and those who work with the returned samples. Considering the recent progress that has been made in understanding the cratering process and the fact that the above problems are so inter-related that the solution to one will provide at least a partial solution to others, answers appear to be achievable.

Acknowledgments—This contribution has benefitted from general discussions on the effects of cratering with M. R. Dence and P. B. Robertson. Reviews by F. Hörz, S. Kieffer and J. L. Warner have resulted in improvements in the text and are gratefully acknowledged.

REFERENCES

Ahrens T. J. and O'Keefe J. D. (1977) Equations of state and impact-induced shock-wave attenuation on the moon. In *Impact and Explosion Cratering* (D. J. Roddy, R. O. Pepin, and R. B. Merrill, eds.), p. 639–657. Pergamon, N.Y.

Ahrens T. J., O'Keefe J. D., and Gibbons R. V. (1973) Shock compression of a recrystallized anorthositic rock from Apollo 15. *Proc. Lunar Sci. Conf. 4th*, p. 2575–2590.

Anders E. (1978) Procrustean science: Indigenous siderophiles in the lunar highlands, according to Delano and Ringwood. *Proc. Lunar Planet. Sci. Conf. 9th*, p. 161–184.

Baldwin R. B. (1971) On the history of lunar impact cratering: The absolute time scale and the origin of planetismals. *Icarus* **14,** 36–52.

Baldwin R. B. (1972) The Tsunami model of the origin of ring structures concentric with lunar craters. *Phys. Earth Planet. Int.* **6,** 327–339.

Baldwin R. B. (1974) Was there a "terminal lunar cataclysm" 3.9×10^9 years ago? *Icarus* **23,** 157–166.

Bickel C. E. and Warner J. L. (1978) Survey of lunar plutonic and granulitic fragments. *Proc. Lunar Planet. Sci. Conf. 9th*, p. 629–652.

Bottomley R. J., York D., and Grieve R. A. F. (1978) ^{40}Ar-^{39}Ar ages of Scandinavian impact structures: 1 Mien and Siljan. *Contrib. Mineral. Petrol.* **68,** 79–84.

Chabai A. J. (1977) Influence of gravitational fields and atmospheric pressure on scaling of explosion craters. In *Impact and Explosion Cratering* (D. J. Roddy, R. O. Pepin, and R. B. Merrill, eds.), p. 1191–1214. Pergamon, N.Y.

Chao E. C. T., Minkin J. A., and Best J. B. (1972) Apollo 14 breccias: General characteristics and classification. *Proc. Lunar Sci. Conf. 3rd*, p. 645–660.

Cooper H. F. (1977) A summary of explosion cratering phenomena relevant to meteor impact events. In *Impact and Explosion Cratering* (D. J. Roddy, R. O. Pepin, and R. B. Merrill, eds.), p. 11–44. Pergamon, N.Y.

Croft S. K. (1979) Proportional vs non-proportional growth of basin-sized excavation cavities: A reconciliation (abstract). In *Lunar and Planetary Science X*, p. 248–250. Lunar and Planetary Institute, Houston.

Delano J. W. and Ringwood A. E. (1978) Siderophile elements in the lunar highlands: Nature of the indigenous component and implications for the origin of the moon. *Proc. Lunar Planet. Sci. Conf. 9th*, p. 111–160.

Dence M. R. (1976) Notes towards an impact model for the Imbrium basin. In *Interdisciplinary Studies by the Imbrium Consortium, Vol. 1* (J. A. Wood, ed.), p. 147–155. LSI Contr. No. 268D. Center for Astrophysics, Cambridge.

Dence M. R. (1973) Dimensional analysis of impact structures (abstract). *Meteoritics* **8,** 343–344.

Dence M. R. (1976) Notes towards an impact model for the Imbrium basin. In *Interdisciplinary studies by the Imbrium Consortium, Vol. 1* (J. A. Wood, ed.). LSI Contr. No. 268D, 147–155. Center for Astrophysics, Cambridge.

Dence M. R. and Grieve R. A. F. (1979) The formation of complex impact structures (abstract). In *Lunar and Planetary Science X*, p. 292–294. Lunar and Planetary Institute, Houston.

Dence M. R., Grieve R. A. F., and Plant A. G. (1974) The Imbrium basin and its ejecta (abstract). In *Lunar Science V*, p. 165–167. Lunar Science Institute, Houston.

Dence M. R., Grieve R. A. F., and Plant A. G. (1976) Apollo 17 grey breccias and crustal composition in the Serenitatis Basin region. *Proc. Lunar Sci. Conf. 7th*, p. 1821–1832.

Dence M. R., Grieve R. A. F., and Robertson P. B. (1977) Terrestrial impact structures: Principal characteristics and energy considerations. In *Impact and Explosion Cratering* (D. J. Roddy, R. O. Pepin and R. B. Merrill, eds.), p. 247–275. Pergamon, N.Y.

Dence M. R. and Plant A. G. (1972) Analysis of the Fra Mauro samples and the origin of the Imbrium basin. *Proc. Lunar Sci. Conf. 3rd,* p. 379–399.

Engelhardt W. V., Arndt J., Stöffler D., and Schneider H. (1972) Apollo 14 regolith and fragmental rocks, their compositions and origins by impacts. *Proc. Lunar Sci. Conf. 3rd,* p. 753–770.

Floran R. J., Grieve R. A. F., Phinney W. C., Warner J. L., Simonds C. H., Blanchard D. P., and Dence M. R. (1978) Manicouagan impact melt, Quebec, 1. Stratigraphy, petrology and chemistry. *J. Geophys. Res.* **83,** 2737–2759.

Gault D. E., Quaide W. L. , and Oberbeck V. R. (1968) Impact cratering mechanics and structures. In *Shock Metamorphism of Nature Materials* (B. M. French and N. M. Short, eds.), p. 87–99. Mono, Baltimore.

Göbel E., Reimold U., Baddenhausen H., and Palme H. (1980) The projectile of the Lappajärvi impact crater. *Zeit. für Natur.* In press.

Grieve R. A. F. (1975) Petrology and chemistry of the impact melt at Mistastin Lake crater, Labrador. *Bull. Geol. Soc. Amer.* **86,** 1617–1629.

Grieve R. A. F. (1978a) The melt rocks at Brent crater, Ontario, Canada. *Proc. Lunar Planet. Sci. Conf. 9th,* p. 2579–2608.

Grieve R. A. F. (1978b) Meteoritic component and impact melt composition at Lac à l'Eau Claire (Clearwater) impact structures, Quebec. *Geochim. Cosmochim. Acta* **42,** 429–431.

Grieve R. A. F., Dence M. R., and Robertson P. B. (1977) Cratering processes as interpreted from the occurrence of impact melts. In *Impact and Explosion Cratering* (D. J. Roddy, R. O. Pepin, and R. B. Merrill, eds.), p. 791–814. Pergamon, N.Y.

Grieve R. A. F., McKay G. A., Smith H. D., and Weill D. F. (1975) Lunar polymict breccia 14321: A petrographic study. *Geochim. Cosmochim. Acta* **39,** 229–246.

Hartmann W. K. (1972) Paleocratering of the Moon: Review of post Apollo data. *Astrophys. Space Sci.* **17,** 48–64.

Hartmann W. K. (1975) Lunar "cataclysm": A misconception? *Icarus* **24,** 181–187.

Hartmann W. K. (1979) Formative conditions controlling structure of planetary highlands (abstract). In *Papers Presented to the Conference on the Lunar Highlands Crust,* p. 42–44. Lunar and Planetary Institute, Houston.

Hartmann W. K., Blasius K., Chapman C. R., Dence M. R., Diaz J., Grieve R. A. F., Jones K., Shoemaker E., Soderblom L., Strom R., Weidenschilling S., and Woronow A. (1980) Chronology of planetary volcanism by comparative studies of planetary cratering. In *Basaltic Volcanism of the Terrestrial Planets,* Ch. 8. Pergamon, N.Y. In press.

Hawke B. R. and Head J. W. (1977) Local *v's* Imbrium basin material at the Apollo 14 site: Some observations and suggestions. In *Interdisciplinary Studies by the Imbrium Consortium, Vol. 2* (J. A. Wood, ed.), p. 91–96. LSI Contr. No. 268D. Center for Astrophysics, Cambridge.

Head J. W., Settle M., and Stein R. S. (1975) Volume of material ejected from major lunar basins and implications for the depth of excavation of lunar samples. *Proc. Lunar Sci. Conf. 6th,* p. 2805–2829.

Hess P. C., Rutherford M. J., and Campbell H. W. (1977) Origin and evolution of LKFM basalt. *Proc. Lunar Sci. Conf. 8th,* p. 2357–2373.

Hodges C. A. and Wilhelms D. E. (1978) Formation of lunar ring basins. *Icarus* **34,** 294–323.

Hörz F. and Banholzer G. S. Jr. (1980) Deep seated target materials in the continuous deposits of the Ries. In *Proceedings of the Conference on the Lunar Highlands Crust.* This volume.

Hörz F., Gall H., Hüttner R., and Oberbeck V. R. (1977) Shallow drilling in the "Bunte Breccia" impact deposits Ries Crater, Germany. In *Impact and Explosion Cratering* (D. J. Roddy, R. O. Pepin, and R. B. Merrill eds.), p. 425–448. Pergamon, N.Y.

Hörz F., Gibbons R. V., Hill R. E., and Gault D. E. (1976) Large scale cratering of the lunar highlands: Some Monte Carlo model considerations. *Proc. Lunar Sci. Conf. 7th,* p. 2931–2945.

Hörz F. and Ostertag R. (1979) The transient cavity of the Ries crater, Germany (abstract). In *Lunar and Planetary Science X,* p. 570–572. Lunar and Planetary Institute, Houston.

Howard K. A., Wilhelms D. E., and Scott D. H. (1974) Lunar basin formation and highland stratigraphy. *Rev. Geophys. Space Phys.* **12,** 309–327.

Irving A. J. (1977) Chemical variation and fractionation of KREEP basalt magmas. *Proc. Lunar Sci. Conf. 8th,* p. 2433–2448.

James O. B. (1973) Crystallization history of lunar feldspathic basalt 14310. *U.S. Geol. Survey Prof. Paper 841*. 29 pp.

James O. B. and Blanchard D. P. (1976) Consortium studies of light-gray breccia 73215: Introduction, subsample distribution data and summary of results. *Proc. Lunar Sci. Conf. 7th,* p. 2131–2143.

Jessberger E. K., Staudbacher Th., Dominik B., Kirsten T., and Schaeffer, O. A. (1978) Limited response of the K-Ar system to the Nordlinger Ries giant meteorite impact. *Nature* **271,** 338–339.

Kieffer S. W. and Simonds C. H. (1980) The role of volatiles in the cratering process. *Rev. Geophy. Space Phys.* **18,** 143–182.

Kirsten T. and Horn P. (1974) Chronology of the Taurus-Littrow region-III: Ages of mare basalts and highland breccias and some remarks about the interpretation of lunar highland ages. *Proc. Lunar Sci. Conf. 5th,* p. 1451–1475.

Lange M. A. (1979) Absolute rock ages and the lunar cratering record (abstract). *EOS (Trans. Amer. Geophys. Union)* **60,** 871.

Lange M. A. and Ahrens T. J. (1979a) Impact melting during the first 1.5 b.y. of lunar history (abstract). In *Lunar and Planetary Science X,* p. 700–702. Lunar and Planetary Institute, Houston.

Lange M. A. and Ahrens T. J. (1979b) Impact melting in early lunar history. *Proc. Lunar Planet. Sci. Conf. 10th,* p. 2707–2725.

Mak E. K., York D., Grieve R. A. F., and Dence M. R. (1976) The age of Mistastin Lake crater, Labrador. *Earth Planet. Sci. Lett.* **31,** 345–357.

Malin M. C. and Dzurisin D. (1978) Modification of fresh crater landforms: Evidence from the moon and Mercury. *J. Geophys. Res.* **83,** 233–243.

Maurer P., Eberhardt P., Geiss J., Grögler N., Stettler A., Brown G. M., Peckett A., and Krähenbühl V. (1978) Pre-Imbrian craters and basins: Ages, compositions and excavation depths of Apollo 16 breccias. *Geochim. Cosmochim. Acta* **42,** 1687–1720.

Melosh H. J. (1977) Crater modification by gravity: A mechanical analysis of slumping. In *Impact and Explosion Cratering* (D. J. Roddy, R. O. Pepin, and R. B. Merrill, eds.), p. 1245–1260. Pergamon, N.Y.

Miller D. S. and Wagner G. A. (1979) Age and intensity of thermal events by fission track analysis: The Ries impact crater. *Earth Planet. Sci. Lett.* **43,** 351–358.

Murray J. B. (1980) Oscillating peak models of basin and crater formation. *The Moon* **22,** 269–292.

Neukum G., König B., Fechtig H., and Storzer D. (1975) Cratering in the earth-moon system: Consequences for age determination by crater counting. *Proc. Lunar Sci. Conf. 6th,* p. 2597–2620.

Oberbeck V. R. (1975) The role of ballistic erosion and sedimentation in lunar stratigraphy. *Rev. Geophys. Space Phys.* **13,** 337–362.

Oberbeck V. R. and Morrison R. H. (1976) Candidate areas for *in situ* ancient lunar materials. *Proc. Lunar Sci. Conf. 7th,* p. 2983–3006.

Oberbeck V. R., Morrison R. H., Hörz F., Quaide W. L., and Gault D. E. (1974) Smooth plains and continuous deposits of craters and basins. *Proc. Lunar Sci. Conf. 5th,* p. 111–136.

O'Keefe J. D. and Ahrens T. J. (1978) Impact flows and cratering scaling on the Moon. *Phys. Earth Planet. Int.* **16,** 341–351.

Onorato P. I. K., Uhlmann D. R., and Simonds C. H. (1977) The thermal history of the Manicouagan impact melt sheet, Quebec. *J. Geophys. Res.* **83,** 2789–2798.

Palme H., Göbel E., and Grieve R. A. F. (1979) The distribution of volatile and siderophile elements in the impact melt of East Clearwater (Quebec). *Proc. Lunar Planet. Sci. Conf. 10th,* p. 2465–2492.

Palme H., Janssens M. J., Takahashi H., Anders E., and Hertogen J. (1978) Meteoritic material at five large impact craters. *Geochim. Cosmochim. Acta* **42,** 313–324.

Phinney W. C., Simmonds C. H., Cochran A., and McGee P. E. (1978) West Clearwater, Quebec, impact structure, Part II: Petrology. *Proc. Lunar Planet. Sci. Conf. 9th,* p. 2659–2693.

Pike R. J. (1974) Ejecta from large craters on the moon: Comments on the geometric model of McGetchin *et al. Earth Planet. Sci. Lett.* **23,** 265–274.

Pike R. J. (1977) Size-dependence in the shape of fresh impact craters in the moon. In *Impact and Explosion Cratering* (D. J. Roddy, R. O. Pepin, and R. B. Merrill, eds.), p. 489–510. Pergamon, N.Y.

Reed V. S. and Wolfe E. W. (1975) Origin of the Taurus-Littrow massifs. *Proc. Lunar Sci. Conf. 6th*, p. 2443–2461.

Reid A. M., Duncan A. R., and Richardson S. H. (1977) In search of LKFM. *Proc. Lunar Sci. Conf. 8th*, p. 2321–2338.

Reimold W. U. and Stöffler D. (1978) Experimental shock metamorphism of dunite. *Proc. Lunar Planet. Sci. Conf. 9th*, p. 2805–2824.

Robertson P. B. and Grieve R. A. F. (1977) Shock attenuation at terrestrial impact structures. In *Impact and Explosion Cratering* (D. J. Roddy, R. O. Pepin and R. B. Merrill, eds.), p. 687–702. Pergamon, N.Y.

Roddy D. J., Pepin R. O., and Merrill R. B. (1977) *Impact and Explosion Cratering*. Pergamon, N.Y. 1301 pp.

Schaeffer G. A. and Schaeffer O. A. (1977) ^{40}Ar-^{39}Ar ages of lunar rocks. *Proc. Lunar Sci. Conf. 8th*, p. 2253–2300.

Schaeffer O. A., Husain L., and Schaeffer G. A. (1976) Ages of highland rocks: The chronology of lunar basin formation. *Proc. Lunar Sci. Conf. 7th*, p. 2067–2092.

Schonfeld E. (1974) The contamination of lunar highland rocks by KREEP: Interpretation by mixing models. *Proc. Lunar Sci. Conf. 5th*, p. 1269–1286.

Settle M. (1979) Volume of impact crater fallback ejecta on the earth, Moon and Venus (abstract). In *Lunar and Planetary Science X*, p. 1113–1115. Lunar and Planetary Institute, Houston.

Settle M. and Head J. W. (1979) The role of rim slumping in the modification of lunar impact craters. *J. Geophys. Res.* **84**, 3081–3096.

Shoemaker E. M. (1960) Penetration mechanics of high velocity meteorites, illustrated by Meteor Crater, Arizona. *Internat. Geol. Congr.*, 21st, Norden, 1960. Pt. 18, p. 418–434.

Simonds C. H., Phinney W. C., Warner J. L., McGee P. E., Geeslin J., Brown R. W., and Rhodes J. M. (1977) Apollo 14 revisited, or breccias aren't so bad after all. *Proc. Lunar Sci. Conf. 8th*, p. 1869–1893.

Simonds C. H., Warner J. L., and Phinney W. C. (1976) Thermal regimes in cratered terrane with emphasis on the role of impact melt. *Amer. Mineral.* **61**, 569–577.

Stewart D. B. (1975) Apollonian metamorphic rocks—the products of prolonged sub-solidus equilibration (abstract). In *Lunar Science VI*, p. 774–776. The Lunar Science Institute, Houston.

Stöffler D., Gault D. E., Wedekind J., and Polkowski G. (1975) Experimental hypervelocity impact into quartz sand: Distribution and shock metamorphism of ejecta. *J. Geophys. Res.* **80**, 4062–4077.

Stöffler D. and Hornemann U. (1972) Quartz and feldspar glasses produced by natural and experimental shock. *Meteoritics* **7**, 371–394.

Strom R. G. and Whitaker E. A. (1976) Populations of impacting bodies in the inner solar system (abstract). In *Reports of Accomplishments of Planetology Programs*, 1975–1976. NASA TMX-3364, p. 194–196.

Stuart-Alexander D. E. and Howard K. A. (1970) Lunar maria and circular basins—a review. *Icarus* **12**, 440–456.

Sutton R. L., Hait M. H., and Swann G. A. (1972) Geology of the Apollo 14 landing site. *Proc. Lunar Sci. Conf. 3rd*, p. 27–38.

Tera F., Papanastassiou D. A., and Wasserburg G. J. (1974) Isotopic evidence for a terminal lunar cataclysm. *Earth Planet. Sci. Lett.* **22**, 1–21.

Turner G. (1977) Potassium-argon chronology of the Moon. *Phys. Chem. Earth* **10**, p. 145–195.

Turner G. and Cadogan P. H. (1975) The history of lunar bombardment inferred from ^{40}Ar-^{39}Ar dating of highland rocks. *Proc. Lunar Sci. Conf. 6th*, p. 1509–1538.

Uhlmann D. R., Klein L. C., and Handwerker C. A. (1977) Crystallization kinetics, viscous flow and thermal history of lunar breccia 67975. *Proc. Lunar Sci. Conf. 8th*, p. 2067–2078.

Ullrich G. W., Roddy D. J., and Simmons G. (1977) Numerical simulations of a 20-ton TNT detonation on the earth's surface and implications concerning the mechanics of central uplift formation. In *Impact and Explosion Cratering* (D. J. Roddy, R. O. Pepin, and R. B. Merrill, eds.), p. 959–982. Pergamon, N.Y.

Vaniman D. T., Lellis S. F., Papike J. J., and Cameron K. L. (1976) The Apollo 16 drill core: Modal petrology and characterization of the mineral and lithic component. *Proc. Lunar Sci. Conf. 7th*, p. 199–239.

Wänke H., Dreibus G., and Palme H. (1978) Primary matter in the lunar highlands: The case of the siderophile elements. *Proc. Lunar Planet. Sci. Conf. 9th,* p. 83–110.

Wänke H., Palme H., Kruse H., Baddenhausen H., Cendales M., Dreibus G., Hofmeister H., Jagoutz E., Palme C., Spettel B., and Thacker R. (1976) Chemistry of lunar highland rocks: A refined evaluation of the composition of primary matter. *Proc. Lunar Sci. Conf. 7th,* p. 3479–3499.

Warner J. L. and Bickel C. E. (1978) Lunar plutonic rocks: A suite of materials depleted in trace siderophile elements. *Amer. Mineral.* **63,** 1010–1015.

Warner J. L., Phinney W. C., Bickel C. E., and Simonds C. H. (1977) Feldspathic granulitic impactites and pre-final bombardment lunar evolution. *Proc. Lunar Sci. Conf. 8th,* p. 2051–2066.

Warren P. H. and Wasson J. T. (1977) Pristine nonmare rocks and the nature of the lunar crust. *Proc. Lunar Sci. Conf. 8th,* p. 2215–2235.

Warren P. H. and Wasson J. T. (1978) Compositional-petrographic investigation of pristine nonmare rocks. *Proc. Lunar Planet. Sci. Conf. 9th,* p. 185–217.

Wilhelms D. E., Oberbeck V. R., and Aggarwal H. R. (1978) Size-frequency distributions of primary and secondary lunar impact craters. *Proc. Lunar Planet. Sci. Conf. 9th,* p. 3735–3762.

Wolf R., Woodrow A. B., and Grieve R. A. F. (1980) Meteoritic material at four Canadian impact structures. *Geochim. Cosmochim. Acta.* In press.

Wolfe S. H. (1971) Potassium-argon ages of the Manicouagan-Mushalagan lakes structure. *J. Geophys. Res.* **76,** 5424–5436.

Woronow A. (1978) A general cratering-history model and its implications for the lunar highlands. *Icarus* **34,** 76–88.

Papike, J.J. and Merrill, R.B., eds.
Proc. Conf. Lunar Highlands Crust (1980), p. 197-209
Printed in the United States of America

Orientations of central peaks in lunar craters: Implications for regional structural trends

Wendy Hale

Department of Geological Sciences, Brown University, Providence, Rhode Island 02912

Abstract—Impact craters are dominant features on the lunar surface and all fresh craters with diameters greater than 35 km possess central peaks. Approximately half of these central peaks (54%) have either linear or arcuate geometry and display some preferential orientation. These orientations were measured for 200 lunar craters of Pre-Nectarian to Copernican age. A pervasive regional north-south trend is identified for central peak orientations, with secondary trends at ± 20–$30°$ from this axis. No evidence for regional east-west trends is found. These orientations are shown to persist in craters formed through the last 4 b.y. of lunar history. A lack of coincidence between impact direction, as determined from ejecta asymmetries and peak orientation, argues for structural control by the target as a source of the linear peak morphology. This is supported by the regional and temporal persistence of preferred peak orientations. It is proposed that linear central peak orientations in lunar craters reflect a pervasive regional structural system which developed early in lunar history and has persisted through geologic time.

INTRODUCTION

Impact craters are pervasive landforms on the lunar surface, with central peaks occurring in all fresh craters larger than 35 km in diameter (Cintala *et al*., 1977; Wood and Andersson, 1978). Virtually all recent workers agree that these central peaks are related to the surrounding impact crater in origin. Studies conducted on central peaks have shown that these features may be classified morphologically by complexity (simple and complex) and geometry (linear, arcuate, or symmetric) (Hale and Head, 1979). Complexity refers to the apparent coherency of the central structures; a simple peak consists of a single massive mountain peak while a complex form consists of peak clusters. Geometry refers to peak orientation, symmetric ones being those which are oriented concentric to the crater center (Fig. 1); 46% of all central peaks in fresh craters display this geometry (Hale and Head, 1979). The remaining 54% may be classified as either linear or arcuate ridges or clusters which display some preferential orientation other than the crater center (Fig. 2).

Central peaks in general may form in response to the high stresses associated with shock and rarefaction waves concentrated at the sub-impact point during the cratering event (Milton and Roddy, 1972; Head, 1978). The origin of linear

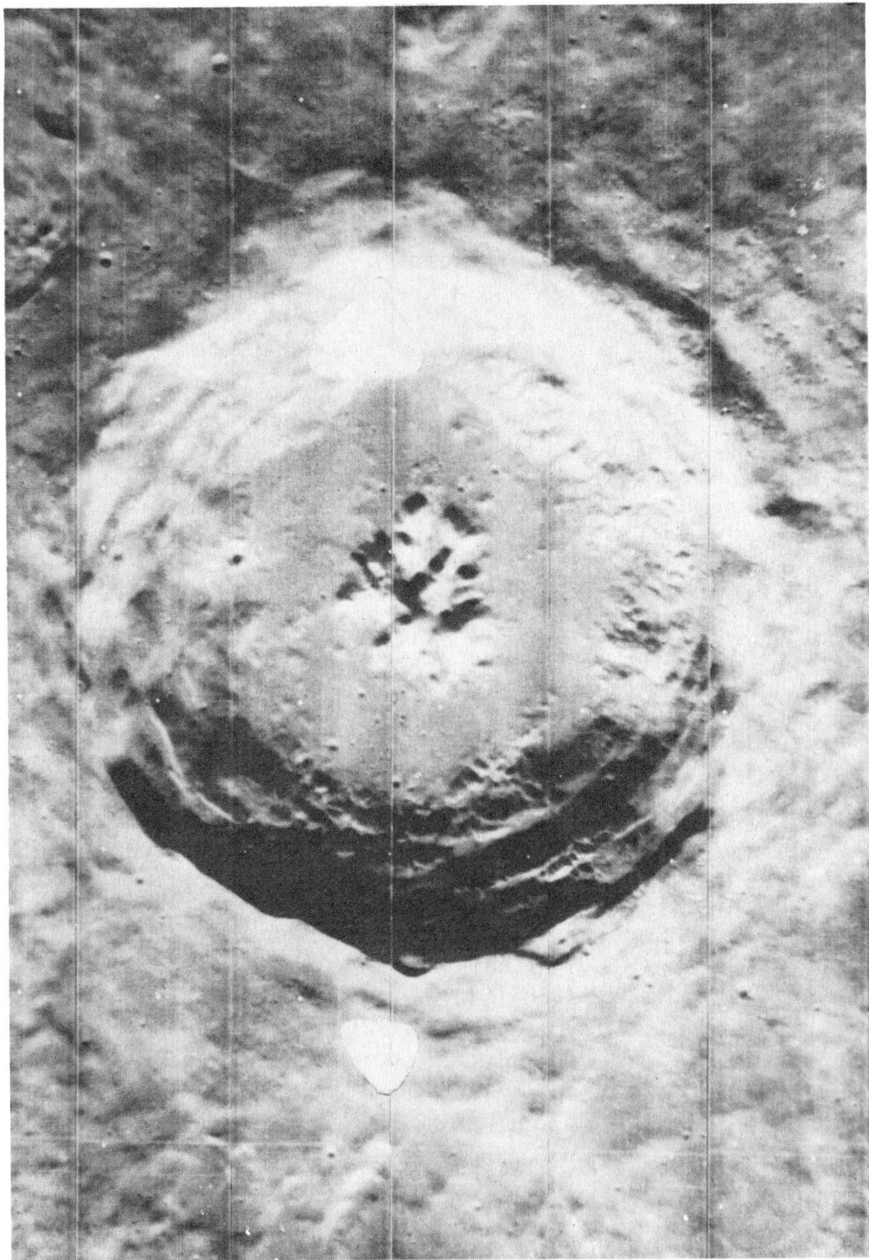

Fig. 1. Aristillus crater (diameter = 55 km). The central structure consists of a cluster of peaks oriented concentric to the crater center. It is thus classified morphologically as a complex, symmetric central peak. 46% of all central peaks are symmetrically oriented. LO IV 110H2.

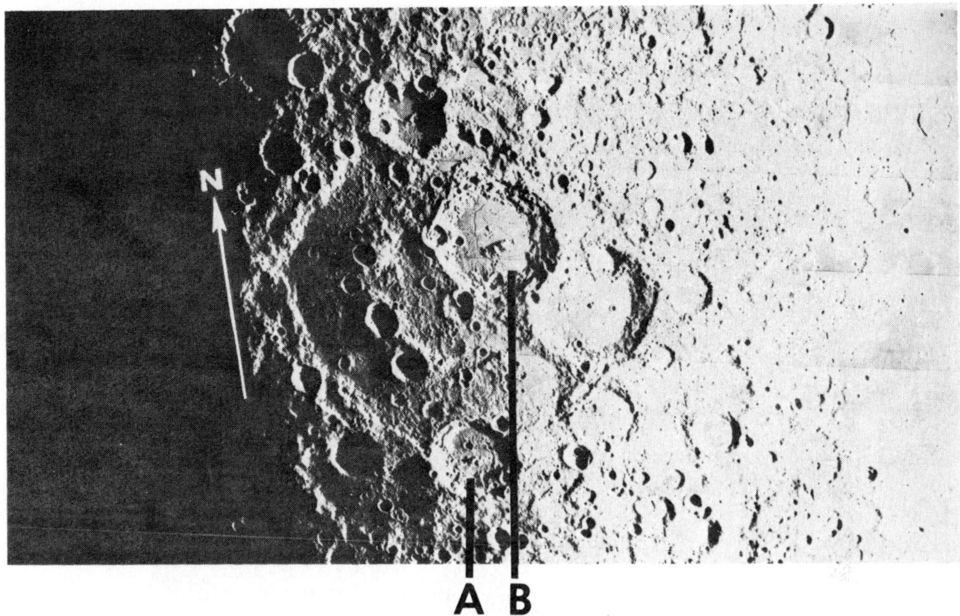

Fig. 2. Two examples of craters with linear central peaks—A) Laue (diameter = 100 km), central peak orientation = 95°, and B) Nernst (diameter = 150 km), central peak orientation = 75°. Both of these central peaks have approximately east-west trends. Note that these are clearly visible despite low sun angle. LO V 189M.

peak morphology is unclear, and two possible mechanisms have been proposed. The first suggests that impact angle may exercise strong controls on central peak geometry. Highly elongate lunar craters, which may show elongated central features, are believed to result from impacts at extremely oblique angles (Schultz, 1972; Gault, 1974; Gault and Wedekind, 1978). Less extreme impact angles are believed to produce the assymetries observed in ejecta blankets, ray patterns and melt ponds around some circular craters (Schultz, 1972; Howard and Wilshire, 1975). Thus it may be possible to produce linear central peaks in essentially circular craters by impacts at slightly oblique angles. Linear central peaks could also form as a result of some type of discrete structural control by the target. Preexisting regional structural trends could act as stress concentrators or barriers to shock wave propagation, controlling the final geometry of the central peak as well as the crater. Indeed, this is thought to be the reason for the polygonal outline of many fresh crater rims (Baldwin, 1963; Shoemaker, 1963; Roddy *et al.*, 1975). Supporting evidence for a structural origin of linear peak morphology can also be found in the frequency of occurrence of this peak geometry relative to terrain type. Linear peaks occur more frequently in craters developed on highland terrain than in those on mare targets (Hale and Head, 1979), and the lunar highlands may possess a relict texture from early tectonism, seen as a criss-crossing

Fig. 3. Moretus crater (diameter = 119 km). An example of the determination of a central peak orientation direction. The N-S axis (heavy white line) is considered to be 0° and the arrow indicates the measured orientation direction in this case, 312°. The crater is Moretus (diameter = 119 km). LO IV 118H2.

pattern of fractures and lineaments termed the lunar grid (Fielder, 1963; Strom, 1964). If such consistent structural patterns do exist in the Moon's crust, they may be portrayed as a recurring direction or set of directions of elongation for linear central peaks in lunar craters. Further, since a qualitative method of assigning ages to craters based on degrees of modification has been developed (Pohn and Offield, 1970; Head, 1975; Schultz and Gault, 1975), it may be possible to detect changes in the regional stress system over geologic time by examining changes in the preferred elongation directions of central peaks in craters of different ages. The purposes of this study were: 1) to examine a large number of craters with linear central peaks of all ages over a broad diameter range, to determine if any preferential orientation(s) exist; 2) to compare directions of elongation to downrange direction of the impacting body, where this can be determined by asymmetrically emplaced ejecta deposits, in order to define the role of impact direction in the formation of linear central peaks; and 3) to assess the degree of change in preferential peak orientations over geologic time.

DATA BASE

In order to be classified as linear, central peaks must exhibit pronounced elongation. The orientation of linear peaks in a given crater was measured in degrees from the north-south axis, on Lunar Orbiter photography (Fig. 3). For arcuate central peaks the direction of maximum elongation was measured. It has been suggested (Strom, 1964) that because of illumination angle, shadowing effects would tend to accentuate north-south trends in lunar surface features while masking east-west ones. While this may apply to subtle regional lineaments of relatively low relief, such as those of the lunar grid, it does not hold for such large-scale, high relief features as central mountain peaks in craters. This study has found examples of E-W peaks under all lighting conditions from normal to high sun elevation angles. Two such examples, at a low sun angle, can be seen in the craters Nernst and Laue (Fig. 2). On the lunar nearside, where sufficient overlapping coverage at multiple illumination angles exists, orientations were measured under differing lighting conditions. No appreciable variations due to illumination angle were observed. Further, no increase in the number of peaks with N-S orientation at high versus equatorial latitudes was observed, strongly suggesting that changing illumination azimuth has no effect on measured peak orientation. Thus no substantial effects due to lighting variations are expected in this data set, which consists of 200 central peaks occurring in craters from Pre-Nectarian to Copernican age. These include 74 from the nearside, 78 from the farside, and 48 from polar regions (latitudes $> \pm 45°$), and include craters 25–200 km in diameter.

RESULTS

A. Orientations of non-symmetric peaks

The distribution of orientations for all central peaks examined in this study can be seen in Fig. 4. In general, there exists a pervasive north-south (0°) trend for central peaks over the entire lunar surface. Also, there exist secondary trends at

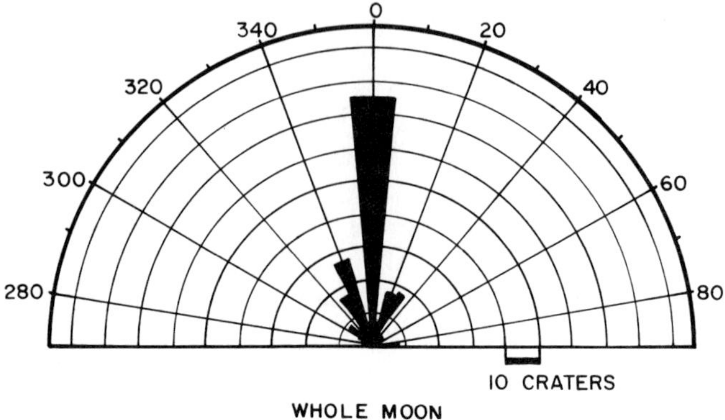

Fig. 4. The rose diagram above depicts the trends in non-symmetric central peak orientations over the entire lunar surface (n = 200). A prominent N-S trend can be seen here, as well as weaker ones at ±20–30°. This includes craters whose ages range from Pre-Nectarian to Copernican and whose diameters range from 25–200 km.

angles of plus and minus 20–30° from the north-south axis. Only 4.5% (N = 7) of all craters examined have central peaks with orientations at 90 ± 10°. Comparison of near equatorial, nearside distributions of linear peak orientations to those at high latitudes or on the farside shows no statistically important variations in these trends. Thus no evidence of a regional east-west trend for linear central peaks is found.

Extending each peak's elongation direction into a trend of regional scale and plotting these on lunar equal area maps has produced Figs. 5 and 6. The persistence of the north-south regional trend is evident here. It dominates even on the nearside (Fig. 5), where patterns appear more complex than on the farside (Fig. 6). The absence of any significant regional east-west trend (90° ±10°) is also conspicuous here. Since east-west peaks can be readily discerned where present, this distribution is thought to reflect a real absence of linear central peaks elongated in this direction. Further, there is no evidence of a strongly radial or concentric pattern for linear central peaks associated with nearby impact basins. Craters with linear peaks which occur near large basins such as Moscovienese or Korolov (Fig. 6) maintain the overall regional trends. A possible exception to this is the Imbrium basin, which may create a locally concentric pattern in central peak trends (Fig. 5). Finally, no distinct change in orientation of central peak trends occurs with increasing latitude. The 0° trend dominates equatorial (±15°), mid-latitude (±15–45°) and polar (> ±45°) data sets.

B. Impact angle effects and temporal variations

Previous workers have noted that ejecta deposits and melt ponds are not always concentrically distributed around fresh craters but may instead possess inherent asymmetries in distribution, despite their association with essentially circular craters (Shoemaker, 1962; Howard and Wilshire, 1975; Hawke and Head, 1977). Such asymmetries may result from slightly oblique impact angles with maximum ejecta and melt deposits occurring in the downrange direction (Howard and Wilshire, 1975; Hawke and Head, 1977; Gault and Wedekind, 1978). Mapping of such deposits has identified downrange directions for 55 lunar craters (Hawke and Head, 1977), of which 28 have central peaks. Of these 28 only 6, or 21.4%, have linear central peaks whose elongation directions coincide (within 20°) to the

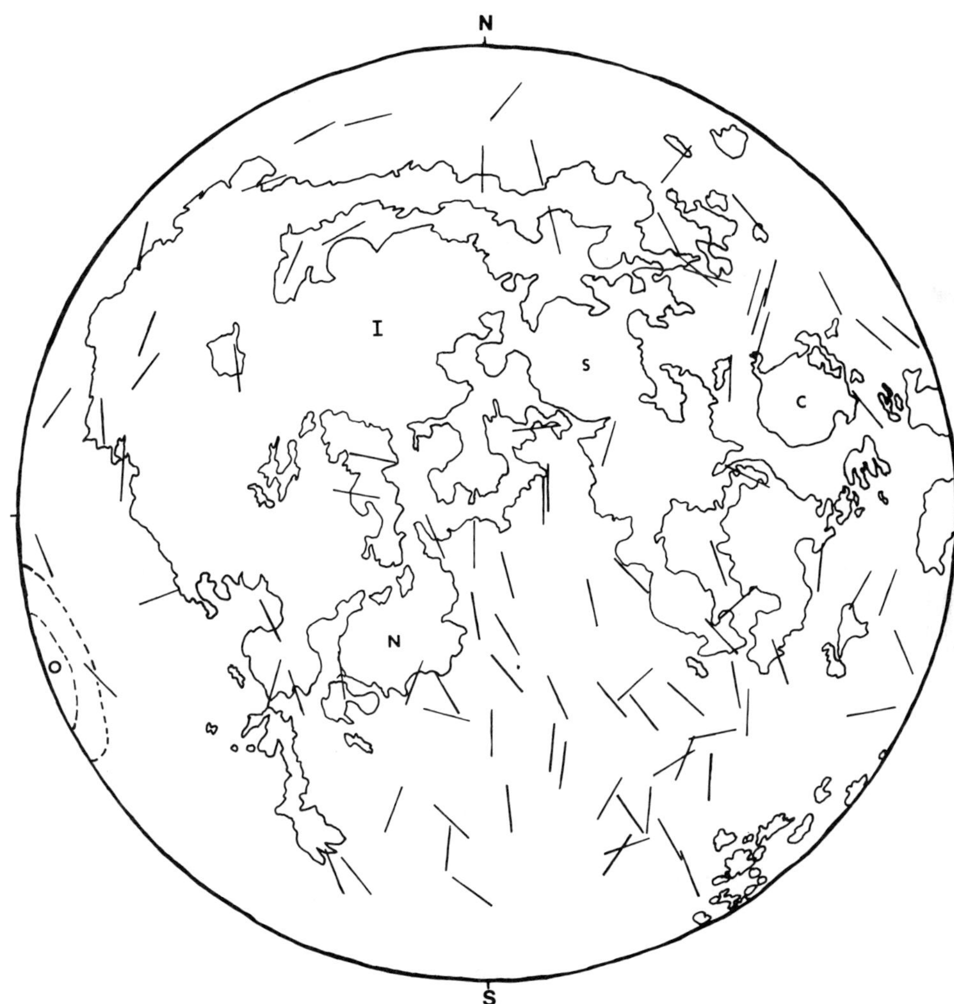

Fig. 5. Equal area projection of the lunar nearside. The elongation direction of each central peak has been extended into a trend of regional scale (short black lines). Letters indicate several large impact basins: C—Crisium, I—Imbrium, N—Nectaris, O—Orientale, S—Serenitatis. Note the more complex pattern relative to Figure 6.

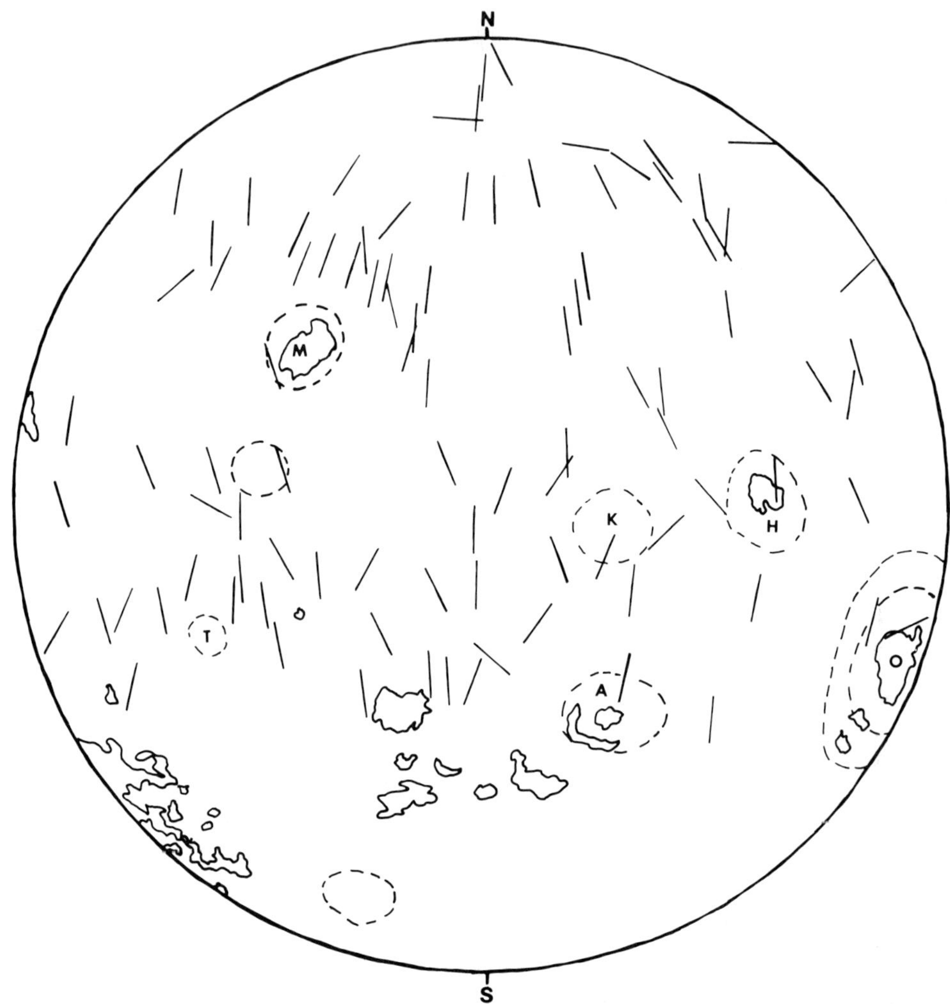

Fig. 6. Equal area projection of the lunar farside. The elongation direction of each central peak has been extended into a trend of regional scale (short black lines). Letters indicate several large basins: A—Apollo, H—Hertzsprung, K—Korolov, M—Moscovienese, O—Orientale. The crater Tsiolkovsky (T) is also indicated.

measured downrange direction. While this is approximately twice the number that would be expected for the totally random case, it is still far fewer than is required to show a strong correlation between impact angle and elongation direction. These data, plus the regional continuity of central peak orientations, thus suggest that impact angle is not the dominant, controlling factor for the elongation direction of linear central peaks in circular to polygonal craters.

The craters examined in this study range in age from Pre-Nectarian to Cop-

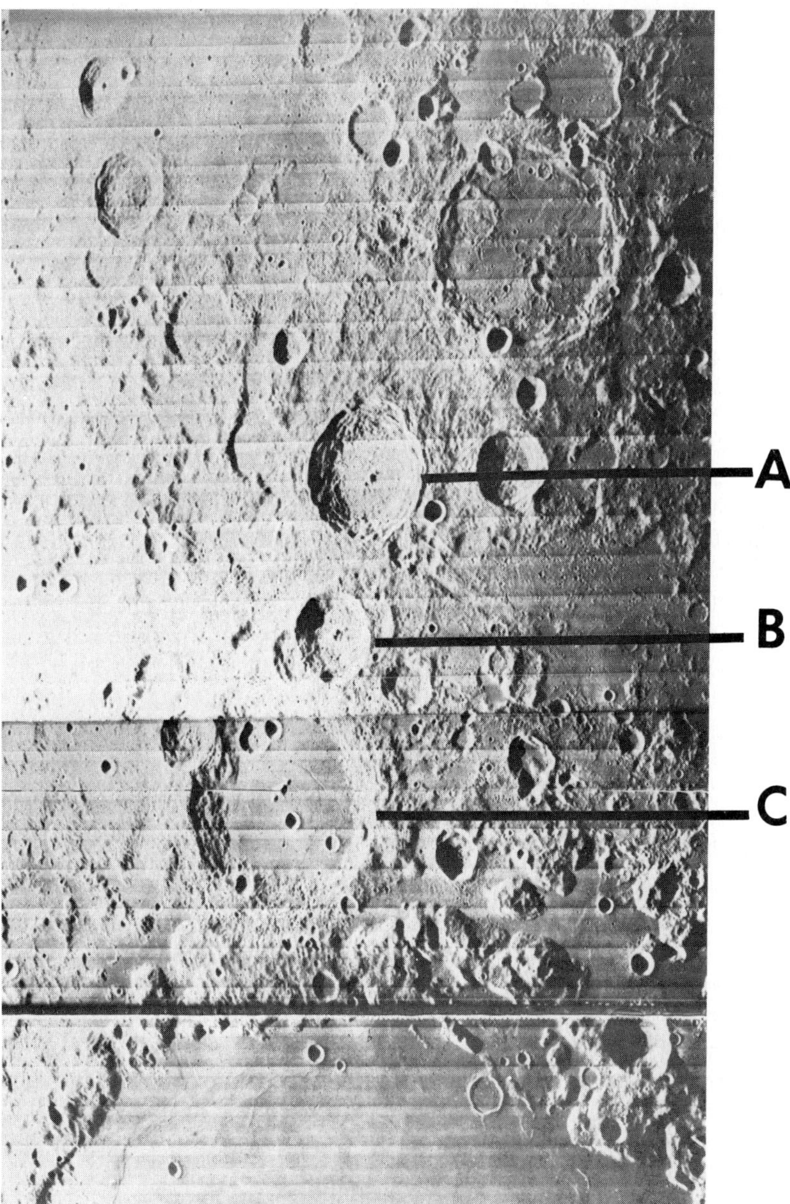

Fig. 7. A region north of Mare Crisium showing temporal persistence of central peak trends in local areas. The craters Geminus (A), Burkhardt (B) and Cleomedes (C) which have Eratosthenian, Imbrian and Pre-Imbrian ages, respectively, all show linear central peak orientations of 30°. The northern rim of Mare Crisium occurs just below Cleomedes. LO IV 192H1 and 2.

ernican, and thus span the last 4 b.y. of lunar history. The trends in central peak orientations discussed in the preceeding sections persist throughout this age range. Indeed, in local areas where craters of several different ages occur together, no appreciable change in orientation is observed over time. An example of this occurs north of Mare Crisium, where the craters Cleomedes (Pre-Imbrian), Burkhardt (Imbrian) and Geminus (Eratosthenian) occur together and all three have elongation orientations of 30° (See Fig. 7). This suggests that a strong continuity through time exists for any structural system exercising control on peak geometry.

DISCUSSION AND CONCLUSIONS

The Earth's Moon is in synchronous rotation. Studies of the regional stress patterns induced by despinning to its present state could provide important clues to the Moon's early tectonic history and the petrogenesis of its surface. Previous attempts to study lunar regional stress systems as revealed by preferred orientations in surface features have indicated the presence of a structural pattern for the Moon's nearside, termed the lunar grid (Fielder, 1963; Strom, 1964). The grid consists of intersecting lineaments which are comprised of definite linear faults, linear ridges and valleys, rilles, joints, crater chains and polygonal crater rims. Fielder (1963) noted that these lineaments intersect and cross each other, and described five distinct systems or families of structural trends, including a SW-NE trend (System A), a SE-NW trend (B), a weak N-S trend (C), a faint E-W trend (D) and a joint system radial to Imbrium basin, termed R_1. The A and B trends are the strongest in this system, are mutually orthogonal and have selenocentric position angles (measured clockwise from the lunar rotation axis) of 45° and 135° respectively. Strom (1964) identified a similar system with 3 major trends, + 45° − 55°, − 45 − 55°, and 0° from the rotation axis. A less clearly delineated trend at + 15 − 20° was also identified here, and no evidence for an E-W system was found.

The examination of 200 linear central peaks in craters of all ages has identified a pervasive, regional N-S (0°) trend for these structures. Secondary trends at ±20–30° from the rotation axis are also identified. These preferred orientations have been shown to exist over the entire lunar surface and do not appear to vary significantly with latitude or near large basins (except, perhaps, Imbrium). The dominant (N-S) central peak trend corresponds to one of the three major trends identified by Strom (1964). However, no evidence is found within the central peak data for the other two lunar grid trends at ±45–55°. Indeed, the secondary central peak trends intersect at angles of 40–60°, much less than the 90° intersections required of an orthogonal system. No evidence for an east-west trend is found in the central peak data, in agreement with earlier studies for the lunar grid.

Allen (1975) has reported that out of a data set of 580 nearside craters, 197 have central peaks with measurable trends. The study indicates that central peak trends match those of the lunar grid as defined by Strom (1964). The lack of agreement between this earlier study and the present one is thought to be due to two factors. First, over 50% of the earlier data set consists of craters less than 35 km in diameter, a size range where many factors, including the transition from simple to complex craters and from scalloped to terraced walls, may combine to make a determination of central peak structure difficult. In particular, craters below 35 km may display extended wall slumps which terminate against the central peak. These may significantly alter a directional measurement if they are incorrectly interpreted to be part of the primary central peak. Second, the earlier data set is restricted to the lunar nearside, where, the present study indicates, central peak trends are somewhat more complex than on the lunar farside (Figs. 5 and 6).

The disparity between the lunar grid patterns and those for linear central peaks may stem from the nature of the two data sets. The lineaments which comprise the lunar grid may include lineated deposits now believed to be basin ejecta (Head, 1976a) and impact crater chains (secondaries). This suggests that at least some of the features identified as structural in origin may instead be superposed on near-surface features not necessarily related to structure at depth. Central peaks in craters, however, may represent material brought up from some depth below the present crater floor by the impact event (Milton and Roddy, 1972; Head, 1978; Hodges and Wilhelms, 1978; Hale and Head, 1979). Studies of terrestrial impact (Dence *et al.*, 1977) and explosion craters (Roddy, 1977) support this view. The craters examined in this study range in diameter from 25–200 km and according to the depth/diameter relationship derived by Pike (1977) for fresh lunar craters, should have depths ranging from 2.7–5 km. Central peaks in these craters should be derived from material originally seated below such depths (Roddy, 1968). The megaregolith has an estimated thickness 2–3 km (Head, 1976b), indicating that central peaks are predominantly derived from the fractured crystalline basement layer which underlies it. The preferred orientations of linear central peaks may thus represent deeper crustal structure preserved in the crystalline basement, making them more effective indicators of regional stress patterns than surface lineaments.

A comparison of independently determined downrange directions to linear central peak orientations, indicating only a 21.4% coincidence (within 20°) between these two directions argues against angle of impact as the most important controlling influence on central peak morphology. The presence of regionally and temporally persistent trends in linear peak orientations also supports this view. Thus structural control by the target is the favored mechanism for the formation of linear central peaks. The continuity of preferred orientations in craters whose ages span the last 4.0 b.y. of lunar history therefore suggests the presence of a pervasive regional structural system which was generated early in lunar history and has persisted through geologic time. The exact origin of such a system is unknown, but could be related to initial crustal cooling or to tidal effects.

Acknowledgments—This work was carried out under NASA Grant NGR-40-002-116 from the Office of Space Science Lunar and Planetary Programs, which is gratefully acknowledged. Thanks are also extended to the National Space Science Data Center for the photographs used in this study, and to S. Bosworth for assistance in preparation of this manuscript.

REFERENCES

Allen C. C. (1975) Central peaks in lunar craters. *The Moon* **12,** 463–474.

Baldwin R. B. (1963) *The Measure of the Moon.* Univ. Chicago Press, Chicago. 488 pp.

Cintala M. J., Wood C. A., and Head J. W. (1977) The effects of target characteristics on fresh crater morphology: Preliminary results for the Moon and Mercury. *Proc. Lunar Sci. Conf. 8th,* p. 3409–3425.

Dence M. R., Grieve R. A. F., and Robertson P. B. (1977) Terrestrial impact structures: Principal characteristics and energy considerations. In *Impact and Explosion Cratering* (D. J. Roddy, R. O. Pepin, and R. B. Merrill, eds.), p. 247–275. Pergamon, N.Y.

Fielder G. (1963) Lunar Tectonics. *Quart. J. Geol. Soc. London* **119,** 65–94.

Gault D. E. (1974) Impact Cratering. In *A Primer in Lunar Geology* (R. Greeley and P. H. Schultz, eds.), p. 137–176. NASA TM X-62359.

Gault D. E. and Wedekind J. A. (1978) Experimental studies of oblique impacts. *Proc. Lunar Planet. Sci. Conf. 9th,* p. 3843–3875.

Hale W. S. and Head J. W. (1979) Central peaks in lunar craters: Morphology and Morphometry. *Proc. Lunar Planet. Sci. Conf. 10th,* p. 2623–2633.

Hawke B. R. and Head J. W. (1977) Impact melt on lunar crater rims. In *Impact and Explosion Cratering* (D. J. Roddy, R. O. Pepin, and R. B. Merrill, eds.), p. 815–841. Pergamon, N.Y.

Head J. W. (1975) Processes of lunar crater degradation: Changes in style with geologic time. *The Moon* **12,** 299–329.

Head J. W. (1976a) Evidence for the sedimentary origin of Imbrium sculpture and lunar basin radial texture. *The Moon* **15,** 445–462.

Head J. W. (1976b) The significance of substrate characteristics in determining morphology and morphometry of lunar craters. *Proc. Lunar Planet. Sci. Conf. 7th,* p. 2913–2929.

Head J. W. (1978) Origin of central peaks and peak rings (abstract). In *Lunar and Planetary Science IX,* p. 485–487. Lunar and Planetary Institute, Houston.

Hodges C. A. and Wilhelms D. E. (1978) Formation of lunar basin rings. *Icarus* **34,** 294–323.

Howard K. A. and Wilshire H. G. (1975) Flows of impact melt at lunar craters. *J. Res. U. S. Geol. Survey 3* No. 2, 237–251.

Milton D. and Roddy D. S. (1972) Displacements within impact craters. *Proc. 29th Int. Geol. Congr.,* Sect. 15, p. 119–124.

Pike R. J. (1977) Size dependence in the shape of large impact craters on the moon. In *Impact and Explosion Cratering* (D. J. Roddy, R. O. Pepin, and R. B. Merrill, eds.), p. 489–509. Pergamon, N.Y.

Pohn H. A. and Offield T. W. (1970) Lunar crater morphology and relative age determination of lunar geologic units, Part 1 Classification. *U.S. Geol. Survey Prof. Paper 700-C,* C-153-162.

Roddy D. J. (1968) The Flynn Creek crater, Tennessee. In *Shock Metamorphism in Natural Materials* (B. M. French and N. M. Short, eds.), p. 291–322. Mono, Baltimore.

Roddy D. J. (1977) Large scale impact and explosion craters: Comparison of morphological and structural analogs. In *Impact and Explosion Cratering* (D. J. Roddy, R. O. Pepin, and R. B. Merrill, eds.), p. 185–246. Pergamon, N.Y.

Roddy D. J., Boyce J. M., Colton G. W., and Dial A. L. (1975) Meteor Crater, Arizona, rim drilling with thickness, structural uplift, diameter, depth, volume and mass balance calculations. *Proc. Lunar Sci. Conf. 6th,* p. 2621–2644.

Schultz P. H. (1972) *Moon Morphology.* Univ. Texas Press, Austin. 626 pp.

Schultz P. H. and Gault D. E. (1975) Seismic effects from major basin formation. *The Moon* **12,** 159–177.

Shoemaker E. M. (1963) Impact mechanics at Meteor Crater, Arizona. In *The Moon, Meteorites and Comets* (B. M. Middlehurst and G. P. Kuiper, eds.), p. 301–336. Univ. Chicago Press, Chicago.

Shoemaker E. M., Batson R. M., Holt H. E., Morris E. C., Rennilson J. J., and Whitaker E. A. (1962) In *Television observations from Surveyor VII, Part II,* p. 9–75. JPL Tech. Rep. 32-1264.

Strom R. G. (1964) Analysis of lunar lineaments I: Tectonic maps of the Moon. *Comm. Lunar and Planetary Lab., Univ. Arizona 39,* 205–216.

Wood C. A. and Andersson L. (1978) New morphometric data for fresh lunar craters. *Proc. Lunar Planet. Sci. Conf. 9th,* p. 3669–3689.

Papike, J.J. and Merrill, R.B., eds.
Proc. Conf. Lunar Highlands Crust (1980), p. 211-231
Printed in the United States of America

Deep seated target materials in the continuous deposits of the Ries Crater, Germany

Friedrich Hörz

NASA Johnson Space Center, Houston, Texas 77058

Gordon S. Banholzer, Jr.

Department of Geology, University of Houston, Houston, Texas 77004

Abstract—A shallow core drilling program in the SW sector of the continuous deposits of the Ries yielded ≈560 m of core materials from 9 different locations. Components derived from the crystalline basement, i.e., from depth greater than 600 m in this 26 km diameter impact structure, are extremely rare and amount to ≲.1% (weight) of the total continuous deposits.

The crystalline samples of the continuous deposits are dominated by granites (≈70%), followed by gneisses (≈22%) and rare amphibolites and lamprophyres. The most highly shocked samples experienced pressures ≲40 GPa, i.e., "maskelynite" grade, observed in ≈10% of all samples; 70% of the samples were shocked between 10 and 25 GPa, while the remaining 20% were either shocked to <10 of the crystalline basement. Sedimentary clasts derived from the mesozoic strata capping the crystalline basement are virtually unshocked as are Tertiary breccia components incorporated into the deposits as "local" materials from the immediate crater environs. It follows that the rare and volumetrically insignificant crystalline clasts are the only shocked lithologies in the entire deposits. Their spatial distribution is random both with regard to radial distance and to vertical profile.

Peak temperatures for the highest shocked materials may have been 700°C; however, since >99% of the entire Bunte Breccia appears unshocked, the important conclusion is that massive, large scale continuous crater deposits are emplaced as relatively cool masses essentially at ambient temperature(s). Most of the thermal energy partitioned during cratering resides with the impact formations inside the final crater structure.

These conclusions force us to take issue with a variety of popular concepts in lunar sample analysis, such as assignment of basin formation ages, the chemistry of basin projectiles and the regular distribution of shock metamorphic effects in a crater's continuous deposits. These Ries studies also illustrate that reconstruction of a firm stratigraphic-structural framework, essential for proper interpretation of lunar highlands samples, is unfortunately still more complex than envisioned previously.

INTRODUCTION

Most lunar highland samples, especially those which may yield clues to the early crustal evolution of the Moon, were collected from impact formations which dominate the heavily cratered lunar terrains. As ejection of crustal materials occurs during impact cratering, these deposits are a mixture of laterally and vertically redistributed lithologies. In addition, most highland samples appear to have undergone multiple impact histories as revealed by, for example, complex breccia-within-breccia relationships. As a result, a firm structural-stratigraphic framework does not exist and interpretation of sample analyses in terms of regional and global geologic and petrogenetic process is seriously handicapped.

The 26 km diameter Ries Crater, Germany, is the largest terrestrial structure with substantial parts of its structural elements and ejecta preserved (Pohl *el al.*, 1977). It is, therefore, a valuable link to extend our cratering knowledge from small scale laboratory experiments, explosive cratering and small (<5 km) terrestrial impact craters to large scale planetary processes culminating in basin forming events of global consequence.

Of particular significance at the Ries is the preservation of a variety of breccia deposits: the "suevites," "crystalline breccias" and "Bunte Breccia" (Gall *et al.*, 1975; Pohl *et al.*, 1977). Since the Ries target stratigraphy is well documented (e.g., Preuss and Schmidt-Kaler, 1969), it is possible, in principle, to relate the petrographic characteristics as well as modes of occurrence of these three major breccia types to PT environments and mass transport regimes during cratering. Suevites and crystalline breccias are almost exclusively derived from the crystalline basement, which is capped by ≈600 m of Mesozoic sediments. In contrast, the Bunte Breccia is predominantly made up of the Mesozoic sediments and also includes up to 80% Tertiary sediments (Hörz *et al.*, 1977, 1979; Ostertag, 1978). Pohl *et al.* (1977) estimate approximately 22–30 km^3 of displaced crystalline material and some ≈150 km^3 displaced sediments of which 100 km^3 were deposited outside the present-day crater rim of 26 km diameter. Adding an average of 50% "local" Tertiary contributions (Hörz *et al.*, 1979) to the Bunte Breccia deposits, the following volumes for the various breccia types emerge: ≈9–11 km^3 suevite; ≈20 km^3 crystalline breccias including megablocks; ≈200 km^3 Bunte Breccia. The suevites and crystalline breccias occur predominantly, although not exclusively, within the crater depression and on the rim vicinity, while the Bunte Breccia resides largely outside the rim area. As a consequence the Bunte Breccia makes up >90% of all crater deposits outside the present-day cavity and it constitutes the "continuous deposits" around the crater structure (Oberbeck, 1975), or the "ejecta blanket" in photogeologic studies of lunar craters.

A shallow drilling program in the best preserved Bunte Breccia deposits towards the SW yielded a cumulative core length of 560 m from 9 localities, illustrated in Fig. 1 (Hörz *et al.*, 1977). Although the major objectives of the drilling program were concerned with the incorporation of local materials into the breccia deposit and the consequences thereof for the emplacement mechanism(s) of large scale crater deposits (Oberbeck, 1975; Chao, 1977), investigations of which are not yet completed, we report here on some results on relatively rare, crystalline clasts within the Bunte Breccia (Banholzer, 1979).

Such crystalline fragments are demonstrably from depths >600 m. They are the most deepseated target materials in general and specifically in the Bunte Breccia. Therefore their radial and vertical position within the Ries' continuous deposits assumes significance when addressing lateral and vertical redistribution of planetary crustal and surface materials during large scale impacts. During

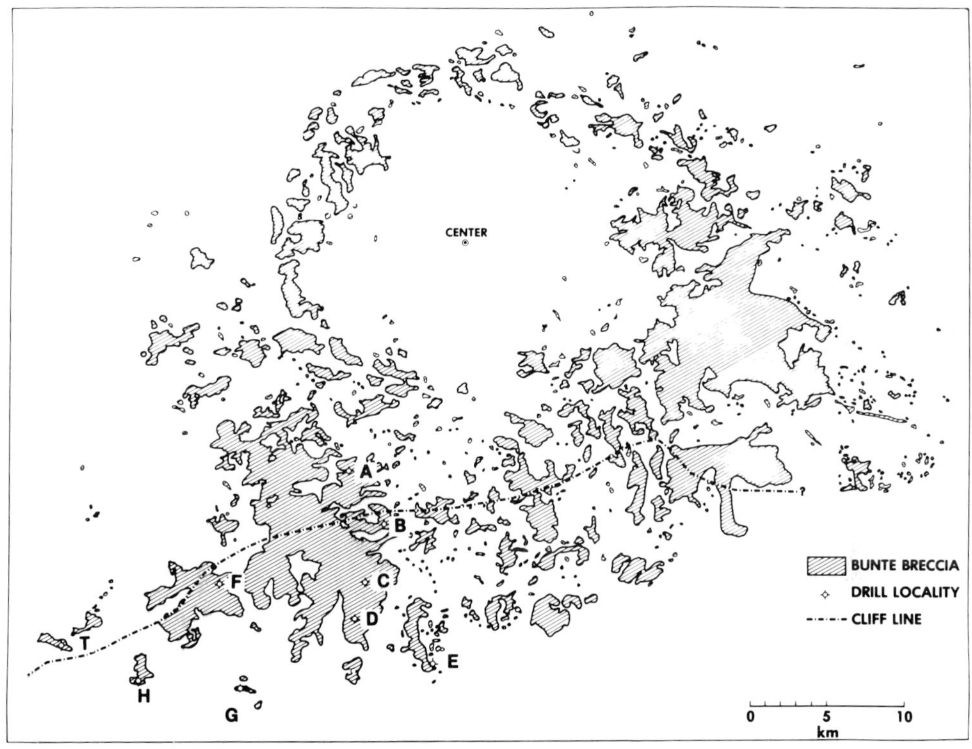

Fig. 1. Distribution of Bunte Breccia deposits around Ries Crater and locations of 3 drill holes, labeled A through I with increasing radial range from impact center.

cratering the vertical and lateral position of any stratigraphic element relative to the expanding shock front defines specific pressure-temperature conditions. Therefore specific shock histories manifested in the dislodged ejecta may also potentially be used to reconstruct stratigraphic information. We therefore determined the degree of shock experienced by the Ries samples. In brief, the well established target stratigraphy (e.g., Preuss and Schmidt-Kaler, 1969), the relatively well known structural elements of the crater (Pohl *et al.*, 1977), and the firm spatial control of the cores samples (Hörz *et al.*, 1977) permit findings in the Ries ejecta to be related to stratigraphic-structural relationships of the target, an exercise all too often attempted on the moon via relatively uncertain, if not speculative, model assumptions.

GENERAL METHODS AND RESULTS

The surfaces of the cylindrical breccia cores were macroscopically mapped via transparent overlays for all clasts >1 cm; materials <1 cm were defined as "matrix." The relative surface areas obtained by weighing of corresponding paper clippings (= planimetric %) were converted via appropriate densities into absolute weight %. The total weight % for crystalline materials varied from .003 to .13% for individual cores. The average spatial frequency for individual cores ranged from .04 to .55 crystalline clasts per core meter. Therefore deepseated crater materials contribute only a minute amount to the continuous deposits. This is an extremely important observation of far reaching consequences in lunar sample analyses and interpretations.

As illustrated in Fig. 2, the spatial distribution of the deepseated clasts appears to be random, both in a vertical as well as a radial sense. Every core contained some crystalline material although core locations (Fig. 1) spanned a considerable radial range. While specific concentrations (N/m) of crystalline clasts may be observed in vertical profile (Fig. 2; cores A, B, F), there are no systematic trends as to their exact location, at least within the limited number of samples available in this study. The vertical distribution of deepseated ejecta is irregular in the Ries' continuous deposits.

All crystalline clasts >1 cm were dislodged from the cores and thin sectioned; the total population consists of 98 specimens. All samples were weathered although to variable degrees, because they have resided in a clay-rich, wet, near-surface environment since deposition some 15 million years ago. Twelve specimens were so badly weathered that they deformed plastically during dislodging and since their primary textures were largely destroyed, they were unsuitable for petrographic analysis. As a consequence, the present study is based on only 86 samples and even many of these were marginally preserved. The small absolute number of samples, as well as their weathered state made it impractical to follow the very detailed subclassification of crystalline Ries ejecta of Graup (1975, 1977). We only differentiated between the principal lithologies: granites, gneisses, amphibolites and lamprophyres. Their respcctive number frequencies are 60, 20, 3 and 3 specimens. Thus the sample suit is dominated (≈70%) by granitic lithologies.

Our "granites", where Graup's classification could be applied, also include diorites and slightly metamorphosed granite gneiss. They are composed typically of quartz (20–40%), potassium feldspar (20–50%), plagioclase (15–25%) and biotite (1–10%), with apatite and opaques as ubiquitous accessories. Secondary alteration is pervasive with feldspars predominantly converted to sericite and clay minerals (kaolin?) and the biotites converted to chlorite as well as iron oxides. Frequent replacement of all major phases by calcite is observed. The "gneisses" could not be distinguished into ortho- and paragneisses and other gneiss varieties described from larger and better preserved sample suites (see Graup, 1975, 1977). They contain quartz (20–40%), feldspars (30–50%), biotite (10–25%), garnet (0–2%), cordierite (0–1%) as well as accessory apatite, zircon and opaques. Interestingly, no sillimanite gneisses appear to be present in our Bunte Breccia sample suite. The 3 amphibolites are extremely weathered and display a weak to moderate foliation. Most of the hornblende and plagioclase are secondarily altered into chlorite and kaolin(?) and significant replacement by fine grained calcite has occurred. Assignment of the 3 "lamprophyres" as minettes is somewhat tenta-

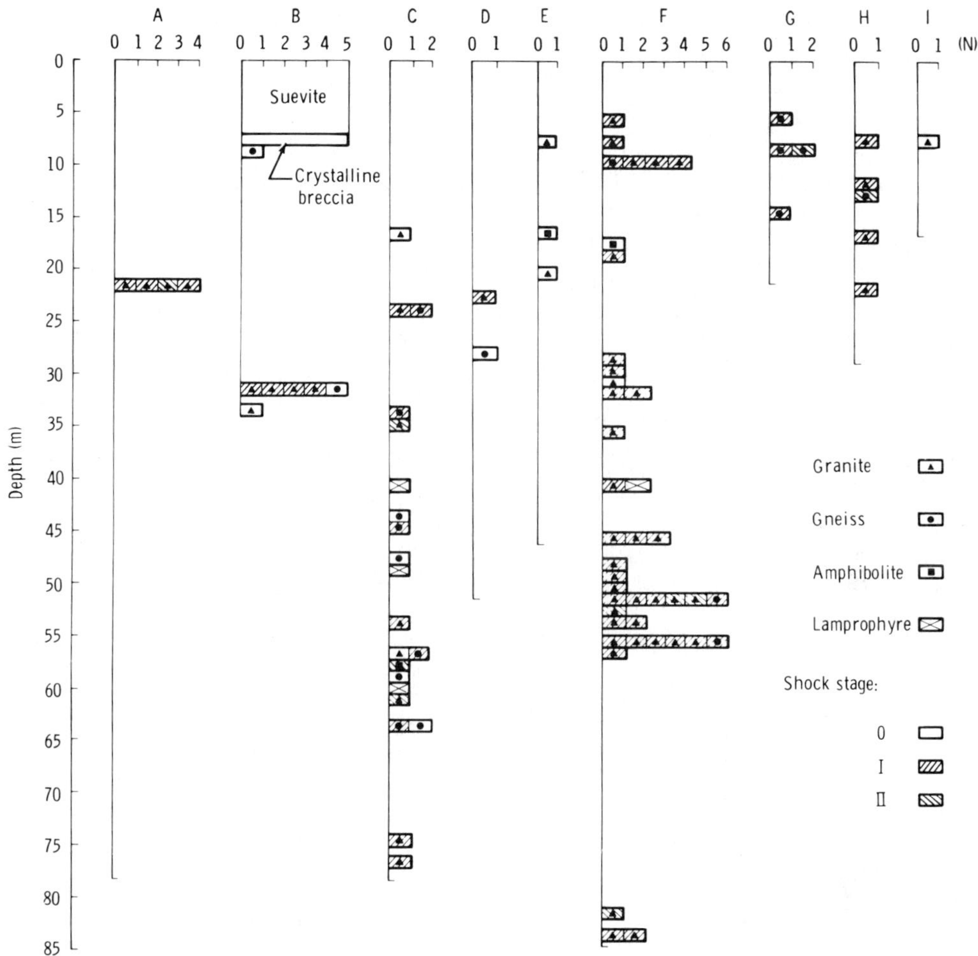

Fig. 2. Spatial distribution of crystalline clasts and the degree of shock in 9 Bunte Breccia cores, which are arranged along a hypothetical, radial traverse. Scale on top of core gives number of samples (N). "Depth" refers to position in core measured from the local top surface. (See Hörz *et al.*, 1977). Each box represents one sample.

tive, but suggested by their seriate biotites, chloritized pyroxene and an aphanitic ground mass containing anhedral feldspar; prominent accessories are apatite and quartz.

The description of shock metamorphic effects was predominantly obtained via detailed plane and universal stage observations of quartz. The feldspars were frequently too weathered to assign reliable shock pressure levels. The shock classification scheme of Stöffler (1972, 1974), which is largely based on laboratory shock recovery experiments of quartz (Hörz, 1968; Müller and Defournaux, 1968) and feldspars (Stöffler and Hornemann, 1972; Gibbons and Ahrens, 1977), was adopted.

Shock stage 0: 0 to ≈10 GPa: Quartz contains sparse to extremely abundant irregular, intragranular fractures that frequently form anastomosing branches. Some grains may also contain single or multiple subplanar fractures that may cross grain boundaries, although no significant offsets within the grains or at grain boundaries are observed. Some grains contain large planar fractures which may or may not be controlled by rational lattice planes; they represent the "cleavages" reported by Hörz (1968). Kinking of biotites is common. Based on detailed studies of similarily deformed crystalline materials from the deep drill core, "Nördlingen 1973," which penetrated ≈600 m of dislodged basement, we follow the conclusions of Graup (1977) and Chao and El Goresey (1977) that most of these deformations, although not necessarily diagnostic for shock, were caused by the Ries event rather than by tectonic activity.

Stage I: 10–25 GPa: Quartz grains display diagnostic shock features in the form of single or multiple sets of "planar features," irregular to patchy extinction, and domains of decreased birefringence. Similar features occur in the feldspars. Universal stage measurements on quartz yielded the typical concentrations along crystallographically controlled planes, predominantly the $\{10\bar{1}3\}$ and $\{10\bar{1}2\}$ rhombohedra. The penetrative "cleavages," though less prominent, cluster around the unit rhombohedron $\{10\bar{1}1\}$ and the basal plane $\{0001\}$ (Banholzer, 1979).

Stage II: 25–≈40 GPa: Stage II is characterized by complete solid state vitrification of the framework silicates, resulting in diaplectic quartz and feldspar glasses (= maskelynite). No inter- or intragranular melt is formed. Classification of Bunte Breccia samples into this stage rests almost exclusively on the presence of diaplectic quartz glass; the presumably coexisting maskelynite is particularly prone to weathering as well as to replacement by calcite, and genuine diaplectic feldspar glasses were rarely observed. Crystalline clasts shocked higher than stage II, i.e., ≳400 kb, were not observed in our collection, a finding consistent with previous studies (Schneider, 1971; Abadian, 1972; Graup, 1975, 1977). In 3 samples, however, we observed veins of colorless to light brown glass, a few mm long and up to .5 mm wide. The original glass is largely weathered and replaced by calcite and clay minerals. The host rocks of these veins, however, belong clearly to shock Stages I or II. Similar veins were described by Dressler *et al.* (1969).

It is important to recapitulate that we had to reject 12 specimens, constituting ≈12% of the total population, because of their advanced degree of weathering. Diaplectic feldspars and diaplectic feldspar glass in shock Stage II samples are observed to be particularly sensitive to secondary alterations. It is possible that many or all of these rejected samples may belong into shock Stage II. We are, however, confident that they were not shocked ≳400 kb, because macroscopic inspection (including hand lens) prior to dislodging from the cores revealed, that their primary igneous or metamorphic textures were relatively well preserved. At pressures ≈400 kb, framework silicates tend to melt selectively, in particular feldspar, thus causing gradual destruction of the primary textures (Stöffler, 1972); this was not the case for our rejected samples.

Following the criteria described above, all clasts were classified into lithologic types and their respective degree of shock. The results are presented in Fig. 2, together with field and laboratory data regarding specific location. Detailed study of Fig. 2 reveals that there are no systematic trends, either in vertical profile or along a hypothetical, radial traverse, in the distribution of rock types. Importantly, there are also no systematic radial and vertical trends in the distribution of the degree of shock. For reasons of clarity and in support of the above conclusions, Figs. 3 and 4 illustrate the petrographic data in normalized form.

DISCUSSION

Crystalline lithologies, their spatial distribution and shock effects in the Ries

Abadian (1972) and, in particular, Graup (1975, 1977) provide the most detailed descriptions of crystalline breccias, including attempts to reconstruct the complex structural relationships of the crystalline basement. As best as can be determined it appears that the crystalline basement immediately below the Mesozoic sediments is formed by a variety of granitic complexes that have intruded the older gneisses. The reconstruction of the crystalline basement according to Graup (1975, 1977) is reproduced in Fig. 5.

This reconstruction is based on structural relations in the crystalline basement outside the crater, in particular the regional trends of the Moldanubikum some 80 km to the east. The spatial distributions of the principal lithologies within the crater and the ejecta appear to reflect such regional trends. In addition, this reconstruction also includes consideration of the degree of shock as an estimate of the proximity to the impact center, as well as the preimpact position of megaclasts relative to the propagating shock front as determined by kink band orientations in biotites (Graup, 1975).

Figure 6 compares the relative frequencies of granites, gneisses and amphibolites plus lamprophyres in the suevites, crystalline breccias and crystalline megaclasts with the relative frequencies observed in the Bunte Breccia cores. The average of the crystalline clasts in our cores from the SW quadrant of the continuous deposits is close to the average of Graup's megaclasts for the entire Ries. The Bunte Breccia distributions are distinctly different from those in suevites which are dominated by gneisses and which also contain more mafic rocks. A $\approx$1200 m deep drill hole, "Nördlingen 1973," some 3.5 km west of the crater center, penetrated the crystalline basement at $\approx$600 m depth; it produced almost exclusively metamorphic gneisses. These findings are compatible with Graup's reconstruction of the basement which entails the granites to occur predominantly at or near the basement surface as somewhat elongated, slab like bodies, similar to those observed in the Moldanubikum. The suevites apparently originated from somewhat deeper levels containing largely metamorphic rocks. Following structural uplift, these metamorphic lithologies now dominate the central crater area

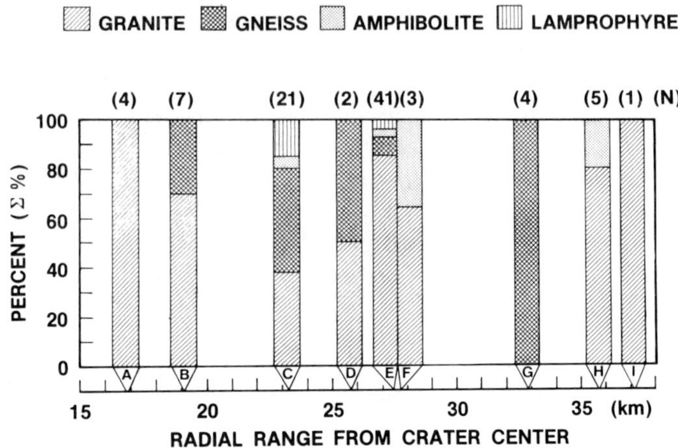

Fig. 3. Relative frequencies of lithologies and their relation to radial range. Note irregular distribution, i.e., cores may exclusively contain gneisses or granites; also note that granites dominate the sample suite.

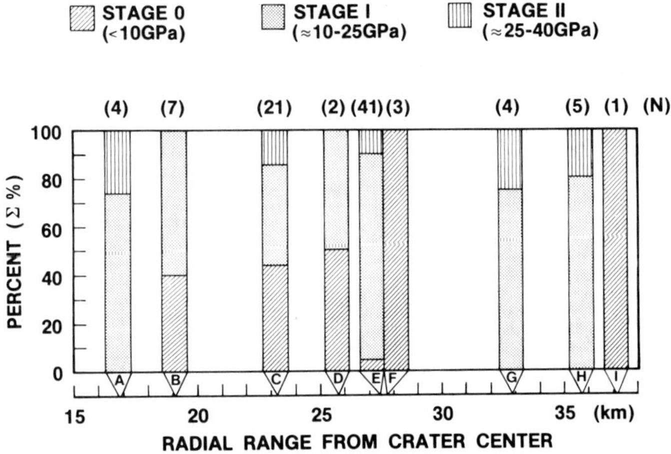

Fig. 4. Relative frequency of various shock levels and their relatin to radial range.

below the genuine fallback materials (Pohl *et al.*, 1977). Therefore the lithologic makeup of the basement clasts in the Bunte Breccia cores appears to reflect the highest stratigraphic, granite-rich, levels of the crystalline basement. The proponderance of granitic inclusions in the Bunte Breccia deposits is a Ries-wide phenomenon (Graup, 1975, 1977) and not confined to the SW-sector, the general location of the drill cores.

The distribution of shock levels in our Bunte Breccia cores is also compatible with the findings of others (Schneider, 1971; Abadian, 1972 and Graup, 1975,

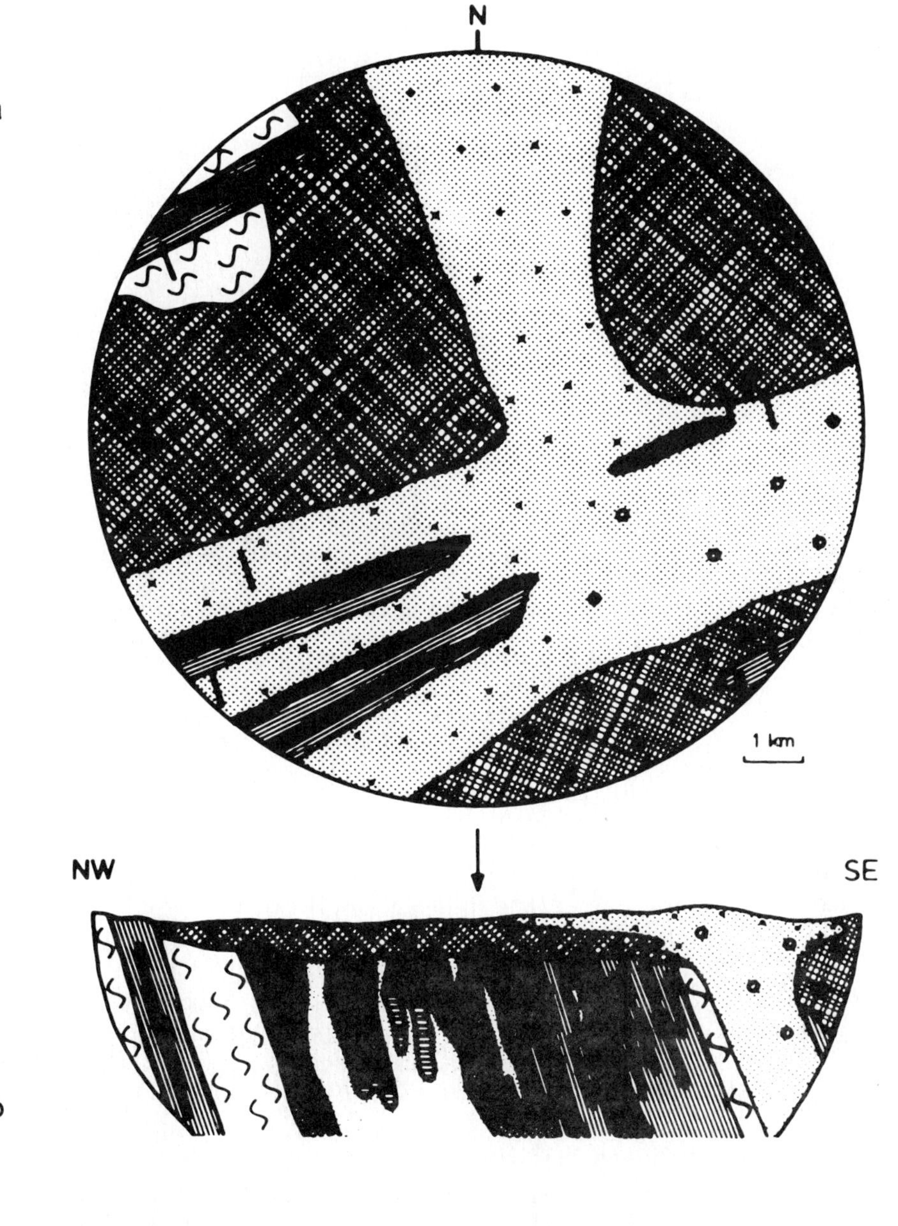

Fig. 5. Schematic reconstruction of Ries crystalline basement according to Graup (1977). a. Plan view, b. cross-section (exaggerated vertical scale). A = cordierite-sillimanite gneiss; B = biotite-plagioclase gneiss; C = mixed gneiss; D = orthogneiss; E = amphibolite; F = biotite-rich granite; G = biotite-poor granite; H = lamprophyre.

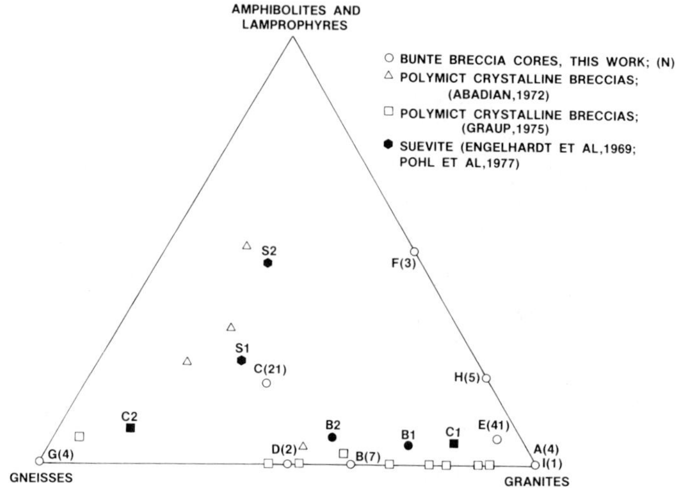

Fig. 6. Distribution of principal crystalline lithologies in various impact deposits of the Ries. Circles: this work; triangles: polymict breccias from outside and inside present rim (Abadian, 1972); squares: polymict breccias outside present rim; representative examples only; a total of 293 localities were investigated, many of which are 100% granite (Graup, 1975). Solid symbols represent averages: B_1: average of 86 Bunte Breccia samples, i.e., total cores; B_2: average of 31 additional samples, 4–10 mm in size, recovered during sieve analyses of Bunte Breccia cores and not further described in this report; C_1: average of all polymict and monomict breccias, especially including megaclasts (Pohl *et al.*, 1977; after Graup, 1975); C_2: average of crystalline encountered in deep drill hole Nördlingen 1973 (Pohl *et al.*, 1977; after Bauberger *et al.*, 1974); S_1: average of 5 suevites outside crater rim and S_2: average of fallback suevite in "Nördlingen 73" (Stöffler *et al.*, 1977).

1977). The highest shock levels found in the Bunte Breccia are ≲400 kb. Higher shock levels occur only in the suevite breccias which contain all levels of shock and in particular whole rock melts in the form of glass bombs. Figure 7 compares the distribution of shock levels in various deposits of the Ries crater. Note the higher shock levels of suevite even after considering only the moderately shocked clasts of Stages O-II. The core samples are generally less shocked than the crystalline breccias (Abadian, 1972 and Pohl *et al.*, 1977). They appear slightly more shocked than the "megaclasts" of Graup (1975) which are almost exclusively confined to the "inner ring" rim zone (Pohl *et al.*, 1977).

Schneider (1971) has performed the most detailed study to date on shock effects in the Bunte Breccia *matrix*, mostly from locations other than the core sites. The most highly shocked material was extremely rare, diaplectic quartz glass of Stage II. Even Stage I quartz is volumetrically insignificant. Most Bunte Breccia matrix-components are genuinely unshocked.

During our core studies we also investigated some Triassic and Lower Jurassic sandstones (20 samples) and found no evidence of shock. Shocked clasts of the sedimentary target strata are also extremely rare even in the suevite deposits.

The fact that most sedimentary lithologies are virtually unshocked (e.g., Englehardt *et al.*, 1969) and particularly those in the Bunte Breccia, has puzzled Ries investigators for a long time.

Adding to these "primary" crater ejecta the volumetrically significant Tertiary components of "local" origin (Schneider, 1971; Hörz *et al.*, 1977, 1979), then $\ll 1\%$ of the Ries' continuous deposits are "shocked" by the optical criteria used here. The shock deformation displayed by our crystalline clasts comprises for all practical purposes the presently traceable shock history of the entire deposit. While future advances may be made to establish and fine tune diagnostic criteria for weakly shocked rocks (e.g., $\ll 10$ GPa), the present optical criteria are sufficiently diagnostic to exclude shock levels >40 GPa for Bunte Breccia components. This limit on the maximum shock levels appears firm; we are also confident that $\ll 1\%$ (by weight) of the entire deposit is shocked to pressures between 10 and 40 GPa. The implications of these observations regarding planetary cratering and lunar sample interpretations will be discussed below.

Comparison with other terrestrial impact craters

Equivalent data for continuous deposits do not exist for any other terrestrial impact structure. Thus comparison is limited to small scale laboratory experiments into Ottawa quartz sand (Stöffler *et al.*, 1975). The following is a summary of their findings: craters of ≈ 30 cm diameter (D) and 6–7 cm depth (t) produce ejecta blankets that originate from the following target positions: materials from the upper target strata shallower than .15 t can be found anywhere in the ejecta

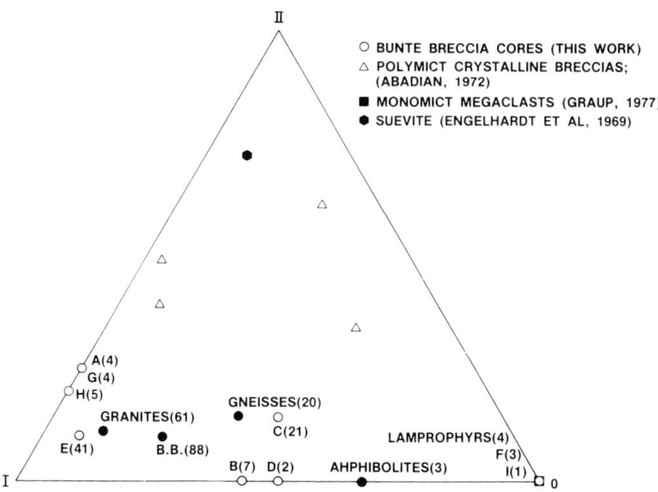

Fig. 7. Distribution of shock metamorphic grades O—II in various Ries deposits. Solid symbols represent averages (B.B. = total population of this Bunte Breccia study; the suevite average excludes all shock stages >II).

blanket; materials from strata deeper than .28 t are deposited inside 2 crater radii and materials from deeper than .33 t do not leave the crater cavity at all. Shock metamorphosed particles of shock Stages I–V, i.e., including impact melts, amount to ≲4% of the total displaced mass and 85% of this shocked material is concentrated in the crater itself. Along a radial traverse within the ejecta the shocked species increase in frequency relative to the unshocked, dominating ejecta mass. Their absolute amount decreases, however, simply because the total mass distribution decays with radial range. The most severely shocked materials, i.e., impact melts, originated from above .15 t and all observable shock effects were generated in the top ≈30% of the target (<.3 t). Furthermore within a radial profile, the relative frequency of severely shocked species increases in comparision to modestly or mildly shocked materials.

Thus the small scale experiments of Stöffler *et al*. (1975) have some interesting analogies, but also significant differences with the Ries results as already pointed out, in part, by Stöffler *et al*. (1975), Pohl *et al*. (1977), and Chao (1977). The major analogies and differences are as follows:

Analogy 1: Only a small fraction of the total displaced mass is recognizably shocked: The small scale Stöffler *et al*. (1975) experiments yielded approximately 4% of the total displaced mass at shock levels >15 GPa. According to Pohl *et al*. (1977), the ejected crystalline basement in the Ries crater amounts to ≈15% of the total ejected mass. With the exception of suevite components, all of these materials are shocked <40 GPa, if shocked at all. Thus a somewhat intuitive estimate yields ≲5% of the total displaced mass in the Ries to be shocked between 10 to 40 GPa.

Analogy 2: Only a small amount of impact melt is generated: The small scale laboratory experiments of Stöffler et al. (1975) yielded <.1% impact melt. The total melt produced during the Ries event is estimated at ≲.2% relative to total displaced mass (Pohl *et al*., 1977). These results differ somewhat from Dence *et al*. (1977) who postulate somewhat larger amounts of melt, 1–5%, for some of the Canadian structures. The latter data are probably representative for impacts into relatively homogeneous, crystalline targets. The exceedingly small melt volume generated in the laboratory craters may be largely due to relatively low impact velocities (≈5–7 km/sec) and the small melt volume in the Ries may be caused by the layered nature of the target, including sedimentary strata containing abundant volatiles. The effects of such lithologies (e.g., carbonates, H_2O saturated clays, etc.) on decreased melt formation has recently been theorized by Kieffer and Simonds (1980).

Analogy 3: Location of shocked materials and impact melts: The small scale laboratory craters had 85% of all shocked components, including melts, residing inside the crater rim, i.e., inside the final crater cavity. Although equivalent measurements do not exist for any natural crater, qualitatively similar conclusions apply to the Ries: the majority of suevite and crystalline breccias reside inside the 26 km rim zone. Similarly, the majority, if not all, of the impact melt studied at Canadian craters, appears to be confined to the actual crater. Photogeologic studies on lunar craters with specific emphasis on spatial distribution of impact melts (Hawke and Head, 1977) yield similar conclusions: impact melts are concentrated inside the crater cavity and only relatively small fractions may be deposited beyond the rim area. The Stöffler *et al.* (1975) small scale experiments together with previous and present Ries investigations indicate that the areal distribution and specifically the total abundance of more moderately shocked components appears to be similarily restricted. The bulk of these materials remains inside the crater cavity and only a small fraction may participate in the make-up of the continuous deposits. Present field and laboratory evidence indicates that the majority of the continuous deposits are for all practical purposes composed of unshocked or mildly shocked materials.

The *major difference* between the Ries findings and the Stöffler *et al.* (1975) experiments appears to be the source area of the highly shocked materials. In the small scale experiments the impact melts derive from the upper 15% of the target (<.15 t) and recognizable shock effects occur as deep as ≈.3 t of the final crater depth. The shocked specimens in the Ries derived from the crystalline basement >600 m deep. Assuming ≈1.5 km for the depth of the transient Ries cavity (Pohl *et al.*, 1977; Hörz and Ostertag, 1979; Pohl, 1979), this corresponds to >.40 t. The impact melts contained in the suevite—following Graup's arguments—may well originate from depth >.5 t. This seeming difference between small and large scale craters is poorly understood. Increasing the depth of the transient Ries crater to 2.5 km, an assumption consistent with present evidence, would make the Ries findings more analogous to the small scale experiments. Production of melts close to the target surface is also to be expected from relatively low velocity experimental impacts (6–7 km/sec) versus "natural" velocities at ≈15 km/sec. This difference may also in part be due to the Ries cap rock consisting of 250 m of limestones (≈.1 to .15 t) and to erosion since formation of the Ries, some 15 million years ago. The factual evidence in the field is that all sedimentary breccia components of the Ries are mildly shocked at most, if shocked at all. Recognizable shocked materials derived from depth >600 m.

A situation similar to the Ries may exist at Meteor Crater, Arizona, where most of the shocked material consists of Coconino sandstone, at least 90 m below the original surface (Roddy, 1978); assuming a total crater depth of ≈180 m (Roddy, pers. comm.), this implies that the bulk of shock metamorphosed samples came from deeper than ≈.5 t. The cap rock at Meteor Crater is also essentially a stratum of limestones (Kaibab); it is abundant in the rim area and continuous deposits, and the bulk of it appears unshocked.

Although comparison of the new Ries data with other terrestrial craters is unfortunately limited, the following generalization is suggested: continuous, large scale crater deposits consist almost exclusively of materials that are unshocked by present criteria. Although such materials may be mechanically comminuted, their pristine textural and specially chemical characteristics are unaffected. Most of the comminution may occur under tensile failure and will not require stresses in excess of ≈.15 GPa (Cohn and Ahrens, 1979). The recognizably shocked materials, which may have undergone textural and chemical modifications, constitute a very small fraction of the ejecta. Genuine melts are exceedingly rare to nonexisting in the continuous deposits outside the rim, especially in the distal parts. Most of the continuous deposits are unshocked.

The specific shock histories for the continuous deposits of the Ries yielded some rare, volumetrically insignificant, clasts which may have experienced peak pressures ≲40 GPa. Such peak pressures in quartz correspond to ≈700°C peak temperature according to Raikes and Ahrens (1980); corresponding temperatures for 30, 20 and 10 GPa shocks are ≈500°, 400° and 200°C respectively. Therefore the important conclusion is reached, that only modest amounts of thermal energy are partitioned into a crater's continuous deposits. The great majority of these deposits is emplaced as a relatively cool mass essentially at ambient temperatures. This conclusion is valid not only for primary ejecta of virtually all stratigraphic target horizons, but especially for the dominant local components which are dislodged from the substrate and incorporated into the deposits by relatively modest kinetic forces (Oberbeck, 1975; Hörz *et al.,* 1977).

While we are confident that these volumetric generalizations are substantially correct, there is however possible complication related to the sequence of ejection and thus sequence of deposition: the most highly shocked materials in the Ries, producing the suevite deposits, were ejected last. Suevite unconformably overlays Bunte Breccia, commonly displaying knife edge sharp contacts. Suevite occurs as far as 8 km from the crater rim in the form of relatively thin deposits and occupies <1% surface area of the Ries deposits beyond the rim. It is unknown to what degree the present distribution is affected by erosion. The possibility that suevite originally formed a significantly more continuous layer cannot be excluded and in particular, that it extended to larger radial ranges. Nevertheless geophysical measurements severely constrain the total amount of displaced crystalline basement and thereby the total volume of suevite; most of it is still contained in the crater fill. Therefore our volumetric generalizations are not seriously affected. A more extensive, albeit thin, suevite cover, may however significantly affect a hypothetical "sample" mission (or a remote sensing experiment) to the original Ries Crater, because emphasis would be placed on a volumetrically unrepresentative surface deposit. Clearly additional field work is required in the Ries crater to more quantitatively assess the total amount of erosion since its formation some 15 million years ago.

Implications for lunar highlands

As stated in the introduction, refined understanding of cratering mechanics and shock effects is essential for reconstruction of a structural-stratigraphic framework of the Apollo sample suite, i.e., to define the source area(s) of specific samples and to characterize petrogenetic processes within that source as opposed to those processes superimposed by impacts.

Our crystalline Bunte Breccia samples allow a better evaluation of the distribution of shocked species in a crater's continuous deposit and they confirmed the long established fact that the continuous Ries deposits contain little shocked material; >99% of the Bunte Breccia appears unshocked (Hüttner, 1969; Englehardt *et al.*, 1969; Schneider, 1971).

We endorse the early conclusions of Dence and Plant (1972) and Chao (1973) that the thermal source of the Apollo 14 rocks, i.e., Fra Mauro formation, is unrelated to the Imbrium event. This has also been postulated more recently and in more detail by Stöffler *et al.* (1977) and Simonds *et al.* (1977), who basically argue that the highly variable modes and textures of Apollo 14 breccias necessitate a variety of cratering events. Photogeologic reinterpretation of the Apollo 14 landing site in the light of Oberbeck's secondary cratering hypothesis is consistent with the above sample interpretations (Morrison and Oberbeck, 1975; Head and Hawke, 1975). There are numerous local and regional craters which are candidate source areas for the Apollo 14 rocks. Their final emplacement, at the Apollo 14 site, however, was accomplished in large part by the secondary debris surge during the Imbrium cratering event, the photogeologic expression of which being defined as "Fra Mauro."

Similar arguments favoring local to regional sample provenance have been advanced for the Apollo 16 site (Head, 1974; Morrison and Oberbeck, 1975), although contrary to Apollo 14, there is uncertainty as to which of the big, basin forming events and associated debris surges contributed (dominated?) the final depositional phase. Imbrium and Nectaris are the most prominent candidates, although Mare Orientale and even some of the older basins cannot positively be excluded.

In any case, the important point we wish to make is that while some big basins either demonstrably or very likely affected the geological history of the Apollo 14 and 16 landing sites, the dominating manifestation is only the deposition of material derived from regional sources in the form of an extensive debris surge. By analogy to the Bunte Breccia, this debris surge was cold. Furthermore, even if we could identify "primary" crater material from these basins, which undoubtedly must be present in small quantities, these materials most likely are modestly shocked according to our crystalline clast studies in the Ries' continuous deposits, if shocked at all.

We are fully aware that such conclusions are contrary to many popular concepts in lunar sample interpretation such as assignment of basin formation chronologies (Schaeffer *et al.*, 1976; Mauer *et al.*, 1979), or the assignment of specific chemistries to specific basin forming impactors (Hertogen *et al.*, 1977); "hot" ejecta to e.g., reset isotope systems for absolute chronology (≈1000°C; Maurer *et al.*, 1979), or even impact melts which thoroughly mix with the projectile are simply not observed in the distal Ries deposits.

Furthermore, it is demonstrably wrong to use indiscriminantly the ejecta decay profiles of McGetchin *et al.* (1973), even after the corrections suggested by Moore *et al.* (1974) and Pike (1974). In principle these calculations yield only mass distributions as a function of radial range. It does not follow that an orderly, layer cake type stratigraphy of various ejecta blankets, exclusively consisting of primary crater materials, results. Most importantly, however, it does not follow that these ejecta stacks are hot, as often implied; such implications are erroneous.

Large scale continuous crater deposits are intimate mixtures of "primary" crater and local materials, deposited at ambient temperatures.

Impact melts and other hot impact deposits are largely confined to crater interiors based on terrestrial cratering experience. This is consistent with lunar photogeologic studies concerning the distribution of impact melts (Hawke and Head, 1977). Many lunar samples for which a specific basin source has been suggested have textures that require relatively rapid quenching to or below the solidus, followed by relatively prolonged subsolidus cooling histories as demonstrated at many terrestrial impact craters (Phinney and Simonds, 1977; Stöffler *et al.*, 1979; Grieve, 1980). Most of the lunar highland textures are duplicated only in terrestrial melt sheets, i.e., large scale ponds of impact melt. They are indicative of various cooling histories which in turn are controlled by the geometry and position of the melt sheet relative to the final crater cavity. The overall diversity of textures and especially those requiring prolonged subsolidus cooling histories appear to argue against cooling during ballistic trajectories (e.g., Onorato *et al.*, 1976). Their prolonged cooling histories make a direct association with distant basins such as Imbrium, Nectaris or Orientale highly improbable; they are not impact melts formed by the above basin events.

In summary, we wish to underscore that, in the absence of better field control, assignment of specific lunar rocks to specific cratering events is extremely speculative at present. Because cratering is a stochastic process, one must invoke probability arguments based on cratering mechanis. The latter indicate that most of the samples were intensly processed by small local to modest size regional events and not by a few large basins. At most, such basins are only responsible for the last depositional phase.

Grieve (1980), in part following earlier conclusions by Stewart (1975), suggested that thermal energy sources other than impact be considered for the abundant thermal metamorphism observed in the highland samples. Lange and Ahrens (1979) calculated that the total thermal energy partitioned by all observable impact craters may be sufficient to account for the sample evidence, but they also pointed out that serious constraints on the cratering flux could be established, if absolute ages of the highland samples are considered. Since most isotope systems were affected between $\approx$3.9 to 4.2 b.y., a correspondingly high cratering rate of truly "cataclysmic" proportions seems to be required, contrary to the model fluxes suggested by Baldwin (1974) or Neukum *et al.* (1975). As a consequence, resolution of the important question whether cratering provided sufficient thermal energy depends primarily on model assumptions regarding cumulative bombardment history. Additional, detailed sample studies testing the suggestions of Grieve (1980) and Stewart (1975) are therefore paramount not only to understanding the thermal history(ies) of lunar highland rocks but also to assess associated implications regarding the early lunar cratering history.

Our Ries studies on deep seated target materials may also have implications regarding interpretation of remotely obtained compositional information, which relies heavily on cratering mechanics for stratigraphic probing (e.g., Andre *et al.*,

1978; Head *et al.*, 1978). Principally two cratering arguments are used: a depth/diameter relationship and the concept of inverted stratigraphy, enabling one to reconstruct relative stratigraphic positions of chemically distinct lithologies observed in the ejecta blanket. As noted above, the crystalline Bunte Breccia samples appear to be derived from near surface environments of the Ries' crystalline basement which lies at 600 m depth; they seem to originate from <1000 m depth although the transient Ries cavity is thought to be 1.5–2.5 km deep (Pohl *et al.*, 1977; Hörz and Ostertag, 1979). Thus the primary Ries ejecta now incorporated into the continuous deposits originated from the upper 50% of the stratigraphic stack penetrated by the Ries event. In fact >90% of all primary crater materials in the continuous deposits were dislodged from depth <.2 t of the total penetration (Hörz *et al.*, 1980), with the Upper Jurassic limestones (<.16 t) dominating the primary crater components. Thus, a hypothetical remote sensing experiment of the present-day Ries would not yield deep seated target materials, much less materials derived from the crater "bottom," occasionally the preferred model assumption in various remote sensing studies. Next to the Tertiary environs, either in situ or the dominant component in the Bunte Breccia, such an experiment would reveal limestones, which formed the cap rock of the target; materials from depths greater than 250 m (≈.15 t) constitute in favorable cases ≲5% of the total continuous deposit and in most cases <2% by volume. These Ries observations are in qualitative agreement with the Stöffler *et al.* (1975) quartz sand craters in so far as no material from depths >.33 t were found outside the rim of these experimental structures.

As stated above, the initial extent of "deep seated" suevite, unfortunately, is poorly understood. The present distribution could be the erosional remnant of a much more extensive, albeit surficial, deposit. Nevertheless we suggest that remotely obtained chemistries not be indiscriminantly placed at or near the crater bottom because of "cratering-mechanics." The latter is not sufficiently understood in general and specifically for large scale structures (e.g., >10 km). There is the distinct possibility that volumetrically significant portions of the continuous deposits, including the rim area, are not from (extremely) deepseated sources but from the upper half if not upper third of the target stack.

Our Bunte Breccia studies have additional implications regarding recent suggestions by Hawke and Head (1978) and Ahrens and O'Keefe (1978) that regular distributions of shock metamorphic effects may be found in large scale crater deposits. In particular, Hawke and Head (1978) suggest very specific shock pressures (≈25 GPa) to be an additional criterion for identifying primary Imbrium ejecta in the Apollo 14 collection. This is based on directly equating compressional peak pressure and associated shock particle velocity with ejection velocity; assuming an ejection angle, a ballistic trajectory may be defined. Following their model an annular distribution results, with increasingly higher shock levels encountered at increasingly greater radial ranges. Ahrens and O'Keefe (1978) predict similar radial trends but, in addition, they postulate vertical zoning within the ejecta blanket with a layer of more heavily shocked material sandwiched between light to moderately shocked bottom and top layers. Neither of the above models is consistent with our observations which yield a more or less random radial and vertical distribution.

We explain these discrepancies by the fact that neither of the models accounts for secondary cratering and the ensuing, turbulent debris surge. While the proposed systematics may indeed exist during ballistic ejection, the ensuing secondary debris surge is turbulent (Hörz *et al.*, 1977); it thoroughly disturbs and masks the predicted trends. Furthermore primary and secondary ejecta may be transported radially for significant distances as evidenced by specific "local" components now found in the Bunte Breccia some 10 km from their original position (Gall *et al.*, 1975). Primary crater materials incorporated into the debris surge are therefore not easily related to the ejection sequence, specifically to their ballistic termination point. Their ultimate location within the continuous deposits is a function of the ballistic transport phase as well as turbulent motion during emplacement of the debris surge. Radial distances traveled in the turbulent surge may be as much as a crater radius from ballistic termination.

CONCLUDING REMARKS

In closing, it is unfortunate that our analyses of deepseated Ries target materials and our general Bunte Breccia investigations cannot provide positive clues for assigning specific highland rocks to specific source areas. Our studies raised many questions concerning popular, lunar interpretations and they illustrate that reconstruction of a stratigraphic-structural framework for lunar sample interpretation is still more complex, than suggested previously. Until significant advances in our understanding of the cratering process are accomplished, reliable field control may not be brought to bear on lunar sample analysis. Assignment of specific rocks to specific source areas, especially individual, large scale basins, is highly speculative.

Acknowledgment—This paper presents parts of a Master's thesis (G.S.B.) during which Prof. M. Carman, University of Houston, provided invaluable encouragement and advice. We also appreciate stimulating discussions with R. Ostertag, R. B. Schaal, D. Stöffler, W. v. Engelhardt and G. Graup. R. A. F. Grieve and R. M. Housley provided substantial reviews. Work at the University of Houston was in part supported by NASA contract NAS-9-14209.

REFERENCES

Abadian M. (1972) Petrographie, Stosswellenmetamorphose und Entstehung polymikter kristalliner Breccien im Nördlinger Ries. *Contrib. Mineral. Petrol.* **35**, 245–262.

Ahrens T. J. and O'Keefe J. D. (1978) Energy and mass distribution of impact ejecta blankets on the Moon and Mercury. *Proc. Lunar Planet. Sci. Conf. 9th,* p. 3787–3802.

Andre C. G., Wolfe R. W., and Adler I. (1978) Evidence for a high magnesium subsurface basalt in Mare Crisium from orbital x-ray fluorescence data. In *Mare Crisium: The View from Luna 24* (R. B. Merrill and J. J. Papike, eds.) p. 1–12. Pergamon, N.Y.

Baldwin R. B. (1974) Was there a "terminal cataclysm" 3.9 × 10^9 years ago? *Icarus* **23,** 157–166.

Banholzer G. (1979) Distribution and Shock Metamorphism of Crystalline Components in the Bunte Breccia, Ries Crater, Germany. M.S. thesis, Univ. Houston, Texas, 118 pp.

Bauberger W., Mielke H., Schmeer D., and Stettner G. (1974), Petrographische Profildarstellung der Forschungsbohrung Nördlingen 1973. *Geol. Bavarica* **72,** 33–34.

Chao E. C. T. (1973) Geologic implications of the Apollo 14 Fra Mauro breccia and comparison with ejecta from the Ries Crater, Germany. *J. Res. U.S. Geol. Survey* **1,** 1–18.

Chao E. C. T. (1977) The Ries Crater of Southern Germany, a model for large basins on planetary surfaces. *Geol. Jahrb.* **43,** Reihe A, 85 pp.

Chao E. C. T. and El Goresy A. (1977) Shock attenuation and the implantation of Fe-Cr-Ni veinlets in the compressed zone of the 1973 Ries research deep drill core. *Geol. Bavarica* **75,** 289–304.

Cohn S. N. and Ahrens T. J. (1979) Dynamic tensile strength of analogs to lunar rocks (abstract). In *Lunar and Planetary Science X,* p. 224–226. Lunar and Planetary Institute, Houston.

Dence M. R., Grieve R. A. F., and Robertson P. B. (1977) Terrestrial impact structures: Principal characteristics and energy considerations. In *Impact and Explosion Cratering,* (D. J. Roddy, R. O. Pepin, and R. B. Merrill, eds.), p. 247–275. Pergamon, N.Y.

Dence M. R. and Plant A. G. (1972) Analysis of Fra Mauro samples and the origin of the Imbrium basin. *Proc. Lunar Sci. Conf. 3rd,* p. 379–399.

Dressler B., Graup G., and Matzke K. (1969) Die Gesteine des kristallinen Grundgebirges im Nördlinger Ries. *Geol. Bavarica* **61,** 201–228.

Engelhardt W. v., Stöffler D., and Schneider W. (1969) Petrologische Untersuchungen im Ries. *Geol. Bavarica* **61,** 229–296.

Gall H., Müller D., and Stöffler D. (1975) Verteilung, Eigenschaften und Entstehung der Auswurfsmassen des Impaktkrates Nördlinger Ries. *Geol. Rundsch.* **64,** 915–1947.

Gibbons R. V. and Ahrens T. J. (1977) Effects of shock pressures on calcic plagioclase. *Phys. Chem. Minerals* **1,** 95–107.

Graup G. (1975) Das Kristallin im Nördlinger Ries. Ph.D. thesis, Univ. Tübingen, Germany. 176 pp.

Graup G. (1977) Die Petrographie der kristallinen Gesteine der Forschungsbohrung Nördlingen 1973. *Geol. Bavarica* **75,** 219–229.

Grieve R. A. F. (1980) Cratering Record: Processes and Effects (abstract). In *Papers presented to the conference on the Lunar Highlands Crust,* p. 27–29. Lunar and Planetary Institute, Houston.

Hawke B. R. and Head J. W. (1977) Impact melt on lunar crater rims. In *Impact and Explosion Cratering* (D. J. Roddy, R. O. Pepin, and R. B. Merrill, eds.), p. 815–841. Pergamon, N.Y.

Hawke B. R. and Head J. W. (1978) Criteria for the identification of the Imbrium ejecta and local components in the Apollo 14 samples (abstract). In *Lunar and Planetary Science IX,* p. 477–479. Lunar and Planetary Institute, Houston.

Head J. W. (1974) Stratigraphy of the Descartes region (Apollo 16): Implications for the origin of samples. *The Moon* **11,** 77–99.

Head J. W., Adams J. B., McCord T. B., Pieters C., and Zisk S. (1978) Regional stratigraphy and geologic history of Mare Crisium. In *Mare Crisium: the View from Luna 24,* (R. B. Merrill and J. J. Papike, eds.), p. 43–74. Pergamon, N.Y.

Head J. W. and Hawke B. R. (1975) Geology of the Apollo 14 region (Fra Mauro): Stratigraphic history and sample provenance. *Proc. Lunar Sci. Conf. 6th,* p. 2483–2501.

Hertogen J., Janssens M. J., Takahashi H., Palme H., and Anders E. (1977) Lunar basins and craters: Evidence for systematic compositional changes of bombarding population. *Proc. Lunar Sci. Conf. 8th,* p. 17–45.

Hörz F. (1968) Statistical measurements of deformation structures and refractive indices in experimentally shock loaded quartz. In *Shock Metamorphism of Natural Materials* (B. M. French, and N. M. Short, eds.), p. 243–254. Mono, Baltimore.

Hörz F., Gall H., Hüttner R., and Oberbeck V. R. (1977) Shallow drilling in the "Bunte Breccia" impact deposits, Ries Crater, Germany. In *Impact and Explosion Cratering* (D. J. Roddy, R. O. Pepin, and R. B. Merrill, eds.), p. 425–448. Pergamon, N.Y.

Hörz F. and Ostertag R. (1979) The transient cavity of the Ries Crater, Germany (abstract). In *Lunar and Planetary Science X,* p. 570–572. Lunar and Planetary Institute, Houston.

Hörz F., Ostertag R., and Rainey D. A. (1980) Target stratigraphy and its manifestation in continuous crater deposits: The "Bunte Breccia" of the Ries Crater, Germany (abstract). In *Lunar and Planetary Science XI,* p. 477–479. Lunar and Planetary Institute, Houston.

Hüttner R. (1969) Bunte Trümmermassen und Suevit. *Geol. Bavarica* **61,** 142–200.

Kieffer S. W. and Simonds C. H. (1980) The role of volatiles and lithology in the impact cratering process. *Rev. Geophys. Space Phys.* In press.

Lange M. A. and Ahrens T. J. (1979) Impact melting early in lunar history. *Proc. Lunar Planet. Sci. Conf. 10th,* p. 2707–2725.

Maurer P., Eberhardt P., Geiss J., Grögler N., Stettler A., Brown G. M., Peckett A., and Krähenbühl U. (1979) Pre Imbrian craters and basins: Ages, compositions and excavation depths of Apollo 16 breccias. *Geochim. Cosmochim. Acta* **42,** 1687–1720.

McGetchin T. R., Settle M., and Head J. W. (1973) Radial thickness variation in impact crater ejecta: Implications for lunar basin deposits. *Earth Planet. Sci. Lett.* **20,** 226–236.

Moore H. J., Hodges C. A., and Scott D. H. (1974) Multiringed basins—illustrated by Orientale and associated features. *Proc. Lunar Sci. Conf. 5th,* p. 71–100.

Morrison R. H. and Oberbeck V. R. (1975) Geomorphology of crater and basin deposits—emplacement of the Fra Mauro formation. *Proc. Lunar Sci. Conf. 6th,* p. 2503–2530.

Müller W. F. and Defourneaux M. (1968) Deformationsstrukturen in Quarz als Indikator für Stosswellen: eine experimentelle Untersuchung an Quarzeinkristallen. *Z. Geophys.* **34,** 483–504.

Neukum G., König B., Fechtig H., and Storzer D. (1975) Cratering in the Earth-Moon system: Consequences for age determination by crater counting. *Proc. Lunar Sci. Conf. 6th,* p. 2597–2620.

Oberbeck V. R. (1975) Role of ballistic erosion and sedimentation in lunar stratigraphy. *Rev. Geophys. Space Phys.* **13,** 337–362.

Onorato P. I. K., Uhlmann D. R., and Simonds C. H. (1976) Heat flow in impact melts: Apollo 17 Station 6 Boulder and some applications to other breccias and xenolith laden melts. *Proc. Lunar Sci. Conf. 7th,* p. 2449–2468.

Ostertag R. (1978) Sedimentologische Untersuchungen der NASA Forschungsbohrung Demmingen in der Ejektadecke des Impaktkraters Nördlinger Ries M.S. thesis, Univ. Münster, Germany. 68 pp.

Phinney W. C. and Simonds C. H. (1977) Dynamical applications of the petrology and distribution of impact melt rocks. In *Impact and Explosion Cratering* (D. J. Roddy, R. O. Pepin, and R. B. Merrill eds.), p. 771–790. Pergamon, N.Y.

Pike R. J. (1974) Ejecta from large craters on the Moon: Comments on the geometric model of McGetchin et al. *Earth Planet. Sci. Lett.* **23,** 265–271.

Pohl J. (1979) A comparison of cratering models for the Ries Crater (abstract). *Meteoritics* **14,** 520.

Pohl J., Stöffler D., Gall H., and Ernstson K. (1977) The Ries impact crater. In *Impact and Explosion Cratering* (D. J. Roddy, R. O. Pepin, and R. B. Merrill, eds.), p. 343–404. Pergamon, N.Y.

Preuss E. and Schmidt-Kaler H. (eds.) (1969) Das Ries; Geologie, Geophysik und Genese eines Kraters. *Geol. Bavarica* **61,** 1–478.

Raikes S. A. and Ahrens T. J. (1979) Post shock temperatures in minerals. *Geophys. J. Roy. Astron. Soc.* In press.

Roddy D. J. (1978) Pre-impact geologic conditions, physical properties, energy calculations, meteorite and initial crater dimensions and orientations of joints, faults and walls at Meteor Crater, Arizona. *Proc. Lunar Planet. Sci. Conf. 9th,* p. 3891–3930.

Schaeffer O. A., Husain L., and Schaeffer G. A. (1976) Ages of highland rocks: The chronology of lunar basin formation revisited. *Proc. Lunar Sci. Conf. 7th,* p. 2067–2092.

Schneider W. (1971) Petrologische Untersuchungen der Bunten Breccie im Nördlinger Ries. *Neues Jahrb. Mineral. Abh.* **114,** 136–180.

Simonds C. H., Phinney W. C., Warner J. L., McGee D. E., Geeslin J., Brown R. W., and Rhodes

M. J. (1977) Apollo 14 revisited, or breccias aren't so bad after all. *Proc. Lunar Sci. Conf. 8th*, p. 1869–1893.

Stewart D. B. (1975) Apollonian metamorphic rock—the product of prolonged subsolidus equilibration (abstract). In *Lunar Science VI*, p. 774–776. The Lunar Science Institute, Houston.

Stöffler D. (1972) Deformation and transformation of rock-forming minerals by natural and experimental shock processes: I. Behavior of minerals under shock compression. *Fortschr. Mineral.* **49,** 50–113.

Stöffler D. (1974) Deformation and transformation of rock forming minerals by natural and experimental shock processes: II. Physical properties of shocked rocks. *Fortschr. Mineral.* **51,** 256–289.

Stöffler D., Gault D. E., Wedekind J., and Polkowski G. (1975) Experimental hypervelocity impact into quartz sand: Distribution and shock metamorphism of ejecta. *J. Geophys. Res.* **80,** 4062–4077.

Stöffler D. and Hornemann U. (1972) Quartz and feldspar glasses produced by natural and experimental shock. *Meteoritics* **7,** 371–394.

Stöffler D., Knöll H. D., and Maerz U. (1979) Genetic classification and nomenclature of lunar highland rocks based on the texture and geologic setting of terrestrial impact breccias (abstract). In *Lunar and Planetary Science X*, p. 1177–1179. Lunar and Planetary Institute, Houston.

Stöffler D., Knöll H. D., Reimold W. U., and Schulien S. (1977) Grain size statistics, composition and provenance of fragmental particles in some Apollo 14 breccias. *Proc. Lunar Sci. Conf. 7th*, p. 1965–1985.

Papike, J.J. and Merrill, R.B., eds.
Proc. Conf. Lunar Highlands Crust (1980), p. 233-252
Printed in the United States of America

An experimental study of the thermal history of fragment-laden "basalt" 77115

Carl R. Thornber and J. Stephen Huebner

U.S. Geological Survey, Reston, Virginia 22092

Abstract—Phase equilibrium and dynamic crystallization experiments using a synthetic analog of the 77115 matrix composition restrict interpretations of the thermal history of the lunar sample. Rapid early cooling (10°–25°C hr^{-1}) of melt containing a large number of a nuclei and slow late stage cooling (<7°C hr^{-1}) are required to crystallize the textures and mineralogy of the 77115 matrix. This result is in agreement with that predicated by the nature and extent of diffusive cation exchange between various xenocrysts phases and the 77115 matrix.

INTRODUCTION

Lunar highlands sample 77115, the "blue-grey breccia" incorporated in the Apollo 17 station 7 boulder, is a fragment-laden, pigeonite-bearing basalt similar to many samples from the highlands region southeast of Mare Serenitatis. James *et al.* (1978) have suggested that lunar rocks with similar texture, lithology and chemistry crystallized from melts of impact origin. Sample 77115 consists of matrix, xenocrysts and minor xenolith components. The matrix, shown in Fig. 1, comprises ~60 vol. % plagioclase (An_{85-88}), ~30 vol. % pigeonite ($Wo_{4-13}En_{77-66}Fs_{19-21}$), ~6 vol. % olivine ($Fo_{66-72}$) and minor or trace amounts of augite, K-Si rich mesostasis, phosphate, troilite and metallic iron (Chao *et al.*, 1975; Minkin *et al.*, 1978). This matrix has a distinctively igneous texture dominated by a microsubophitic intergrowth of plagioclase and pigeonite. Xenocrysts that are predominant in the matrix are plagioclase, olivine and pyroxene. The sizes of xenocrysts range from several millimeters down to tens of microns. Small xenocrysts are virtually indistinguishable from matrix phases, introducing some error into the calculation of a matrix composition. Xenocryst rims show varying amounts of diffusive cation exchange with the matrix (Huebner, in Chao *et al.*, 1975). Those orthopyroxene, pigeonite, plagioclase, and olivine xenocrysts having compositions similar to phases in the matrix are homogeneous from core to rim. Notably the margins of these orthopyroxene and pigeonite xenocrysts show no evidence of dissolution/resorption in the melt/matrix. Xenocrysts that have compositions different from those of analogous matrix phases have zoned rims which compositionally approach those of corresponding phases in the matrix

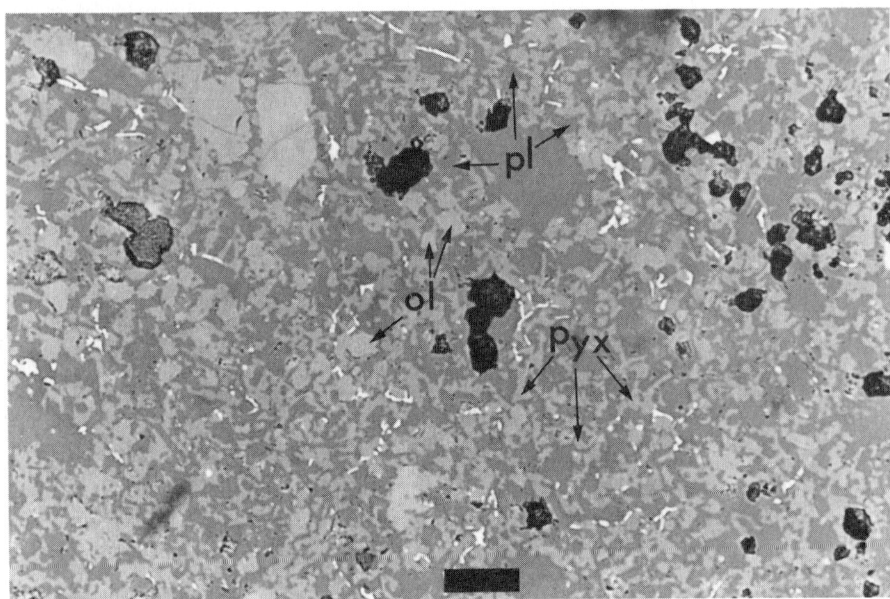

Fig. 1. Reflected light photomicrograph of matrix of thin section 77115,53 showing a matrix of microsubophitic intergrowth of calcic plagioclase (pl) with clinopyroxene (pyx). Matrix olivine (ol) grains are randomly dispersed. Ilmenite, mesostasis and metallic iron occur interstitially in the submicrophitic network. Bar scale = 0.05 mm.

(Chao *et al.*, 1975, Tables 2a, 2b and 3). This sample affords an excellent opportunity to test theoretical cooling models of the kinds proposed by Huebner (1976), Walker *et al.* (1977), Taylor *et al.* (1978) and Sanford and Huebner (1979).

In this investigation, we have experimentally evaluated the thermal conditions required to reproduce, by crystallization from a melt, the texture and mineralogy of the 77115 matrix. In addition, we show that the extent of cation exchange between xenocrysts and matrix restricts the possible interpretations of the thermal history of the lunar sample. To the extent that 77115 is similar to other highland samples, our attempts to evaluate its thermal history are pertinent to the interpretation of the nature and duration of individual events which generated the lunar highlands crust (Sanford and Huebner, 1980).

EXPERIMENTAL TECHNIQUE

Isothermal phase equilibrium experiments and dynamic crystallization experiments were conducted on a synthetic analogue (Table 1) of the 77115 matrix composition reported by Chao *et al.* (1975). Starting material was prepared by the following procedure:

1. Mixing of weighed oxide and carbonate components.
2. Repeated fusing (in air) and crushing of the glass product.

3. Crystallizing at ~1000°C for 24 hours in a 1:1 CO_2/H_2 mixture (fO_2 = $10^{-14.5}$ atm) to form plagioclase, olivine, clinopyroxene and ilmenite (as determined by x-ray powder diffraction).
4. Grinding of crystallized matrix.

All experimental charges in this study were run in cylindrical iron capsules milled dry from ¼" diameter ARMCO electromagnetic iron rod (Corey Steel Co.). These capsules came from the same lot of capsules that was used by Lipin (1978). A typical mill analysis (Table 2) shows that this rod contains no more than 100 ppm carbon; subsequent carbon analysis shows that the pre-treatment (see below) of the iron capsules reduces the carbon content to ≤30 ppm, the value of the analytical blanks. The effect of this amount of carbon on reduction of ferrous iron in these basaltic charges is negligible. "In vacuo heating" was instituted as part of the pre-treatment procedure to insure the removal of hydrogen absorbed by the capsules during their initial reduction at high temperature in a hydrogen atmosphere.

Table 1. Compositions of starting material and the 77115 matrix.

	Glass (fused) twice in air)	Crystallized starting material fused in Fe capsules (in vacuo)							7715 matrix composition reported by Chao *et al.* (1975)
Run no.		1	2	3	5	7			
Temp °C	1300	1368	1293	1300	1258	1264			
Time (hrs)	1.5	4.25	1.50	5	17.5	21.5			
No. Analyses	10	4	6	7	9	8			
Weight %							*AVG*	*SD*	
SiO_2	45.21	46.50	45.77	46.67	46.18	47.11	46.44	0.51	46.1
TiO_2	2.81	2.95	2.82	2.70	2.80	2.76	2.81	0.09	3.03
Al_2O_3	18.15	18.36	17.81	17.95	17.84	18.00	17.99	0.22	18.7
"FeO"	9.02	7.68	9.29	8.17	9.84	7.46	8.48	1.04	9.7
MgO	11.19	11.13	11.30	11.23	10.99	11.41	11.21	0.16	10.7
CaO	9.87	10.56	10.29	10.36	10.30	10.53	10.40	0.13	11.0
Na_2O	1.07	0.98	1.11	1.13	0.86	1.13	1.04	0.12	0.65
K_2O	0.07	0.18	0.12	0.12	0.06	0.06	0.11	0.05	0.16
TOTAL	97.39	98.34	98.51	98.33	98.87	98.46	98.48		100.04
Cations/6 oxys									
Si	1.699	1.688	1.674	1.697	1.680	1.704	1.689	0.012	1.663
Ti	0.076	0.081	0.078	0.074	0.077	0.075	0.077	0.003	0.082
Al	0.775	0.785	0.768	0.769	0.765	0.768	0.771	0.008	0.794
Fe^{2+}	0.228	0.233	0.284	0.249	0.300	0.226	0.258	0.032	0.293
Mg	0.620	0.602	0.616	0.610	0.597	0.615	0.608	0.008	0.545
Ca	0.412	0.410	0.402	0.404	0.402	0.407	0.405	0.003	0.425
Na	0.050	0.035	0.080	0.079	0.079	0.078	0.070	0.020	0.022
K	0.004	0.008	0.004	0.004	0.004	0.004	0.005	0.002	0.004
TOTAL	3.864	3.877	3.906	3.886	3.904	3.879	3.883		3.828
Wo	32.7	32.9	30.9	32.0	30.9	32.7	31.9	0.96	33.7
En	49.2	48.4	47.3	48.3	46.0	49.2	47.8	1.23	43.1
Fs	18.1	18.7	21.8	19.7	23.1	18.1	20.3	2.11	23.2

Table 2. Analysis of "typical milling" of ARMCO Electromagnetic Iron used for Fe capsules in this study (provided by the Corey Steel Co.).

C	0.010 % maximum
Mn	0.08 % maximum
S	0.02
P	0.008
Si	0.08
Al	0.05
Ti	0.07
Fe	remainder

Capsule and charge preparation involved the following:

1. Reduction of capsules and lids in H_2 at ~ 800°C (15 min).
2. In vacuo heating of capsules and lids at ~ 800°C (15 min), with cooler (~700°C) Fe-sponge between the capsules and pump.
3. In vacuo heating of loaded capsules at 800°C (15 min), with cooler (~700°C) Fe-sponge between the capsules and pump.
4. Sealing with iron lid by impact (in air).
5. Enclosure in evacuated silica tubes.

Isothermal phase equilibria experiments were conducted both by melting or partial melting of crystallized starting material and by crystallizing charges initially quenched as glasses from 1270°C. Charges in iron capsules were contained in evacuated silica tubes which were hung in conventional quenching furnaces. Runs were 24–600 hours duration. Temperature was continuously monitored throughout each run using a Pt/Pt-Rh_{10} thermocouple (calibrated at the melting point of gold) adjacent to the charge. Reversals were accomplished for plagioclase and olivine "phase-in" boundaries, which occur at temperatures at which significant (>70 vol.%) melt is present. Reactions involving pyroxene and ilmenite took place at lower temperatures in the presence of ≤30 vol.% melt; attainment of equilibrium in these runs was indicated by similarity of the phase assemblages in both crystallization and melting experiments conducted under identical conditions.

ANALYTICAL TECHNIQUES

Modal percentages of crystalline matrix phases were estimated by incident and transmitted light microscopy of grain mounts and polished thin sections of each charge. Sections were cut through each capsule parallel and close to the axis of the crucible providing reprsentative cross sections of each charge. This technique minimized sampling error, especially of high temperature runs in which crystal settling may have taken place. Grain mounts were particularly useful in determining the volume percent of glass in finely crystalline charges heated isothermally at near solidus temperatures.

All optical phase determinations were checked by X-ray powder diffraction methods. This technique could not be used reliably to distinguish the clinopyroxenes pigeonite and augite in charges consisting largely of plagioclase and olivine. Microprobe analysis of calcium in the clinopyroxenes was also not definitive because the pyroxene grains were very small and adjacent to grains of olivine and calcic plagioclase. A single pyroxene-bearing run product (#70; 1162°C; Table 3) was examined by transmission electron microscopy and found to contain pigeonite but not augite (G. L. Nord, pers. comm.).

Where grain size permitted, glasses and crystalline phases were routinely analyzed by electron microprobe analysis methods. The microprobe technique used compositionally similar standards of olivine, pyroxene, plagioclase and glass; analyses were performed at 15 KV and at a beam current of 0.10 microamperes, both by wavelength dispersive and by integrated wavelength/energy dispersive techniques. Count data were converted to weight % oxides using the matrix correction scheme of Bence and Albee (1968).

RESULTS AND DISCUSSION

77115 phase equilibria

Equilibrium (isothermal) experiments provide a standard with which the results of dynamic crystallization runs can be compared. The equilibrium phase relations determined for the 77115 matrix composition are shown in Fig. 2 and Table 3. The bulk composition intersects the liquidus at 1255° ± 5°C and the solidus at 1055° ± 20°C. Plagioclase crystallizes at the liquidus and is joined by olivine at

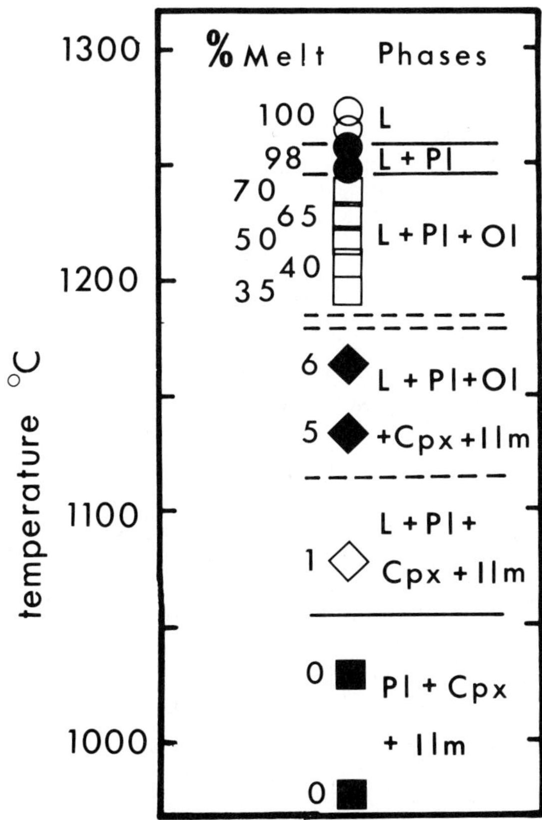

Fig. 2. Phase equilibria of 77115 matrix composition as determined by isothermal melting experiments. L, liquid; Pl, plagioclase; Ol, olivine; Cpx, clinopyroxene; Ilm, ilmenite.

Table 3. Summary of isothermal phase equilibria experiments.

Run	Temp. (°C)	Time (hrs)	Starting material	Phases Present (volume %): L	Pl	Ol	Cpx	Ilm
7	1264	21.5	CM	100	—	—	—	—
3	1256	5	CM	98	2	—	—	—
53a	1241	71	GL	X	X	X	—	—
53b	1241	71	CM	75	15	10	—	—
64	1227	69	CM	65	20	15	—	—
62	1226	113	CM	65	20	15	—	—
35a	1219	48	GL	50	35	15	—	—
35b	1219	48	CM	50	40	10	—	—
44	1194	51.5	GL	35	40	25	—	—
70	1162	40.5	CM	6	X	X	X	tr
55	1134	24	CM	5	X	tr	X	X
58	1133	93	GL	5	X	tr	X	X
41a	1080	330	GL	tr	X	—	X	X
41b	1080	330	CM	tr	X	—	X	X
24	1035	600	CM	—	X	—	X	X
26	1035	600	GL	—	X	—	X	X
43	975	313	GL	—	X	—	X	X

CM = crystallized matrix
GL = Glass (fused in Fe capsules, in vacuo)
tr = Trace
X = phase present, volume % not determined
L = liquid

Pl = plagioclase
Ol = olivine
Cpx = clinopyroxene
Ilm = Ilmenite

1243°C. Between 1180°C and 1110°C clinopyroxene and ilmenite coexist with plagioclase and olivine, and the equilibrium melt proportion diminishes from 30 to 5 vol. %. Below 1110°C, olivine is no longer a stable phase and the equilibrium mineral assemblage is plagioclase, clinopyroxene and ilmenite. These data are in agreement with the paragenetic sequence of 77115 matrix crystallization estimated petrographically by Chao *et al.* (1975, p. 502) and are consistent with equilibrium phase relations determined for the similar 77135 matrix composition by Storey *et al.* (1974) in a CO_2/H_2 atmosphere. The sequence and temperatures of mafic silicate crystallization are also similar to those predicted from a model (Huebner, 1975, 1979) for basalt crystallization in the system $CaO\text{-}MgO\text{-}FeO\text{-}SiO_2\text{-}Al_2O_3$ at plagioclase saturation (Fig. 3a,b). According to this model, plagioclase and olivine should precipitate first from 77115 bulk composition, followed by calcium-poor pyroxene (either orthopyroxene or pigeonite, depending on the Fe/Mg ratio of the melt). Augite, if it were to crystallize, would do so only very late in the crystallization sequence, when the Fe/Mg and/or the Ca/(Ca+Mg+Fe) of the residual melt was enriched by fractionation. The slight differences between the composition of starting material and the lunar sample (Table 1) would not significantly affect the crystallization path.

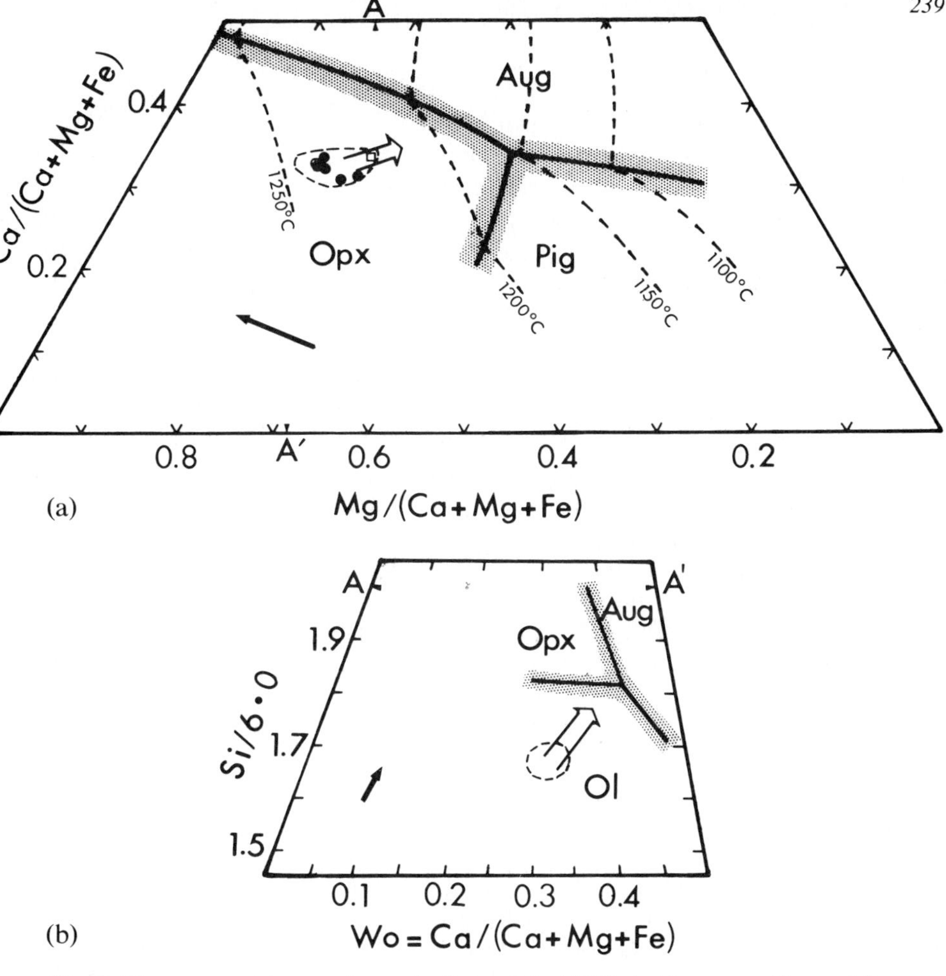

Fig. 3. Figures a and b comprise projections of the 77115 matrix composition and its synthetic analogs along with saturation surfaces in the system CaO-MgO-FeO-SiO_2-Al_2O_3 (at 1 bar) onto two different compositional planes in the system CaO-MgO-FeO-SiO_2 at plagioclase saturation:

Opx = orthopyroxene	Aug = augite
Pig = pigeonite	Ol = olivine

a) Projection onto the pyroxene quadrilateral plane: $Mg_2Si_2O_6$-$CaMgSi_2O_6$-$CaFeSi_2O_6$-$Fe_2Si_2O_6$. Dashed isotherms lie in olivine-pyroxene saturation surfaces.

Filled circles	-	analog matrix fused in Fe capsules
Open circle	-	analog matrix fused in air (adjacent to three overlapping filled circles)
Open square	-	Chao et al (1975) matrix (see Table 1)

b) Projection onto the "Wo"-silica plane along section A-A' in Fig. a. The region enclosed by a dashed line corresponds to that in Fig. a. "arrow-bands" in each figure represent suggested crystallization path as estimated from modes of olivine and plagioclase observed in isothermal experiments, projected onto the compositional plane of each diagram. The vectors in the lower left part of each diagram indicate the magnitude and direction of maximum possible bulk compositional changes due to "Fe-loss".

Minor variation of iron content is observed in charges containing 77115 analog matrix material that was fused in iron capsules (Table 1). In addition, the iron/magnesium ratios of liquid and of coexisting olivine in isothermal runs between 1241° and 1194°C, do not always vary consistently with temperature. These compositional discrepancies may be the result of minor reduction of the silicate charge and precipitation of coarse metallic iron blebs. We have yet to resolve whether or not this apparent iron-loss is due to impurities in the capsule material or to a reducing agent introduced during capsule preparation. Coarse blebs of metallic iron were observed only in those charges which were indicated by electron microprobe analysis to have been depleted in ferrous iron. In the lower temperature isothermal experiments, which were too fine grained to be analyzed by electron microprobe, we did not observe the metallic iron blebs and therefore we conclude that these runs did not experience "iron-loss". As the data from isothermal experiments are in agreement with independent estimates of phase equilibria of 77115 (and similar compositions), we assume that "iron-loss" was minor in the sense that it did not significantly affect the determination of "phase in-out" boundaries for the 77115 matrix composition. This conclusion is substantiated by calculations of the possible error in temperature of "phase-in" boundaries that could be induced by the observed variations in iron content of experimental charges. The "iron-loss" vector projected in Fig. 3a, originates at the composition FeO and represents the maximum possible "iron-loss" of runs fused above the liquidus and phase equilibria runs, relative to the composition of the lunar sample (Table 1). The composition change associated with this vector would result in an increase of 25°-30°C in the temperature at which olivine begins to precipitate (Roeder and Emslie, 1970, Fig. 7). An increase of 25°C in the temperatures at which pyroxene crystallization begins, is similarly estimated from the isotherms that lie on the olivine-pyroxene saturation surfaces projected in Fig. 3a. "Iron-loss" affects similar changes in olivine and pyroxene "phase-in" boundaries. This temperature error for "phase-in" boundaries does not change the relative order in appearance of phases crystallizing from a melt of 77115 matrix composition.

DYNAMIC CRYSTALLIZATION EXPERIMENTS

Dynamic crystallization experiments were performed with two interdependent objectives in mind: 1) to determine the thermal history necessary for reproduction of the mineralogy and textures of the 77115 matrix and 2) to reproduce, by use of analog "xenocrysts," the apparent diffusive exchange reactions between xenocrysts and matrix in the lunar sample.

We determined the thermal history of the lunar 77115 matrix by examining the texture and the mineralogy of charges crystallized along a variety of cooling curves (Fig. 4). All charges were brought to their initial temperature by placing them in a furnace preheated to that temperature; we estimate that the initial

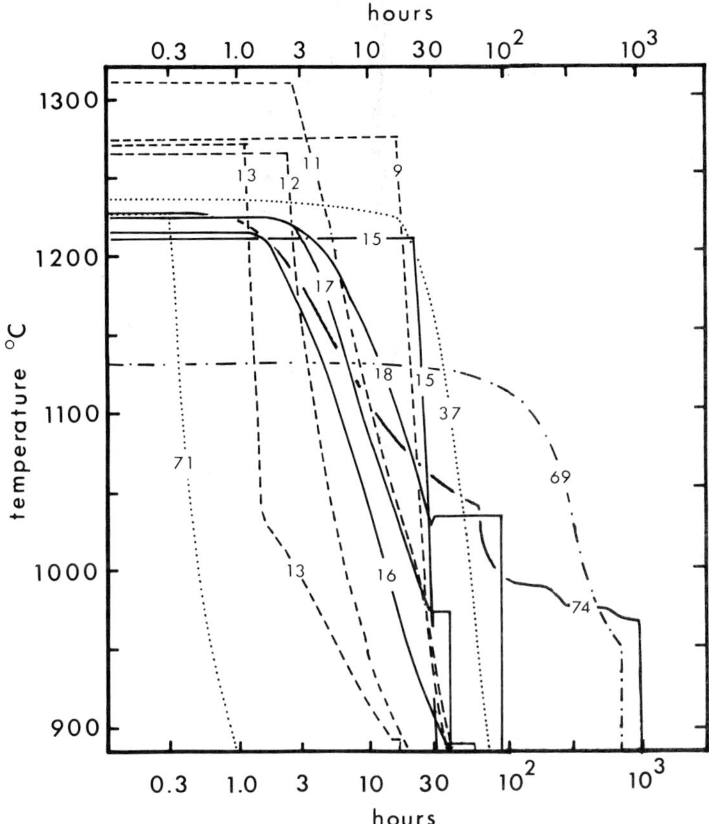

Fig. 4. Cooling curves of dynamic crystallization experiments.

temperature was reached within 60 seconds. The initial temperature was the highest of any particular run.

All runs in which starting material was completely melted and then cooled at various rates (curves 9, 11, 12 and 13 in Fig. 4) resulted in coarse radial growths of plagioclase and poikilitic olivine (Fig. 5 a,b,c, and d). We interpret this texture as having resulted from undercooling, formation of a few nucleii, and subsequent rapid outward growth from these few nucleation centers. Presumably, coarse textures developed because interference of crystal growth was minimized by the presence of relatively few growth centers.

Aphanitic texture similar to, but finer than, that of the 77115 matrix was attained in some runs crystallized from temperatures at which crystals and liquid were present (curves 15, 16, 17 and 71 in Fig. 4 and Fig. 5 e,f,g, and h). The mineralogy of each of the run products, as determined by x-ray powder diffraction, is plagioclase, olivine and ilmenite (clinopyroxene was observed in run 17). In these runs, finely disseminated crystalline residua of partially melted starting

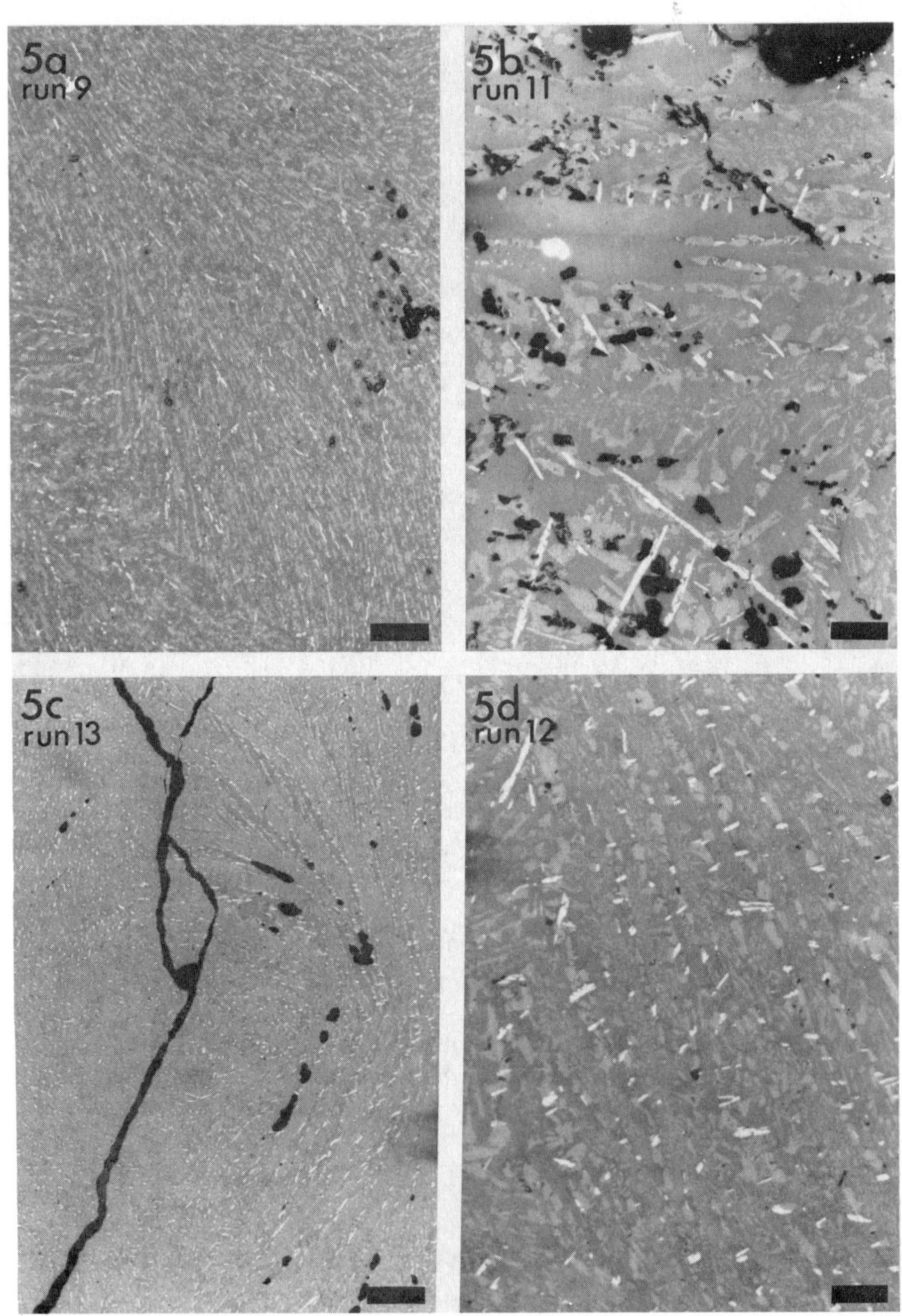

5a
run 9
5b
run 11
5c
run 13
5d
run 12

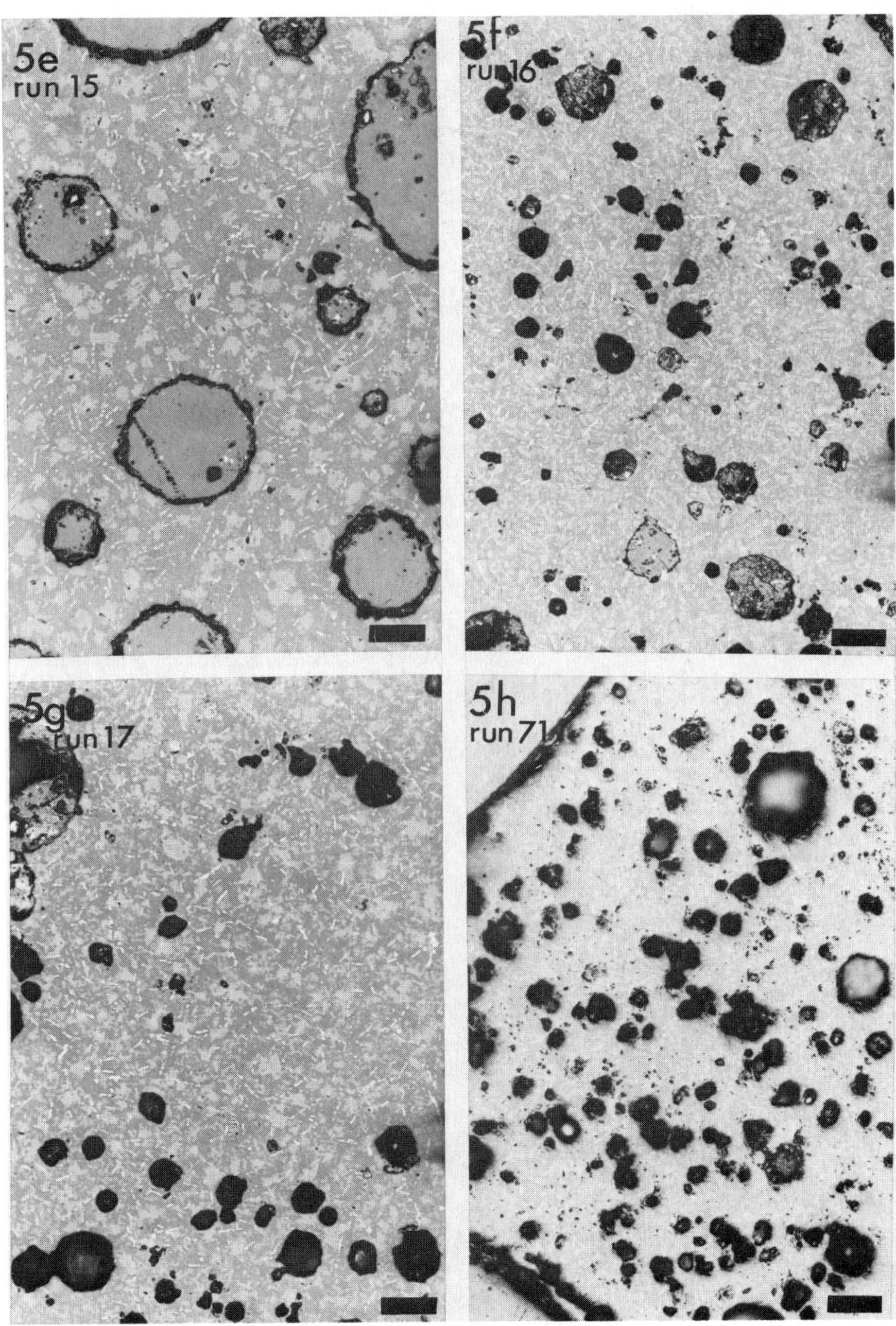
5e
run 15
5f
run 16
5g
run 17
5h
run 71

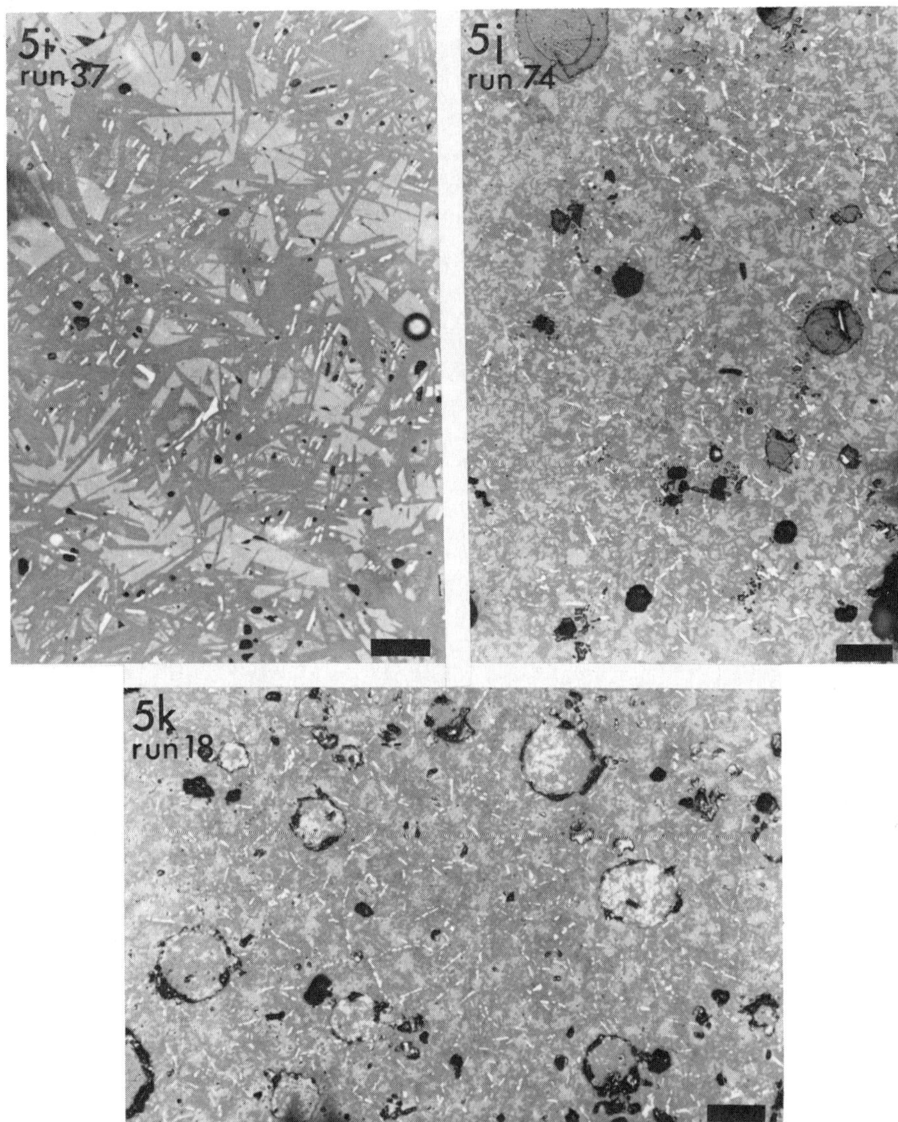

Fig. 5. Reflected light photomicrographs showing textures observed in dynamic crystallization experiments. Run numbers correspond to those of cooling curves (Fig. 4). Bar scale = 0.05 mm.

material served as nucleii for crystallization. Run 37 (curve 37, Fig. 4 and Table 4) was slowly cooled from a higher subliquidus temperature and crystals grew for a longer time at higher temperatures (1235°–1180°C). The result was a relatively coarse-grained intergrowth of plagioclase, olivine, and minor clinopyroxene and ilmenite (Fig. 5i). A prerequisite for the development of an aphanite texture appears to be a large number of nuclei per unit volume initially within the melt followed by rapid crystallization. Lofgren (1979) reached similar conclusions in a study of terrestrial olivine-tholeiite crystallization. Lofgren (1978) also observed that aphanitic textures developed from crystallization of melted lunar soil. In those kinetic melting experiments that were maintained above their liquidus temperature sufficiently long to cause complete melting, coarse, radial textures developed, whereas shorter melting times resulted in aphanitic textures. In our experiments, we assured the presence of nuclei for crystallization by heating the crystalline starting material to an initial temperature within the plagioclase + olivine stability field. A large number of nucleii were observed in isothermal runs quenched from similar temperatures.

The texture of the lunar sample was best mimicked by run 18 (curve 18, Fig. 4) which was initially cooled rapidly (~ 10°C hr^{-1}) from 1225°C, a temperature at which abundant olivine and plagioclase is present, then held at 1040°C for 80 hours. This run produced an intergrowth of plagioclase, olivine and minor clinopyroxene + ilmenite with traces of Si-K rich mesostasis (Fig. 5k) which closely resembles the igneous *texture* of the 77115 matrix but has a different modal mineralogy.

Cooling rate also influences the mineralogy and mode of the crystallized product. Pigeonite, the major mafic phase observed in the lunar sample, was absent

Table 4. Summary of thermal histories and results of runs containing "xenocrysts"

Thermal Histories of Runs Containing "Xenocrysts"

Run	Cooling curve	Initial temp. (T_i, °C)	Time at T_i (hrs)	Final temp. (T_f, °C)	Cooling Rate (°C hr^{-1}): From T_i to 1180°C (>35% melt)	From 1180 to 1110°c (olivine - out interval)	From 1110°C to T_f
19	18	1225	1.8	1040	10	10	10–0
37	37	1235	1.6	580	2–3	7	7
74	74	1226	1.0	965	25	15–8	8–0

Results of Runs Containing "Xenocyrsts"

Run	Matrix olivine composition (%Fa) Avg.	S.D.	Range	Olivine "xenocryst" rim composition range (%Fa)†	Extent of Fe/Mg diffusive exchange (%Fa-μ)*: ~parallel to a (avg)	~parallel to b (avg)	~parallel to c
19	31.6	2.1	23.7–34.1	20.5–23.6	24.4	—	19.5
37	18.3	3.0	13.8–22.9	15.6–20.9	4.8	8.0	—
74	23.5	3.4	16.5–29.7	21.4–22.5	11.0	—	13.2

† extrapolated values of outermost rim compositions as shown in Fig. 7.
* as defined by the areas under respective curves in Fig. 7.

in all runs cooled through the 1180–1110°C interval (in which olivine reacts with liquid to form pigeonite) at a rate greater than 25°C per hour and is present only as a minor phase in runs 17, 18, and 37 which were colled at 17°, 10°, and 7°C per hour through this interval. However, pigeonite was the mafic silicate phase (olivine was absent) in the very fine-grained product of run 69 which was cooled very slowly from 1130°C (curve 69, Fig. 4). These results suggest that the rate at which olivine reacts with melt to form pyroxene is an important consideration in an interpretation of the cooling history of 77115. Cooling at rates greater than 7°C hr^{-1} does not appear to provide sufficient time for growth of pyroxene that is the stable mafic silicate below 1110°C and is the major mafic phase in the lunar sample.

Restrictions on the interpretations of the thermal history of the 77115 matrix, as provided by dynamic crystallization experiments, are enumerated below.

1. *Upper thermal limit of 1220°–1240°C*. This is the maximum temperature at which plagioclase and olivine residua of partially melted starting material are present in sufficient quantity to provide the high nucleation density required for aphanitic crystallization. If the melt that is now represented by the matrix was heated to a higher temperature, the melt did not remain sufficiently long at that temperature to digest all of the fragmental debris. Such rapid cooling could occur if the superheated (above its liquidus temperature) melt were rapidly mixed with cold fragmental debris (Onorato *et al.*, 1976).

2. *Rapid early-stage cooling (10°–25°C hr^{-1}) from 1230° to 1180°C*. Cooling rates slower than 10°C hr^{-1} through the early crystallization interval result in coarse plagioclase and olivine crystals. Rates of 10° to 25°C hr^{-1} produce aphanitic textures similar to that of the lunar sample and cooling at rates >25°C hr^{-1} result in finer grained textures. Rapid initial cooling of the lunar sample is consistent with the survivial, in 77115, of orthopyroxene xenocrysts that are conspicuously unresorbed and homogeneous from core to rim (Fig. 6). As illustrated by liquidus relations in a model basaltic system at 1 bar (Fig. 3b), these pyroxenes could not have been in equilibrium with a melt of 77115 matrix composition at low pressures. The lack of resorption of these phases indicates that these xenocrysts could not have remained long in contact with a melt of the 77115 matrix composition. The 77115 melt must have cooled rapidly through the high-temperature interval during which the melt would have had a composition similar to that of the matrix. Slower cooling would be possible only after extensive crystallization and fractionation caused the melt to evolve to the more siliceous (and iron-rich) compositions required for Ca-poor pyroxene stability. This experimental observation, combined with the fact that unreacted xenocrysts were compatible with the lunar 77115 matrix assemblage during late stage cooling, substantiates the suggestion by Huebner (in Chao *et al.*, 1975 and Huebner, 1976) that diffusive cation exchange observed between matrix and other xenocrysts phases took place after the melt was nearly or completely crystalline.

3. *Slow late-stage cooling (<7°C hr^{-1} from 1190–1110°C)*. The 77115 sample must have cooled more slowly than 7°C hr^{-1} in order to obtain the large proportion of pigeonite to olivine that exists in the lunar 77115 matrix.

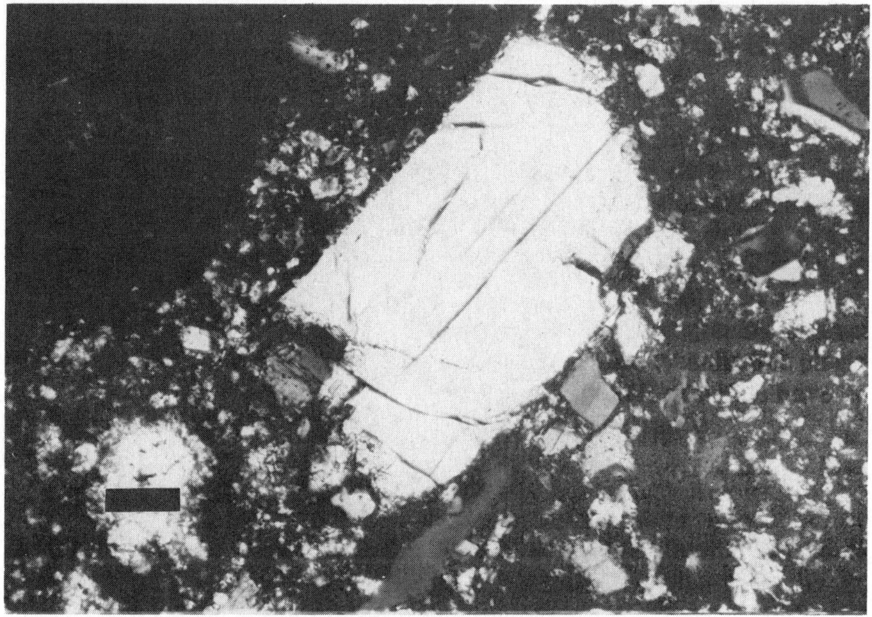

Fig. 6. Photomicrograph of orthopyroxene xenocryst in polished thin section 77115,50. Bar scale = 0.05 mm, transmitted light, crossed polarizers.

DYNAMIC CRYSTALLIZATION EXPERIMENTS USING ANALOG XENOCRYSTS

Based on the observation that rim compositions of 77115 xenocrysts reflect an approach to equilibrium with crystalline matrix phases, previous investigators (Huebner, 1976 and in Chao *et al.*, 1975) interpreted the zoning as due to diffusive cation exchange between matrix and xenocrysts at near solidus temperatures. One objective of the current investigation has been to model experimentally the reaction between xenocrysts and matrix in a cooling system, thereby placing additional restrictions on the thermal history of the lunar sample. These experiments involved dynamic crystallization of analog 77115 matrix about olivine plates ("xenocrysts") of Fa_8 composition. Cooling curves were chosen to comply with experimentally determined constraints (described in a previous section) yet also to provide sufficient time for iron-magnesium cation diffusion in olivine as calculated by Sanford and Huebner (1980).

Run 19, cooled initially at a "fast" rate along curve 18 (Fig. 4), then held isothermally ("slowly" cooled) at a temperature where a significant proportion of residual melt persisted, crystallized matrix olivines with an average composition of $Fa_{31.6}$ nearly identical with those in 77115 matrix. Diffusive exchange of iron and magnesium between xenocryst and matrix is extensive in this run and closely resembles that observed in 77115. The microsubophitic texture of the matrix in run 19 (Fig. 7) is identical with that of run 18 (Fig. 5k) and closely

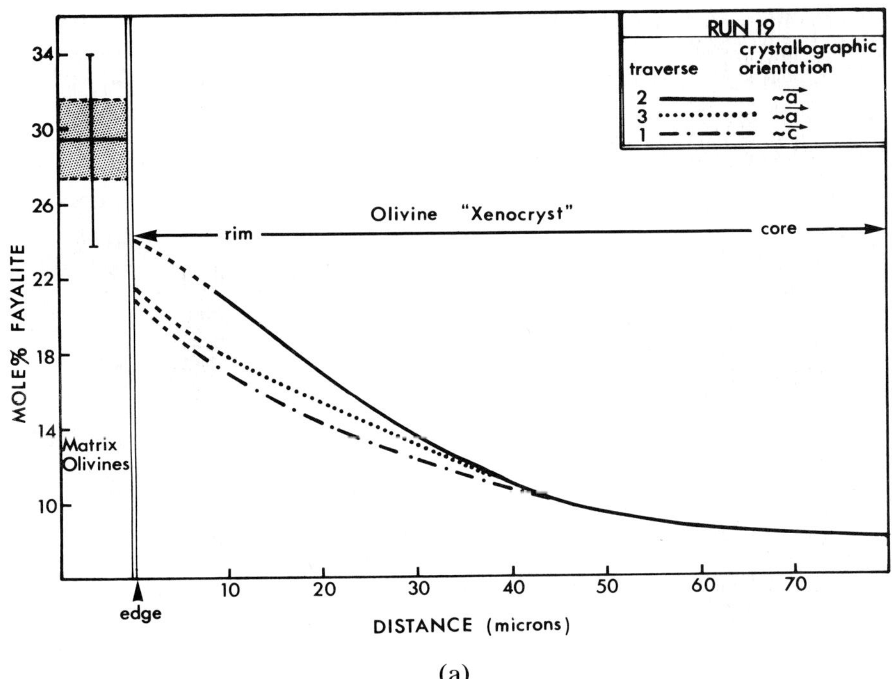
RUN 19
traverse
crystallographic orientation
2 ~a
3 ~a
1 ~c
Olivine "Xenocryst"
rim
core
Matrix Olivines
MOLE % FAYALITE
34
30
26
22
18
14
10
edge
10
20
30
40
50
60
70
DISTANCE (microns)

(a)

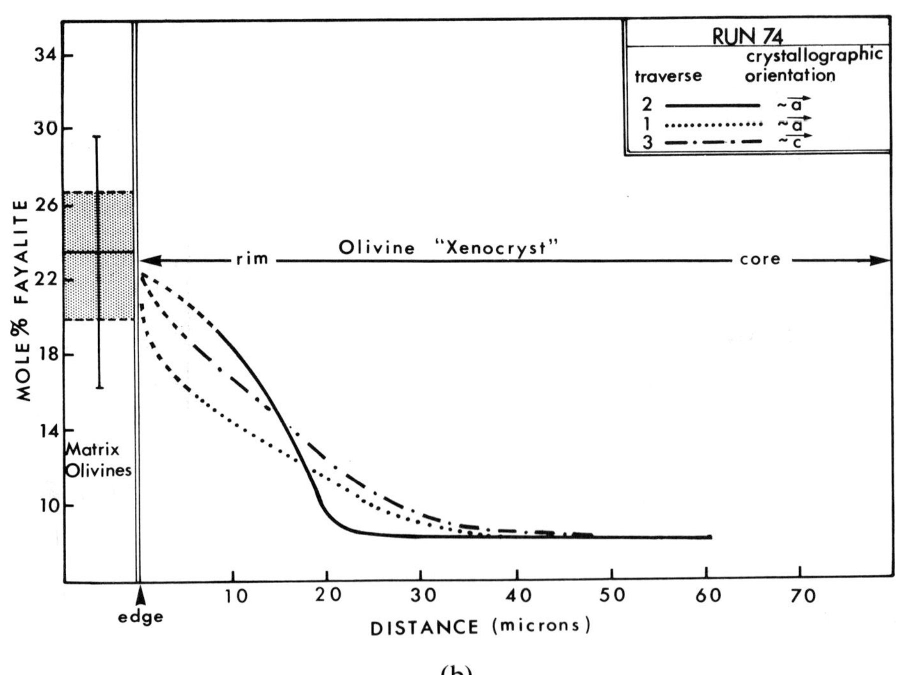
RUN 74
traverse
crystallographic orientation
2 ~a
1 ~a
3 ~c
Olivine "Xenocryst"
rim
core
Matrix Olivines
MOLE % FAYALITE
34
30
26
22
18
14
10
edge
10
20
30
40
50
60
70
DISTANCE (microns)

(b)

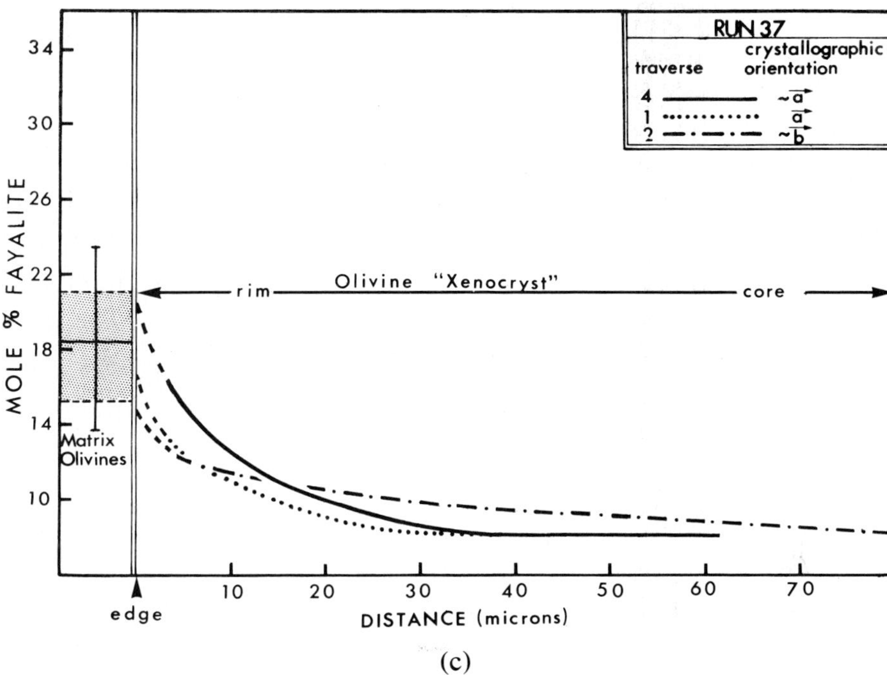

(c)

Fig. 7. Diffusion profiles of olivine "xenocrysts" in dynamically crystallized 77115 analog matrix. Dashed portions of curves represent extrapolated values of % fayalite component within 5μ of the crystal edge (electron microprobe analysis in this region is hampered by the relatively large excitation volume of the electron beam). The composition of matrix olivines (% fayalite) is shown in the left-hand margin of each diagram; the shaded region illustrates the compositional range within one standard deviation (lσ) about the mean and the vertical lines correspond to the range of matrix olivine compositions.
a - run 19; b - run 74; c - run 37.

resembles that of the lunar 77115 matrix (Fig. 1). Matricies in run 19 and other runs containing olivine plates, show textural uniformity throughout the charges and adjacent to "xenocrysts" (Fig. 8). The thermal histories and results of Run 19 and two different analog xenocryst experiments (37,74), are summarized in Table 4. Diffusion profiles of olivine xenocrysts in these experiments are illustrated in Fig. 7 and can be compared with data for 77115 olivine xenocrysts presented by Sanford and Huebner (1980). In these experiments, olivine "xenocryst" rim compositions approach that of matrix olivine. All three runs (Fig. 4) were cooled from a similar subliquidus temperature (1235–1225°C), at which a large number of nuclei existed. Run 74 crystallized an aphanitic texture (Fig. 5j) similar to that of run 19 (Fig. 7) and to the lunar sample (Fig. 1). Run 37 which, as mentioned previously, was cooled slowly from a higher temperature (curve 37, Fig. 4), produced a relatively coarse grained texture (Fig. 5i).

Runs 37, 74, and 19, have matrix olivine of average composition 18.3, 23.5 and

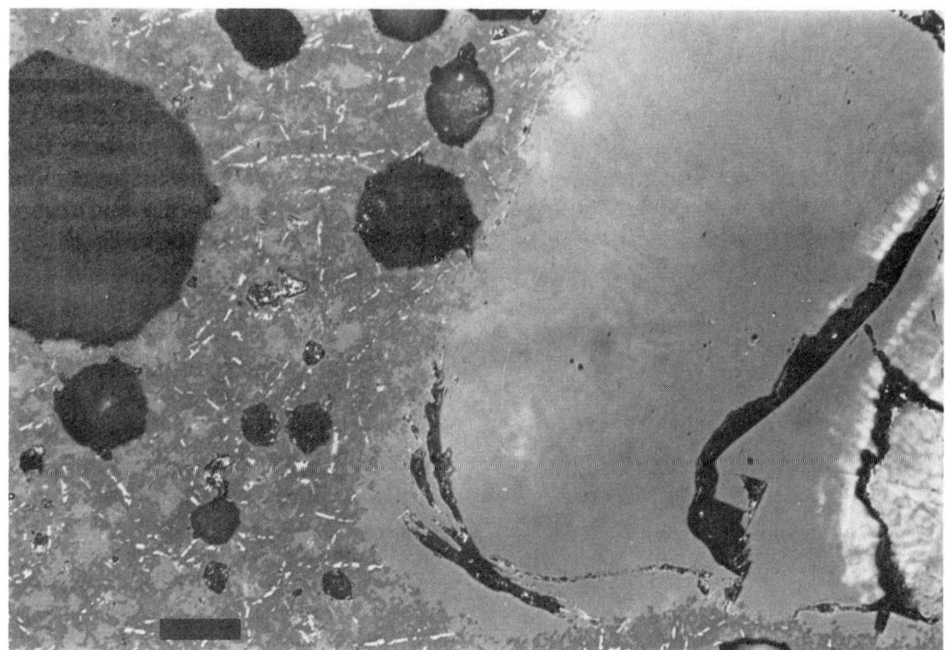

Fig. 8. Reflected light photomicrograph showing matrix texture and olivine "xenocryst" in run 19. Bar scale = 0.05 mm.

31.6 mole percent fayalite, respectively. None of these runs contained the coarse metallic iron blebs characteristic of "iron-loss". As might be predicted from the difference between the analogue xenocrysts' composition (Fa_8) and the composition of respective matrix olivines, the xenocrysts show an increasing extent of Fe-Mg exchange in runs 37, 74, and 19, respectively (Table 4). In experiments 37 and 74, average matrix olivines are more magnesian than those of the 77115 matrix (Fa_{31}). A possible explanation is that in run 37, slow initial cooling resulted in the growth of coarse forsteritic (early) olivine. During more rapid cooling at lower temperatures, these magnesian olivine grains did not have time to react with residual melt, causing that melt to fractionate to a more iron-rich composition than during equilibrium cooling. Melt (or its quenched products) is interstitial to plagioclase in this run. Although good microprobe analyses could not be obtained on this fine-grained material, these areas are enriched in silicon, potassium, and iron, relative to other melts analyzed in this study. Run 74, which was rapidly cooled initially, crystallized olivines that had a higher iron content than those of runs 37, but was not cooled slowly enough at lower temperatures to crystallize olivine with the Fe/(Mg+Fe) characteristic of the lunar sample matrix silicates.

The dynamic crystallization experiments reinforce our earlier conclusions about the cooling history of the lunar 77115 matrix. The cooling curve (#18) with which we best simulated the lunar matrix texture, also caused compositional zoning to develop in analog xenocrysts (in run 19) which reproduces well the zoning observed in the lunar sample. The cooling experiments using analog "xenocrysts" which failed to simulate the lunar sample provide clues to the conditions required to form the features observed in the lunar sample. Comparison of runs 74, 37 and 19 suggests that 1) diffusive cation exchange between xenocryst and matrix is not significantly affected by those differences in early thermal history which drastically affect matrix texture, and 2) that late stage cooling history is the predominant factor controlling the extent of olivine xenocrysts diffusion in the lunar sample. In order for xenocrysts in a rock of 77115 composition to form rims similar to those observed in the lunar sample and for the rock to have cooled within limits suggested by dynamic crystallization experiments, it had to have been maintained at a moderate temperature where a small but significant melt proportion was present. Presumably, this residual melt provides a conduit for cation migration through the matrix, toward and away from the xenocrysts. This observation is consistent with constraints on 77115 late-stage matrix crystallization. The sample must have remained for an extended period of time at a temperature within the olivine-out interval (1180°–1110°C) where residual liquid reacts with olivine to form pigeonite.

SUMMARY

Previous investigations of 77115 and similar highland rock types (Chao *et al.*, 1975; James, 1976; and Minkin *et al.*, 1978) collectively point out difficulties in assigning either an endogenous or impact melting hypothesis to their origin. Chao (Chao *et al.*, 1975) suggested the possibility of endogenous intracrustal melting (igneous partial melting) of 77115 and that slow initial and rapid late stage cooling were required to generate compositionally zoned xenocryst rims and subsequent aphanitic matrix. This interpretation is inconsistent with our interpretation of the thermal history of 77115. Phase equilibria and dynamic crystallization experiments using a synthetic analog of 77115 matrix composition have led to the following conclusions regarding the thermal history of the 77115 fragment-laden lunar basalt.

1. The data illustrate the necessity of a high nucleation density prior to cooling and crystallization of the melt. This places an upper interval limit of 1220°–1240°C on the sample and is consistent with a fragment-laden impact-generated melt sheet hypothesis, where residual crystal fragments serve as nucleii for crystallization.
2. The melt had to crystallize rapidly initially (i.e., 10°–25°C hr^{-1}) through the early-stage crystallization interval (1230°–1180°C) which is consistent with the survival of unrimmed orthopyroxene—which could not have been in equilibrium with the melt at initial stages of cooling.
3. 77115 must have cooled more slowly than 7°C hr^{-1} through the "olivine out" interval (1190°–1110°C) in order to obtain the existing proportion of pigeonite to olivine in the matrix and to form the observed compositional zonation of xenocryst rims.

Acknowledgments—R. Helz and J. McGee of the U.S. Geological Survey are gratefully acknowledged for their critical reviews of this manuscript. Also, R. Wendlandt and O. James provided reviews of a preliminary short paper on this subject which were helpful in writing the current paper. Discussions with R. Sanford were much appreciated. We thank L. B. Wiggins for ably maintaining our ageing microprobe, J. McGee for technical advice in obtaining microprobe analyses and also Nanzetta Merriman for assistance with photomicrography. This research was largely supported by the National Aeronautics and Space Administration (contract T-2356A).

REFERENCES

Bence A. E. and Albee A. L. (1968) Empirical correction factors for the electron microanalysis of silicates and oxides. *J. Geol.* **76,** 382–403.

Chao E. T. C., Minkin J. A., Thompson C. L., and Huebner J. S. (1975) The petrogenesis of 77115 and its xenocrysts: Description and preliminary interpretation. *Proc. Lunar Sci. Conf. 6th,* p. 493–515.

Huebner J. S. (1975) Origin of the SiO_2 variation of mare basalt melts (abstract). In *Lunar Science VI,* p. 411–413. The Lunar Science Institute, Houston.

Huebner J. S. (1976) Diffusively rimmed xenocrysts in 77115 (abstract). In *Lunar Science VII,* p. 396–398. The Lunar Science Institute, Houston.

Huebner J. S. (1979) The system CaO-MgO-FeO-SiO_2-"Others": possible model for the crystallization of basaltic magmas (abstract). *Abstracts With Programs,* Geol. Soc. Amer. **11,** 447 pp.

James O. B. (1976) Petrology of aphanitic lithologies in consortium breccia 73215. *Proc. Lunar Sci. Conf. 7th,* p. 2145–2178.

James O. B., Hedenquist J. W., Blanchard D. P., Budahn J. R., and Compston W. (1978) Consortium breccia 73255: Petrology, major- and trace-element chemistry and Rb-Sr systematics of aphanitic lithologies. *Proc. Lunar Planet. Sci. Conf. 9th,* p. 789–819.

Lipin B. R. (1978) The system Mg_2SiO_4-Fe_2SiO_4-$CaAl_2Si_2O_8$-SiO_2 and the origin of Fra Mauro basalts. *Amer. Mineral.* **63,** 350–364.

Lofgren G. E. (1978) Dynamic crystallization and kinetic melting of the lunar soil. *Proc. Lunar Planet. Sci. Conf. 9th,* p. 959–975.

Lofgren G. E. (1979) The effect of nucleation on basaltic textures (abstract). *Abstracts with Programs,* Geol. Soc. Amer. **11,** 467–468.

Minkin J .A., Thompson C. L., and Chao E. T. C. (1978) The Apollo 17 Station 7 boulder: Summary of study by the international consortium. *Proc. Lunar Planet. Sci. Conf. 9th,* p. 877–903.

Onorato P. I. K., Uhlmann D. R., and Simonds C. H. (1976) Heat flow in impact melts: Apollo 17 Station 6 boulder and some implications to other breccias and xenolith laden melts. *Proc. Lunar Sci. Conf. 7th,* p. 2449–2467.

Roeder P. L., and Emslie R. F. (1970) Olivine-liquid equilibrium. *Contrib. Mineral Petrol.* **29,** 275–289.

Sanford R. F., and Huebner J. S. (1979) Reexamination of diffusion processes in 77115 and 77215 (abstract). In *Lunar and Planetary Science X,* p. 1052–1054. Lunar and Planetary Institute, Houston.

Sanford R. F., and Huebner J. S. (1980) Model thermal history of 77115 and implications for generation of fragment-laden basalts. *Proc. Conf. Lunar Highlands Crust.* This volume.

Storey W. C., Humphries D. J., and O'Hara M. J. (1974) Experimental petrology of 77135. *Earth Planet. Sci. Lett.* **23,** 435–438.

Taylor L. A., Onorato P. I. K., Uhlmann D. R., and Coish R. A. (1978) Subophitic basalts from mare crisium: cooling rates. Mare Crisium: The View from Luna 24 (R. B. Merrill and J. J. Papike, eds.), p. 433–481. Pergamon, N.Y.

Walker D., Longhi J., Lasaga A. C., Stolper E. M., Grove T. L., and Hays J. F. (1977) Slowly cooled microgabbros 15555 and 15065. *Proc. Lunar Sci. Conf. 8th,* p. 1521–1547.

Papike, J.J. and Merrill, R.B., eds.
Proc. Conf. Lunar Highlands Crust (1980), p. 253-269
Printed in the United States of America

Model thermal history of 77115 and implications for the origin of fragment-laden basalts

Richard F. Sanford and J. Stephen Huebner

U.S. Geological Survey, Reston, Virginia 22092

Abstract—Calculations of the cooling history of fragment-laden lunar pigeonite basalt 77115 are based on experimental data of Thornber and Huebner (1980), observed diffusion profiles in 77115 olivine xenocrysts, and idealized heat-flow models. The results indicate that 77115 was subjected to 1) initial rapid cooling (10°–25°C/hr from 1230° to 1180°C) caused by local thermal equilibration on a scale of about 1–10 cm; 2) slow cooling (<7°C/hr) in the temperature range 1180°–1050°C resulting from thermal equilibration with hot host rock; and 3) rapid termination of the cooling process after a few weeks to months. These conclusions narrow the range of possible interpretations on the nature of the impact that is believed to have produced the fragment-laden melts. Because of the similarity of 77115 to other fragment-laden basalts of impact origin from the Apollo 17 landing site, the conditions inferred for the origin of 77115 are considered to be typical, in general respects, of conditions responsible for the formation of many such rocks. Among these rocks, variations in nucleation density, grain size, texture, pyroxene/olivine ratio, and extent of Fe–Mg diffusion in olivine can be used to estimate differences in thermal histories relative to that of 77115.

INTRODUCTION

Studies of Apollo 17 samples are many, but the main features of the rocks and the various interpretations are summarized in a few papers. Regional geology in the vicinity of the Apollo 17 landing site was described by Apollo Field Geology Investigation Team (1973), Wolfe and Reed (1976) and Wolfe *et al*. (1977). Petrogenesis of samples, including 77115, from the Station 7 boulder has been discussed by Chao *et al*. (1974, 1975, 1976) and Minkin *et al*. (1978) among others. Also relevant to the petrology of these rocks are studies by Simonds (1975) and James (1976).

Lunar basalt sample 77115 is typical of fragment-laden pigeonite basalts found at the Apollo 17 landing site. Interpretations about its origin and history have hinged on precise knowledge of its cooling history. Recent experimental work described by Thornber and Huebner (1980) places tight restrictions on interpretations of the cooling history of 77115 and aids in choosing among the various models proposed for the history of this rock. Because 77115 is representative of a major rock type in the lunar highlands, understanding its thermal history will help in understanding the formation of lunar crustal materials. Several controversial issues will be addressed in this report:

1. *Single- versus multiple-impact events*. Most workers agree on the relative ages of the rocks in the Station 7 boulder. From oldest to youngest, these rock types are light-gray noritic breccia, 77215; dark dikelets represented by 77075, which cut 77215; blue-gray fragment-laden basalt 77115, which encloses a meter-wide "clast" of 77215; and greenish-gray fragment-laden basalt 77135, which appears free of fractures that penetrate the other rock types. Most workers agree that the noritic clast 77215 represents older (4.4 b.y., Nakamura *et al.*, 1976; Nakamura and Tatsumoto, 1977) lunar crust that had crystallized well before the enclosing fragment-laden basalts formed. However, controversy exists over the time intervals between emplacement of the individual basalt units. Multiple-impact events are responsible for the different basalt units, according to Wolfe and Reed (1976), Chao *et al.* (1974, 1975, 1976), Nunes *et al.* (1974a and b), Stettler *et al.* (1974, 1975, 1978), Nakamura *et al.* (1976) and Nakamura and Tatsumoto (1977). Alternatively, the various Apollo 17 fragment-laden basalts could have been produced by a single event, the South Serenitatis meteor impact, as advocated by Simonds (1975), Ryder and Wood (1977), Winzer *et al.* (1977), Wolfe *et al.* (1977), James *et al.* (1978), Jessberger *et al.* (1978) and McGee *et al.* (1980).

2. *Source of melt*. Disagreement exists concerning the source of the melt in these rocks. Some workers consider that the fragment laden melts were derived by mixing of clasts and superheated, impact-generated melt (Simonds *et al.*, 1973, 1974; Grieve *et al.*, 1974; Simonds, 1975; James *et al.*, 1975; and Dence *et al.*, 1976). Other workers have suggested that all or part of the melt was derived from igneous activity in the lunar interior, then mixed with debris at or near the lunar surface (Chao *et al.*, 1974, 1975, 1976; Crawford and Hollister, 1974; Hollister, 1975; James, 1976; and Minkin *et al.*, 1978).

3. *Mechanism of cooling*. The fine grain size of the lunar basalts has been interpreted to result from rapid cooling. Chao *et al.* (1974, 1975) and Minkin *et al.* (1978) suggested that this rapid cooling occurred near the surface. Simonds (1975), Simonds *et al.* (1976) and Onorato *et al.* (1978) attributed the fine grain size to rapid equilibration of small cold clasts and superheated melt, and they suggested that the rock crystallized at depth in the ejecta blanket.

RESTRICTIONS ON THERMAL MODEL

Thornber and Huebner (1980) cooled a synthetic analogue of 77115 along various cooling curves to simulate, as well as possible, the actual cooling history of the lunar 77115 sample. These results can be summarized in terms of three quantitative restrictions on the cooling history of 77115. Diffusion profiles in zoned olivine xenocrysts in 77115 provide a fourth restriction.

1. The microsubophitic igneous texture and high density of crystal nuclei indicate that crystallization began from a mixture of clasts and melt within the field of olivine + plagioclase + liquid at a temperature of about 1230°C.

2. Lack of reaction between orthopyroxene xenocrysts and melt, which could

not have been in equilibrium, indicates that these crystals could not have remained long in contact with the melt. That is, initial cooling must have been rapid. Experimental reproduction of the actual grain size in the matrix indicates that the rock cooled through the 1230°–1180°C interval at 10°–25°C/hr. This cooling rate is partly dependent on the previous thermal history and the number of preexisting small clasts that serve as nucleation sites.

3. The rate of reaction of olivine with liquid to form pyroxene indicates that the cooling rate below 1180°C must have been less than 7°C/hr. Some liquid is necessary for this reaction to proceed. The solidus is at about 1050°C, but during nonequilibrium crystallization, some liquid may be present at lower temperatures. Slow cooling in the temperature range 1180°–1050°C is also suggested by the close approach of xenocryst olivine rim compositions to average matrix olivine composition. This approach appears to be aided by the presence of a small amount of liquid that acts as a conduit for diffusion.

4. The actual, integrated time-temperature path of 77115 must have permitted growth of the observed olivine diffusion profiles. Growth of these profiles is calculated for various thermal models.

Any model for the thermal history of 77115 must satisfy these four restrictions. In addition, we assume that the actual cooling curve corresponds in its general form to heat-flow equations for geometrically simple arrangements of hot melt and cooler host rock.

CALCULATION OF DIFFUSION PROFILES IN OLIVINE

Previous calculations of olivine diffusion profiles (Huebner, 1976; Sanford and Huebner, 1979) used an analytical solution (Crank, 1975, Eq. 6.22) to the diffusion equations. This solution used the simplifications of constant temperature and diffusion coefficient to model the profiles observed in the lunar olivine xenocrysts. In the present calculations, finite-difference techniques are used, and diffusion coefficients are allowed to vary with temperature (a function of time), composition, and oxygen fugacity. Instead of assuming that the grains are spherical, the present calculations assume that the olivine grains have planar surfaces, because many margins of olivine xenocrysts are planar, and the diffusion distance is small relative to the grain dimensions. (Deviations from this geometry are found at sharp corners of grains, within small grains, and near fractures in grains, all of which were avoided during data collection.)

The computer program used for calculating olivine diffusion profiles was modified from a finite-difference program for the carburization of steel (Goldstein and Moren, 1978). Diffusion of Fe in olivine is governed by Fick's second law:

$$\frac{\partial C}{\partial t} = \frac{\partial}{\partial x}\left(D\left(\frac{\partial C}{\partial x}\right)\right) \tag{1}$$

where C is the concentration of Fe (mol% Fe_2SiO_4) in olivine, t is time, x is distance from the grain edge, and D is the diffusion coefficient for Fe-Mg exchange. Diffusion of Mg is assumed to be equal to that of Fe but opposite in direction. Eq. (1) can be differentiated to yield:

$$\frac{\partial C}{\partial t} = D\frac{\partial^2 C}{\partial x^2} + \frac{\partial D}{\partial x}\cdot\frac{\partial C}{\partial x} \tag{2}$$

Because D is not a strong function of x, the last term can be neglected. Substitution of the second order correct Crank-Nicholson finite-difference approximations yields:

$$\frac{C_i^{n+1}-C_i^n}{\Delta t} = D_i^n\cdot\frac{1}{2}\left\{\frac{C_{i+1}^{n+1}-2C_i^{n+1}+C_{i-1}^{n+1}}{(\Delta x)^2}+\frac{C_{i+1}^{n}-2C_i^{n}+C_{i-1}^{n}}{(\Delta x)^2}\right\} \tag{3}$$

where i is the number of the grid point along the space coordinate, n is that along the time coordinate, Δt is the time between grid points along the time coordinate, and Δx is the distance between points along the space coordinate. Collecting terms for the nth and (n+1)th grid points on opposite sides of the equation yields:

$$C_{i-1}^{n+1} + C_i^{n+1}\left[-2-2\left(\frac{(\Delta x)^2}{D_i^n\Delta t}\right)\right] + C_{i+1}^{n+1} = -C_{i+1}^{n} + C_i^n\left[2-2\left(\frac{(\Delta x)^2}{D_i^n\Delta t}\right)\right] - C_{i+1}^{n}$$

Knowing the concentration (C_i^n) at every space point i at time n, we can calculate the concentrations (C_i^{n+1}) at time n+1 using the Thomas algorithm for a tridiagonal matrix (Rosenberg, 1975).

Diffusion coefficients used in the calculations were determined from an analytical fit to the data of Buening and Buseck (1973) and are a function of composition, oxygen fugacity, crystallographic direction and temperature. The analytical expression for D of Fe-Mg exchange in olivine was found as follows. First the data of Buening and Buseck (1973) were fit by least-square methods to linear curves relating 1nD, log fO_2, and T(K),

$$1nD = \left(\frac{a_1}{T} + a_2\right)\log f_{O_2} + \left(\frac{a_3}{T} + a_4\right). \tag{5}$$

Coefficients a_i were obtained for each composition in each of the three principal crystallographic directions. Separate expressions were obtained from 1000°–1100°C data and 1150°–1200°C data. Diffusion coefficients for the b axis at 1150°–1200°C, not measured by Buening and Buseck (1973), were estimated by multiplying the a axis coefficient at 1150° and 1200°C by the ratio of the coefficients (D_b/D_a) at lower temperatures, which was found to be 0.765 ($\sigma = 0.034$). Because the diffusion coefficient was measured only at one oxygen fugacity at 1150°–1200°C, we assumed that the dependence of the diffusion coefficient on oxygen fugacity at these temperatures is the same as at lower temperatures. Compositional dependence of D was accounted for by fitting the a_i coefficients to equations of the type

$$a_i = a_{i1}X + a_{i2}(1-X) + a_{i3}X^2(1-X) + a_{i4}X(1-X)^2 \tag{6}$$

where X is the mole fraction of Fe_2SiO_4, and a_{ij} are constants whose values are listed in Table 1. This equation has the same form as that of a regular solution model for the variation of an intensive thermodynamic property with composition. For a given crystallographic direction, the diffusion coefficient is calculated from the expression

$$D = D_a D_b D_c / (D_b^2 D_c^2 \cos^2\phi \sin^2\Theta + D_a^2 D_c^2 \sin^2\phi \sin^2\Theta + D_a^2 D_b^2 \cos^2\Theta)^{1/2} \quad (7)$$

where D_a, D_b, and D_c are diffusion coefficients parallel to the three principal crystallographic directions.

The olivine grain is assumed to be initially chemically homogeneous and to have a constant surface composition. Depending on the type of calculation being performed, different surface and initial core compositions were used as input parameters. For the surface composition in calculations designed to simulate

Table 1. Values of coefficients (a_{ij} in Eq. 6) used in analytical expression for olivine diffusion coefficients.

i	a_{i1}	a_{i2}	a_{i3}	a_{i4}
T ≤ 1100°C				
		a axis		
1	2736.7973	434.0056	−4674.4602	2454.7248
2	−2.2199	−0.0843	5.5594	−0.6927
3	19957.2676	−15006.9124	−53348.1572	27601.0566
4	−18.5904	12.5208	80.4410	−11.2099
i		*b axis*		
1	8474.5363	−1909.7915	−30596.8352	12808.0345
2	−6.3701	1.7525	24.1022	−8.6977
3	110084.2217	−47725.4951	−464592.8750	165542.1992
4	−84.4844	37.8926	377.2997	−118.1000
i		*c axis*		
1	1713.7215	962.0448	−453.9722	−1216.6809
2	−1.0665	−0.2779	1.2212	0.6702
3	16761.6470	−4654.6365	−11711.3651	−13544.9336
4	−10.1416	8.6889	34.4670	2.4768
T > 1100°C				
i		*a axis*		
1	2736.7973	434.0056	−4674.4602	2454.7248
2	−2.2199	−0.0843	5.5594	−0.6927
3	−98031.9053	−33447.5742	237092.0156	43985.6782
4	68.4812	23.1688	−152.7719	−34.9835
i		*c axis*		
1	1713.7215	962.0448	−453.9722	−1216.6809
2	−1.0665	−0.2779	1.2212	0.6702
3	−19901.3042	−33178.8794	−3693.5697	11519.0043
4	14.7212	24.3142	15.7336	−12.9028

diffusion profiles in 77115 olivine xenocrysts, we used the average composition of olivine in the 77115 matrix (Fo_{69}). The core compositions (Fo_{82}-Fo_{91}) of natural olivine grains that have homogeneous cores were used as the starting composition of olivine xenocrysts. For calculations intended to reproduce olivine diffusion profiles produced experimentally, the surface composition in the calculations was set to equal the composition of olivine in the surrounding matrix (which was powdered olivine grains in some runs and a synthetic mixture having the composition of the 77115 matrix in others). The initial compositions of the analogue "xenocrysts" that were modeled were set equal to the composition of the olivine analyzed before the experiment.

Oxygen fugacity in all calculations was set at one log unit below that of the iron-wüstite buffer assemblage (Eugster and Wones, 1962; Huebner, 1975). This oxygen-fugacity range approximates that of the measured lunar samples and of experiments conducted in iron capsules, in vacuo (see Huebner *et al.,* 1976, Fig. 2). Crystallographic directions (ϕ and Θ) of each olivine grain whose profile was calculated were determined optically by using the universal stage.

Model temperatures were varied according to the calculations being performed. For reproduction of experimentally produced diffusion profiles, the actual temperatures in the experiments were used as input. The calculated profile was then compared with the actual profile to determine the accuracy of the model in predicting diffusion profiles. For simulating natural 77115 olivine xenocryst diffusion profiles, various temperature-time paths were tested. In these calculations, simulated diffusion was terminated when a "best fit," as determined by visual comparison of measured and calculated profiles, was obtained. The result is an estimate of the time necessary to produce the actual diffusion process on the moon.

EXPERIMENTALLY PRODUCED DIFFUSION PROFILES

Accuracy of the finite-difference calculations was tested by comparing calculated olivine diffusion profiles with profiles obtained in isothermal and "cooling-rate" experiments. Isothermal experiments were performed upon natural olivine grains ($Fo_{87.5}$) from lunar troctolite 76535 in a powdered synthetic olivine matrix (Fo_{68}). The samples were run for 312, 672 and 1004 hrs at 1050°C and log fO_2 (bars) = -14.5, then quenched in mercury. Excellent agreement was obtained for many profiles, as illustrated in Fig. 1a for a 312 hr experiment. The dip in fayalite content within 5–10 μm of the grain edge is due to "edge effects" associated with the size of the sample excited by the electron microprobe beam and does not indicate true compositional variations. Where found, poor agreement between calculated and observed profiles, in nearly all cases, can be attributed to nearby cracks that could act as a conduit for Fe and Mg and/or to deviation of grain-matrix contacts from the planar surface assumed in the diffusion equations. These observations show that the diffusion data of Buening and Buseck (1973) and the finite-difference approximations are sufficiently accurate to be used in calculating

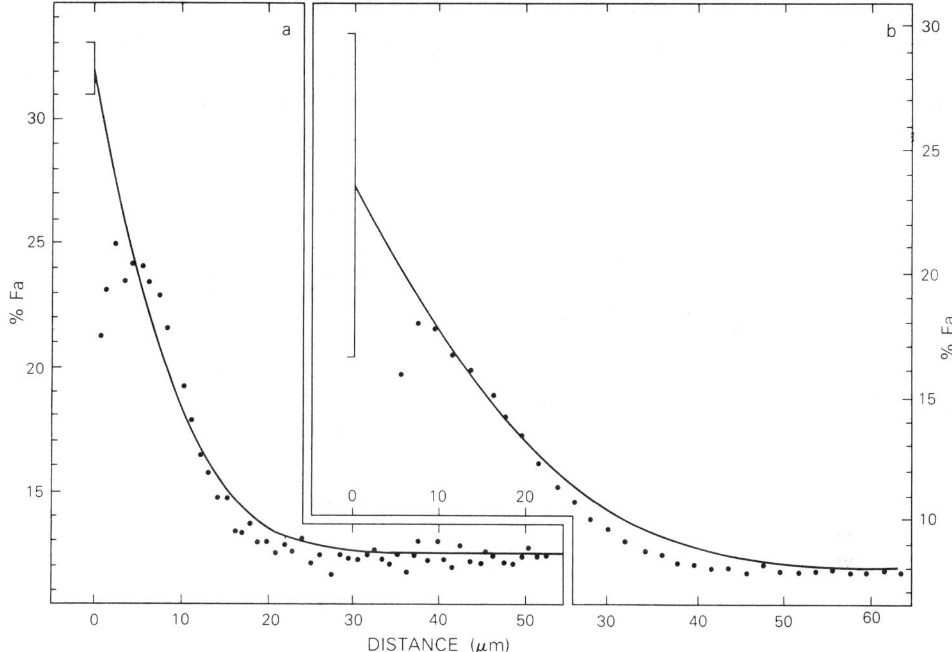

Fig. 1. Experimentally produced and calculated olivine diffusion profiles, mol% Fe_2SiO_4 versus distance. Dots represent analyzed compositions; solid curve represents the calculated profile. Grain edge is at x=0. The measured range of matrix olivine compositions is indicated by the vertical bar at x=0. Dip in Fe_2SiO_4 within 5–10 μm of edge is due to microprobe beam "edge effects." The vertical (composition) axes in the two parts of the figure are drawn to the same scale but are displaced relative to one another. a) Olivine grain ($Fo_{87.5}$) placed in powdered olivine matrix (Fo_{68}) run for 312 hrs at 1050°C and log fO_2 (bars) = −14.5. b) Olivine grain (Fo_{92}) placed in synthetic 77115 matrix composition and run in an iron crucible, in vacuo. The averaged analyzed matrix olivine composition in experiment is $Fo_{76.5}$. Temperature was controlled by equation (8) in which T_o = 1230°C, T_h = 670°C, A = 100 and $0 \leq t \leq 2.5722 \times 10^6$ sec (714.5 hrs). Analyzed profile is the same as traverse 3 run 74 of Thornber and Huebner (1980), Fig. 7b.

the times necessary to produce observed profiles in naturally occurring olivine grains, provided that the grains are free of cracks or other "defects."

Olivine diffusion profiles produced in "cooling rate" experiments by Thornber and Huebner (1980) were compared with curves calculated using the experimental run conditions as boundary conditions. For samples in which rim and matrix olivine compositions coincide, the calculated and experimentally produced profiles agree well as shown in Fig. 1b. For the profile in Fig. 1b, the same temperature-time function was used in the calculations as in experimental run 74 of Thornber and Huebner (1980). The average analyzed matrix olivine composition ($Fo_{76.5}$) in the experiment was used as the rim composition in the calculations, and the core composition (Fo_{92}) was that of the "xenocryst" placed in the experimental charge. Real time for the experiment and calculated time for the model were equal, 714.5 hrs. The agreement between calculated and experimentally produced profiles further validates the methods of calculation.

THERMAL MODELS

During this investigation progressively more complex models for the thermal history of 77115 were developed. Because results of preliminary models were used in later models, we will present a brief synopsis of the models as they became more refined and more closely approximated the experimental conditions.

Initial calculations were performed to find the time necessary to form the profiles observed in 77115 by an isothermal process. Results are shown in Table 2. The fractional sorption M_t/M_∞ (Crank, 1975, Eq. 6.22) was corrected by applying

Table 2. Olivine xenocryst data for specimen 77115.

PTS	Traverse	Core (Fo %)	Profile	M_t/M_∞	a (cm)	Dt, x10^6 (cm^2)	Θ (°)	Φ (°)	D, x10^{11} (cm^2/sec)	t, x10^{-6} (secs)
			Fe_{975}	0.826		4.54				5.62
50	1.2	82.8	Fe_{925}	0.715	0.006	2.82	70	58	0.0808	3.49
			Mg	0.716		2.83				3.47
	3.3	90.8	Fe	0.576	0.0150	9.78	60	22	0.0762	12.78
			Mg	0.512		7.38				9.69
	3.4	79.5	Fe	0.680	0.0060	2.45	24	28	0.2148	1.14
			Mg	0.743		3.18				1.48
	3.10	84.7	Fe	0.627	0.0065	2.29	82	64	0.0704	3.25
			Mg	0.617		2.19				3.11
51	1	85.1	Fe	0.505	0.0110	3.79	83	44	0.0755	5.02
			Mg	0.480		3.37				4.46
	2	91.2	Fe	0.615	0.0115	6.87	39	2	0.1045	6.58
			Mg	0.666		8.60				8.23
	3	82.4	Fe	0.645	0.0130	5.90	23	14	0.2105	2.81
			Mg	0.652		6.05				2.87
	4	88.9	Fe	0.426	0.0115	3.24	74	69	0.0606	5.35
			Mg	0.396		2.76				4.56
53	1	87.8	Fe	0.650	0.0105	6.62	34	37	0.1190	5.56
			Mg	0.642		6.40				5.38

Abbreviations: PTS, polished thin section number; M_t/M_∞, fractional sorption; a, grain radius; D, diffusion coefficient; t, time; Θ, Φ, polar coordinates of crystallographic axes, Fo in mole %.

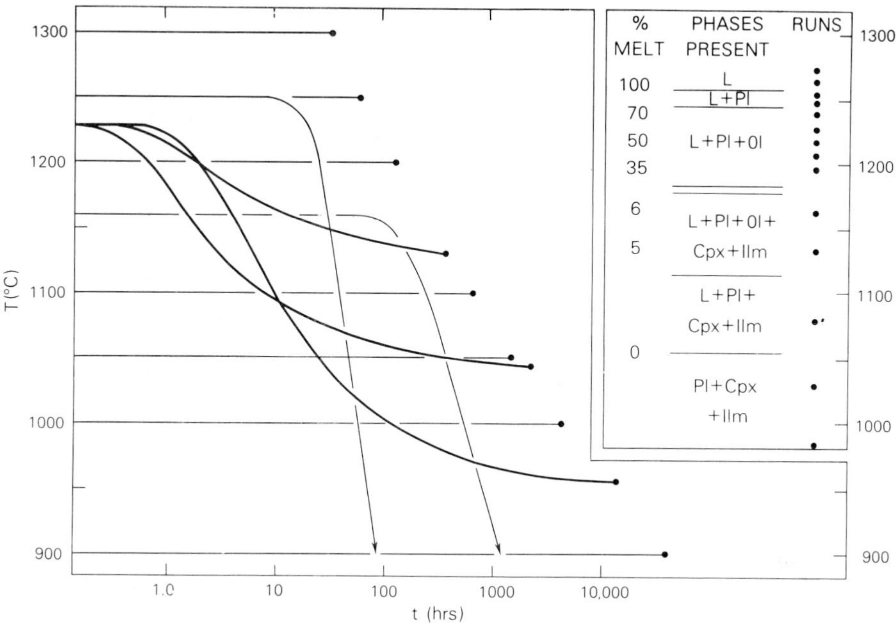

Fig. 2. Typical calculated cooling curves that yielded observed olivine diffusion profiles. Horizontal lines show times for isothermal diffusion. Curves ending at arrows pointing below 900°C are for intrusion into cold host rock according to model 3. Bold curves, for intrusion into hot host rock, satisfy restrictions from experimental data. Equilibrium phase relations (Thornber and Huebner, 1980) are shown in the inset to the right. Abbreviations: L, liquid; Pl, plagioclase; Ol, olivine; Cpx, clinopyroxene; Ilm, ilmenite.

the factor r^2/a^2 (r is the distance from the core of the grain; a is the grain radius) to the microprobe count data to account for the exponential increase in volume from core to rim of the roughly spherical grains. Microprobe traverse orientations with respect to olivine crystallographic axes are expressed in polar coordinates (Θ and Φ). For each traverse, a diffusion coefficient (D) was calculated from the diffusivities for the principal crystallographic axes at 1050°C and log $fO_2 = -14$, using the xenocryst core compositions (Table 2). Values of Dt, calculated from the fractional sorption, were used to obtain values of time (t). The value of t indicates the time required for growth of the rims if the process had taken place isothermally at 1050°C. Estimates for different grains range from 1×10^6 to 13×10^6 sec (13 to 148 days) and are not greatly dependent on the profile (Fe or Mg) used to obtain the fractional sorption. Larger values of t are associated with traverses that do not pass precisely through the centers of the grains and with grains that deviate from the spherical geometry assumed in these calculations. Thus, we prefer the smallest values of t (Table 2), about 1×10^6 sec. In Fig. 2, the family of horizontal lines terminating at 30 hrs (1300°C) to 40,000 hrs (900°C) shows the variation in computed time as a function of temperature for a particular traverse (77115,50, profile 3.10).

Next, an approximate cooling rate was obtained by assuming linear cooling. The values of the diffusion constant at lower temperatures were obtained from Eq. (6) at 1000°–1100°C. For an initial temperature on the liquidus of 1250°C, cooling at 50°C/day to <400°C would produce the observed diffusion profile.

In order to approximate the actual shape of the cooling curve and to evaluate the effect of the geometry of the melt body on the cooling rate, three models having different geometrical arrangements of hot melt and cold rock were tested. These models are illustrated and solutions to the heat flow equations are shown in Fig. 3. Initial calculations for each model were performed by assuming that host-rock temperature (T_h) was 0°C. For a given initial temperature of the melt, only one parameter (A) in the heat flow equations is unknown. The corresponding distance from the sample to the contact (x) can be found by the relation $x = 2A\kappa^2$, where κ is the thermal diffusivity of the rock. Using a value of $\kappa = 0.002$ cm^2/sec determined on a similar lunar "breccia," 77035,44 (Horai and Winkler, 1976) and a starting temperature of 1250°C, we obtain 45, 45 and 27 cm, respectively, for the distance of the sample from the contact with host rock for the three models. The different geometries yield roughly the same distance, on the order of decimeters. Two representative cooling curves using the third model and initial temperatures of 1250°C and 1160°C are shown in Fig. 2 as the curves that terminate below 900°C.

Comparison with the experimental restrictions 2 and 3 above shows that these calculated cooling curves involve too much time at high temperatures (1250°–1180°C) and not enough time at intermediate temperatures (1180°–1050°C). A cooling curve that satisfies these restrictions therefore must have a faster cooling rate at high temperatures and a slower cooling rate at intermediate temperatures. In order to have extended slow cooling at the intermediate temperatures, the adjacent host rock must be either hot or a very good insulator. Because the rocks

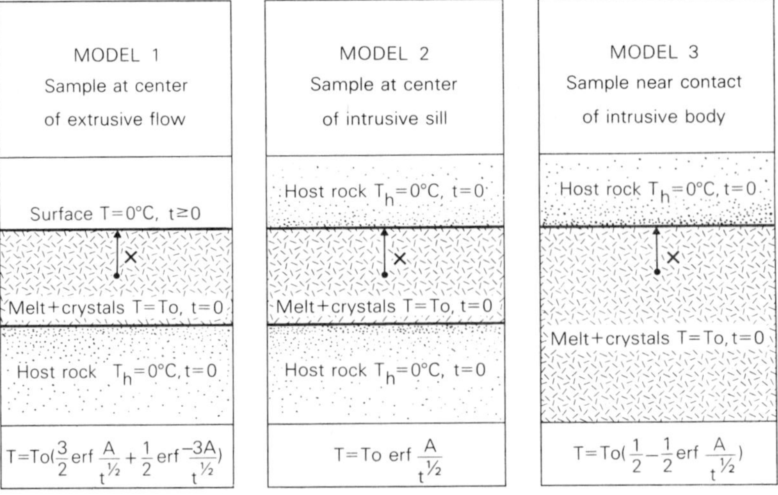

Fig. 3. Three models used to test effect of geometry on cooling rates and sample location.

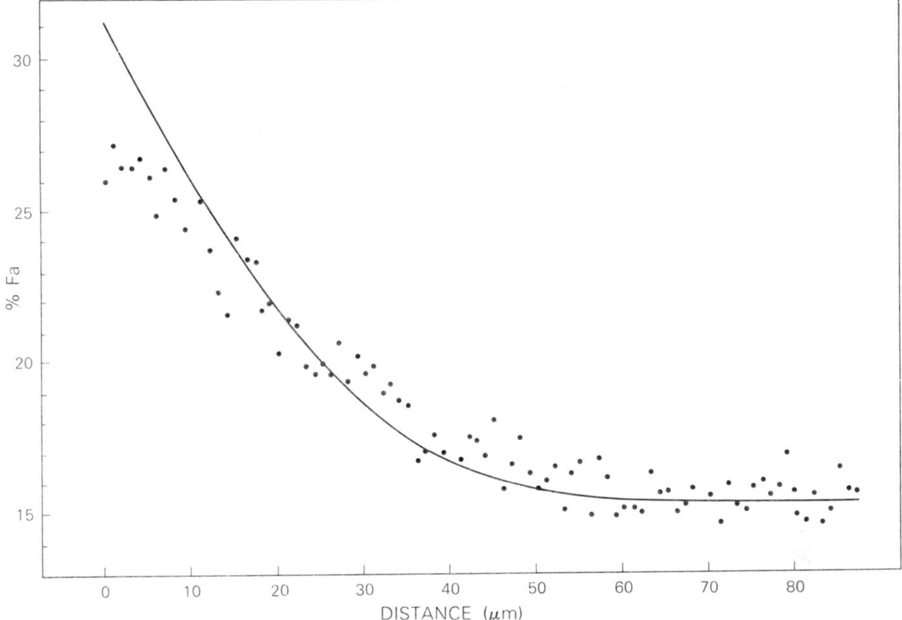

Fig. 4. Composition of olivine xenocryst in thin section 77115,50 as a function of distance from rim (at origin). Traverse number 3.10. Dots represent microprobe analyses; solid line represents the calculated curve. The calculated curve was found by using the boundary conditions of constant rim composition equal to the average matrix olivine composition (Fo_{69}) and the initial xenocryst composition equal to the homogeneous core composition ($Fo_{84.7}$) in the natural grain. Temperature was reduced according to Eq. (8) in which T_o = 1230°C, T_h = 1030°C, and A = 60. The simulated cooling process was terminated at 346.6 hr (14.4 days), when the composition 20 μm from the rim reached $Fa_{22.2}$.

adjacent to 77115 are very similar to 77115, their thermal diffusivities were probably similar. Therefore, the host rock was probably hot while 77115 was cooling.

The equation for heat flow from a large (semi-infinite) intrusive body in hot host rock (geometry as in model 3) can be written,

$$T = (T_o - T_h)\left(\tfrac{1}{2} - \tfrac{1}{2}\,\mathrm{erf}\frac{A}{t^{1/2}}\right) + T_h, \tag{8}$$

where T is the temperature at time t, T_o and T_h are initial temperatures of intrusion and host rock, respectively, and A is a constant, defined above. This is the final form of the equation which we fit to the experimental restrictions by varying T_h and A (T_o was set at 1230°C). Because the final temperature approaches a constant high value (1050°–1180°C) where olivine grains would homogenize in 0.3–5.6 yrs, the cooling process must be terminated when the observed amount of Fe-Mg diffusion is achieved. Typical curves of this type are shown in Fig. 2 as bold lines. Fig. 4 shows a calculated olivine diffusion profile and the measured composition profile 3.10 across an olivine xenocryst in thin section 77115,50.

Our calculations of temperature as a function of time (according to equations in Fig. 3 and Eq. 8) assume that the thermal diffusivity remained constant during cooling, and that the latent heat of fusion is negligible. Although data are insufficient to calculate the magnitude of these approximations, we can evaluate the effects of variable thermal diffusivity and latent heat of fusion in terms of their tendency to increase or decrease the cooling rate. Thermal diffusivities of solids are higher than that of their respective melts at the same temperature. (See, for example, Powell and Childs, 1963.) This effect will tend to increase the cooling rate during crystallization. However, latent heat given off during crystallization will tend to decrease the cooling rate. In addition, we must consider the effect of liberated gases that form vesicles, probably in the late stages of crystallization. The increase in porosity would lower the thermal diffusivity (Robertson, 1979), but the increase in gas pressure would raise it (Horai and Winkler, 1976). Decrease in radiative heat transport as temperature declines (Shankland *et al.*, 1979; Robertson, 1979) both during crystallization and at subsolidus temperatures would lower the thermal diffusivity and tend to decrease the cooling rate. Thus, we have five conflicting effects. In view of the fact that measured diffusivities of lunar rocks are available only to 327°C (Horai and Winkler, 1976), that diffusivities in olivine single crystals (determined to 950°C by Kanamori *et al.*, 1968) are an order of magnitude greater than those in lunar rocks, and that diffusivities in basaltic melts are unknown, it seems appropriate to assume simply that the thermal diffusivity is a constant.

INTERPRETATION

At the beginning of crystallization, a mixture of melt and crystals was present. The fact that many xenocrysts (the more Fe-rich olivine and all the pyroxene xenocrysts) could not be in equilibrium with a melt having the composition of the matrix indicates that the xenocrysts could not be residua from partial melting. The xenocrysts must have been added to the melt, which was probably already superheated (Simonds *et al.*, 1973, 1974; Grieve *et al.*, 1974; Simonds, 1975; James *et al.*, 1975; Dence *et al.*, 1976). Early cooling was rapid enough so that xenocrysts unstable in the melt were not resorbed.

The fine grain size of 77115 can be reproduced by cooling from 1230° to 1180°C in about 4 hrs; this cooling rate is consistent with the sample being within about 1–10 cm of the cooler rock that served as a heat sink. Different geometries of melt and host rock have little effect on this calculated distance. In contrast, thermal equilibration of superheated melt and small, cold clasts would occur in the first 100 sec. (Simonds *et al.*, 1976; Onorato *et al.*, 1978). Such rapid cooling would cause a much finer texture than observed in 77115, for similar initial clast/melt ratios. Although thermal equilibration of melt and small clasts must have taken place during this earliest time period, we see no evidence of this process, such as very fine grained or radial textures about the small clasts. The grain size and microsubophitic texture appear to be due to slower cooling at about 10°–25°C/hr. We suggest that equilibration of the melt-crystal mixture with neighboring melt sheets (such as 77135) or large clasts (such as 77215) was responsible for the observed texture. Our conclusion that the melt sheet equilibrated with *hot* rock also makes it unlikely that the melt crystallized at the Moon's surface as suggested by Chao *et al*. (1974, 1975) and Minkin *et al*. (1978).

An extended time at temperatures slightly above the solidus is necessary for pigeonite to form at the expense of matrix olivine (cooling restriction 3) and for olivine xenocryst rims to form (restriction 4 and Thornber and Huebner 1980). This nearly constant temperature for a period of about a week or more suggests that thermal equilibrium on the scale of decimeters to meters was closely approached during this time, which followed rapid initial cooling and preceded rapid final cooling. Latent heat of fusion probably tended to buffer the temperature and prolong the time spent within the crystallization interval. The exact temperature of thermal equilibrium would be a function of the initial temperatures of melt and clasts and the relative proportions of each.

Termination of the cooling process when the observed amount of olivine diffusion has taken place requires an increase in cooling rate after a week to several months at 1180°–1050°C. Rapid cooling to below 900°C would drastically slow olivine diffusion. Some orthopyroxene xenocrysts lack any optically visible exsolution lamallae, which are nearly ubiquitous in pyroxenes from metamorphic and plutonic environments (e.g., see Huebner *et al*., 1975; Ross and Huebner, 1979). McGee *et al*. (1980) have found submicroscopic (≤25.0 nm) augite lamellae in matrix pigeonite from 77075, 77115 and 77135. The lack of visible exsolution lamellae in orthopyroxene is consistent with rapid final cooling at low temperatures. Such cooling could be caused by 1) effects of radiation from the surface of the ejecta pile, 2) exposure at the surface because of subsequent meteor impact, and 3) exposure at the surface because of late tectonic adjustments in the crust.

1. The effects of radiation can be evaluated by considering field evidence for depth of burial. By analogy with the station 6 boulder, which has a track extending ≈ 500 m up the 2200-m-high North Massif to its probable source, Muehlberger *et al*. (1973) and Apollo Field Geology Investigation Team (1973) inferred that the station 7 boulder also came from the same unit. If the height from the top of the massif to the presumed source of the boulder is an indication of the original depth of burial, then boulder 7 was buried about 1700 m deep. The time required for radiative cooling at the surface to affect rocks at this depth is orders of magnitude greater than the time required to form the observed olivine diffusion profiles and would have resulted in homogeneous olivine xenocrysts. Unless this burial depth is grossly overestimated, rapid final cooling probably was not due to surface radiation while 77115 was at its original depth immediately after impact.

2. We can evaluate whether a subsequent meteor impact is a likely explanation by considering the probability of two impacts taking place weeks to a year apart. Estimates of impact frequency (e.g., Gault *et al*., 1974; Duennebier, *et al*., 1975; Hörz *et al*., 1976; Hartmann, 1979) suggest that two impacts in the same area so close in time would be extremely improbable. However, the ejecta blanket was probably piled in a very unstable arrangement immediately after impact, and a large meteor impact anywhere on the moon might have triggered large-scale slumping of the crater walls.

3. Contraction of the cooling ejecta blanket could similarly cause large-scale

fracturing and slumping at the crater rim within weeks to a year after the impact. Such fracturing is characteristic of basaltic flows, and fallen blocks of fractured lava are common in the volcanic craters of Kilauea, Hawaii.

In contrast to the slow initial cooling followed by rapid cooling of melt suggested by Chao *et al.* (1975), rapid initial cooling followed by slow cooling is suggested by our results. Rapid initial cooling, which helped create the fine-grained texture, is necessary to preserve unreacted orthopyroxene xenocrysts. Later, slow cooling, while only a small fraction of melt remained, produced the zoning observed in olivine and other xenocrysts. If the olivine zoning had formed while at near liquidus temperatures, then orthopyroxene xenocrysts would have been resorbed, olivine nuclei would have been resorbed, the average grain size would have been much coarser, and plagioclase would have formed large euhedral crystals, as demonstrated by the experimental results. Also, as pointed out by Huebner (Alternative Interpretation, *in* Chao *et al.,* 1975), olivine and pyroxene xenocryst rims are too Fe-rich to have equilibrated with a melt having the composition of the 77115 matrix.

Formation of olivine rims at subsolidus temperatures, as originally suggested by Huebner (Alternative Interpretation, *in* Chao *et al.,* 1975) is also unlikely. Apparently a small amount of melt is necessary in order to maintain olivine xenocryst rim compositions (and the compositions of adjacent matrix olivine) at a constant value close to the average matrix olivine composition. However, Huebner's earlier conclusion, that the xenocryst rims represent approach to equilibrium with matrix crystals, is consistent with a small proportion of melt being present in the dominantly crystalline matrix.

On the basis of compositional and textural similarity of matrix material in the fragment-laden basalts from the Apollo 17 landing site, Simonds (1975), Ryder and Wood (1977), Winzer *et al.* (1977), Wolfe *et al.* (1977), James *et al.* (1978) and Jessberger *et al.* (1978) suggested that the basalts formed from the same event, presumably the South Serenitatis impact. A difficulty with this interpretation lies in the fact that 77115 has fractures, which must have formed after the rock was nearly completely crystalline, but 77135 does not. That is, 77115 was cool enough to have fractured at a time that 77135 could not be fractured, either because 77135 was still fluid, or because it had not yet formed. Our results, suggesting that a few percent of melt was still present a week or more after impact, indicates that the fractures in 77115 could not have formed *before* this time. However, 77115 and 77135 could still have originated from the same impact. 77135 differs from 77115 in having larger vesicles and coarser grain size (Chao *et al.*, 1974, 1975; Minkin, 1978). These data suggest that 77135 spent more time at higher temperatures than 77115 or that 77135 initially contained *fewer nucleii* in the form of small clasts. Therefore, 77115 may have cooled, crystallized and fractured, while 77135 was still partly molten and unable to form fractures.

Constraints determined for the cooling history of 77115 can be used as a standard for comparing the cooling histories of other fragment-laden basalts. The crystal form, grain size, and matrix texture are mainly a result of the earliest cooling interval and abundance of preexisting nucleii. Large euhedral crystals are associated with an initially large proportion of melt and few nucleii. Numerous fine-grained crystals indicate a substantial number of nuclei for crystal growth. The microsubophitic texture indicates rapid cooling of a melt and is not formed by metamorphic processes.

The fine grain size resulting from cooling at about 10–25°C/hr through the crystallization interval can be compared with the grain size in other samples to set limits on cooling rates. For example, we deduce that parts of 77135 probably cooled slower than 10°C/hr in the 1230°–1180°C interval, because the grain size is coarser than that of 77115. Local variations in cooling rate, as indicated by variations in grain size, may be related to distance from cooler rock or to initial melt/clast ratios. In contrast, 77075 probably cooled faster than 25°C/hr at high temperatures. This cooling rate is probably due to the small size of the dikelets and a lower temperature for 77215 which the dikelets intrude. The reaction of olivine and melt to pyroxene and Fe-Mg diffusive exchange in olivine xenocrysts probably took place at temperatures slightly above the solidus for a period of about a week. More extensive pyroxene growth, shown by the poikilitic texture of pyroxene oikocrysts in 77135 (Chao *et al.,* 1974; Minkin *et al.,* 1978) and Apollo 16 rocks (Simonds *et al.,* 1973), probably represents slightly higher temperature (but not higher than the upper stability limit of pyroxene), a larger proportion of melt and/or longer times above the solidus. More extensively exchanged olivine rims may also be expected to form in xenocrysts that remain longer in the melt and/or in xenocrysts held at higher temperatures.

Acknowledgments—Carl Thornber generously made available his experimental data and insights on the cooling history of 77115. Stimulating discussions with O. B. James and R. Wendlandt were particularly helpful. E. Robertson and G. Nord of the U.S. Geological Survey reviewed the manuscript and made many constructive suggestions. J. I. Goldstein and A. Romig of Lehigh University kindly provided copies of their finite difference computer programs. This research was supported by NASA Contract T-2356A.

REFERENCES

AFGIT (Apollo Field Geology Investigation Team) (1973) Geologic exploration of Taurus-Littrow: Apollo 17 landing site. *Science* **182,** 672–680.

Buening D. K. and Buseck P. R. (1973) Fe-Mg lattice diffusion in olivine. *J. Geophys. Res.* **78,** 6852–6862.

Chao E. C. T., Minkin J. A., and Thompson C. L. (1974) Preliminary petrographic description and geologic implications of the Apollo 17 station 7 boulder consortium samples. *Earth Planet. Sci. Lett.* **23,** 413–428.

Chao E. C. T., Minkin J. A., and Thompson C. L. (1976) The petrology of 77215, a noritic impact ejecta breccia. *Proc. Lunar Sci. Conf. 7th,* p. 2287–2308.

Chao E. C. T., Minkin J. A., Thompson C. L., and Huebner J. S. (1975) The petrogenesis of 77115 and its xenocrysts: Description and preliminary interpretation. *Proc. Lunar Sci. Conf. 6th,* p. 493–515.

Crank J. (1975) *The mathematics of diffusion.* 2nd ed., Clarendon, Oxford. 414 pp.

Crawford M. L. and Hollister L. S. (1974) KREEP basalt: A possible partial melt from the lunar interior. *Proc. Lunar Sci. Conf. 5th,* p. 399–419.

Dence M. R., Grieve R. A. F., and Plant A. G. (1976) Apollo 17 gray breccias and crustal composition in the Serenitatis Basin region. *Proc. Lunar Sci. Conf. 7th,* p. 1821–1832.

Duennebier F., Dorman J., Lammlein D., Latham G., and Nakamura Y. (1975) Meteoroid flux from passive seismic experimental data. *Proc. Lunar Sci. Conf. 6th,* p. 2417–2426.

Eugster H. P. and Wones D. R. (1962) Stability relations of the ferruginous biotite, annite. *J. Petrol.* **3,** 82–125.

Gault D. E., Hörz F., Brownlee D. E., and Hartung J. B. (1974) Mixing of the lunar regolith. *Proc. Lunar Sci. Conf. 5th,* p. 2365–2386.

Goldstein J. I. and Moren A. E. (1978) Diffusion modeling of the carburization process. *Met. Trans. A* **9,** 1515–1527.

Grieve R. A. F., Plant A. G., and Dence M. R. (1974) Lunar impact melts and terrestrial analogs: Their characteristics, formation and implications for crustal evolution. *Proc. Lunar Sci. Conf. 5th,* p. 261–273.

Hartmann W. K. (1979) Formative conditions controlling structure of planetary highlands (abstract). In *Papers Presented to the Conference on the Lunar Highlands Crust,* p. 42–44. Lunar and Planetary Institute, Houston.

Hollister L. S. (1975) Outline of a model for lunar evolution (abstract). In *Lunar Science VI,* p. 381–383. The Lunar Science Institute, Houston.

Horai K. and Winkler J. L. Jr. (1976) Thermal diffusivity of four Apollo 17 rock samples. *Proc. Lunar Sci. Conf. 7th,* p. 3183–3204.

Hörz F., Gibbons R. V., Hill R. E., and Gault D. E. (1976) Large scale cratering of the lunar highlands: Some Monte Carlo model considerations. *Proc. Lunar Sci. Conf. 7th,* p. 2931–2945.

Huebner J. S. (1975) Oxygen fugacity values of furnace gas mixtures. *Amer. Mineral.* **60,** 815–823.

Huebner J. S. (1976) Diffusively rimmed xenocrysts in 77115 (abstract). In *Lunar Science VII,* p. 396–398. The Lunar Science Institute, Houston.

Huebner J. S., Lipin B. R., and Wiggins L. B. (1976) Partitioning of chromium between silicate crystals and melts. *Proc. Lunar Sci. Conf. 7th,* p. 1195–1220.

Huebner J. S., Ross M., and Hickling N. (1975) Significance of exsolved pyroxenes from lunar breccia 77215. *Proc. Lunar Sci. Conf. 6th,* p. 529–546.

James O. B. (1976) Petrology of aphanitic lithologies in consortium breccia 73215. *Proc. Lunar Sci. Conf. 7th,* p. 2145–2178.

James O. B., Brecher A., Blanchard D. P., Jacobs J. W., Brannon J. C., Korotev R. L., Haskin L. A., Higuchi H., Morgan J. W., Anders E., Silver L. T., Marti K., Braddy D., Hutcheon I. D., Kirsten T., Kerridge J. F., Kaplan I. R., Pillinger C. T., and Gardiner L. R. (1975) Consortium studies of matrix of light gray breccia 73215. *Proc. Lunar Sci. Conf. 6th,* p. 547–577.

James O. B., Hedenquist J. W., Blanchard D. P., Budahn J. R., and Compston W. (1978) Consortium breccia 73255: Petrology, major- and trace-element chemistry, and Rb-Sr systematics of aphanitic lithologies. *Proc. Lunar Planet. Sci. Conf. 9th,* p. 789–819.

Jessberger E. K., Staudacher T., Dominik B., and Kirsten T. (1978) Argon-argon ages of aphanite samples from consortium breccia 73255. *Proc. Lunar Planet. Sci. Conf. 9th,* p. 841–854.

Kanamori H., Fujii N., and Mizutani H. (1968) Thermal diffusivity measurement of rock-forming minerals from 300° to 1100°K. *J. Geophys. Res.* **73,** 595–605.

McGee J. J., Nord G. L., Jr., and Wandless M. V. (1980) Comparative thermal histories of matrix pyroxenes from Apollo 17 boulder 7 fragment-laden melt rocks (abstract). In *Lunar and Planetary Science XI,* p. 700–702. Lunar and Planetary Institute, Houston.

Minkin J. A., Thompson C. L., and Chao E. C. T. (1978) The Apollo 17 station 7 boulder: Summary of study by the International Consortium. *Proc. Lunar Planet. Sci. Conf. 9th,* p. 877–903.

Muehlberger W. R., Batson R. M., Cernan E. A., Freeman V. L., Hait M. H., Holt H. E., Howard K. A., Jackson E. D., Larson K. B., Reed V. S., Rennilson J. J., Schmitt H. H., Scott D. H., Sutton R. L., Stuart-Alexander D., Swann G. A., Trask N. J., Ulrich G. E., Wilshire H. G., and Wolfe E. W. (1973) Preliminary geologic investigation of the Apollo 17 landing site. In *Apollo 17 Prelim. Sci. Rep.* NASA SP-330, p. 6–1 to 6–91.

Nakamura N. and Tatsumoto M. (1977) The history of the Apollo 17 station 7 boulder. *Proc. Lunar Sci. Conf. 8th,* 2301–2314.

Nakamura N., Tatsumoto M., Nunes P. D., Unruh D. M., Schwab A. P., and Wildeman T. R. (1976) 4.4 b.y.-old clast in boulder 7, Apollo 17: A comprehensive chronological study by U-Pb, Rb-Sr and Sm-Nd methods. *Proc. Lunar Sci. Conf. 7th,* p. 2309–2333.

Nunes P. D., Tatsumoto M., and Unruh D. M. (1974a) U-Th-Pb and Rb-Sr systematics of Apollo 17 boulder 7 from the north massif of the Taurus-Littrow valley. *Earth Planet. Sci. Lett.* **23,** 445–452.

Nunes P. D., Tatsumoto M., and Unruh D. M. (1974b) U-Th-Pb systematics of some Apollo 17 lunar samples and implications for a lunar basin excavation chronology. *Proc. Lunar Sci. Conf. 5th,* p. 1487–1514.

Onorato P. I. K., Uhlmann D. R., and Simonds C. H. (1978) The thermal history of the Manicouagan impact melt sheet, Quebec. *J. Geophys. Res.* **83,** 2789–2798.

Powell R. L., and Childs G. E. (1963) Thermal Conductivity. In *American Institute of Physics Handbook,* p. 142–162 3rd ed., section 4g. McGraw-Hill, N.Y.

Robertson E. C. (1979) Thermal conductivities of rocks. U.S. Geol. Survey open-file report 79–356, 52 pp.

Rosenberg D. U. von (1975) *Methods for the Numerical Solution of Partial Differential Equations.* Elsevier, N.Y. 128 pp.

Ross M. and Huebner J. S. (1979) Temperature-composition relationships between naturally occurring augite, pigeonite, and orthopyroxene at one bar pressure. *Amer. Mineral.* **64,** 1133–1155.

Ryder G. and Wood J. A. (1977) Serenitatis and Imbrium impact melts: Implications for large-scale layering in the lunar crust. *Proc. Lunar Sci. Conf. 8th,* p. 655–668.

Sanford R. F. and Huebner J. S. (1979) Reexamination of diffusion processes in 77115 and 77215 (abstract). In *Lunar and Planetary Science X,* p. 1052–1054. Lunar and Planetary Institute, Houston.

Shankland T. J., Nitsan U., and Duba A. G. (1979) Optical absorption and radiative heat transport in olivine at high temperature. *J. Geophys. Res.* **84,** 1603–1610.

Simonds C. H. (1975) Thermal regimes in impact melts and the petrology of the Apollo 17 Station 6 boulder. *Proc. Lunar Sci. Conf. 6th,* p. 641–672.

Simonds C. H., Phinney W. C., and Warner J. L. (1974) Petrography and classification of Apollo 17 non-mare rocks with emphasis on samples from the Station 6 boulder. *Proc. Lunar Sci. Conf. 5th,* p. 337–353.

Simonds C. H., Warner J. L., and Phinney W. C. (1973) Petrology of Apollo 16 poikilitic rocks. *Proc. Lunar Sci. Conf. 4th,* p. 613–632.

Simonds C. H., Warner J. L., Phinney W. C., and McGee P. E. (1976) Thermal model for impact breccia lithification: Manicouagan and the moon. *Proc. Lunar Sci. Conf. 7th,* p. 2509–2528.

Stettler A., Eberhardt P., Geiss J., and Grögler N. (1974) ^{39}Ar-^{40}Ar ages of samples from the Apollo 17 station 7 boulder and implications for its formation. *Earth Planet. Sci. Lett.* **23,** 453–461.

Stettler A., Eberhardt P., Geiss J. Grögler N., and Guggisberg S. (1975) Age sequence in the Apollo 17 station 7 boulder (abstract). In *Lunar Science VI,* p. 771–773. The Lunar Science Institute, Houston.

Stettler A., Eberhardt P., Geiss J., Grögler N., and Guggisberg S. (1978) Chronology of the Apollo 17 Station 7 boulder and the south Serenitatis impact (abstract). In *Lunar and Planetary Science IX,* p. 1113–1115. Lunar and Planetary Institute, Houston.

Thornber C. R., and Huebner J. S. (1980) An experimental study of the thermal history of fragment-laden "basalt" 77115. *Proc. Conf. Lunar Highlands Crust.* This volume.

Winzer S. R., Nava D. F., Schuhmann P. J., Lum R. K. L., Schuhmann S., Lindstrom M: M., Lindstrom D. J., and Philpotts J. A. (1977) The Apollo 17 "melt sheet": Chemistry, age and Rb/Sr systematics. *Earth Planet. Sci. Lett.* **33,** 389–400.

Wolfe E. W., Bailey N. G., Lucchitta B. K., Muehlberger W. R., Scott D. H., Sutton R. L., and Wilshire, H. G. (1977) The geological investigation of the Taurus-Littrow valley: Apollo 17 landing site. U.S. Geol. Survey open-file report 77–783, 800 pp.

Wolfe E. W., and Reed V. S. (1976) Geology of the massifs at the Apollo 17 landing site. *U.S. Geol. Survey J. Res.* **4,** 171–180.

Papike, J.J. and Merrill, R.B., eds.
Proc. Conf. Lunar Highlands Crust (1980), p. 271-337
Printed in the United States of America

Lunar highland melt rocks: Chemistry, petrology and silicate mineralogy

D. T. Vaniman[1] and J. J. Papike[2]

[1] Los Alamos Scientific Laboratory, Geoscience Division, Los Alamos, New Mexico 87545
[2] Department of Earth and Space Sciences, State University of New York, Stony Brook, New York 11794

Abstract—A selected suite containing several of the largest samples of lunar highland melt rocks includes impact melt specimens (anorthositic gabbro, low-K Fra Mauro) and volcanic specimens (intermediate-K Fra Mauro). Although previous assumptions of LKFM volcanism have fallen into disfavor, no fatal arguments against this hypothesis have been presented, and the evidence of a possibly "inherited igneous" olivine-plagioclase cosaturation provides cause for keeping a volcanic LKFM hypothesis viable. Comparisons of silicate mineralogy with melt rock compositions provide information on the specimen's composition and cooling history. Plagioclase-rock compositions can be matched to the experimentally determined equilibria for appropriate samples to identify melt rocks with refractory anorthitic clasts. Olivine-rock compositions indicate that melt rock vitrophyres precipitate anomalously Fe-rich olivine; the cause of this anomaly is not immediately evident. The Al-Ti and Ca-Fe-Mg zonation in pyroxene provide information on relative cooling rates of highland melt rocks, but Cr- and Al-content (where Al-rich low-Ca pyroxene cores are preserved in rapidly cooled samples) can be correlated with composition of the host rock.

INTRODUCTION

Lunar highland melt rocks are samples of feldspathic composition (wt.% ratio $CaO/Al_2O_3 < 0.7$) with $TiO_2 + FeO + MgO < 22$ wt.% and with textures that reflect crystallization from thin, mostly-molten bodies. The relative concentrations of incompatible elements in these highland melt samples provide a more precise distinction from the less feldspathic, more femic mare basalts: the trace-element signature of lunar highland melt rocks is that of the lunar highlands crust in general, being marked by a constant negative slope on chondrite-normalized REE plots (wt. La/wt. Yb $\simeq$ 3). This constant slope reflects the "spiking" of most highland samples with variable amounts of incompatible-element (REE) rich, LIL-enriched (K), and phosphorus-uranium-thorium enriched compositions ("KREEP").

Highland melt rocks include samples which are volcanic, and samples which crystallized from impact melt-sheets. At present the most generally accepted criterion for distinguishing between these two origins is based on siderophile contamination: samples with each of the elements Ir, Au, Re and Os in concentrations >1 ppb are inferred to have obtained these high siderophile concentrations by meteoritic impact contamination (Warren and Wasson, 1977). Irving (1975) uses the presence of metals high in Ni (>1.5 wt.% Ni) to identify samples which are meteorite contaminated, although low-Ni metals may also occur in impact melt rocks. Irving (1975) and other workers suggest that impact melt rocks may occur with little siderophile contamination; extrusive texture and low siderophile contents alone are not proof of a volcanic origin. A suggestion of volcanic origin requires some assumption of petrologically acceptable source and magma compositions.

The problem of distinguishing highland volcanic rocks from highland impact melts has been a focus of much research. At present there are sufficient data to identify many samples which are impact-melt mixtures of a variety of highland cumulus materials, to identify a very few samples of highland volcanic rocks, and to suggest that there may be a suite of highland melt rocks which remelted ancient lunar igneous compositions by impact.

The purpose of this paper is to review the geochemistry and petrography of several large-sample (>1 cm) highland melt rocks, and to apply this review to a comparison of mineral/rock systematics by concentrating on silicate mineralogy. An important part of this review will be the distinction of mineral-composition characteristics which depend on melt-rock composition from characteristics which are modified by crystallization and cooling histories.

CLASSIFICATIONS OF HIGHLAND MELT ROCKS

Several studies have already synthesized a great deal of the available data on highland melt rocks. Irving (1975) summarizes the chemical and textural systematics of highland melt rocks, discussing mineral compositions in his classification scheme. Other studies (Naney *et al.*, 1977; Floran *et al.*, 1976) group lunar highland melt rocks on the basis of chemical data alone. A six-category classification by Naney *et al.* (1977), based on cluster analysis using 9 major elements, is shown in Fig. 1. The present paper uses Naney *et al.* (1977) as a basis for selecting a representative suite of 17 highland melt rocks that span the major compositional and textural groups, that are large enough to ensure reliable analysis of major- and trace-element composition, and that include samples of undoubted impact-melt origin as well as probable volcanic samples. Available analyses of these 17 samples are plotted on the Walker *et al.* (1972) pseudoternary in Fig. 2; chemical data for these samples are summarized in Table 1, and modal data are listed in Table 2.

We are confident that the analyses listed in Table 1 are representative and adequate for the applications in this paper. One may argue that clasts in highland melt rocks will be a factor in biasing chemical analyses, but in only four of the samples discussed here is the clast content greater than 5%, and the greatest clast content is ~16% (Table 2). Moreover, the detailed study by Simonds (1975) on

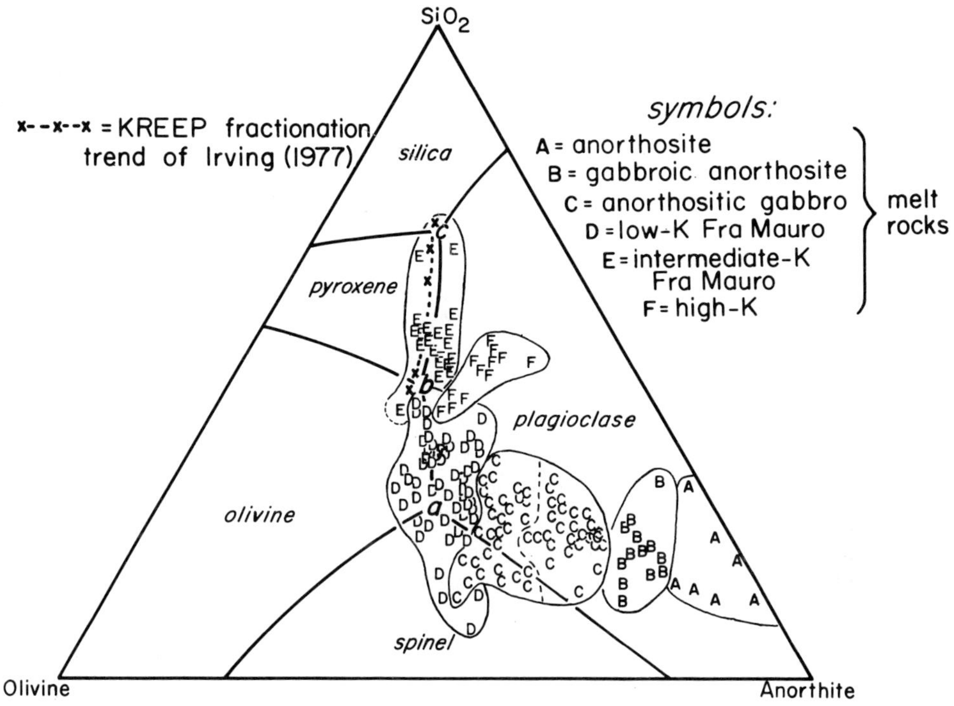

Fig. 1. 173 melt rocks plotted in the olivine-anorthite-silica pseudoternary (projection method of Walker *et al.*, 1972). Dashed line in group C represents a lower-significance split of that group into low-An and high-An anorthositic gabbro.

the Apollo 17 Station 6 boulder, a poikilitic-ophitic melt-rock, indicates that the matrices of impact melt rocks are very homogeneous. Figure 2 plots multiple analyses (where available) for each of the seventeen samples chosen for this paper. Some scatter is observed (e.g., 60335, 14310), possibly due to analyses of clasts plus matrix. Wherever possible we have chosen an analysis by X-ray fluorescence to be representative, since analysis by this method generally requires a larger sample size than the other methods used to obtain the analyses cited in Table 1 (INAA, atomic absorption, isotope dilution). In the case of sample 66095, the three analyses plotted in Fig. 2a differ considerably from the single analysis of 66095 plotted in one of our previous papers (Vaniman and Papike, 1978). The single analysis plotted previously may have included a sizeable gabbroic anorthosite clast (or clasts); the petrographic descriptions of the Lunar Receiving Laboratory (document MSC 03210, 1972) indicate that ~10% of this sample is composed of gabbroic anorthosite clasts, which we attempted o avoid by concentrating on the matrix. For an alternative explanation, the reader is referred to work by Garrison and Taylor (1979) indicating heterogeneities within the melt matrix of 66095.

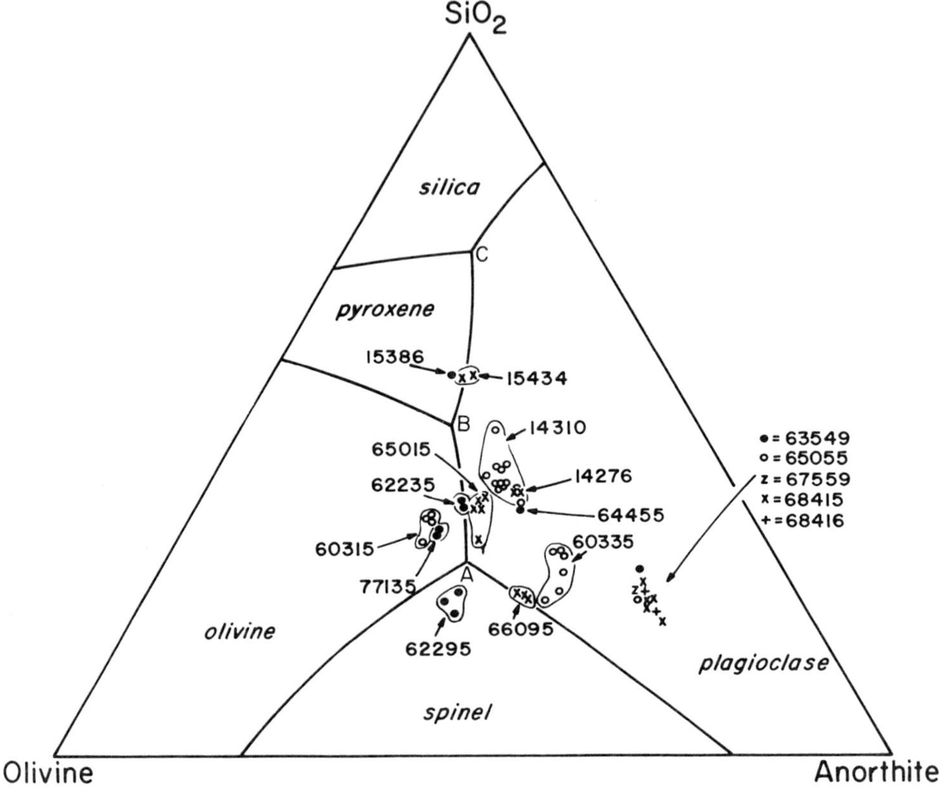

Fig. 2. a. Compilation of all available whole-rock analyses for 17 representative large highland melt rock samples. Note the separate symbols for 14310 (o) and 14276 (x).

Readers of Irving (1975), of Naney *et al.* (1977), and of several studies on lunar "quartz monzodiorite" and other granitic-type compositions (Ryder, 1976; Irving, 1977; Taylor *et al.* 1979), will note that some melt-rock compositions are not represented among the 17 samples discussed in this paper. These compositions include the troctolitic melt rock group of Irving (1975), the high-K melt rock group of Naney *et al.* (1977), and the KREEP-rich granitic-type compositions related to intermediate-KREEP melt rocks by either crystal/liquid fractionation (Irving, 1977) or liquid immiscibility (Hess *et al.,* 1978). The troctolitic melt rocks are treated as part of a larger compositional group designated as Low-K Fra Mauro (LKFM) in the cluster analysis of highland melt rocks by Naney *et al.* (1977). Although the traditional usage of the term "LKFM" is reserved for samples with ~11 wt.% MgO and olivine-plagioclase cosaturation (Reid *et al.,* 1972), the cluster analysis performed by Naney *et al.* (1977) on 148 melt-rock compositions using 9 major elements shows a continuous gradation between troctolitic basalts (17–13 wt.% MgO) and those samples which fit the original definition of LKFM with a range of 13 to 9 wt.% MgO (Reid *et al.,* 1972). Other workers have suggested a possible petrogenetic relationship between Mg-rich and Mg-poor

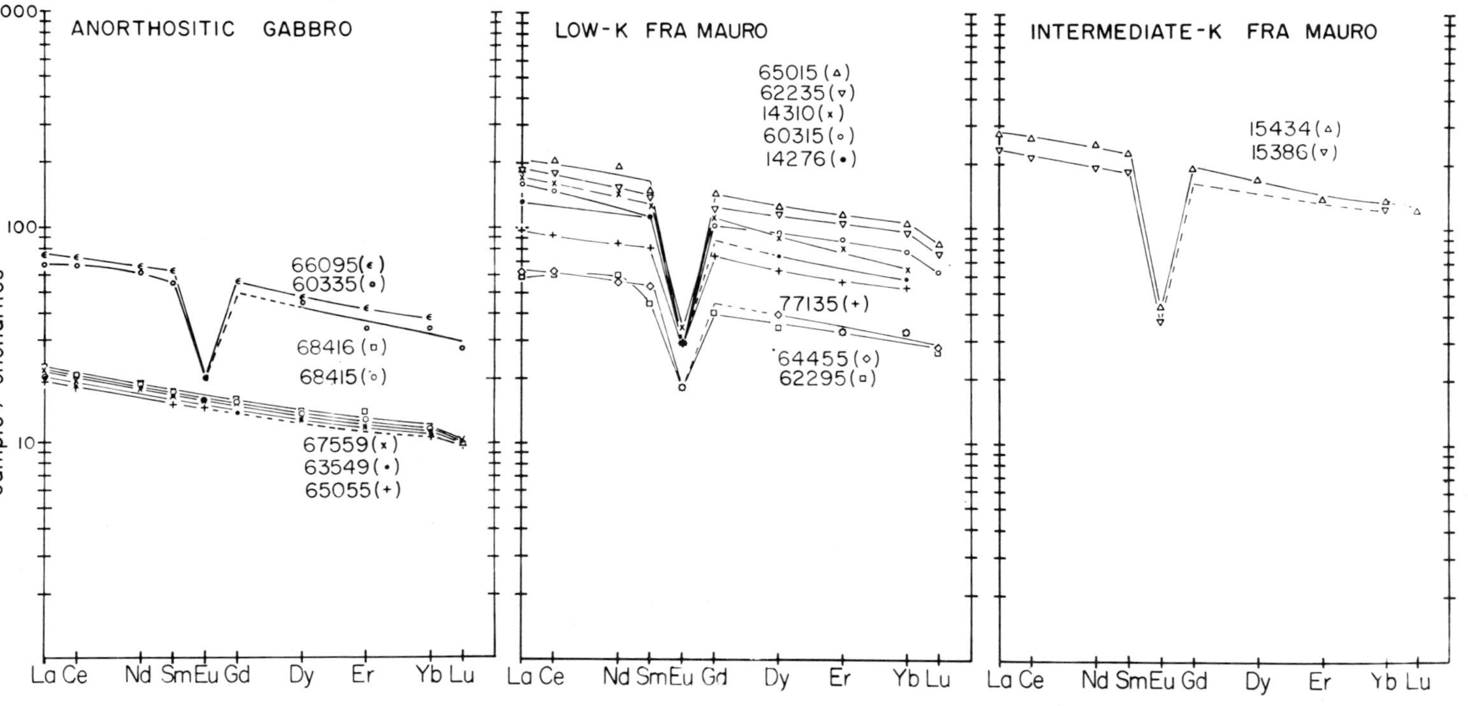

Fig. 2. b. Chondrite-normalized REE data for the samples plotted in Fig. 2a.

Table 1.

	High-An Anorthositic gabbro					Low-An Anorthositic gabbro	
	67559	68415	68416	63549	65055	60335	66095
SiO_2	45.3	45.40	45.60	45.68	45.41	46.19	44.47
TiO_2	0.26	0.32	0.30	0.30	0.28	0.58	0.71
Al_2O_3	27.42	28.63	28.40	28.59	28.46	25.27	23.55
FeO	4.31	4.25	4.22	4.27	3.90	4.51	7.16
MnO	0.054	0.06	0.06	0.05	0.05	0.07	0.08
MgO	4.47	4.38	4.64	4.33	4.81	8.14	8.75
CaO	16.40	16.39	16.32	15.20	16.13	14.43	13.69
Na_2O	0.50	0.41	0.44	0.43	0.44	0.52	0.42
K_2O	0.078	0.06	0.08	0.07	0.13	0.23	0.15
Cr_2O_3	0.08	0.10	0.13	0.09	0.08	0.13	0.15
P_2O_5	0.113	0.07	0.07	0.07	0.13	0.19	0.24
Σ	99.0	100.07	100.26	99.08	99.82	100.26	99.37
Mg/(Mg+Fe)	0.65	0.65	0.66	0.64	0.69	0.76	0.69
CaO/Al_2O_3	0.60	0.57	0.57	0.53	0.57	0.57	0.58
La	7.2	6.81	7.24	6.39	6.2	19.9	18.5
Ce	18	18.3	18.4	16.6	16	51.5	50.0
Pr						7.8	
Nd	11	10.9	11.5	10.6		32.4	40.0
Sm	3.1	3.09	3.28	2.99	2.6	9.05	8.8
Eu	1.20	1.11	1.11	1.03	1.0	1.28	1.63
Gd		3.78	4.07	3.67		10.6	14.0
Tb	0.66			0.58	0.55	1.90	1.3
Dy	4.3	4.18	4.29	3.90		11.5	9.4
Ho						2.5	1.3
Er		2.57	2.86	2.40		6.77	4.1
Tm							
Yb	2.4	2.29	2.42	2.23	2.1	6.23	4.9
Lu	0.34	0.34		0.26	0.29	0.675	0.90
Y		22	21		19	57	70.0
La/Yb	3.00	2.97	2.99	2.87	2.95	3.19	3.78
Cr	~500	~700	~900	~600	~500	910	~1000
V	21	20	21	13	35	26	110
Sc	8.8	8.2	9.2	6.8	7.2	8.53	6.8
Ni	257	184	205	192	392	720	710.0
Cu		12	14		2.4	7.44	3.9
Zn	≤ 5.6	4.8		1.12	0.56	4.27	92.0
Ga	3.8	2.0	1.7	2.54	3.0	3.12	3.8
V/Ni	0.08	0.11	0.10	0.07	0.09	0.04	0.15
Cr/V	~24	~35	~43	~46	~14	35	~9
U	0.39	0.32	0.34	0.329	0.311	0.92	1.0
Th	1.10	1.26	1.24	0.9	1.18	2.75	2.2
Zr		97.5	80		72	292	322
Hf	2.4	2.4		1.8	2.1	6.86	5.0
Sn							
Nb		5.6	<10		<10	10	20.6
Mo							
Th/U	2.82	3.94	3.65	2.74	3.79	2.99	2.20

Table 1. *(Continued)*

Low-K Fra Mauro								Intermediate-K Fra Mauro	
14276	14310	64455	62295	60315	62235	65015	77135	15386	15434
47.60	47.2	48.5	45.16	46.75	47.04	46.99	46.13	50.83	52.14
1.20	1.24	0.65	0.70	1.38	1.21	1.26	1.54	2.23	2.23
21.34	20.1	22.4	20.05	17.10	18.69	19.68	18.01	14.77	16.16
7.94	8.38	5.47	6.40	8.64	9.45	8.59	9.11	10.55	10.82
0.12	0.11	0.074	0.09	0.11	0.11	0.112	0.13	0.16	0.17
7.10	7.87	9.29	14.85	13.42	10.14	9.31	12.63	8.17	7.84
13.18	12.3	13.4	11.85	10.50	11.52	11.90	11.03	9.71	9.67
0.72	0.63	0.57	0.48	0.61	0.48	0.55	0.53	0.73	0.64
0.48	0.49	0.245	0.11	0.49	0.34	0.36	0.30	0.67	0.61
0.26	0.18	0.16	0.19	0.24	0.20	0.19	0.20	0.35	0.36
0.40	0.34		0.15	0.48	0.41	0.41	0.28	0.70	
100.34	98.8	100.8	100.03	99.72	99.59	99.35	99.89	98.87	100.64
0.61	0.63	0.75	0.81	0.73	0.66	0.66	0.71	0.58	0.56
0.62	0.61	0.60	0.59	0.61	0.62	0.60	0.61	0.66	0.60
43	56.4	21.1	19.2	49.6	60.3	65.5	32.1	83.5	90
	144	56.0	52.2	136	160	185	81.2	211	228
	17		7.4	19.2	21.5	26.0			
	87.0	34.0	33	88	94	118	51.6	131	150
20.0	24.0	9.8	8.10	20.6	25.2	26.7	14.6	37.5	40.4
1.9	2.15	1.23	1.21	1.95	1.93	1.97	1.99	2.72	3.00
	28.1		10.4	26.0	31.7	36.0	18.5	45.4	48.0
4.8	5.1	1.97	1.76	4.67	5.56	6.28		7.9	7.9
23	32.7	12.2	11.0	28.5	35.0	38.1	19.1	46.3	51.7
7.3	6.5	2.7	2.4	6.0	6.8	8.3			11
	19.7	6.7	6.9	18.0	21.0	23.8	11.4	27.3	27
	3.0			1.4					
11.7	18.4	6.6	6.29	15.8	19.4	21.1	10.5	24.4	27.2
	2.5	0.96	0.905	2.10	2.59	2.84		3.4	4.12
200	174		59	131	166	174	109		
3.68	3.07	3.20	3.05	3.14	3.11	3.10	3.06	3.42	3.31
~1800	1260	1100	1300	1600	1400	1280	1400	2400	2400
25	36		23	47	65				
18.2	20	7.8	10.3	15.1	16.5	17.1		23.6	38.7
113	271	540	330	878	830	730	86	12.5	
38.0	5		18.1	18.6	24.9	5.06			
1.7	1.8	4.0	22.1	2.37	11	1.46	4	3.5	
3.8	3.2	3.05	2.74	3.85	3.82	3.81			
0.22	0.13		0.07	0.05	0.08				
~72	35		~57	~34	~22				
2.5	3.10	1.430	0.76	2.19	2.08	2.22	1.42	2.8	
8	10.42		2.70	6.84	7.83	8.92	5.5	10	
620	842		283	720	858	940	501	970	
16	21.0	7.8	6.55	17.1	20.7	21.6		31.6	40
33	52		18	31	43	48	33		
3.20	3.35		3.55	3.12	3.76	4.02	3.87	3.57	

Table 1. *(Continued)*

	High-An Anorthositic gabbro					Low-An Anorthositic gabbro	
	67559	68415	68416	63549	65055	60335	66095
Zr/Hf		40.6			34.3	42.6	64.4
Zr/Nb		17.4			>7	29.2	15.6
Cs						0.28	0.40
Rb		1.704	1.705	1.764		6.18	4.15
K%		0.065	0.05	0.0594	0.11	0.19	0.12
Ba	76	76.2	78.2	74	80	193	237
Pb		0.784				2.0	15.0
Sr		182.4	130	170.2	140	162.5	154
K/Rb			~300	337		~300	~290
Rb/Cs						22.0	10.4
Ba/Rb		44.7	45.9	42.0		31.2	57.1
Rb/Sr		0.0093	0.0131	0.0104		0.038	0.027
Sr/Eu		164	117	165	140	127	94
K/Ba		~8.5	~6.4	8.03	~14	~10	~5
ppb Ir	11	4.58		7.8	10.2	17	
ppb Au	5.0	2.7		3.1	5.0	16.8	18
ppb Re		0.434				3.8	
ppb Os							
Ta	0.28			0.3	0.3	0.81	0.44
Co	20.6	11	10	16.9	29.0	36.8	44.0
ppb-Ru	26				26		
ppb-Ge	320	73		96	240	700	
ppb-In	≤11			6.8	6.4		680.0
ppb-Cd		2.75		5.8	1.9		
ppb-W						301	
ppb-As						193	
ppb-Se		98				230	
Li		5.1	4.4	4.8	2.2	13.1	10.7
F						48	
Cl						15.8	220
B						0.055	
Be						2.2	
Sb		0.53					

Oxides in weight percent, all other data in ppm unless stated otherwise. Data sources are as follows. 67559: Nava, 1974; Wasson *et al.*, 1977. 68415: LSPET, 1972; Hubbard *et al*, 1974. 68416: Rose *et al.*, 1973; Hubbard *et al.*, 1974. 63549: Hubbard *et al.*, 1974. 65055: Christian *et al.*, 1976; Boynton *et al.*, 1976. 60335: LSPET, 1972; Wänke *et al.*, 1976. 66095: LSPET, 1972; Brunfelt *et al.*, 1973. 14276: Brunfelt *et al.*, 1972; Rose *et al.*, 1972. 14310: Hubbard *et al.*, 1972. 60315: LSPET, 1973a; Rose *et al.*, 1973; Wanke *et al.*, 1976. 62235: LSPET, 1973a;

LKFM samples (note the incorporation of the olivine-spinel saturated basalt 62295 in studies of olivine-plagioclase cosaturated compositions by Walker *et al.*, 1973). Since we use a modified classification scheme based on Naney *et al.* (1977), we retain troctolitic basalts within the LKFM group.

Of the six groups of highland melt rocks identified by Naney *et al.* (1977), only three are discussed here: anorthositic gabbro, LKFM, and intermediate-K Fra

Table 1. *(Continued)*

Low-K Fra Mauro								Intermediate-K Fra Mauro	
14276	14310	64455	62295	60315	62235	65015	77135	15386	15434
38.8	40.1		43.2	42.1	41.4	43.5		30.7	
18.8	16.2		15.7	23.2	19.9	19.6	15.2		
0.57	0.54	0.27	0.560	0.560	0.350	0.423			
14	12.8	6.0	5.58	9.54	9.32	10.2	7.32	18.46	15.77
0.40	0.4075	0.203	0.09	0.41	0.28	0.30	0.25	0.56	0.51
540	630		197	479	530	570	337	837	
9.0	6.18		1.9	5.2		4.66			
170	250		132	159	160.7	146	172.2	187.4	140.7
~290	318	340	~160	~430	~300	~290	~340	~300	~320
24.6	23.7	22.2	10.0	17.0	26.6	24.1			
38.6	49.2		35.3	50.2	56.9	55.9	46.0	45.3	
0.08	0.0512		0.0423	0.0600	0.0580	0.0699	0.0425	0.0985	
90	116		110	82	83	74	87	69	47
~7	6.47		~4.6	~8.6	~5.3	~5.3	~7.4	~6.7	
	10.5	2.25	5.5	9.4	17	17		0.061	
0.3	4.31	1.56	7.0	16.7	17.6	10.9		0.22	
	1.02	0.284	1.2	1.3	1.7	1.8			
	10.5								
2.1	2.3		0.78	1.96	2.17	2.46		4.6	
13	17.0	31.1	23.0	35.8	54.2	42.6		17.9	36
	14								
	110	62	642	625	800	600		61	
56	30			0.07	10.0				
<20	8.4	5.3	4.9	5.0		9.25		10	
2.5	1200		372	488	994	883			
160	30		110	340	250	170			
	120	190	186	520	260	290			
21	22		14.9	24		20.9	19.3	27.2	
			50	52		23			
23	22		7.6	9.7	21.0	7.4			
				12					
4.0	4.2		2.1	5.2					
0.024	0.004	0.00045	0.00088	0.011		0.00434			

Wanke *et al.*, 1976. 62295: Walker *et al.*, 1973; Wanke *et al.*, 1976; Rose *et al.*, 1973. 64455: Haskin *et al.*, 1973. 65015: Haskin *et al.*, 1973; Duncan *et al.*, 1973; Wanke *et al.*, 1973. 77135: Rhodes *et al.*, 1974; LSPET, 1973b; Hubbard *et al.*, 1974. 15386: Rhodes and Hubbard, 1973; Blanchard (unpublished); Warren and Wasson, 1978. 15434: Helmke *et al.*, 1973; Helmke (unpublished).

Mauro (IKFM). The three groups which are not discussed (anorthosite, gabbroic anorthosite, and high-K) are considered to be minor, inadequately sampled, or local variants restricted to a single occurrence. The granite-type compositions constitute a special sub-group of IKFM that may represent liquid immiscibility, introducing complexities of liquid/liquid interaction which we deliberately avoid in this paper in order to restrict our data to problems of crystal/liquid interaction.

Table 2. Modal Data for the Highland Melt-Rock Suite.

	67559	68415*	68416	63549	65055	60335	66095	14276*
MATRIX								
feldspar	74.7	82.2	75.5	62.4	68.5	56.5	45.0	64.6
pyroxene								
clinopyroxene	(20.1)	11.9	13.6	(35.0)	(27.6)	13.4	(30.6)	13.3
orthopyroxene		—	2.3			0.6		15.7
olivine	2.1	3.1	6.7	—	—	9.0	10.2	—
silica	—	<0.1	—	—	—	—	—	—
ilmenite	0.2	0.1	<0.1	—	—	0.3	—	1.2
ulvöspinel	—	<0.1	—	—	0.7	—	—	0.1
Mg,Al spinel	—	—	—	—	—	—	—	—
metal	0.2	0.2	0.1	0.1	0.7	0.1	1.2	0.3
troilite	0.3	<0.1	0.3	0.1	0.4	0.8	0.1	0.1
phosphate	—	—	—	—	—	<0.1	—	0.5
mesostasis	1.4	2.1	1.5	2.4	2.0	3.6	—	3.5
glass	—	—	—	—	0.2	—	1.4	—
accessories (<0.1%)		schreibersite armalcolite				schreibersite armalcolite	schreibersite cohenite sphalerite akagareite	schreibersite armalcolite tranquillityite K-Ba feldspar
CLASTS								
xenocrysts								
feldspar	—	—	—	—	—	15.0	8.5	?
olivine	—	—	—	—	—	—	—	—
pyroxene	—	—	—	—	—	—	0.1	—
spinel	—	—	—	—	—	—	—	—
xenoliths	—	—	—	—	—	POIK(0.8)	POIK(0.3) ANT(1.6) LMB(1.0)	—

* (68415, 14276 and 14310 from Gancarz *et al.*, 1972; 62295 from Walker *et al.*, 1973) (65015 from Albee *et al.*, 1973)
Xenolith types are poikilitic (POIK) and ophitic (F.B.) melt rocks, light-matrix breccias (LMB), and anorthositic rocks (ANT).

Table 2. *(Continued)*

	14310*	64455	62295*	60315	62235	65015*	77135	15434	15386
MATRIX									
feldspar	59.0	59.9	35.5	43.2	34.1	57.1	41.1	44.0	37.5
pyroxene									
clinopyroxene	16.6	(32.7)	—	4.9	6.0	6.4	(30.0)	18.9	18.4
orthopyroxene	16.3		—	39.0	36.3	28.9		16.0	20.0
olivine	—	6.5	28.3	6.1	6.0	1.1	15.1	—	—
silica	—	—	—	—	—	—	—	3.9	4.4
ilmenite	1.8	—	—	2.1	2.8	1.2	1.4	2.7	3.5
ulvöspinel	0.1	—	—	—	—	—	—	0.2	—
Mg,Al spinel	—	—	6.5	—	—	—	—	—	—
metal	0.1	0.2	<0.1	0.7	1.2	0.4	0.2	—	0.4
troilite	0.1	0.3	—	—	0.1	0.1	0.1	<0.1	0.3
phosphate	0.3	—	—	—	—	0.7	—	<0.1	—
mesostasis	4.4	0.4	n.a.	—	—	3.6	—	—	15.3
glass	—	—	30	—	—	—	—	14.1	—
accessories (<0.1%)	schreibersite armalcolite tranquillityite	schreibersite akaganeite	schreibersite	K feldspar					
CLASTS									
xenocrysts									
feldspar	?	—	0.1	3.0	9.0	(<1)	6.2	—	—
olivine	—	—	<0.1	1.0	—	—	—	—	—
pyroxene	—	—	—	—	0.5	—	2.0	—	—
spinel	—	—	<0.1	—	—	—	—	—	—
xenoliths	—	—	—	—	ANT(3.4) F.B.(0.6)	—	ANT(2.9) F.B.(1.0)	—	—

METHODS

Electron microprobe analyses of feldspar, pyroxene, olivine, ilmenite, armalcolite and spinel were collected from specimens for which such data were scant or unavailable. Analyses were collected to a uniform standard of 9 elements in pyroxene and olivine (Si, Al, Ti, Fe, Mn, Mg, Ca, Cr, Na), 7 elements in feldspar (Si, Al, Fe, Mg, Ca, Na, K), and 7 elements in the opaque oxide minerals (Si, Al, Ti, Fe, Mn, Mg, Cr). During this study electron microprobe data were collected from polished thin sections 67559,9; 68416,5; 68416,69; 63549,5; 63549,8; 65055,16; 65055,32; 60335,70; 60335,73; 60335,75; 66095,81; 66095,139; 66095,140; 64455,36; 62295,65; 62295,76; 60315,61; 60315,66; 60315,76; 62235,6; 62235,67; 62235,69; 77135,17; 77135,23; 77135,128; 15386,3; 15434,116; 15434,131. Approximately 1100 mineral analyses were collected from these specimens. Analyses were made on an ARL EMX-SM microprobe with three automated positionable spectrometers and one semi-fixed spectrometer used for determination of Si. Counting intervals were standarized by beam integration; counting times for each element were 20–25 seconds. The accelerating potential was maintained at 15 KV with sample current of 0.015 μA (brass sample).

Representative mineral analyses are summarized in Tables A-1 to A-4, including analyses which were kindly provided by D. Beaty of California Institute of Technology from data collected by A. Albee, B. Dymek, A. Gancarz, and A. Chodos. In the following pages we discuss in some detail the minor-element characteristics of pyroxene, the major-element features of olivine, and the Na-Ca composition of plagioclase. We have not included a discussion of minor elements (Fe, Mg) in plagioclase even though other workers have found that Fe/Mg ratios can be useful in identifying plagioclase clasts within melt rocks (Walker *et al.*, 1973). Our reason for avoiding this topic lies in the low content of Mg in plagioclase clasts and in many plagioclase crystals which appear to have grown from melt. In the great majority of plagioclase xenocrysts, Mg was below our detection limits (~0.005 cations/6 oxygens). Moreover, Fe contents often were also below detection limits. Our goal was to obtain as many accurate analyses as we could by running in the automated mode; we found that accurate measurement of Mg in many highland plagioclase samples at such low levels would have required manual operating conditions that would not meet our goal. (The Fe + Mg content of plagioclase may be useful in distinguishing clasts, even though both elements may be below detection limits in the clasts. This is the case in sample 66095, where both Fe and Mg are below detection limits in xenocryst plagioclase but well above detection limits in the groundmass plagioclase. However, this distinction is not generally applicable to other samples.)

The same polished specimens used for microprobe analysis were point counted in reflected light with transmitted light capability for identifying silicate minerals. A total of 500 points were identified on each of the polished specimens (14,000 for all samples totalled).

PETROGRAPHIC DESCRIPTIONS: SEVENTEEN REPRESENTATIVE HIGHLAND MELT ROCKS

The samples described here are arranged in categories based on those of Naney *et al.* (1977), modified by variation in chemical or petrographic character.

Naney Group c: anorthositic gabbro
1. An-rich anorthositic gabbro (67559,68415,68416)
2. olivine-free, An-rich anorthositic gabbro (65055, 63549)
3. An-poor anorthositic gabbro (60335, 66095)

Naney Group d: low-K Fra Mauro
1. basaltic-textured, high-Al_2O_3 LKFM (14276, 14310, 62295, 64455)
2. poikilitic, low-Al_2O_3 LKFM (60315, 62235, 65015, 77135)

Naney Group e: intermediate-K Fra Mauro
1. clast-free low-siderophile IKFM (15434, 15386)

Modal data for these samples are summarized in Table 2. Photomicrographs are presented in Fig. 5.

Anorthositic gabbro

An-rich anorthositic gabbros

Anorthositic gabbros with >50–55 mol.% An/(An + SiO_2 + Ol) include 67559, 68415 and 68416. The two samples 68415 and 68416 were collected about 20–30 cm apart on a boulder of ~0.5 m diameter (LSPET, 1973). This boulder is one of the largest melt rocks at the Apollo 16 site. Sample 67559 is similar to 68415 and 68416 in texture and mineral chemistry (Steele and Smith, 1973). These rocks are ophitic/intergranular to intersertal with groundmass feldspar laths ≤0.5 mm and feldspar phenocrysts of ~1 mm. LSPET (1972) reports irregular 1 cm clasts of plagioclase in 68415 and 68416 which are not seen in thin section; most of the large feldspar crystals seen in thin section are probably phenocrysts (Gancarz *et al.*, 1972) though rare, irregular plagioclase grains of An_{98-93} might be xenocrysts (Helz and Appleman, 1973). Feldspars crystallized from melt have calcic cores (An_{98}) and zone outward to sodic rims (~An_{77}) though reversals have been noted and attributed to autointrusion or convection overturn in a cooling lava lake (Gancarz *et al.*, 1972), or to late stage volatile loss or migration (Walker *et al.*, 1973). Olivines in these three rocks cover approximately the same compositional range (Fo_{73-61}). These olivines are less magnesian than those from the more mafic anorthositic gabbro 60335 (Fo_{83-78}).

Although 68415 and 68416 are almost identical in bulk composition (Table 1), there are significant differences in their pyroxene structural and compositional ranges and in the makeup of their residual minor-phase assemblages. Sample 68416 is somewhat coarser-grained and has orthopyroxenes ($En_{82}Wo_4$) zoned to pigeonite (~$En_{67}Wo_8$) and augite (as calcic as $En_{46}Wo_{40}$) with little Fe-enrichment (minimum Mg'^{*}_{px} = 0.55). In comparison, 68415 is slightly finer-grained and lacks orthopyroxene. In contrast to 68416, 68415 has pronounced Fe enrichment in augite and pigeonite grains included within the mesostasis (Fig. 3). Pyroxenes in both 68415 and 68416 have initial low-Ca compositions poor in Ti (<0.01 cations/6 oxygens in orthopyroxene, slightly more Ti in low-Ca pigeonite) with heterogeneous Al content (0.05 to 0.15 cations/6 oxygens) and are zoned to Ca and/or Fe-rich pyroxenes with Al:Ti ≃ 2. Minor phases in 68416 include metal, troilite and ilmenite with residual alkali-rich glass. The more evolved minor-phase assemblage in 68415 includes metal with minor amounts of schreibersite, troilite, ilmenite, ulvöspinel, armalcolite, and cristobalite. Walker *et al.* (1973) note that the occurrence of metal within early-formed plagioclase laths suggests early liquid immiscibility and oxygen fugacity well below the Fe/FeO buffer. In his compilation of metal Ni-Co data, Irving (1975) notes that 68415 has metals with a much broader range of Ni + Co content than the metals in 68416. We note that the metals in 68416 tend to concentrate as large single grains (up to 0.3 mm) and are not as abundant or as dispersed as in 68415.

Olivine-free, An-rich anorthositic gabbro.

The two basaltic-textured melt rocks 63549 and 65055 are similar to 67559, 68415 and 68416 in bulk composition, but show a much more pronounced pyroxene

*Mg′ = Mg/(Mg + Fe) atomic.

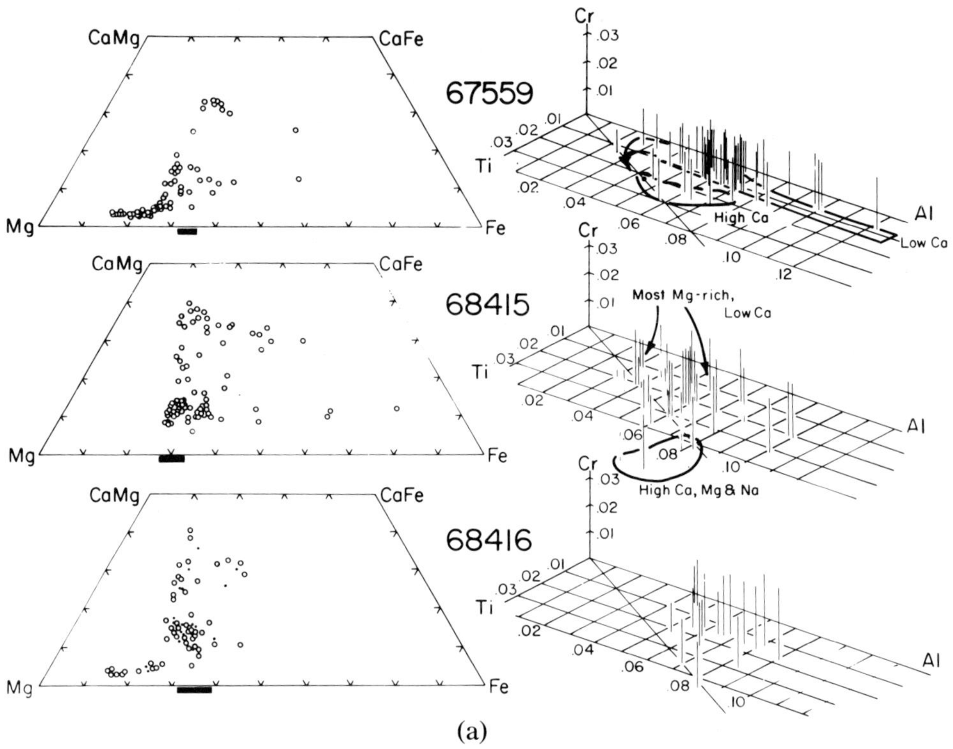

(a)

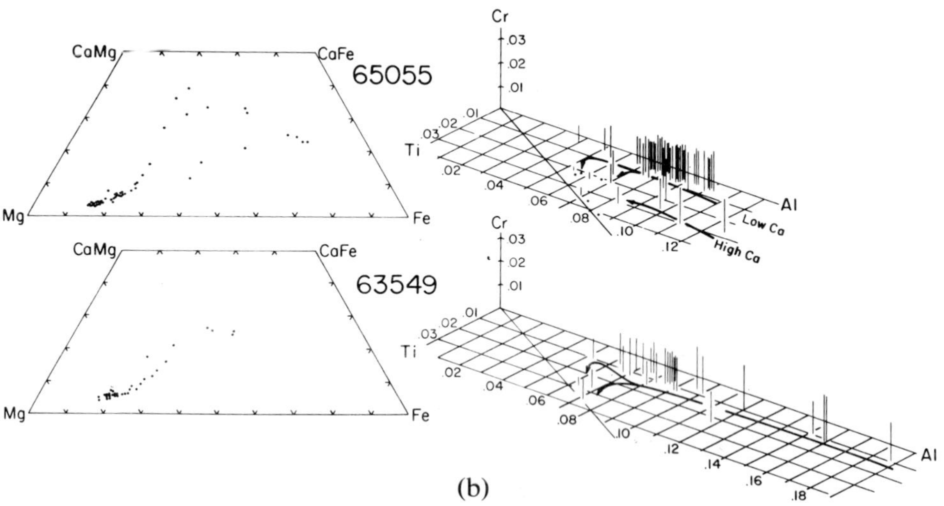

(b)

Fig. 3 (a-f). Ca-Fe-Mg composition of pyroxenes, Cr-Ti-Al composition of pyroxenes, and Fe-Mg composition of olivines in samples from the melt rock suite. In addition to data collected for this study, data from the following studies is included: 14068; Helz, 1972 and Warner, 1972; 14310: Brown *et al.*, 1972, Crawford and Hollister, 1974, Kushiro *et al.*, 1972 and Ridley *et al.*, 1972; 60315: Bence *et al.*, 1973, Hodges and Kushiro,

LOW-AN ANORTHOSITIC GABBRO

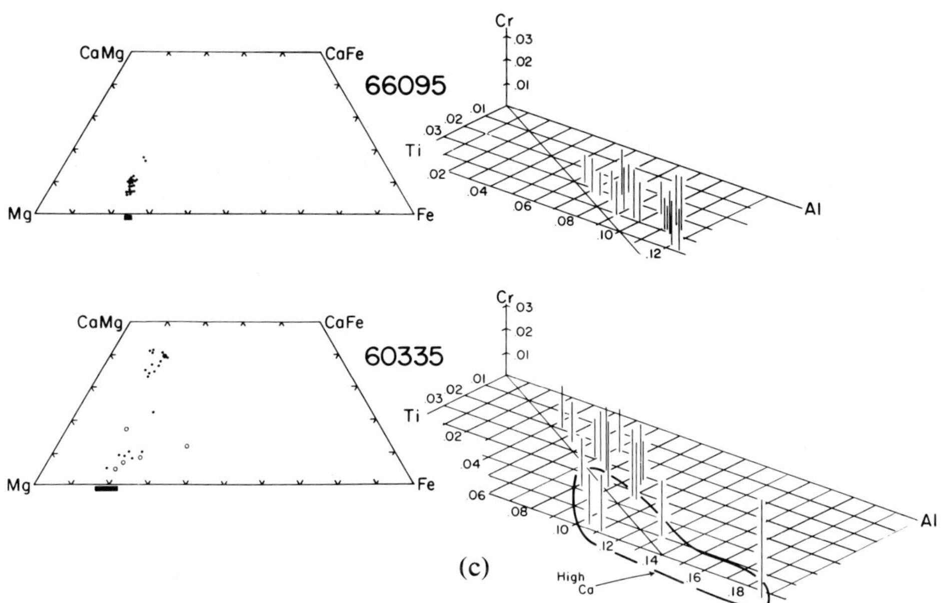

LOW-K FRA MAURO (LKFM): SUBOPHITIC/INTERGRANULAR

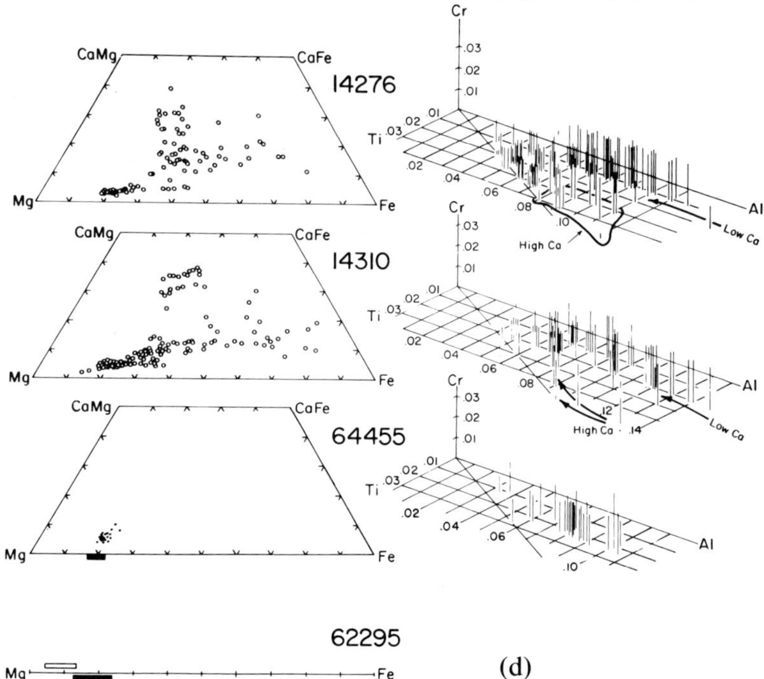

(d)

1973 and Walker *et al.,* 1973; 60335: Walker *et al.,* 1973; 62235: Crawford and Hollister, 1974; 62295: Walker *et al.,* 1973; 65015: Albee *et al.,* 1973; 68415: Helz and Appleman, 1973, Gancarz *et al.,* 1972 and Walker *et al.,* 1973; 68416: Hodges and Kushiro, 1973; 77135: Bence *et al.,* 1973. Symbols (x) specify ophitic textural patches in otherwise poikilitic samples.

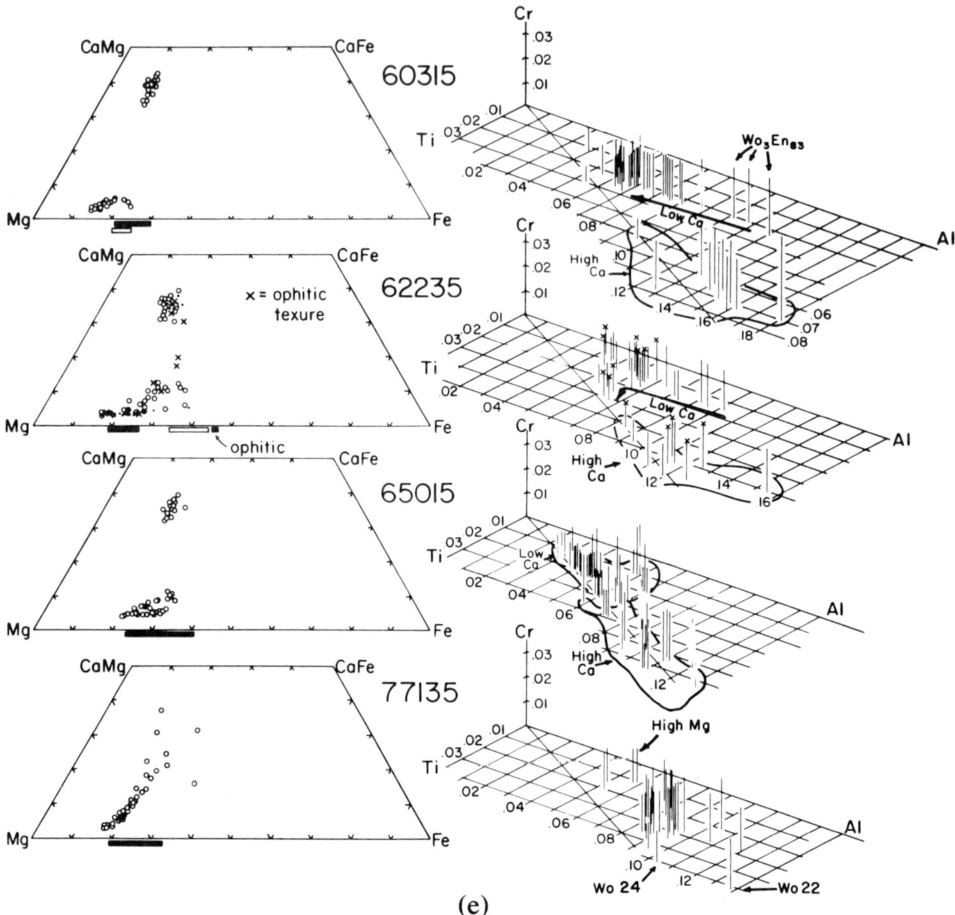

(e)

zonation pattern and lack olivine. The textural interpretations and experimental work of Walker *et al.* (1973) suggest that plagioclase is the liquidus phase* in crystallization of this composition at low pressure, followed by orthopyroxene and olivine. The effects of cooling rate on the crystallization sequence for this composition are not known, but we note that the matrix grain-size in olivine-free anorthositic gabbro melt rocks (~0.5 to 0.05 mm) is finer than in coarse olivine-bearing anorthositic gabbro melt rocks (≥ 0.5 mm, excluding sparse micro-ophitic patches). Perhaps rapid cooling rates, manifested by the development of finer-grained rocks, resulted in an overstepping of the olivine stability interval for these bulk compositions.

*J. Longhi suggests that the observed phase appearances may be affected by Fe-loss from the experimental charges.

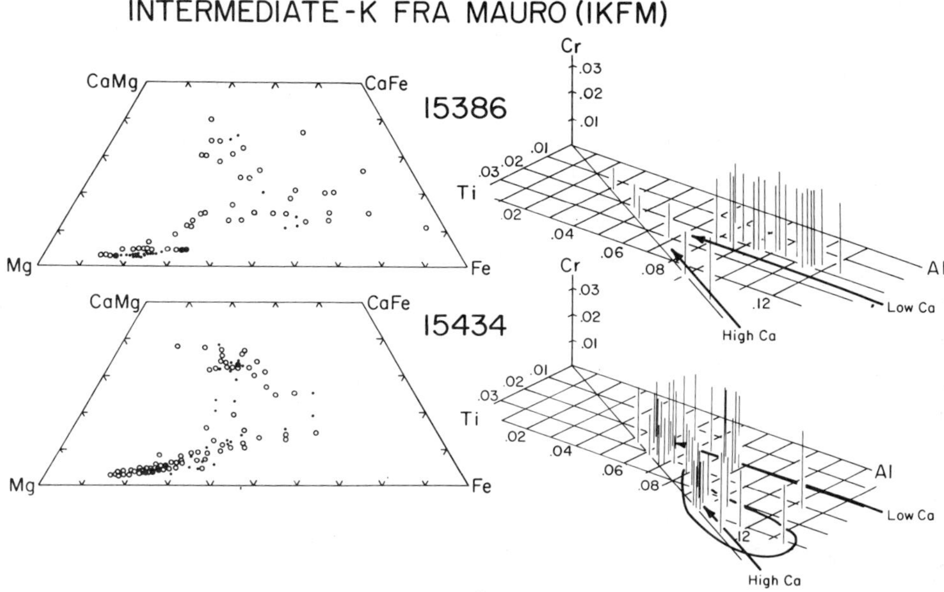

(f)

Of the two olivine-free melt rocks 65055 and 63549, 65055 is more coarse-grained, has a much broader range of feldspar zonation ($An_{97.5}$ to $An_{70}Ab_{28}Or_2$, compared to $An_{97.5}$ to An_{88} in 63549), and attains a much more extensive range of Ca and Fe enrichment in pyroxene zonation (Fig. 3). Both rocks have an orthopyroxene → pigeonite → augite pyroxene crystallization sequence; orthopyroxene cores in 65055 are more magnesian and less calcic than the initial orthopyroxene compositions in 63549 ($En_{82}Wo_3$ vs. $En_{79}Wo_4$).

Pyroxene minor-element zonation in 65055 is similar to that observed in olivine-bearing anorthositic gabbros 68415 and 68416, with low Ti-content and heterogeneous Al content in initial orthopyroxene. However, the very fine-grained rock 63549 (~0.1 to 0.05 mm plagioclase) has initial orthopyroxene and Mg-rich pigeonite compositions with 0.2 Al/6 oxygens and a distinct decrease in Al with Ca, Fe enrichment. Final pyroxene minor-element compositions in olivine-free anorthositic gabbro melt rocks are similar to compositions in olivine-bearing anorthositic gabbro melt rocks, with Al:Ti $\simeq$ 2, though the more evolved Fe-rich pyroxenes in 65055 are unique in that many are Cr-free. Both 65055 and 63549 contain metal and FeS, but 63549 is considerably richer in glassy mesostasis (Table 2).

An-poor anorthositic gabbro.

Rock 60335 is an intergranular ophitic-textured specimen which falls into the An-poor subgroup of anorthositic gabbro (Naney, pers. comm., 1978), with <50–55 mol.% An/(An + SiO_2 + Ol). Sample 60335 is also more Mg-rich (Mg′ = 0.76)

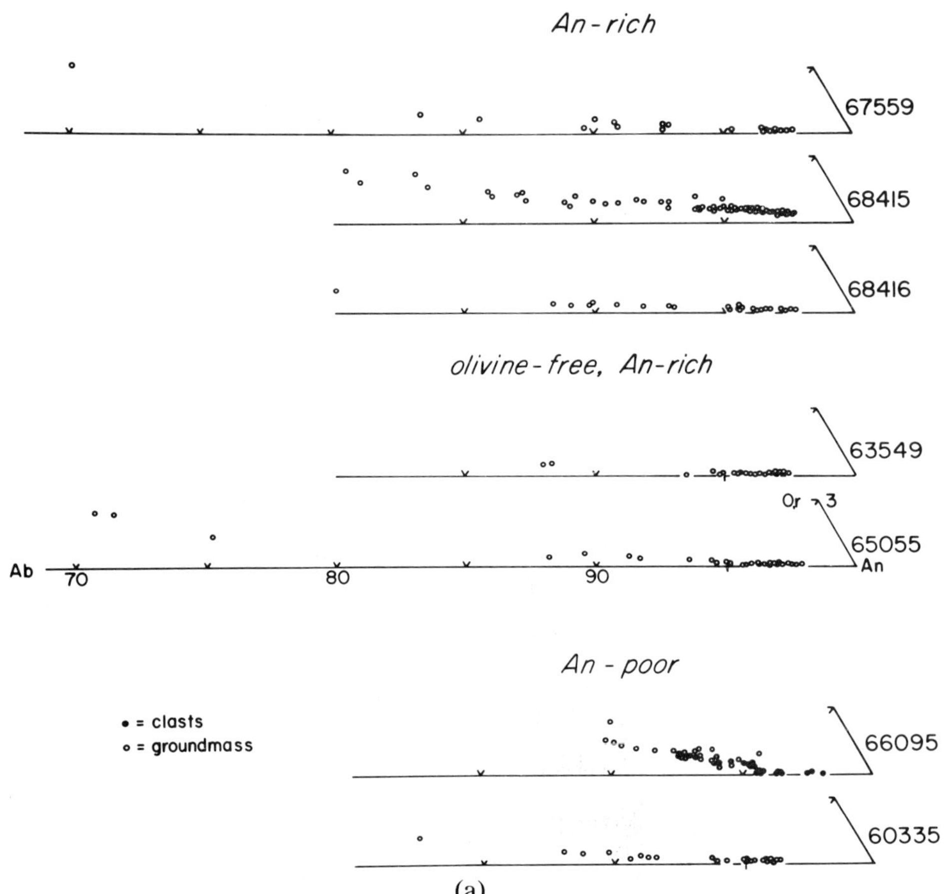

(a)

Fig. 4 (a-c). Ab-An-Or composition of feldspars in samples of the melt rock suite. Sources of data from the literature are the same as those listed for Fig. 3.

than the other large rock samples in the anorthositic gabbro group (Mg′ = 0.64 to 0.69). This higher Mg:Fe content is reflected in a more magnesian initial olivine composition (Fo_{83} vs. Fo_{77-69} in other olivine-bearing anorthositic gabbro melt rocks). Orthopyroxene ($En_{78}Wo_5$) is rare, and is rimmed by clinopyroxene zoned from $En_{72}Wo_9$ to $En_{48}Wo_{41}$ with little Fe-enrichment (Mg'_{px} ranges from 0.82 to 0.73; Walker *et al.*, 1973, report one analysis of a more Fe-rich pyroxene with Mg'_{px} = 0.6). The more calcic pyroxenes are enriched in both Al and Ti (Al content up to 0.2 cations/6 oxygens, Al:Ti variable but ranging to values less than 2.0, indicating the presence of some Ti^{3+} as $R^{2+}Ti^{3+}SiAlO_6$); the presence of Na indicates that some Ti^{3+} may be charge-compensated by Na^+ in a $NaTi^{3+}Si_2O_6$ component.

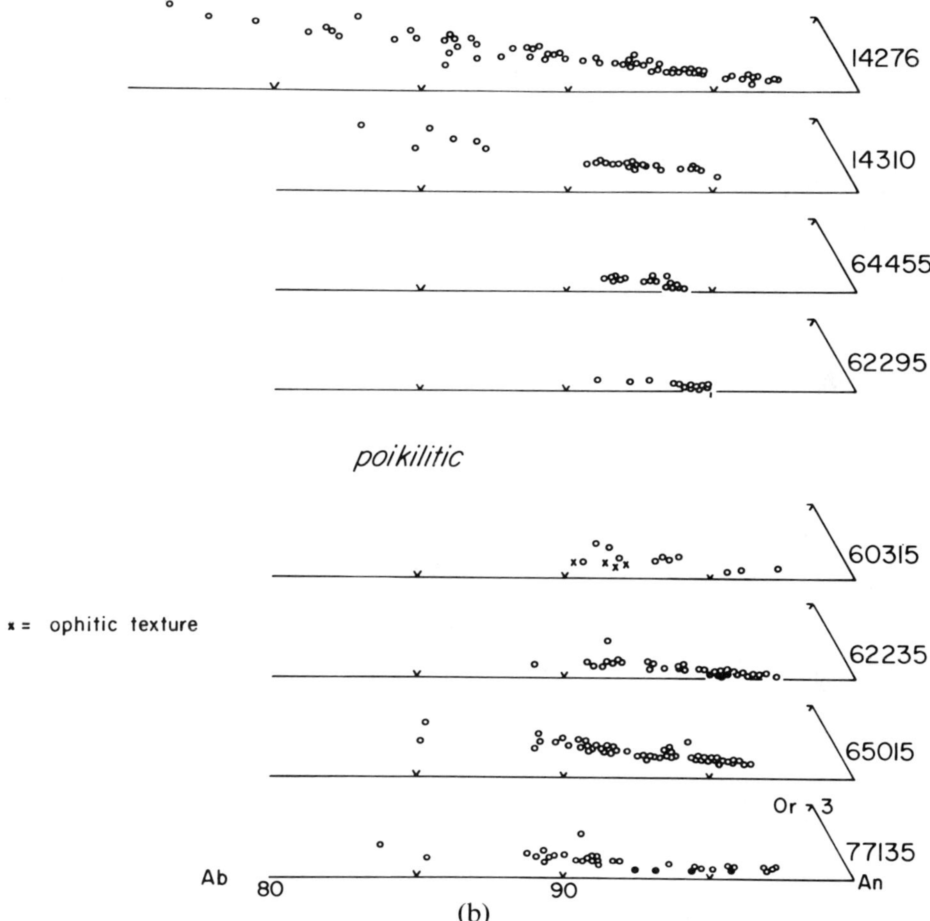

(b)

Feldspar cores have compositions as calcic as An_{97}; Walker *et al.* (1973) report rim and matrix compositions as sodic as An_{73}. Sample 60335 contains a high percentage (~15%; Table 2) of fractured, shocked or polycrystalline anorthoclase that is xenocrystic. Walker *et al.* (1973) note that although some plagioclase xenocrysts (An_{95-97}) can be distinguished by habit (equant) or texture (shocked or fractured), it is often difficult to distinguish feldspar xenocrysts from feldspar phenocrysts. This is a problem which is often encountered in highland melt rocks with coarse basaltic texture (>0.5 mm grain size). The dominant opaque phase is a metal with moderate amounts of Ni (~2–4 wt.%) and Co (<0.5 wt.%). This

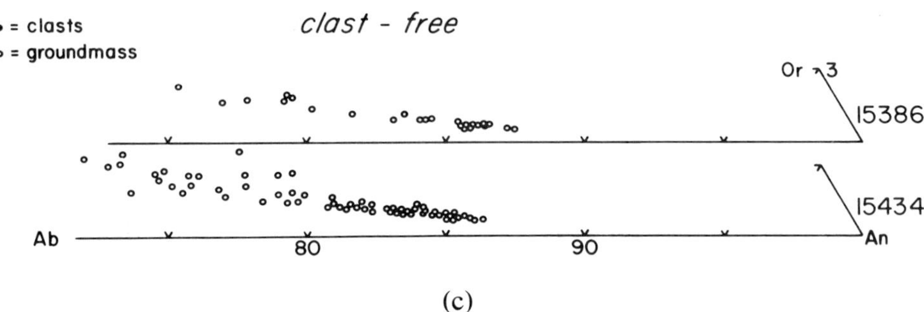

(c)

rock is host to metal spheres with a striking graphic intergrowth of schreibersite and troilitic rims. Minor ilmenite, armalcolite and phosphate are found within glassy portions of the mesostasis.

Sample 66095 is a 1185 gm specimen from the plains northeast of South Ray crater at the Apollo 16 site. Early sample studies focused on 66095 as a representative of one of the "rusty rocks" which were found at every station visited during the Apollo 16 mission. Taylor *et al.* (1973) note that much of the rusty stain in 66095 was due to the development of an oxyhydrate (akaganeite: β FeOOH; Taylor *et al.*, 1974). The β FeOOH is associated with Cl concentrations, suggesting that lawrencite ($FeCl_2$) had been oxidized when the sample was exposed to earth's atmosphere. Nunes *et al.* (1973) note a high concentration of excess Pb in 66095; the similar concentration of Cl and Pb in volatile coatings on volcanic lunar glass spheres (Chou *et al.*, 1975) supports a volcanic, lunar origin for the original Cl-rich component prior to oxyhydration. The restriction of "rust" to metallic portions of Cl-rich lunar rocks suggests that akaganeite formed in the earth's atmosphere and not on the moon (Taylor *et al.*, 1973).

Rock 66095 is micro-ophitic ($<$0.05 mm grain size) with restricted mineral zonation. Pyroxene ranges from $En_{72}Wo_7$ to $En_{67}Wo_{17}$. Olivine (Fo_{75-77}) and fine-grained laths of plagioclase are zoned from $An_{94.5}Ab_5Or_{0.5}$ to $Ab_{89}Ab_9Or_2$. This restricted zonation persists despite an apparent lack of bulk-rock homogeneity (Garrison and Taylor, 1979); in this regard 66095 is more like a breccia than like a clast-poor melt rock. Blocky feldspar xenocrysts are more calcic (An_{98-95}). These mineral compositions are summarized in Figs. 3 and 4. Other abundant constituents of 66095 are xenoliths, glass, Fe-Ni metal and FeS (Table 2). Minor constituents include schreibersite, cohenite, and sphalerite; FeS and Fe,ZnS occur as rims on metallic grains. The Ni-Co contents of metal grains fall entirely outside the range of metal compositions in volcanic lunar mare basalts, a criterion used by Irving (1975) to identify impact melts contaminated by meteoritic metal.

Low-K Fra Mauro

We have subdivided the low-K Fra Mauro melt rocks into two groups: those with >20% Al_2O_3 and ophitic or intersertal texture, and those with <20% Al_2O_3 and poikilitic texture. This division of textures based on Al-content is not a rigid distinction, but exceptions are rare (Simonds *et al.*, 1973).

Ophitic/intersertal-textured (high Al_2O_3) LKFM.
Three ophitic-textured LKFM samples (14276, 14310 and 64455) plot together on the pseudoternay projection (Fig. 2a); a fourth intersertal sample (62295) is troctolitic and plots within the spinel primary field. Samples 14276 and 14310 may be considered together, for they are similar in composition, texture and paragenesis. Although 64455 plots with the Apollo 14 KREEP basalts in Fig. 2a, it is significantly poorer in iron and alkalis (Table 1), strikingly restricted in mineral zonation, and contains minor olivine (which is absent in the Apollo 14 KREEP basalts).

The Apollo 14 KREEP basalt 14310 has been intensively studied. The reader is referred to Kushiro *et al.* (1972), Longhi *et al.* (1972), Brown *et al.* (1972), Ridley *et al.* (1972), James (1973) and Crawford and Hollister (1974) for detailed descriptions. Samples 14276 and 14310 are intergranular to sub-ophitic basalts with extensive plagioclase zonation (An_{95} in 14310 or $An_{97.5}$ in 14276, to $An_{75}Ab_{22}Or_3$; Ridley *et al.* (1972) report rare rim compositions as sodic as An_{58}). Pyroxenes are highly zoned, ranging from an initial aluminous orthopyroxene ($En_{80}Wo_4$, with ~0.14 Al/6 oxygens) through pigeonite and augite to Fe-rich low-Ca augite (Fig. 3). Kushiro *et al.* (1972) put the orthopyroxene to pigeonite transition at $Mg'_{px} \sim 0.65$. The absence of olivine may be explained by experimental crystallization studies of 14310 at atmospheric pressure where olivine (Fo_{89}) has an early but brief appearance at ~1230°C (Green *et al.*, 1972) or ~1200°C (Walker *et al.*, 1972) before reaction with melt and resorption at ~1190°C. The dominant accessory phase is ilmenite, with minor amounts of troilite, ulvöspinel, armalcolite, metal, schreibersite, and tranquillityite ($Fe_8(Y,Zr)_2Ti_3Si_3O_{24}$). Gancarz *et al.* (1972) also describe apatite, whitlockite, and K-Ba feldspar in 14276.

In contrast to 14276 and 14310, sample 64455 is strikingly restricted in mineral zonation (Figs. 3, 4) and contains abundant olivine (Fo_{81-78}). Texturally, this rock is a highly fractured subophitic basalt, but with its texture and relatively coarse grain size (~0.5 mm feldspar laths) one would expect a greater range in mineral zonation. Re-equilibration may explain the limited chemical zonation of olivine, pyroxene and feldspar in 64455.

Rock 64455 is a unique specimen, for it is both a "rusty rock" (similar to 66095; see Taylor *et al.*, 1973) and a glass-coated, aerodynamically rounded broken spheroid. Grieve and Plant (1973) found that the glass coating of 64455 does not have the same composition as the crystalline base to which it adheres, but is instead a different highland composition (anorthositic gabbro: $Al_2O_3 \sim 25.2\%$

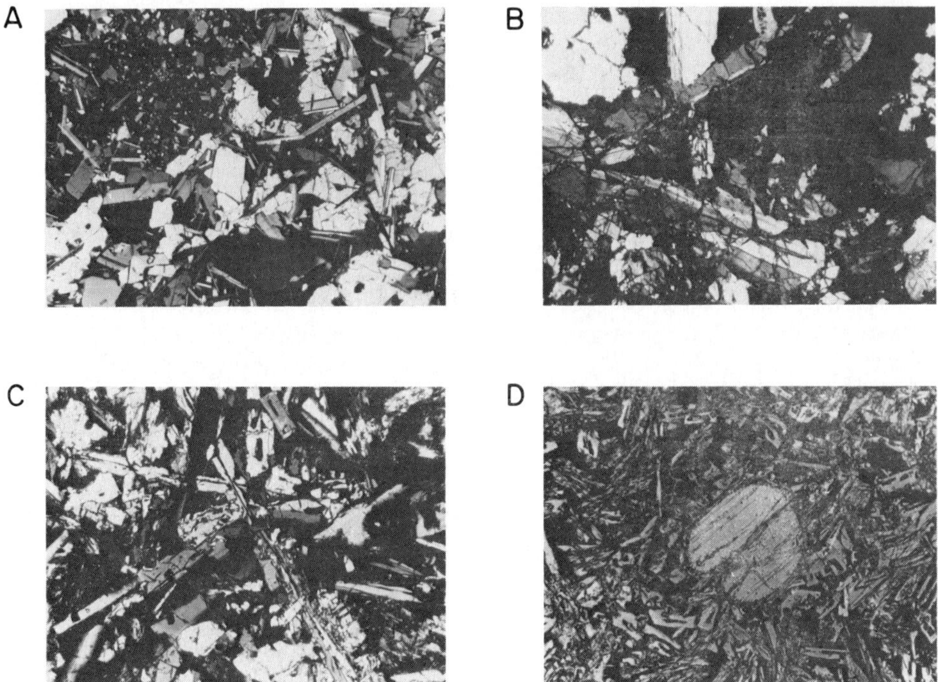

Fig. 5. Photomicrographs of representative highland melt-rock samples. The long dimension in each photograph is 1.0 mm. (A) Low-An anorthositic gabbro sample 60355,70 in transmitted light, crossed nicols; (B) IKFM sample 15434,116 in transmitted light, crossed nicols; (C) sample 62295,15 in transmitted light, crossed nicols; (D) sample 62295,76 in plane polarized transmitted light. The large, rounded grain is a spinel xenocryst. (E,F,G) sample 65015,143 under crossed nicols, in plane polarized light, and in reflected light. This poikilitic LKFM sample contains a large FeNi-metal spherule with a feldspar rim. (H,I,J) sample 14310,214 under crossed nicols, in plane polarized light, and in reflected light. This subophitic/intergranular LKFM sample contains an especially fine-grained zone around a small metal/sulfide fragment.

and CaO ~ 14.5%). At the contact with the glass rind, the crystalline part of 64455 has been partially melted; here the remaining olivine grains are more Mg-rich (Fo_{83-82}) than in the crystalline portion of the rock and the metal, troilite and schreibersite grains which are angular in the crystalline portion of the rock become rounded. The restricted mineral zonation of 64455 shows that the thermal event in which the glass rind was emplaced and surface melting occurred was evidently pervasive enough to re-equilibrate olivine and pyroxene compositions within the crystalline sample.

The low-K Fra Mauro group includes relatively Si-poor and Mg-rich troctolitic compositions represented by sample 62295 (Mg′ = 0.81). This rock is a fine-grained (<0.2 mm) olivine (Fo_{87-77}), feldspar (An_{89-87}), Mg,Al spinel vitrophyre with rounded xenocrysts of more magnesian olivine (Fo_{95-90}), spinel (subhedral

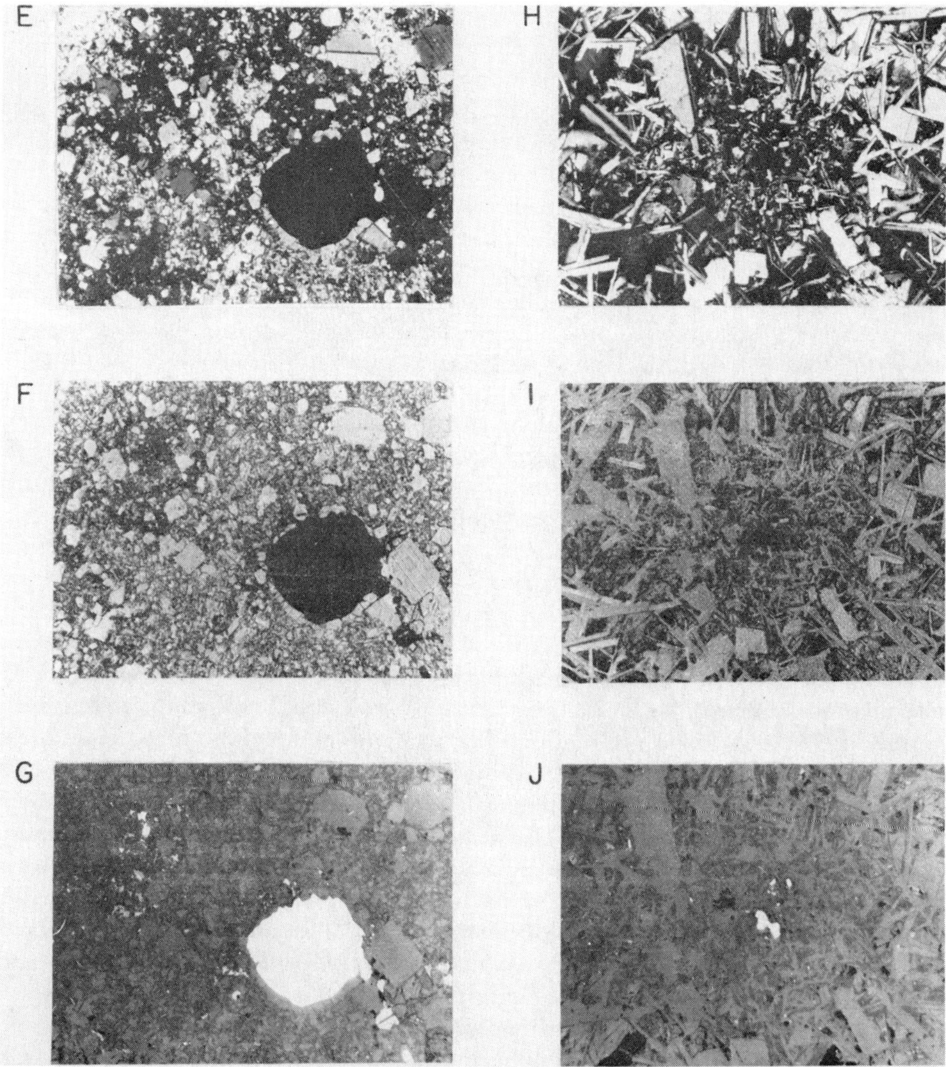

and higher in Cr_2O_3 and FeO than the spinel inclusions within plagioclase), plagioclase and metal + schreibersite. Walker *et al.* (1973) suggest that 62295 is a low-pressure spinel-plagioclase-olivine peritectic partial melt of a source similar in composition to pink-spinel (Mg,Al spinel) troctolite (Apollo 16 sample 67435). However, the high Ir and An contents of 62295 (Table 1) are evidence for an impact-melting origin.

Poikilitic-textured low-Al_2O_3 LKFM.

Four poikilitic rocks are included in Table 1 (60315, 62235, 65015 and 77135). All four of these rocks plot near the feldspar-olivine cotectic (Fig. 2a) and all four have generally similar ranges of mineral composition (Figs. 3 and 4). The Apollo 16 poikilitic rocks have well developed separation of high-Ca and low-Ca pyroxene pairs with the low-Ca pyroxene component consisting of orthopyroxene intergrown with or rimmed by pigeonite. In 77135 the pyroxenes are similar, but the separation between high-Ca and low-Ca pyroxenes is less pronounced. Low-Ca pyroxenes are low in Ti (~0.01 cations/6 oxygens) with heterogeneous Al content; high-Ca pyroxenes are richer in Ti (~0.05 – 0.06 cations/6 oxygens) with Al:Ti ≃ 2. Both 60315 and 62235 have minor diabasic patches with pyroxene compositions that span the solvus gap between low-Ca and high-Ca pyroxenes. All of these samples include xenolithic and xenocrystic clasts; in 77135 the feldspar xenocrysts are notably more calcic than the groundmass feldspar. Minor phases in the four poikilitic IKFM rocks are:

60315: metal, troilite, ilmenite, K-feldspar (sub-ophitic areas)
62235: metal, troilite, ilmenite, pleonaste spinel
65015: metal, troilite, ilmenite, whitlockite (CaP_2O_6)
77135: metal, troilite

The metal Ni contents are >1.5 wt.% (Irving, 1975), in the range of samples affected by meteoritic contamination.

Intermediate-K Fra Mauro

Clast-poor low-siderophile IKFM.

Samples 15386 and 15434 are members of the Apollo 15 KREEP crystal-liquid equilibration series of Irving (1977). Irving has proposed that this series represents a fractionation trend of partial melt from a troctolitic source deep in the lunar crust or upper mantle. The Ni content and Ni/Co ratio of intermediate-K Fra Mauro (Ni ~12 ppm, Ni/Co < 1) contrasts sharply with impact melt-rocks (Ni = 60–1400 ppm, Ni/Co = 4.5 to 20) and suggests a low level of meteoritic contamination. The metal compositions in 15382, a small rock which is very similar to 15386 and 15434, are similar and indicate that IKFM samples do not contain high-Ni metals of meteoritic composition (Irving, 1975).

Pyroxene and feldspar compositions of 15386 and 15434 are summarized in Figs. 3 and 4. Both samples are ophitic to intersertal; the grain size is ~0.5 to 1 mm. Textural evidence and microprobe data indicate that both 15434 (thin section 116) and 15386 are nearly co-saturated with orthopyroxene ($En_{80}Wo_3Fs_{17}$) and plagioclase (An_{87}). The ilmenite in these rocks is particularly distinctive for it occurs as acicular grains (length/width > 25) that contrast sharply with the stubby, anhedral ilmenite grains in other melt rocks. Feldspars are zoned from An_{92} to $An_{70}Ab_{27}Or_3$.

Pyroxene zonation trends follow a pattern which is found in clast-free feldspathic impact melts (Vaniman and Papike, 1979) as well as IKFM basalts: orthopyroxene cores become more Fe-rich toward their rims and are mantled by low-Ca pigeonite (Wo_7) and augite (as calcic as Wo_{35-40}) at Mg'_{px} = 0.8–0.6. A broad range in Wo content is maintained as pyroxene becomes more Fe-rich. In minor-element composition the initial orthopyroxenes are Al-rich (~0.12 cations/6 oxygens) and become less aluminous as the pyroxenes become more Fe-rich. Al/Ti ratios are high (>4) in the orthopyroxene crystallization series. The first Ca-rich augites to form have an Al/Ti ratio of 2, indicating an increased content of Ti as $CaTiAl_2O_6$. With Fe-enrichment the Ti, Al and Cr contents of pyroxene decrease with the Al/Ti ratio maintained at 2. The residual phases in 15386 include apatite, cristobalite, metal, troilite, and a more stubby ilmenite with higher Fe/Mg ratio (Mg'_{ilm} ranges from ~0.03 in early-formed acicular grains to ~0.00 in late-formed grains). Small pyroxene grains isolated in the glassy mesostasis approach pyroxferroite composition. Dowty *et al.* (1976) find tranquillityite, armalcolite, baddeleyite, whitlockite and chromian ulvöspinel in the mesostasis of a smaller but similar Apollo 15 IKFM basalt, 15382. Crawford and Hollister (1977) also find zircon in the groundmass of 15434, and report that the glassy residue has segregated into two immiscible fractions. Note the studies by Hess *et al.* (1977, 1978) on immiscible liquids resulting from crystallization of an Apollo 15 LKFM composition.

HIGHLAND MELT ROCKS IN SPACE AND TIME

Highland melt rocks have crystallization histories reflecting one of two distinct origins: impact melting or internal lunar melting. These two processes must be considered separately.

Impact melt rocks

Impact melts can be generated by a broad range of projectile sizes, from the "moonlets" responsible for large, multi-ring basins such as Orientale (600 km diameter transient cavity; Moore *et al.*, 1974) to small meteorites that create craters <1 km in diameter. Hawke and Head (1979) report impact melt deposits in craters of ≤0.75 to 12 km diameter, although such deposits are small, thin, and often masked by mass wasting of the crater walls. Lange and Hawke (1979) quantify the amounts of impact melt generated as $(1.8 \times 10^{-4}) \times D^{3.4}$, where D is the crater diameter. They note that their calculated impact melt volume for most craters is greater than that calculated by photogeologic investigations, and suggest that much of the melt is widely and thinly dispersed, or intermixed with bulk ejecta. In Lange and Ahrens (1979) it is proposed that the total amount of impact melt generated from 81 large lunar craters ($D \geq 161$ km) is 5.3×10^6 km^3. This is a minimum volume, but it is comparable to the proposed total volume of mare basalts (Head, 1976).

Photogeologic studies of the relatively young Orientale Basin reveal some of the characteristics of the spatial distribution of impact melt sheets around lunar multi-ringed basins. Moore *et al.* (1974) map a *central basin* material, interpreted as impact melt, which fills the transient crater (600 km-diameter), mantling hills and filling depressions; beyond the central basin but within the crater ring (900 km-diameter) is a more hummocky expanse of *knobby basin material* which is interpreted as being largely composed of impact melt. Similar distributions of major impact melt sheets are observed in terrestrial craters, where impact melt sheets are generally confined within the stable crater structure (including the outer crater formed by collapse of the transient cavity; Phinney and Simonds, 1977). Phinney and Simonds (1977) point out that part of the impact melt sheet is injected downward within the crater as dikes and sills. Another part of the impact melt is ejected upward to land within and beyond the crater where it cools rapidly as a breccia component (thus forming *suevites*), rather than following the slower cooling process that occurs in the thick, continuous melt sheets within the crater. Dikes, sills, and suevite ejecta may account in part for the discrepancy between theoretical calculations of impact melt volumes (Lange and Ahrens, 1979) and melt volumes estimated photogeologically from observations of melt sheets within craters. As one example of this discrepancy, the Lange and Ahrens equation estimates a melt volume of $\sim 5 \times 10^5$ km^3 for the Orientale basin, contrasted with photogeologic estimates of only 2×10^5 km^3 of impact melt in the impact melt sheets within the Orientale crater (Head, 1979). These differing estimates suggest that in the large multi-ringed lunar basins, over half of the impact melt is ejected beyond the crater or injected within the crater as dikes and sills; less than half of the melt remains ponded within the crater to cool as an impact melt sheet.

None of the relatively intact intra-crater impact melt sheets have been sampled directly by any of the lunar sampling missions. The closest approach to direct observation of a lunar melt sheet has been through the sampling of boulders from a thin ($\sim$10 m) impact melt sheet uplifted in the South Massif at Apollo 17 (Simonds, 1975). Boulders at the foot of the massif provide a fairly complete section through the melt sheet, including melt-matrix breccias as well as melt rocks. All samples from the melt sheet, whether clast-rich melt-matrix breccias or melt rocks, have approximately the same composition represented by sample 77135 (Table 1).

The homogenization of inhomogeneous target materials by impact mixing and melting has been documented by Simonds *et al.* (1978) in a terrestrial occurrence. The homogenization proceeds by generating a relatively small volume of total melt near the point of impact and turbulently mixing this total melt with clastic debris as the melt travels down and then out along the walls of the transient impact crater (Grieve and Floran, 1978; Simonds *et al.,* 1978; Grieve, 1975; Simonds, 1979). The intermixed clasts are melted as they draw heat from the surrounding superheated liquid; melting of clasts may be partial or nearly complete, but the result in either case is a rather homogeneous mixture of the rock types found throughout the area of the transient crater. This characteristic homogenization of melt sheet composition, plus rapid viscous cooling that prevents crystal/liquid fractionation (Simonds, 1979), provides a means of distinguishing different melt sheets in a chaotically mixed cratered terrain.

Floran *et al.* (1976) distinguish three chemical types among Apollo 16 impact melt rocks. These categories correspond closely to our groups of An-rich anorthositic gabbro (67559, 68416, 68415, 63549, 65055), An-poor anorthositic gabbro (60335, 66095), and to a subset of our group LKFM (60315, 62235, 65015). The implication of melt-sheet homogenization is that these three melt rock groups correspond to three distinct melt sheets. Fig. 6a shows the tight compositional range of each of these melt sheets. Fig. 6b shows the distribution of representative fragments from these three melt sheets at the Apollo 16 landing site. Clearly, samples from all three melt sheets are scattered throughout the landing site, and it is not possible to map the areal distributions of these melt sheets with any resolution finer than ~10 km. This chaotic distribution at Apollo 16 is probably characteristic of reworked melt sheets throughout the lunar highlands.

The $^{40}Ar/^{39}Ar$, Rb–Sr isochron, and concordant U–Pb ages of impact melt rocks fall between 3.80 and 4.01 G.y. This spread of ages is seen in the broad sample base of Irving (1975), and in the samples described here (Table 3). The consistency of impact melt rock ages in the span 3.9 ± 0.1 G.y. reflects the pervasive character of impact reworking at or until ~3.9 G.y. These age data, which are susceptible to thermal resetting, may represent reheating of previously formed melt-rock fragments in impact ejecta blankets, as well as melt rocks actually formed in the age span of 3.9 ± 0.1 G.y. There is thus no reliable measurement of the actual age of impact melting in any of the impact melt rocks.

Lugmair and Scheinin (1975) and Unruh *et al.* (1977) show that Sm–Nd internal isochron ages are relatively resistant to metamorphic resetting. Carlson and Lugmair (1979) have used this persistence of high-T° (igneous) Sm–Nd isochrons to show that the 3.94 ± 0.04 G.y. Rb–Sr isochron of 15386 is indeed an igneous and not a metamorphic age, for the reported Rb–Sr isochron age is comparable (3.87 ± 0.12 G.y.: Table 3). Particular concern has been given to the true age of melt crystallization in sample 15386, because this sample is one of the few highland melt rocks which has consistent evidence for origin as an internal lunar melt (discussion below). Unfortunately there are no Nd–Sm data for samples which are undoubted impact melt rocks, although such data might reveal the true crystallization ages of impact melt rocks, and not the age of the latest reheating event.

Volcanic melt rocks

As mentioned above, the consistent Nd–Sm and Rb–Sr isochron ages of sample 15386 (Table 2) suggest that ~3.9 G.y. is the true crystallization age of this sample (Carlson and Lugmair, 1979). At the present time, this sample is the best characterized lunar highland volcanic rock; its lack of meteoritic (impact) contamination, its lack of impact-intermixed clastic debris, and its position in the volcanic KREEP fractionation trend of Irving (1977) all point to a volcanic origin. The similar petrographic and chemical features of a number of loosely fragmented IKFM chips in the "pedestal" sample 15434, with an Rb–Sr isochron age of 3.91 ± 0.04, indicate that this sample is a related volcanic rock. Warren and Wasson (1978) provide evidence from lack of siderophile contamination in 15382, another

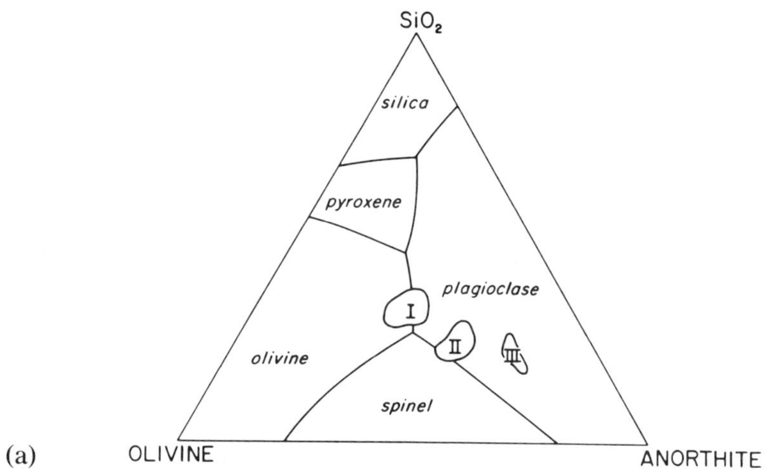

Fig. 6. a. Melt sheets identified by Floran *et al.* (1977) on the basis of chemical discrimination between samples at the Apollo 16 landing site.

very similar IKFM basalt to show that it too is a pristine basalt. The differing initial $^{87}Sr/^{86}Sr$ ratios of 15386 and 15434 (0.70038 vs. 0.70070, Table 3) might be taken to suggest that the volcanic samples 15386 and 15434 are not from a single flow (Meyer, 1977), although there are terrestrial occurrences where a single basalt flow has an even broader range of initial $^{87}Sr/^{86}Sr$ ratios (0.7040-0.7084; Laughlin *et al.*, 1972) due to small amounts of soil assimilation at the flow base (Brookins *et al.*, 1975). Although 15386 and 15434 may be from a single flow, the occurrence of small fragments of another type of IKFM basalt* in Apollo 17 breccias (Ryder *et al.*, 1977) is evidence of widespread and varied KREEP volcanism.

Lugmair and Carlson (1978) show that IKFM samples from Apollo 12 and 14 as well as Apollo 15 have consistent T_{ICE} model ages within the range 4.36 ± 0.06 G.y. Rb–Sr model ages for the same samples, however, are somewhat younger (4.25–4.30 G.y. for those Apollo 15 IKFM basalts which are probably volcanic samples). Within the calculated errors, both Nd–Sm T_{ICE} and Rb–Sr T_{BABI} ages overlap at $\simeq$4.3 G.y. The isotopic systematics of the IKFM source region apparently were "set" at $\simeq$4.3 G.y. and IKFM-type volcanism therefore could have occurred prior to the ~3.9 G.y. IKFM volcanism of Apollo 15. Some evidence for this is found at Apollo 17, where the IKFM-type clasts described by Ryder *et al.* (1977) have an Rb–Sr isochron age of 4.01 ± 0.04 G.y. (Compston *et al.*, 1975). This may be a minimum age reflecting the breccia formation.

The Apollo 17 samples indicate that IKFM-type volcanism occurred prior to 3.9 G.y., at ages $\geq$4.0 G.y. Model ages based on Rb–Sr and Nd–Sm systematics suggest that the upper age limit for IKFM volcanism is $\simeq$4.3 G.y. Another

*similar to A–15 IKFM basalts but more Fe,Mg,Cr-rich and relatively Na,K-poor.

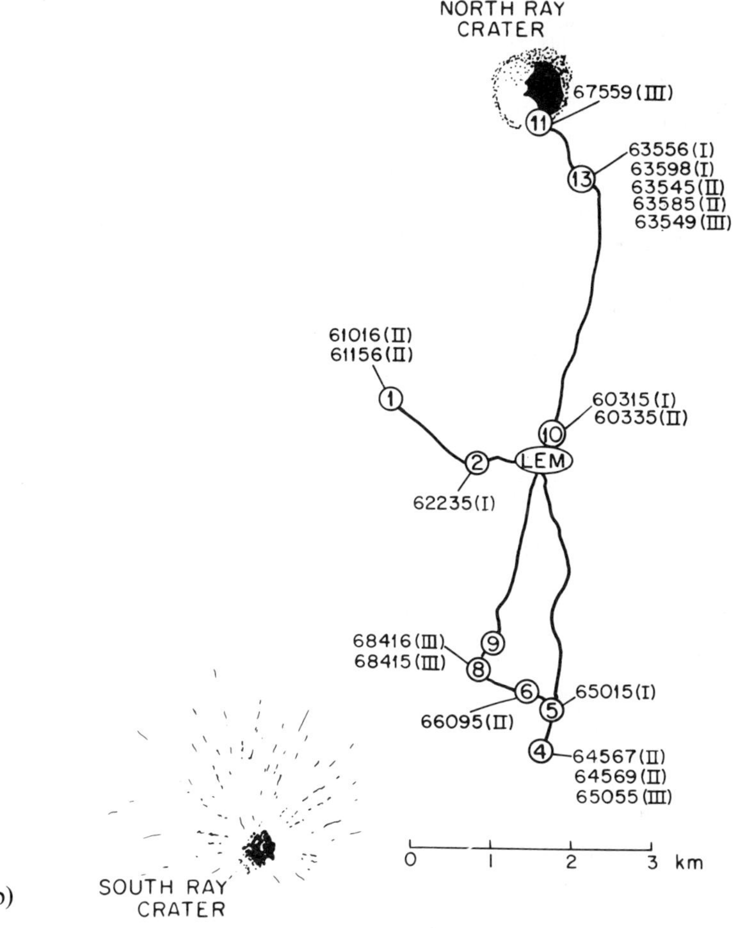

b. Distribution of large melt-rock samples at the Apollo 16 landing site. Note that samples all of the three melt sheets of Fig. 6a are scattered throughout the Apollo 16 site.

maximum age limit is based on the occurrence of KREEP-free granulitic impactites at Apollo 17 with ages ≥4.17 G.y. (Warner *et al.*, 1977 and Oberli *et al.*, 1979). Since highland samples of younger impact age all contain KREEP contamination, Warner *et al.* (1977) and Ryder and Spudis (1979) infer from the KREEP-free granulitic impactites that KREEP was not brought to the lunar surface (or at least, was not widespread) before ≃4.2 G.y. Further age data from granulitic impactites of other localities may eventually confirm an upper age limit for KREEP volcanism, provided that the impact distribution of KREEP volcanics was almost universal once those volcanics appeared on the lunar surface.

Chemical, petrologic and isotopic studies of IKFM basalts at Apollo 15 and Apollo 17 provide the only directly sampled evidence for localities of highland

Table 3. Age and isotopic data for a suite of lunar highland melt rocks.

	$^{40}Ar/^{39}Ar$ whole rock	$^{40}Ar/^{39}Ar$ feldspar	$(^{87}Sr/^{86}Sr)_I$	Rb-Sr internal isochron	"T_{BABI}" model age (0.69910 basis)	$^{207}Pb/^{204}Pb$ $^{206}Pb/^{204}Pb$, concordance	Nd-Sm isochron	$^{143}Nd/^{144}Nd$ T_{ICE} model age
Anorthositic Gabbros								
60335					4.19±0.06[19]			
68415	3.85 ±0.04[3]	4.09[3]	0.69920[22]	3.84±0.01[19]	4.44±0.20[9]			
	3.85 ±0.06[5]							
	3.80 ±0.04[13]							
	4.00 ±0.05[5]							
68416								
66095					3.91±0.27[9]			
Low-K Fra Mauro								
14310	3.88 ±0.06[13]		0.70045[2]	3.93±0.04[2]	4.49±0.13[8]			
	3.88 ±0.05[14]	3.89±0.04[1]	0.70033[6]	3.93±0.06[6]				
	3.92 ±0.01[12]							
62295	3.89 ±0.05[19]		0.69956[16]	4.00±0.06[16]	4.28±0.07[9]			
60315	4.03 ±0.03[5]				4.41±0.06[9]	3.93[7]		
	3.907±0.010[12]							
62235					4.48±0.07[9]			
65015	3.98[4]	4.47[4]	0.69917[19]	3.93[19]	4.45±0.06[9]			4.32±0.12[20]
	3.92 ±0.04[5]				4.41±0.08[10]			
77135	3.90 ±0.03[17]							
Intermediate-K Fra Mauro								
15382	3.90[13]				4.30[9]			4.38±0.14[20]
								4.42±0.11[20]
15386			0.70038[11]	3.94±0.04[11]	4.25±0.04[11]		3.87±0.12[21]	
15434			0.70070[10]	3.91±0.04[10]	4.27±0.05[10]			

References: (1) Turner *et al*. (1971); (2) Compston *et al*. (1972); (3) Huneke *et al*. (1973); (4) Jessberger *et al*. (1974); (5) Kirsten *et al*. (1973); (6) Murthy *et al*. (1972); (7) Nunes *et al*. (1973); (8) Nyquist *et al*. (1972); (9) Nyquist *et al*. (1973); (10) Nyquist *et al*. (1974); (11) Nyquist *et al*. (1975); (12) Schaeffer *et al*. (1976); (13) Stettler *et al*. (1973); (14) Turner *et al*. (1972); (15) Turner *et al*. (1973); (16) Mark *et al*. (1974); (17) Stettler *et al*. (1975); (18) Schaeffer *et al*. (1976b); (19) Papanastassiou and Wasserburg (1972); (20) Lugmair and Carlson (1978); (21) Carlson and Lugmair (1979);

volcanism. The Apollo 15 IKFM basalts can be correlated with the Apennine Bench Formation (Hawke and Head, 1978). The Apollo 17 IKFM-type basalts are found in breccias of the Serenitatis impact ejecta, and thus come from some source within the Serenitatis target zone. Photogeologic studies by Head and McCord (1978) and Hawke and Head (1978) point to further highland basalt features at Gruithuisen in Mare Procellarum (volcanic domes) and throughout the Fra Mauro and Imbrium basins. Metzger and others (1979), using finely tuned reprocessing of orbital gamma ray data to map Th concentration, conclude that the highlands surrounding and underlayering the western mare contain near-surface remnant flows of volcanic KREEP. Their results support the conclusion of Hawke and Head (1978) that the Apennine Bench Formation is an IKFM flow. Moreover, mixing models used by Metzger *et al.* (1979) indicate that sub-Imbrium highland flows brought to the surface by the Lambert and Timocharis craters, to the west of the Apollo 15 site and toward the interior of Mare Imbrium, are also of IKFM composition. Thus both sample and remote-sensing data suggest that IKFM basalts underlie the mare basalts of the Imbrium basin.

Possible concentrations of other highland basalt types are suggested by higher Th values around the crater Archimedes (quartz monzodiorite?), and by lower Th values SE of the Apennine Mountains and in the Haemus Mountains. The lower Th values are subject to the number of interpretations including the possible occurrence of LKFM-composition volcanics. The concept of LKFM-composition igneous activity elicits a wide range of responses, from strong support (Hess *et al.,* 1978) to strong denial (Warren and Wasson, 1977) and including arguments that there is no conclusive evidence to support either position (Reid *et al.,* 1977).

Pristine highland rocks are dominated by ferroan anorthosites, by rocks that plot near point B in the Walker pseudoternary (Fig. 1), and by troctolites, norites and dunites that plot close to the olivine-anorthite join in the pseudoternary (Warren and Wasson, 1978). Warren and Wasson (1977) suggest that all of the LKFM samples on the moon may be formed by mixing anorthosites, a KREEP igneous component, and troctolitic/dunitic material with normative Ol/An > 0.5. This argument is supported by the absence of any pristine LKFM samples.

All of the LKFM samples discussed here (Table 1) are rich in siderophile contaminants and commonly contain impact-intermixed clasts; therefore these samples are impact melts and are not volcanic rocks. Every LKFM specimen in the lunar sample collection is impact modified (Reid *et al.,* 1977) even though many LKFM specimens have textures similar to those of extrusive basalt. Samples 14310 and 62295 both were originally interpreted as volcanic rocks before their siderophile contamination was reported (both have high Ir and Au contents: Table 1). However, despite the evidence for impact reworking, the phase chemistry of some LKFM compositions with apparent multiple saturation provides reason to believe that LKFM basalts may have been widely distributed on the moon.

Walker *et al.* (1973) point out an apparent olivine-anorthite-liquid control in some LKFM compositions. Those melt rocks that have bulk compositions with olivine-anorthite cosaturation also have evidence of meteoritic contamination and thus of an impact melt origin, yet it would be an unlikely coincidence for impact melts which are mixtures of highland rock types to consistently end up with compositions that indicate equilibrium igneous cosaturation (Walker *et al.*, 1973; Reid *et al.*, 1977). This "suspicious coincidence" is replicated (Fig. 7) in the Apollo 14 soil analysis, and in the LKFM glass averages obtained in four different studies (Reid *et al.*, 1972; Reid, 1974; Ridley *et al.*, 1973; Naney *et al.*, 1976). The occurrence of an *apparent* igneous co-saturation in soil or soil-derived glasses, as well as glasses and melt rocks derived by larger deep-melting impacts, may suggest that there were at one time large bodies of LKFM igneous rock at the lunar surface. The suggestion of LKFM igneous activity has recently been reinforced by the discovery of Blanchard and Budahn (1979) that relic mineral clasts in breccia 73255 were once in equilibrium with a magma having LKFM trace-element composition.

Both the impact-origin and igneous LKFM hypotheses have been considered and judged unacceptable by Reid *et al.* (1977), the first because of its fortuitous nature and the second because of the lack of any pristine igneous LKFM samples. The argument may also be raised, that such samples as the Apollo 14 regolith (Fig. 7) and Apollo 14 breccias are obviously impact mixtures despite their apparent olivine-plagioclase cosaturation. This is certainly true; Ryder and Taylor (1976), for instance, have noted the occurrence of rare mare basalt fragments in Apollo 14 samples: the mare-type compositions could not have been derived from an LKFM precursor. On the other hand, if such contaminants were abundant, the LKFM soil would not have its apparent "cosaturation" composition. It is important to consider the *dominant* components of the Apollo 14 samples. Simonds *et al.* (1977) note that the Apollo 14 soil is similar in composition to vitric-matrix breccias that may be locally derived from the soil itself. Pyroxene clast Ca-Fe-Mg compositions and feldspar clast Ab–Or compositions in the vitric-matrix breccias are comparable to the range of compositions in LKFM melt rocks. Grieve *et al.* (1975) find that breccia 14321 ("Big Bertha") is composed primarily of three major components: LKFM basalt fragments, a light binding matrix of LKFM composition, and a set of an older microbreccias and poikilitic materials that probably are also of LKFM composition. Taken together, the three major components in 14321 can be interpreted as impact products from a LKFM terrain. That terrain could have been volcanic.

If small LKFM basalt flows at one time occurred at or near the lunar surface, it is unlikely that impact reworking and remelting would have preserved the original LKFM compositions. For instance, melt rock 14276 contains plagioclase clasts which could account for its deviation from exact plagioclase-olivine-pyroxene cosaturation by impact mixture with anorthositic rocks. The possibility of a similar two-end-member mixing, of LKFM igneous rocks and ferroan anorthosites, is demonstrated at Apollo 16, where soil and breccia samples can be described as simple mixtures of Apollo 16-type LKFM samples (e.g. 60315, 62235) and Apollo 16 anorthosites (LSPET, 1972). However, if LKFM igneous rocks did occur near the lunar surface at one time, one must account for the present lack of pristine samples as well as the apparent ability of many "LKFM

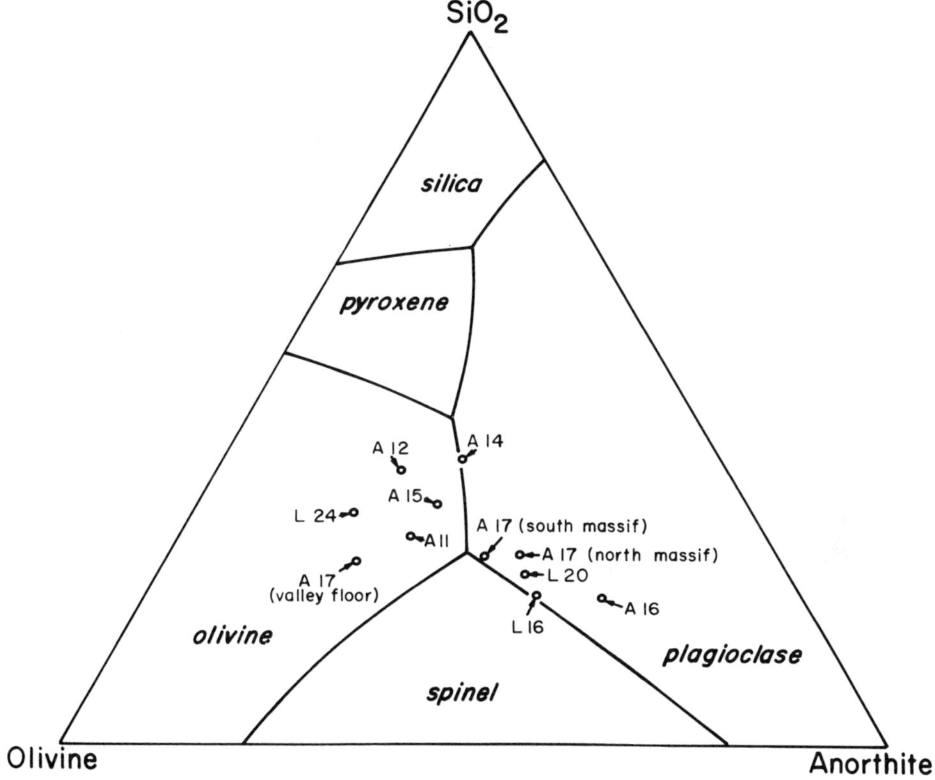

Fig. 7. Apollo (A) and Luna (L) average soil compositions plotted in the Walker *et al.* (1972) pseudoternary.

basalts'' to survive impact remelting without significant intermixture of other rock types. To satisfy these two constraints, the processes of impact melting place certain restrictions on the ages and volumes of ''LKFM basalts'' which could have occurred at the lunar surface.

In impact processes the area of the transient crater is largely homogenized during the production of impact breccias and melt rocks. In a large-scale impact of the type that would produce a multi-ringed basin, the transient crater may be many tens of kilometers in diameter and perhaps ~5 km deep (Moore *et al.*, 1974). The more numerous, smaller craters of a few tens of kilometers will have shallower excavation depths. Studies of the Canadian impact craters of Manicouagan, West Clearwater, and Mistastin provide evidence of how impact homogenization occurs. The turbulent mixing of melt and clasts, as the melt travels down and out along the walls of the transient crater, results in melt sheets which are compositional analogs of the averaged rock type encompassed by the transient crater (O'Keefe and Ahrens, 1975). Clasts intermixed with the impact total-melt phase may be partially or completely melted as they draw heat from the surrounding superheated liquid, but the result in either case is a homogeneous mixture of the rock types found throughout the area of the transient crater. In the discussion above, this homogenization was invoked in the "mapping" of melt-rock types at Apollo 16; this type of homogenization can also be used to place lower limits on the extent and volume of an LKFM volcanic pile if it has to be large enough to survive impact melting without acquiring a large amount of foreign debris.

Assuming the large-crater limits described above, one would need to postulate an LKFM volcanic terrain ~100 km in diameter and perhaps a few km thick. These requirements are large, but they do not exceed the present scale of mare basalt flooding in lunar basins (Head, 1976). Similar large volumes of basalt occur on earth, in the sea-floor igneous regieme and in tholeiitic flood basalts that form flow series over areas as great as 5×10^5 km^2 and ~2 km thick. It is not unreasonable to expect LKFM basaltic floods of this volume during the past history of the moon; Schonfeld and Meyer (1973) suggest basalt floods of LKFM composition in the Imbrium Basin region at 4.4–4.3 G.y. The restrictions of Warner *et al.* (1977) on KREEP appearance would push this date up to ~4.2 G.y. If LKFM basaltic flood(s) appeared at the lunar surface at ~4.1–4.2 G.y., they would have been exposed for 200–300 million years to cataclysmic meteorite bombardments. This exposure could explain the absence of pristine LKFM samples in the lunar collection: we note that none of the known pristine IKFM volcanic samples have volcanic ages >3.9 G.y. The occurrence of LKFM basalt on a broad scale at ~4.2 G.y. would help account for the common occurrence of a KREEP component in all lunar melt rocks, breccias, and soils <4.2 G.y. old.

CRYSTALLIZATION HISTORIES

Although LKFM basalts may have occurred at one time on the lunar surface, the criteria of siderophile contamination indicate that every igneous-textured LKFM sample yet analyzed is in fact crystallized not from a volcanic rock, but from impact melt. Thus, of the samples listed on Table 1, only the IKFM basalts can be thought of as crystallizing from extrusive basaltic flows; all of the others, LKFM and Gabbroic Anorthosite, shared the distinctive cooling histories of impact melt sheets.

Several equilibrium experiments on highland melt rock compositions are summarized in Table 4. For anorthositic gabbro compositions, plagioclase appears at ~1400–1300°C, followed by orthopyroxene or olivine at ~1250°C. Mg-rich LKFM compositions ($Mg' = 0.8$) at point A (Fig. 1) crystallize plagioclase and olivine at ~1270°C, and precipitate pyroxene at ~1230°C. The more typical LKFM samples ($Mg' \sim 0.66$) form plagioclase and olivine at ~1230°C and precipitate pyroxene at ~1180°C. The modified 14310 composition of Hess *et al.* (1977) is ~10% less feldspathic and ~50% more Mg-rich than natural 14310; these modifications result in persistence of olivine from ~1230°C to temperatures less than 1150°C, whereas the natural composition probably attains olivine saturation only briefly at ~1230°C before the olivine is resorbed. In IKFM samples, plagioclase and pyroxene appear at ~1180°C and are joined by ilmenite at ~1080°C. These data are quoted for comparison of equilibrium cooling histories with those inferred from studies on natural samples; many such studies have shown that the natural processes do not follow equilibrium predictions, especially in the impact-melting regime.

Simonds (1975) deduces a two-stage impact melt cooling history which has proved applicable to terrestrial as well as lunar impact melts. The first stage of cooling is a rapid loss of heat from the superheated impact melt to the enclosed clasts, followed by slower second-stage cooling by conduction and radiation, as in a lava flow emplaced by normal volcanic processes. The first stage of cooling is extremely rapid in all impact melt sheets; for example, first-stage cooling rates may be on the order of tens of degrees per second (Onorato *et al.*, 1978). Extrapolated to degrees/hour, the calibration parameter for most cooling rate experiments, the first-stage impact-melt cooling rates fall in the phenomenal range of 10^4–10^6 °C/hour. First stage cooling rates are governed by the size and distribution of clasts in the melt, and therefore these rates do not depend on position within the melt sheet. The second stage of cooling, however, is several orders of magnitude slower and depends greatly on the thickness of the melt sheet, its emplacement cover (if any), position within the melt sheet, and other factors relating the melt sheet to its geologic environment. In thicker impact melts insulated by covering debris, grain sizes tend to be coarser and the clastic debris are more thoroughly digested, leaving fewer relic clasts. Simonds (1975) has noted that in lunar highland melt sheets there is a preferential sequence of clast digestion, starting with Na–rich feldspar and Fe-rich ferromagnesian mineral clasts before the more refractory ferromagnesian minerals, and leaving anorthitic plagioclase as the last to melt. Thus, in a melt sheet which is chemically homogeneous in its clast-plus-melt composition, a sample near the rapidly cooled melt sheet margin may have a melt phase which has assimilated only the most sodic feldspar and ferrous mafic minerals, resulting in a melt phase which is different from that of the melt sheet interior where almost all of the clasts are digested.

Temperature ranges for first-stage and second-stage crystallization in lunar highland melt sheets are suggested by theoretical modelling (Onorato *et al.*, 1978) and petrologic studies (Simonds *et al.*, 1976). Probable first-stage cooling histories fall in the range of ~1900°–1200°C. Second-stage cooling may operate from ~1200°C to solidus. Grove and Bence (1977) have shown in eperiments on Apollo 15 QNB that the temperature for plagioclase saturation is drastically reduced at higher cooling rates (>50°C/hour), and pyroxene and olivine saturation temperatures are also reduced but to a lesser degree. The exact effects are not known in highland-composition impact melts, but should be comparable in sense if not in magnitude. In highland impact melts it is likely that once the suppression of crystallization in first-stage cooling is relaxed, second-stage cooling may take place in a melt phase which is oversaturated with respect to plagioclase, olivine, and pyroxene (Table 4).

Table 4. Experimental data on highland melt rock compositions.

	68415 (a)		63545 (e)*		62295 (a)		14310 (c)	
SiO_2	45.3	42.0	43.8	39.8	45.3	43.0	47.0	43.2
TiO_2	0.35	0.25	1.00	0.68	0.8	0.6	1.2	0.83
$AlO_{1.5}$	28.7	31.4	23.1	24.7	20.8	23.3	19.0	20.6
FeO	3.69	2.86	6.5	4.9	5.99	4.76	9.7	7.4
MnO	—	—	0.03	0.02	—	—	—	—
MgO	4.44	6.14	12.3	16.6	14.4	20.4	11.3	15.4
CaO	16.3	16.2	13.0	12.6	11.5	11.7	11.8	11.6
$NaO_{0.5}$	0.5	0.9	0.21	0.37	0.4	0.7	0.4	0.7
$KO_{0.5}$	0.12	0.14	0.09	0.10	0.14	0.17	0.1	0.1
$CrO_{1.5}$	0.14	0.10	0.10	0.07	0.2	0.2	0.2	0.1
	99.5		100.1		99.5		100.7	
	1399°C	g+pl	1299°C	g+sp	1325°C	g+sp	1250°C	g+sp
	1300°	g+pl	1289°	g+ol+sp	1290°	g+ol+sp	1245°	g+sp
	1275°	g+pl	1275°	g+pl+sp	1271°	g+pl+ol+sp	1235°	g+ol+pl+sp
	1250°	g+pl+opx	1265°	g+ol+pl+sp	1250°	g+pl+ol+sp	1190°	g+ol+pl
	1225°	g+pl+opx+ol	1236°	g+ol+pl+sp	1228°	g+opx+pl+ol+sp	1175°	g+px+ol+pl
			1225°	g+ol+pl	1208°	g+px+pl+ol+sp	1150°	g+px+ol+pl
			1214°	g+ol+pl+px				
An%	15.2		12.1		11.2		9.9	

Notes: for each composition, a first column lists oxide weigh percent and a second column lists cation composition normalized to 100%. A star (*) indicates melting experiments (as opposed to cooling experiments). References (b) and (d, last columns) quote temperatures at phase appearance.

References: (a) Walker *et al.* (1973); (b) Green *et al.* (1972); (c) Hess *et al.* (1977); (d) Hess *et al.* (1978); (e) Delano (1977).

Phases indicated are glass (g), plagioclase (pl), olivine (ol), pyroxene (px) or orthopyroxene (opx), ilmenite (ilm) and spinel (sp).

The selective origins of intersertal, intergranular, ophitic to sub-ophitic, or poikilitic textures are not necessarily related to cooling rate. Lofgren (1977) suggests that poikilitic melt-rock texture may arise from a high density of nucleation sites, the condition one would expect from a dust- and fragment-charged impact melt. However, Simonds (1975) proposes that poikilitic melt-rock texture is a reflection of the two-stage cooling history, with small feldspar laths forming rapidly (disequilibrium) at higher cooling rates before the impact melt begins slower (equilibrium) crystallization of enclosing pyroxene (± olivine) oikocrysts. Both effects, enhanced nucleation on seed debris and rapid disequilibrium plagioclase growth, are likely to have occurred in lunar impact melts.

Those IKFM samples which have been recognized as volcanic rocks (e.g., 15382, 15386, 15434) are in part recognized as such because of their lack of clastic debris. The lack of clastic debris reflects crystallization from a basalt, rather than from a clast-laden impact melt sheet (note however that the lack of clasts does

Table 4. *(Continued)*

	76055 (e)*		15382 (d)		KREEP (b)		LKFM derivation (d)	
SiO_2	45.4	40.9	52.3	49.6	48.5	45.6	48.8	47.4
TiO_2	1.33	0.90	1.8	1.3	2.0	1.4	4.3	3.1
$AlO_{1.5}$	16.6	17.6	17.7	19.8	16.7	18.5	12.2	14.0
FeO	9.5	7.2	8.6	6.8	11.1	8.7	15.3	12.4
MnO	0.11	0.09	—	—	0.0	0.0	—	—
MgO	16.7	22.4	7.2	10.2	8.3	11.6	6.7	9.7
CaO	9.4	9.1	9.8	10.0	11.5	11.6	11.3	11.8
$NaO_{0.5}$	0.68	1.19	0.9	1.6	0.6	1.1	0.4	0.8
$KO_{0.5}$	0.39	0.45	0.6	0.7	1.2	1.4	0.7	0.9
$CrO_{1.5}$	0.17	0.12	—	—	—	—	—	—
	100.3		98.9		99.9		99.7	
	1300°C	g+ol	1200°C	g	at 1225°C	g	at 1150°C	px+pl
	1244°	g+ol	1185°	g+sp	at 1225°	g+ol+pl		
	1225°	g+pl+ol	1180°	g+px+pl				
	1198°	g+pl+ol	1100°	g+px+pl	at 1180°	g+px+pl		
			1180°	g+ilm+px+pl				
An%	8.0		8.8		8.0		6.2	

not prove a volcanic origin: sample 14310 apparently originated in a melt sheet with exceptionally slow second-stage cooling in which all clasts were incorporated into the melt).

SILICATE MINERALOGY OF HIGHLAND MELT ROCKS

Plagioclase

Plagioclase Ca-Na-K compositions are summarized in Fig. 4 for 17 representative melt rocks. In sample 66095, plagioclase clasts are distinguished from groundmass plagioclase on the basis of grain size as well as composition (Fig. 4). The plagioclase clasts in this sample are an order of magnitude larger than the groundmass crystals (~1 mm vs. <0.05 mm). The other samples of impact melt rocks lack such a sharp bimodality of clasts vs. groundmass, and no attempt has been made to distinguish clast/groundmass compositions for these samples (Fig. 4). In many melt-rock samples, texture or grain size is inadequate to identify clastic fragments. Walker *et al.* (1973) note that sample 60335 contains blocky, fractured plagioclase grains that are clearly xenocrystic, yet there is a continuous gradation in texture between these xenocrysts and plagioclase laths, of the same composition, that appear to have grown from melt. In 60335 it is possible that some of the lath-shaped plagioclase may have unrecognized xenocrystic cores. The presence of xenocrystic plagioclase in 60335 implies that the bulk composition of this sample is too plagioclase-rich to represent the melt composition from which crystallization actually occurred in the impact melt. Alternatively, where zoning profiles in clasts may be lost by annealing at near-liquidus temperatures during second-stage cooling in an impact melt sheet, it is possible that the clasts may effectively equilibrate with the melt phase but not dissolve. In order to examine these possibilities, it is helpful to compare experimental and theoretical predictions of plagioclase crystallization temperatures.

Figure 2 shows that highland melt rocks, plotted on the simple (pseudoternary) system, are plagioclase saturated or they are approximately multiply saturated with plagioclase plus a ferromagnesian phase. One may visualize a path from An-rich anorthosic gabbro melt rocks, through the An-poor anorthositic gabbro specimens, toward point A, and then proceeding toward point B and finally into the plane of the projection but toward point C along surfaces of multiple saturation. In practice this is not a real crystal fractionation path, but it serves to systematize plagioclase-melt relationships within the highland melt rock suite. Since calcic plagioclase would constantly be removed along this path, any measure of normative anorthite content must decrease as the path recedes from the An apex. The cation normative anorthite content* in highland melt rocks decreases from ~16% in An-rich anorthositic gabbro samples to ~8% in IKFM basalts. Although this path is hypothetical and does not imply any relationship between samples, it is a useful basis for comparing crystal/melt relationships.

In Fig. 8 the results of several relevant experiments are plotted against catanorm anorthite content, for comparison with plagioclase crystallization temperatures calculated assuming near liquidus plagioclase saturation (or cosaturation with a mafic phase). Crystallization temperatures were calculated using the experimentally calibrated distribution of $NaAlO_2 + SiO_2$ in liquid versus albite in plagioclase (relation K_8 of Drake, 1976). The appearance of plagioclase near the liquidus in all of these samples allows the assumption that the bulk rock analysis approximates the composition of the liquid which was in equilibrium with the most calcic plagioclase observed in the sample. If the rock includes xenocrysts, this assumption requires either that clasts are so sparse as to be insignificant in affecting the melt composition, or that the clasts have largely equilibrated with the melt. By comparing predicted plagioclase/rock relations with those observed, it is possible to view anomalously refractory clasts or re-equilibrated plagioclase on the same diagram.

The experimentally determined plagioclase liquidii are plotted against catanorm % anorthite in Fig. 8. The vertical temperature bars for each experimental charge represent the temperature range within which plagioclase appears, as determined by the temperature intervals used in the relevant experiment (Table 4). Intervals

*Calculated by combining CaO with remaining Al_2O_3, after forming $KAlO_2$ and $NaAlO_2$.

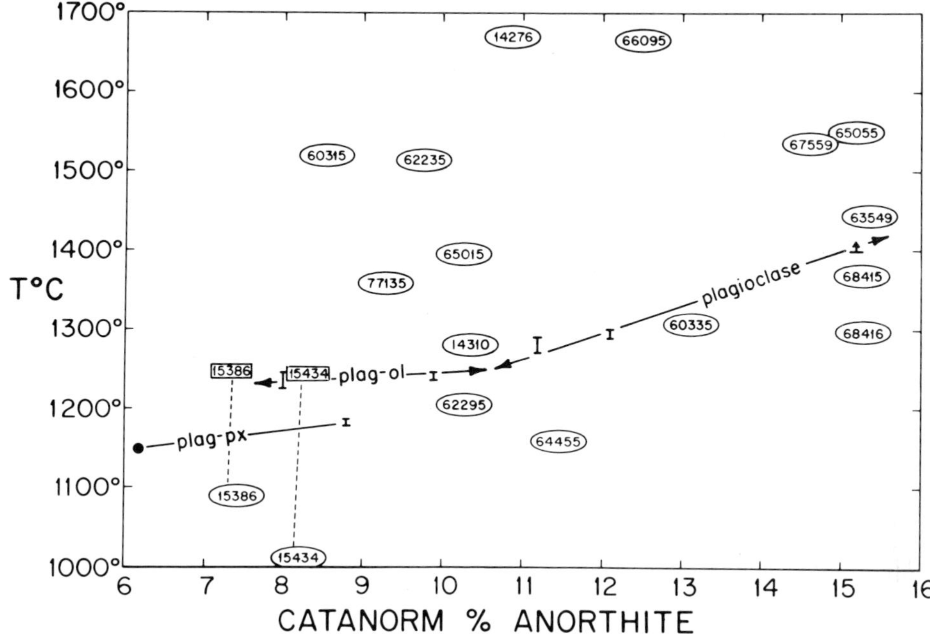

Fig. 8. Experimentally determined feldspar saturation temperatures plotted against cation normative anorthite content (data from Table 4). Vertical bars represent constraints on the temperature range within which plagioclase appears in the equilibrium experiments. Compositions of plagioclase, plagioclase-olivine, and plagioclase-pyroxene saturation are indicated.
Circled sample numbers represent calculated plagioclase/rock equilibration temperatures. Calculated temperatures more than ~100°C above the saturation lines are interpreted as deviations due to the presence of refractory unmelted plagioclase clasts in impact melt rocks; *note* that such anomalous temperatures are *not* real. Calculated temperatures more than 100°C below the saturation lines may reflect low-temperature equilibration, undercooling in an impact melt sheet, or non-representative plagioclase analysis. The deviation between calculated plagiocase and pyroxene equilibration temperatures in 15386 and 15434, samples which should be cosaturated with both phases, and the apparent displacement of plagioclase-olivine and plagioclase-pyroxene saturation lines from continuity are discussed in the text.

of plagioclase-only saturation, plagioclase-olivine cosaturation, and plagioclase-pyroxene cosaturation are indicated. The plagioclase-pyroxene cosaturation trend is discontinuous with and below the plagioclase-olivine cosaturation trend. In part, this discontinuity may be due to the precipitation of another Ca-bearing phase (low-Ca pyroxene) in addition to plagioclase; in part the discontinuity may be due to differences between the experimental charges and the natural samples (the composition 15382, Table 4, is less, Fe,Mg-rich and more sodic than the natural samples 15386 and 15434). Whatever the causes of discontinuity, the calculated temperatures of feldspar appearance in 15386 and 15434 are lower than the experimentally determined temperatures. The textures of these two samples (example in Fig. 5) are intergranular/subophitic to intersertal, and indicate little sub-solidus equilibration. We note however that the lowest albite contents we have found in our study of 15386 and 15434 are in the range $Ab_{12.5-13.5}$; Basu and Bower (1976) find plagioclase grains as Na-poor as Ab_9 in IKFM basalt samples from Apollo 15 soils, a composition close to that needed (Ab_8) in order to bring our estimate of the temperature of plagioclase appearance up to that found experimentally. It is quite possible that our least sodic plagioclase determinations for 15386 and 15434 (Fig. 4) do not represent liquidus plagioclase compositions. This interpretation is supported by higher calculated pyroxene liquidus temperatures of ~1225°–1250°C for 15386 and 15434 (using the K_{24} relationship of Nielsen and Drake, 1979).

Interpretations of the more anorthitic samples (plagioclase, or plagioclase-olivine saturated) are more straightforward, and correlate with petrographic characteristics. The An-rich anorthositic gabbro samples 63549 and 68415 plot close to the predicted plagioclase liquidus temperature (~1400°C), whereas samples 67559 and 65055 apparently contain unequilibrated refractory clastic plagioclase which lead to an anomalously high temperature calculation. Simonds (1975) demonstrates that the more refractory plagioclase clasts in Apollo 17 impact melt rocks are selectively preserved in advanced stages of second-phase absorbtion by the melt phase. On the other hand, sample 68416 contains anomalously sodic plagioclase calculated at ~100°C below the predicted liquidus temperature. The anomalously sodic plagioclase in 68416 is accompanied by anomalously Fe-rich initial olivine composition. Both effects may be due to undercooling during the first-stage cooling (clast/melt thermal equilibration) in an impact melt sheet. Donaldson *et al*. (1975) note an Ab-increase at higher cooling rates in mare basalt compositions. However the plagioclase geothermometer is based on Na contents, and at the low Na contents of plagioclase and melt in 68416 we find that the compound uncertainties indicated by Drake (1976) for the geothermometer and imposed by the detection limits of microprobe analysis lead to a confidence of only ± 110°C.

The An-poor anorthositic gabbro melt rock 60335 plots close to the predicted liquidus temperature (interpolated) of ~1340°C. This close match occurs despite the observation of texturally distinct fractured plagioclase clasts (Walker *et al.*, 1973), because the clastic plagioclase has the same maximum anorthite content (An_{97}) as the cores of plagioclase laths crystallized from melt. Sample 66095, though similar in composition to 60335, is laced with ~8% clasts of distinctively large, anorthitic plagioclase (Fig. 4). These obviously unequilibrated clasts lead to an anomalousy high temperature estimate.

The LKFM sample 14310 plots very close to its predicted plagioclase liquidus temperature of ~1275°C. The concordance of experimental and calculated temperatures of plagioclase appearance in 14310 is in accord with the conclusions of many workers who find no evidence for clastic plagioclase in this sample (Gancarz *et al.*, 1972; Crawford and Hollister, 1974; Meyer, 1977). Although sample 14276 is very similar to 14310 in both texture and composition (Table 1), the extraordinarily high calculated temperature of plagioclase appearance is a strong indication that the more anorthitic subhedral, blocky plagioclase crystals are unequilibrated clasts.

Samples 62295 and 64455 have calculated plagioclase appearance temperatures below predicted values. Although the deviation recorded in 62295 is within the error limits of the calculation method, sample 64455 is over 100°C below its predicted temperature. This sample contains equilibrated plagioclase, olivine and pyroxene with greatly curtailed ranges of zonation (Figs. 3,4), as well as a coating of impact-melt glass and a partially remelted rim, all of which suggest a thermal event which caused re-equilibration of the plagioclase in 64455 to a lower temperature (<1250°C). Prediction is consistent with the petrographic character of the poikilitic LKFM samples, all of which contain blocky anhedral plagioclase crystals which appear to be unequilibrated clasts, and all of which have anomalously high calculated temperatures of plagioclase appearance.

Olivine

Several studies have shown that the Fe-Mg exchange distribution coefficient between olivine and liquid is temperature independent (Roedder and Emslie, 1970; Roedder, 1974; Longhi *et al.*, 1978). Written as the relation $K_D^{ol-liq} = X_{Fe}^{ol}X_{Mg}^{liq}/X_{Mg}^{ol}X_{Fe}^{liq}$, where X represents mole fraction, the equilibrium value of K_D^{ol-liq} is constrained within the range 0.31 ± 0.05 for terrestrial, lunar and meteoritic basalts of a great range in composition. The constancy of K_D^{ol-liq} is maintained throughout a variety of fO_2, P_{H_2O} and P_{total} environments (Longhi *et al.*, 1978; Bender *et al.*, 1978).

Table 5 lists calculated values of K_D^{ol-liq} for the olivine-bearing samples treated in this paper. Using whole-rock compositions to represent melt, the calculated K_D^{ol-liq} values are high and cluster at ~0.6 ± 0.1 (Fig. 9). We have stated in the discussion of *classification* our reasons for believing that the melt rock compositions cited in Table 1 are representative, and close to the actual melt compositions. Though clasts may bias the analysis away from true melt composition, the bias will be towards anorthite (leaving Fe/Mg ratios unaffected), or in the worst case toward other highland xenolithic types with Mg′ compositions which are very much like the melt phase (Mg′ ~ 0.7).

Re-equilibration is a very likely cause for some of the high K_D^{ol-liq} values. Where crystal growth is slow enough that the first-formed olivine is able to re-equilibrate with a lower-temperature melt, or where the cooling history is protracted and Fe-Mg diffusion within olivine has destroyed the initial zoning pattern, the calculated K_D^{ol-liq} using the bulk rock composition will be anomalously high. This effect is certainly seen in some of the melt rocks described here (Table 5). For example, in sample 64455 the lack of silicate zonation (Figs. 3,4) and the anomalously low calculated temperaure of plagioclase/rock equilibration indicate re-equilibration at low temperature. The lack of silicate zonation is also seen in sample 66095. Re-equilibration of olivine with the remaining melt may also account for the high

Table 5. Olivine K_D^{ol-liq} compositions.

	Mg/(Mg+Fe) rock	maximum Mg/(Mg+Fe) olivine	K_D^{ol-liq}
ANG			
68415	0.65	0.73	0.69
60335	0.76	0.84	0.60
66095	0.69	0.77	0.66
LKFM subophitic/intergranular			
64455	0.75	0.83	0.61
62295	0.81	0.88	0.58
poikilitic			
62235	0.66	0.81	0.46
65015	0.66	0.77	0.58
77135	0.71	0.80	0.61
60315	0.73	0.79	0.72
Mg-rich vitrophyre, A-16 drill core (Vaniman *et al*., 1976)			
T.S.			
60004,474	0.95	0.97	0.59

K_D^{ol-liq} calculated values in poikilitic samples (62235, 65015, 77135, 60315) and in subophitic anorthositic gabbro samples (68415, 60335); however, the appearance of a similarly high K_D^{ol-liq} in vitrophyre 62295 can not be explained in this way: extensive olivine-melt re-equilibration is unlikely in a sample with skeletal olivine crystals (Figs. 5c,d) in 30% residual glass (Table 2).

Sample 62295 is not alone among highland melt rock vitrophyres in maintaining a high apparent K_D^{ol-liq}. Vaniman *et al*. (1976) describe another magnesium-rich melt rock vitrophyre from the Apollo 16 drill core with a high K_D^{ol-liq} (0.59; Table 5). For this sample and for 62295, re-equilibration at low temperature is very unlikely. Other factors that may affect K_D^{ol-liq} must be considered.

Mg-enrichment effect?

The current olivine-liquid experimental data does not extend to Mg′ values greater than 0.75. As Mg'_{liquid} approaches 1.0, the K_D^{ol-liq} must also approach 1.0. It is possible that the equilibrium K_D^{ol-liq} is actually between 0.3 and 1.0 in samples of very high Mg'_{rock} (~0.95; Table 5). However, we doubt that this effect would be significant in samples such as 62295 with less extreme Mg'_{rock} values.

Cooling rate, and undercooling effects?

Donaldson *et al*. (1975) report increased K_D^{ol-liq} with increased cooling rate in a low-Ti mare composition: at 86°C/hr the K_D^{ol-liq} is 0.4, and at cooling rates greater than ~400°C/hr the K_D^{ol-liq} approaches 0.5. A similar effect could occur in highland compositions, although experimental data are lacking. Although the textures in most highland melt rocks suggest much slower cooling rates (e.g., ~10°C/hr for 14310; Nabelek *et al*., 1978) it is important to remember that the clast-melt cooling history of an impact melt rock is a two-stage process. Certainly, the initially phenomenal first-stage cooling rates ($\simeq 10^4$ °C/hr) are not evident in the textures described here, which are dominantly those of slower second-stage cooling comparable to a "normal" lava body. Although it is unlikely that crystals could form during rapid first-stage cooling, nucleation will be enhanced by the abundance of particulate mater preserved in impact melts (Nabelek *et al.*, 1978). Nucleation may occur as cooling rates drop abruptly at the transition between first- and second-stage cooling. This transition probably occurs at $\simeq$1200°C (Simonds 1975; Simonds *et al.*, 1976). The initial nucleation and growth of plagioclase, olivine and pyroxene may thus take place below their equilibrium stability (undercooling) and over a wide range of cooling rates. Cooling rates during nucleation could vary from hundreds of degrees per hour at the end of first-stage cooling, to a few degrees or tens of degrees per hour at the inception of second-stage cooling, depending on location in the melt sheet. Thus the textures observed in melt-rock specimens are dominated by second-stage growth, but the cores of many crystals may have experienced rapid cooling rates. There is little direct experimental evidence to elucidate the crystal-melt interactions in highland melt rocks during this process.

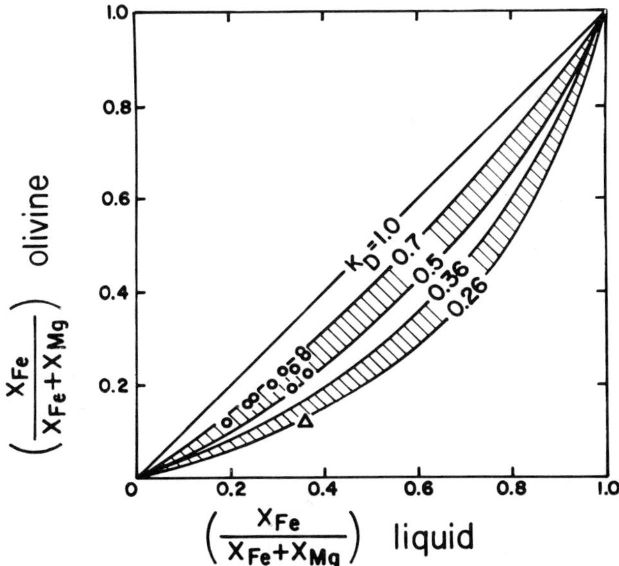

Fig. 9. Data of K_D^{ol-liq} for highland impact melt rocks (Table 5) compared with all lunar, meteoritic, and terrestrial basalts. K_D^{ol-liq} in volcanic basalts is temperature-independent and constrained within the range 0.26–0.36 (Longhi *et al.*, 1979). Highland impact melts, however, have K_D^{ol-liq} values of ~0.6 ± 0.1 (open circles); discussion in text. Note that the triangle represents olivine from equilibrium experiments on 14310 composition (Walker *et al.*, 1973; Green *et al.*, 1972), indicating that equilibrium olivine growth is within the "normal" range of $K_D^{ol-liq} \simeq 0.3$.

Other factors?
Data for cooling-rate effects on crystal/melt equilibria are only available for mare compositions (Donaldson *et al.*, 1975; Grove and Bence, 1977; Bianco and Taylor, 1977). These data can not be easily extrapolated to highland melt-rock compositions or cooling histories. What is the effect of the higher Al and alkali content in highland compositions? This compositional difference should be reflected in a much higher melt viscosity (Cukierman *et al.*, 1973). How much will the higher viscosity be increased by rapid first-stage undercooling? In analogous terrestrial impact melts the viscosity must be great enough to account for suspension of denser blocks of country rocks, many meters in diameter, within thick melt sheets where melt-composition viscosity calculations suggest that the settling velocity should be greater than 5 cm/second (Simonds *et al.*, 1976). Finally, would the melt-phase structure in an impact melt rock differ from that of an identical melt generated volcanically? It is possible that motion in silicate magmas may have a great affect on the growth of polymer-type structures. Since an impact melt is emplaced as a cooling sheet before a large portion of the melt has formed by melting of clasts, there will be a sizeable portion of the impact melt phase which has experienced no motion at all. Moreover, the formation of melt rapidly and in-situ may be incomplete, resulting in the retention of molecular structures that would not occur in melts with a longer and more active history of origin at depth and transport to the surface.

These factors of composition, viscosity, and melt structure may influence cation exchange between olivine and liquid. However, at present we can not predict the results in terms of K_D^{ol-liq}, but can only point to the observed anomalies in melt-rock vitrophyres.

Pyroxene

Variations in pyroxene major-element composition and Ti-Al-Cr composition of 16 highland melt rocks are summarized in Fig. 10. Walker *et al.* (1973) report the occurrence of clinopyroxene in sample 62295, but stoichiometric pyroxene analyses were not obtained from 62295 in our study. The common zonation trend in highland melt rock pyroxenes is initiated with the formation of a Pbca orthopyroxene structure in the compositional range $Wo_{2-4}En_{86-81}Fs_{12-15}$. Orthopyroxene cores of this compositional range are mantled by $P2_1/c$ pigeonite of a composition which probably depends on both the bulk rock composition and the cooling rate. In some samples (e.g., intergranular sample 68415, annealed sample 64455, and poikilitic sample 77135) only pigeonite occurs, and apparently the orthopyroxene phase either is recrystallized as a more Fe,Ca-rich pigeonite or never formed because of undercooling of the melt rock during the final part of first-stage cooling in a melt sheet. In highland melt-rock compositions (Table 4), temperatures for the first appearance of pyroxene are ~1250–1210°C in anorthositic gabbro melt rocks and ~1175–1180°C in LKFM and IKFM samples, though calculated pyroxene temperatures for some IKFM samples are in excess of 1200°C. Crystal-

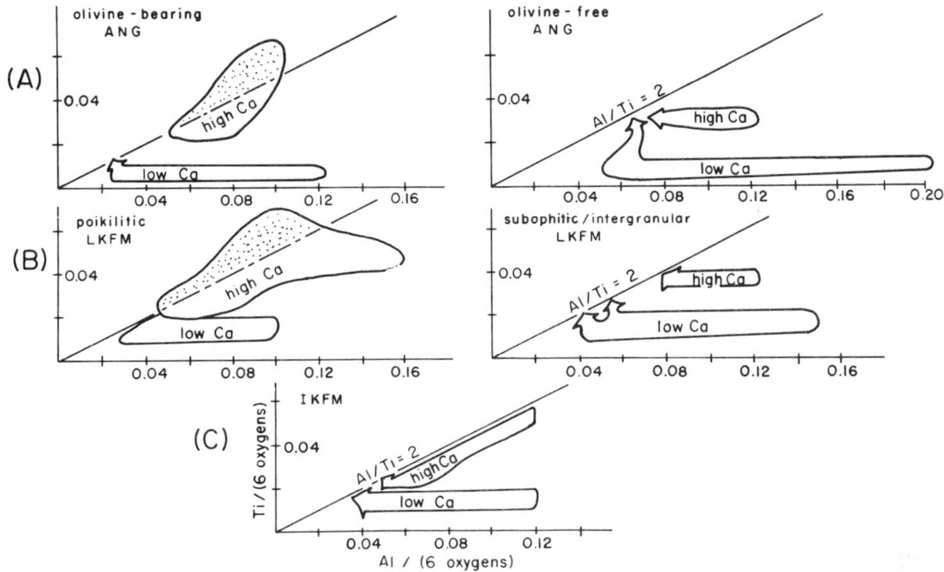

Fig. 10. Summary of Al-Ti zonation in pyroxenes of highland melt rocks. Olivine-bearing ANG is more slowly cooled than olivine-free ANG; poikilitic LKFM is more slowly cooled than subophitic/intergranular LKFM. Stippled areas indicate augites with significant Na content at electron microprobe detection limits (~0.01 cations/6 oxygens). Discussion in text.

lization temperatures near 1200°C are at the approximate lower limit of rapid first-stage cooling by conductive heat loss to cold clasts (Simonds, 1975; Simonds *et al.,* 1976). The persistence of rapid first-stage cooling to temperatures below pyroxene saturation in some melt rocks, but not in others, may account for the variable appearance of orthopyroxene cores in pigeonite grains. This trend is supported by the data of Donaldson *et al.* (1975) and Grove and Bence (1977) indicating that suppression of pyroxene crystallization in mare basalts results in more Fe,Ca-rich initial pyroxene compositions.

Augite (C2/c) appears with continued crystallization. In some samples augite forms by metastable growth on pigeonite, continuing on a zoning trend of constant Ca,Fe-enrichment (e.g., sample 63549, Fig. 3). A more typical occurrence is for high-Ca augite to form closer to a stable composition on the high-Ca limb of the pyroxene solvus (e.g., sample 14276) along with or leading to metastable intrasolvus C2/c pyroxenes of lower Ca content.

Major-element pyroxene crystallization paths may be summarized as follows: Orthopyroxene forms with Mg/(Fe + Mg) atomic, or Mg′, of 0.88 to 0.82. Orthopyroxene may persist to lower Mg content (e.g., to Mg′ ~ 0.65 in 14310; Kushiro *et al.,* 1972) than the Mg′ at which pigeonite becomes stable. Augite forms with initial Mg′ = 0.85–0.65. Both pigeonite and augite may be zoned toward decreasing Mg′, though pigeonite generally disappears at or before Mg′

= 0.60, and final pyroxene crystallization occurs as metastable Fe-rich clinopyroxene which may persist toward pyroxferroite (e.g., 68416, 14310, and 15386).

Three-axis plots in Fig. 3 show detailed zonation of Al, Ti and Cr in each of 16 highland melt-rock samples. The caption for Fig. 3 explains the special symbols used in several of the diagrams. A generalized synopsis of Al, Ti pyroxene zonation is presented in Fig. 10.

Pyroxene Al, Ti content is particularly sensitive to variations in cooling rate. Several studies (Donaldson *et al.*, 1975; Papike *et al.*, 1976; Grove and Bence, 1977) have emphasized the increased Ti, Al, and (to a lesser extent) Cr content of pyroxenes grown with rapid cooling rates. Grove and Bence (1977) find that as cooling rates increase from equilibrium conditions to 150°C/hour, the Al content of first-formed pyroxene is tripled, the Ti content is doubled, but the Cr content is increased by less than half. Although Grove and Bence (1977) studied an Apollo 15 quartz-normative mare basalt, their conclusions appear generally applicable to pyroxene compositions in highland melt rocks. Although initial pyroxene growth could be suppressed by rapid first-stage cooling of an impact melt sheet, the slower second-stage cooling controls the growth and zonation of pyroxene to solidus temperatures.

In Fig. 10a, both An-rich and An-poor anorthositic gabbro (ANG) are considered together, and samples are compared on the basis of cooling rate. The olivine-free ANG melt rocks (63549, 65055) are interpreted as having the most rapid cooling rates among the ANG samples. Sample 63549 is exceptional in its transition from pigeonite directly to low-Ca augite without the formation of a high-Ca augite zonation trend, and in its high initial pyroxene Al-content (0.19 cations/6 oxygens). Sample 65055 does form a separate high-Ca augite trend starting at greater Ti content than the low-Ca orthopyroxene(?)-pigeonite-augite trend; however, 65055 is similar to 63549 in texture, in grain size, and particularly in absence of modal olivine. The absence of olivine in these samples is interpreted as resulting from a higher cooling rate, analogous to the absence of olivine in natural samples of 14310 and 14276 despite olivine appearance in equilibrium experiments (≃ infinitely slow cooling rates) of the 14310/14276 composition (Walker *et al.*, 1973; Green *et al.*, 1972).

The olivine-bearing ANG melt rocks include one sample (67559) with initial pyroxene Al composition as high as 0.12 cations/6 oxygens. Sample 67559 may represent a transition from rapid to slow cooling rates. The olivine-bearing ANG samples 68415 and 68416 have a smaller range of Al-zonation in low Ca,Ti pyroxene and have a distinctive high Ca,Ti pyroxene trend that projects above the line of Al/Ti ratio = 2. Pyroxene compositions above the Al/Ti = 2 line must either contain Ti^{3+} as the component $R^{2+}Ti^{3+}(SiAl)O_6$ or charge-balance additional Ti^{4+}, beyond that accounted for by the formula $R^{2+}Ti^{4+}Al_2O_6$, by incorporating a monovalent cation, e.g., $Na^{+}Ti^{4+}(SiAl)O_6$ or $Na^{+}(Ti_{0.5}R^{2+}_{0.5})Si_2O_6$. Na occurs at or slightly above microprobe detection limits (~0.01 cations/6 oxygens) in augites that plot in the stippled region of Fig. 10a, providing evidence for the Na^{+}–Ti^{4+} coupled substitution in the relatively "slowly" cooled samples 68415 and 68416. Although the microprobe data strongly suggest that Na is involved in augite compositions that plot above the Al:Ti = 2 line, the data are not precise enough to rule out the possibility of an accompanying $Ti^{3+}-{}^{IV}Al$ substitution in the stippled region.

Fig. 10b compares poikilitic and subophitic/intergranular LKFM melt rocks. As in the anorthositic gabbro melt rocks, higher cooling rates are characterized by relatively high initial Al contents in the first-formed low-Ca pyroxene (~0.15 cations/6 oxygens). The poikilitic samples experienced relatively slower cooling rates, as evidenced by solvus-controlled separation into partially equilibrated low-Ca and high-Ca pyroxenes clusters with restricted Fe/Mg-zonation, and by augite Al/Ti compositions above the Al/Ti = 2 line that contain microprobe-detectable amounts of Na (stippled region, Fig. 10b). Note that poikilitic sample 77135, which has not partially equilibrated into high-Ca and low-Ca pyroxene clusters, is also distinctive in having a poor separation of high-Ca and low-Ca clusters in Al,Ti space (Fig. 3) and in lacking any augite compositions that plot above the Al/Ti = 2 line.

Pyroxene Al,Ti contents of the highland IKFM volcanic samples 15434 and 15386 are summarized in Fig. 10c. The augite trend begins at high Ca, Al, Ti and is depleted steadily in all three of these components as the augite composition is enriched in Fe, approaching pyroxferroite. The Al/Ti zonation trend parallels the Al/Ti = 2 line but does not pass above it. Initial Al content in the first-formed orthopyroxene is ~0.12 cations/6 oxygens. These data suggest a relatively rapid cooling history in a lava flow on the lunar surface.

Pyroxene Cr content is not greatly affected by variations in cooling rate, but does drop below detection in the very Fe-rich pyroxenes. At the Cr-rich end of the pyroxene Cr spectrum, however, there is a slight distinction between pyroxenes of the impact melt rocks (≤0.02 cations Cr/6 oxygens) and pyroxenes of the IKFM volcanic rocks (0.02 cations Cr/6 oxygens). The sporadically lower Cr content of pyroxenes from impact melt rocks reflects the bulk rock compositions: the Cr_2O_3 content in IKFM is ~0.35 wt.%, as opposed to ~0.08–0.26 wt.% in other melt rocks (Table 1). Another characteristic which reflects the bulk rock composition is the maximum content of Al attained in first-formed low-Ca pyroxenes. Measured in cations/6 oxygens, the maximum Al content is 0.20 in ANG melt rocks (24–28 wt.% Al_2O_3), 0.15 in LKFM melt rocks (17–22 wt.% Al_2O_3), and 0.12 in IKFM basalts (~15–16 wt.% Al_2O_3). Since Al content in low-Ca pyroxenes can be correlated with cooling rate as well as host rock composition, the Al content of low-Ca pyroxene in rapidly cooled or undercooled melt rocks may approach a fixed maximum which is not exceeded for that composition of melt rock, even at considerably higher cooling rates. Some support for this concept of a limiting maximum can be found in the data of Donaldson *et al*. (1975) on low-Ti mare basalt, where initial Al contents in pyroxene rise continuously as cooling rate increases from equilibrium to 86°/hour but do not rise appreciably as the cooling rate is increased to 380°/hour and 650°/hour. Thus the maximum Al content observed in pyroxenes of highland melt rocks may be a direct reflection of Al content in the host rock. At present this statement can only be based on analogy with experiments on mare basalt compositions, for comparable experiments on highland melt rock compositions have not been performed.

CONCLUSIONS

Highland melt rocks lacking meteoritic siderophile contamination and with isotopic, chemical and textural systematics that are apparently volcanic plot between points B and C in the Walker *et al.* (1972) pseudoternary (Fig. 1). Combined Nd/Sm and Rb/Sr isotopic data suggest that these basalts formed by partial melting of the lunar lower crust or upper mantle at ~3.9–4.0 G.y. No other highland melt rocks have been proven to be volcanic. However, a possible candidate which must still be considered is low-K Fra Mauro (LKFM), even though no pristine volcanic samples of this composition have been found. The fact that glasses, soils, breccias and melt rocks of LKFM composition are abundant on the lunar surface suggests that if LKFM volcanics were at one time on the lunar surface, then they formed thick and widespread lava sequences that could be impact reworked without inevitably suffering an extensive intermixture with other highland rock types. This hypothesis may explain the observed plagioclase-olivine cosaturation in the LKFM composition, a property which would be fortuitous if it must be explained by a repetitively exact impact mixture, and remixture, of other lunar crustal components. The point has already been well made (Ryder and Spudis, 1979) that we must keep an open mind about ancient, pre-4.0 G.y. lunar volcanism.

When large size (>1 cm) is used as a criterion for selecting representative highland melt rocks, samples can be found that range from anorthositic gabbro composition (24–28 wt.% Al_2O_3), LKFM composition (17–22 wt.% Al_2O_3), and the volcanic series of intermediate-K Fra Mauro (IKFM) composition (~15–16 wt.% Al_2O_3). All of these samples share the characteristic of being plagioclase saturated or plagioclase co-saturated. This characteristic allows plagioclase-melt geothermometry to be applied to plagioclase-whole rock compositions. For the anorthositic gabbro and LKFM compositions, these data can be compared with experimentally determined liquidus temperatures to identify samples with refractory plagioclase clasts, and to indicate samples with sub-solidus equilibration histories. Olivine compositions yield anomalously high Mg,Fe K_D^{ol-liq} values of 0.6 ± 0.1 in the anorthositic gabbro and LKFM impact melt rocks. The appearance of Fe-rich olivine reflects re-equilibration in most samples except the Mg-rich melt rock vitrophyres (e.g., 62295). The cause(s) of anomalously Fe-rich olivine in melt rock vitrophyres are unknown.

First-stage cooling in an impact melt sheet may involve variable cooling rates from as high as ~10^6 to 10^5 °C/hour to a few tens of degrees, or only a few degrees, per hour at the lower thermal limit of first-stage crystallization (≃1200°C; Simonds *et al.*, 1976 and Onorato *et al.*, 1978). Thus a great variety of cooling rates may be attained through the range of plagioclase saturation (1400°–1250°C), olivine saturation (1300°–1225°C), and pyroxene saturation (1250°–1175°C). Modification of the equilibrium cooling sequence is most noticeable in olivine-melt relations, but is also reflected in pyroxene-melt relations. Most of the cooling histories of pyroxene in highland impact melts take place below the range of first-stage cooling (below ~ 1200°C); the second-stage cooling which governs pyroxene-melt relations is characterized by a variety of lower cooling rates, from >100°C/hr to approximate equilibrium at $<2 \times 10^{-6}$ °C/hour depending on the samples location in the melt sheet (Onorato *et al.*, 1976). Variations in second-stage cooling histories are reflected by a variety of Al, Ti, Cr zonation profiles in pyroxene (Fig. 10). Taken together, the silicate mineralogy of highland impact melts can yield considerable information about relic pre-impact xenocrysts, about undercooling during rapid loss of heat from the melt phase to enclosed clasts, and about the subsequent cooling of the melt sheet under a thermal regime comparable to that of conduction through and radiation from an extruded lava flow. Only the latter type of cooling is observed in IKFM basalts.

Although general trends for all of these silicate/melt thermal paths can be deciphered, the estimation of temperatures and times must be based on analogies with rate-dependent cooling experiments on mare basalts. There are experimental data on the effects of cooling rates on textural characteristics in highland basalts (Lofgren, 1977 and Nabelek *et al.*, 1978), but there are few data on experimental mineral chemistry. Mineral data from controlled cooling-rate experiments on both impact-melted and volcanic highland compositions hold great potential for a closer calibration of the thermal histories of highland melt rocks.

Acknowledgments—This paper was substantially improved by the thorough reiews of J. R. Garrison, J. Longhi and associate editor C. T. Herzberg. S. Simon collected the new modal data and his efforts are appreciated. Thanks go to Robin Spencer for typing the manuscript and to Pauline Papike for help with technical aspects of putting this paper together. This research was supported by NASA Grant NSG9044 which we gratefully acknowledge.

REFERENCES

Albee A. L., Gancarz A. J., and Chodos A. A. (1973) Metamorphism of Apollo 16 and 17 and Luna 20 metaclastic rocks at about 3.95 AE: Samples 61156, 64423, 14-2, 65015, 67483, 15-2, 76055, 22006, and 22007. *Proc. Lunar Sci. Conf. 4th,* p. 569–595.

Basu A. and Bower J. F. (1976) Petrography of KREEP basalt fragments from Apollo 15 soils. *Proc. Lunar Sci. Conf. 7th,* p. 659–678.

Bence A. E., Papike J. J., Sueno S., and Delano J. W. (1973) Pyroxene poikiloblastic rocks from the lunar highlands. *Proc. Lunar Sci. Conf. 4th,* p. 597–611.

Bender J. F., Hodges F. N., and Bence A. E. (1978) Petrogenesis of basalts from the Project Famous area: Experimental study from 0 to 15 kbars. *Earth Planet. Sci. Lett.* **41,** 277–302.

Bianco A. S., and Taylor L. A. (1977) Applications of dynamic crystallization studies: Lunar olivine-normative basalts. *Proc. Lunar Sci. Conf. 8th,* p. 1593–1610.

Blanchard D. P. and Budahn J. R. (1979) Remnants from the ancient lunar crust: Clasts from consortium breccia 73255. *Proc. Lunar Planet. Sci. Conf. 10th,* p. 803–816.

Boynton W. V., Chou C.-L., Robinson K. L., Warren P. H., and Wasson J. T. (1976) Lithophiles, siderophiles, and volatiles in Apollo 16 soils and rocks. *Proc. Lunar Sci. Conf. 7th,* p. 727–742.

Brookins D. G., Carden J. R., and Laughlin A. W. (1975) Additional note on the isotopic composition of strontium in McCarty's Flow, Valencia County, New Mexico. *Earth Planet. Sci. Lett.* **25,** 327–330.

Brown G. M., Emeleus C. H., Holland J. G., Peckett A., and Phillips R. (1972) Mineral-chemical variations in Apollo 14 and Apollo 15 basalts and granitic fractions. *Proc. Lunar Sci. Conf. 3rd,* p. 141–157.

Brunfelt A. O., Heier K. S., Nilssen B., and Sundvoll B. (1972) Distribution of elements between different phases of Apollo 14 rocks and soils. *Proc. Lunar Sci. Conf. 3rd,* p. 1133–1147.

Brunfelt A. O., Heier K. S., Nilssen B., and Sundvoll B. (1973) Geochemistry of Apollo 15 and 16 materials. *Proc. Lunar Sci. Conf. 4th,* p. 1209–1218.

Carlson R. W. and Lugmair G. W. (1979) Sm–Nd study of pristine KREEP basalt 15386 (abstract). In *Lunar and Planetary Science X,* p. 178–180. Lunar and Planetary Institute, Houston.

Chou C.-L., Boynton W. V., Sundberg L. L., and Wasson J. T. (1975) Volatiles on the surface of Apollo 15 green glass and trace-element distributions among Apollo 15 soils. *Proc. Lunar Sci. Conf. 6th,* p. 1701–1727.

Christian R. P., Berman S., Dwornik E. J., Rose H. J., and Schnepfe M. M. (1976) Composition of some Apollo 14, 15, and 16 lunar breccias and two Apollo 15 fines. In *Lunar Science VII,* p. 138–140. The Lunar Science Institute, Houston.

Compston W., Foster J. J., and Gray C. M. (1975) Rb–Sr ages of clasts from within Boulder 1, Station 2, Apollo 17. *The Moon* **14,** 445–462.

Compston W., Vernon M. J., Berry H., Rudowski R., Gray C. M., Ware N., Chappell B. W., and Kaye M. (1972) Apollo 14 mineral ages and the thermal history of the Fra Mauro formation. *Proc. Lunar Sci. Conf. 3rd,* p. 1487–1501.

Crawford M. L. and Hollister L. S. (1974) KREEP basalt: A possible partial melt from the lunar interior. *Proc. Lunar Sci. Conf. 5th,* p. 399–419.

Crawford M. L. and Hollister L. S. (1977) Evolution of KREEP: Further petrologic evidence. *Proc. Lunar Sci. Conf. 8th,* p. 2403–2417.

Cukierman M., Klein L., Scherer G., Hopper R. W., and Uhlmann D. R. (1973) Viscous flow and crystallization behavior of selected lunar compositions. *Proc. Lunar Sci. Conf. 4th,* p. 2685–2696.

Delano J. W. (1977) Experimental melting relations of 63545, 76015, and 76055. *Proc. Lunar Sci. Conf. 8th,* p. 2097–2123.

Donaldson C. H., Usselman T. M., Williams R. J., and Lofgren G. E. (1975) Experimental modeling of the cooling history of Apollo 12 olivine basalts. *Proc. Lunar Sci. Conf. 6th,* p. 843–869.

Dowty E., Keil K., Prinz M., Gros J., and Takahashi H. (1976) Meteorite-free Apollo 15 crystalline KREEP. *Proc. Lunar Sci. Conf. 7th,* p. 1833–1844.

Drake M. J. (1976) Plagioclase-melt equilibria. *Geochim. Cosmochim. Acta* **40,** 457–466.

Duncan A. R., Erlank A. J., Willis J. P., and Ahrens L. H. (1973) Composition and inter-relationships of some Apollo 16 samples. *Proc. Lunar Sci. Conf. 4th,* p. 1097–1113.

Floran R. J., Phinney W. C., Blanchard D. P., Warner J. L., Simonds C. H., Brown R. W., Brannon J. C., and Korotev R. L. (1976) A comparison between the geochemistry and petrology of Apollo 16—terrestrial impact melt analogs (abstract). In *Lunar Science VII,* p. 263–265. The Lunar Science Institute, Houston.

Gancarz A. J., Albee A. L., and Chodos A. A. (1972) Comparative petrology of Apollo 16 sample 68415 and Apollo 14 samples 14276 and 14310. *Earth Planet. Sci. Lett.* **16,** 307–330.

Garrison J. R. and Taylor L. A. (1979) Petrology of lunar rock 66095: Implications for the genesis of highland basalt and the stratigraphy of the Apollo 16 landing site (abstract). In *Papers Presented to the Conference on the Lunar Highlands Crust,* p. 18–20. Lunar and Planetary Institute, Houston.

Green D. H., Ringwood A. E., Ware N. G., and Hibberson W. O. (1972) Experimental petrology and petrogenesis of Apollo 14 basalts. *Proc. Lunar Sci. Conf. 3rd,* 197–206.

Grieve R. A. F. (1975) Petrology of the impact melt at Mistastin Lake Crater, Labrador. *Bull. Geol. Soc. Amer.* **86,** 1617–1629.

Grieve R. A. F. and Floran R. J. (1978) Manicouagan impact melt, 2, chemical interrelations with basement and formational processes. *J. Geophys. Res.* **83,** 2761–2771.

Grieve R. A., McKay G. A., Smith H. D., and Weill D. F. (1975) Lunar polymict breccia 14321: A petrographic study. *Geochim. Cosmochim. Acta* **39,** 229–245.

Grieve R. A. F. and Plant A. G. (1973) Partial melting on the lunar surface, as observed in glass coated Apollo 16 samples. *Proc. Lunar Sci. Conf. 4th,* p. 667–679.

Grove T. L. and Bence A. E. (1977) Experimental study of pyroxene-liquid interaction in quartz-normative basalt 15597. *Proc. Lunar Sci. Conf. 8th,* p. 1549–1579.

Haskin L. A., Helmke P. A., Blanchard D. P., Jacobs J. W., and Telander K. (1973) Major and trace element abundances in samples from the lunar highlands. *Proc. Lunar Sci. Conf. 4th,* 1275–1296.

Hawke B. R. and Head J. W. (1978) Lunar KREEP volcanism: Geologic evidence for history and mode of emplacement. *Proc. Lunar Planet. Sci. Conf. 9th,* p. 3285–3309.

Hawke B. R. and Head J. W. (1979) Impact melt volumes associated with lunar craters (abstract). In *Lunar and Planetary Science X,* p. 510–512. Lunar and Planetary Institute, Houston.

Head J. W. III (1976) Lunar volcanism in space and time. *Rev. Geophys. Space Phys.* **14,** 265–300.

Head J. W. III (1979) Lateral crustal variations determined from geological studies (abstract). In *Papers Presented to the Conference on the Lunar Highlands Crust,* p. 59–60. Lunar and Planetary Institute, Houston.

Head J. W. III and McCord T. B. (1978) Imbrian-age highland volcanism on the moon: The Gruithuisen and Mairan domes. *Science* **199,** 1433–1436.

Helmke P. A., Blanchard D. P., Haskin L. A., Telander K., Weiss C., and Jacobs J. W. (1973) Major and trace elements in igneous rocks from Apollo 15. *The Moon* **8,** 129–148.

Helz R. T. (1972) Rock 14068: An unusual lunar breccia, *Proc. Lunar Sci. Conf. 3rd,* p. 865–886.

Helz R. T. and Appleman D. E. (1973) Mineralogy, petrology, and crystallization history of Apollo 16 rock 68415. *Proc. Lunar Sci. Conf. 4th,* p. 643–659.

Hess P. C., Rutherford M. J., and Campbell H. W. (1977) Origin and evolution of LKFM basalt. *Proc. Lunar Sci. Conf. 8th,* p. 2367–2373.

Hess P. C., Rutherford M. J., and Campbell H. W. (1978) Ilmenite crystallization in nonmare basalt: Genesis of KREEP and high-Ti mare basalt. *Proc. Lunar Planet. Sci. Conf. 9th,* p. 705–724.

Hodges F. N. and Kushiro I. (1973) Petrology of Apollo 16 lunar highland rocks. *Proc. Lunar Sci. Conf. 4th,* p. 1033–1048.

Hubbard N. J., Gast P. W., Rhodes J. M., Bansal B. M., Wiesmann H., and Church S. E. (1972) Nonmare basalts: Part II. *Proc. Lunar Sci. Conf. 3rd,* p. 1161–1179.

Hubbard N. J., Rhodes J. M., Weismann H., Shih C.-Y., and Bansal B. M. (1974) The chemical definition and interpretation of rock types returned from the non-mare regions of the moon. *Proc. Lunar Sci. Conf. 5th,* p. 1227–1247.

Huneke J. C., Jessberger E. K., Podosek F. A., and Wasserburg G. J. (1973) $^{40}Ar/^{39}Ar$ measurements in Apollo 16 and 17 samples and the chronology of metamorphic and volcanic activity in the Taurus-Littrow region. *Proc. Lunar Sci. Conf. 4th,* p. 1725–1756.

Irving A. J. (1975) Chemical, mineralogical and textural systematics of nonmare melt rocks: Implications for lunar impact and volcanic processes. *Proc. Lunar Sci. Conf. 6th,* p. 363–394.

Irving A. J. (1977) Chemical variations and fractionation of KREEP basalt magmas. *Proc. Lunar Sci. Conf. 8th,* p. 2433–2448.

James O. B. (1973) Crystallization history of lunar feldspathic basalt 14310. *U.S. Geol. Survey Prof. Paper 841.* 29 pp.

Jessberger E. K., Huneke J. C., Podosek F. A., and Wasserburg G. J. (1974) High resolution argon analysis of neutron-irradiated Apollo 16 rocks and separated minerals. *Proc. Lunar Sci. Conf. 5th,* p. 1419–1449.

Kirsten T., Horn P., and Kiko J. (1973) ^{39}Ar-^{40}Ar dating and rare gas analysis of Apollo 16 rocks and soils. *Proc. Lunar Sci. Conf. 4th,* p. 1757–1784.

Kushiro I., Ikeda Y., and Nakamura Y. (1972) Petrology of Apollo 14 high-alumina basalt. *Proc. Lunar Sci. Conf. 3rd,* 115–129.

Lange M. A. and Ahrens T. J. (1979) Impact melting during the first 1.5 B.Y. of lunar history (abstract). In *Lunar and Planetary Science X,* p. 700–702. Lunar and Planetary Institute, Houston.

Lange M.A. and Hawke B. R. (1979) The generation of lunar impact melts: A comparison of theoretical and observational results (abstract). In *Papers Presented to the Conference on the Lunar Highlands Crust,* p. 99–101. Lunar and Planetary Institute, Houston.

Laughlin A. W., Brookins D. G., and Carden J. R. (1972) Variations in the initial strontium ratios of a single basalt flow. *Earth Planet. Sci. Lett.* **14,** 79–82.

Lofgren G. E. (1977) Dynamic crystallization experiments bearing on the origin of textures in impact-generated liquids. *Proc. Lunar Sci. Conf. 8th,* p. 2079–2095.

Longhi J., Walker D., and Hays J. F. (1972) Petrography and crystallization history of basalts 14310 and 14072. *Proc. Lunar Sci. Conf. 3rd,* p. 131–139.

Longhi J., Walker D., and Hays J. F. (1978) The distribution of Fe and Mg between olivine and lunar basaltic liquids. *Geochim. Cosmochim. Acta* **42,** 1545–1558.

LSPET (1972) *The Apollo 16 Prelim. Sci. Rep.* NASA SP-315, Washington, D.C.

LSPET (1973a) The Apollo 16 lunar samples: Petrographic and chemical description. *Science* **179,** 23–34.

LSPET (1973b) The Apollo 17 lunar samples: Chemical and petrographic description. *Science* **182,** 659–662.

Lugmair G. W. and Carlson R. W. (1978) The Sm–Nd history of KREEP. *Proc. Lunar Planet. Sci. Conf. 9th,* p. 689–704.

Lugmair G. W. and Scheinin N. B. (1975) Sm–Nd systematics of the Stannern meteorite. *Meteoritics* **10,** 447–448.

Mark R. K., Chin-Nan L.-H., and Wetherill G. W. (1974) Rb–Sr age of lunar igneous rocks 62295 and 14310. *Geochim. Cosmochim. Acta* **38,** 1643–1648.

Metzger A. E., Haines E. L., Etchegaray-Ramirez M. I., and Hawke B. R. (1979) Thorium concentrations in the lunar surface: III. Deconvolution of the Apenninus region. *Proc. Lunar Planet. Sci. Conf. 10th,* p. 1701–1718.

Meyer C. Jr. (1977) Petrology, mineralogy and chemistry of KREEP basalt. *Phys. Chem. Earth* **10,** 239–260.

Moore H. J., Hodges C. A., and Scott D. H. (1974) Multiringed basins—illustrated by Orientale and associated features. *Proc. Lunar Sci. Conf. 5th,* p. 71–100.

Murthy V. R., Evensen N. M., Jahn B.-M., and Coscio M. R. Jr. (1972) Apollo 14 and 15 samples: Rb–Sr ages, trace elements, and lunar evolution. *Proc. Lunar Sci. Conf. 3rd,* p. 1503–1514.

Nabelek P. I., Taylor L. A., and Lofgren G. E. (1978) Nucleation and growth of plagioclase and the development of textures in a high-alumina basaltic melt. *Proc. Lunar Planet. Sci. Conf. 9th,* p. 725–741.

Naney M. T., Crowl D. M., and Papike J. J. (1976) The Apollo 16 drill core: Statistical analysis of glass chemistry and the characterization of a high alumina—silica poor (HASP) glass. *Proc. Lunar Sci. Conf. 7th,* p. 155–184.

Naney M. T., Papike J. J., and Vaniman D. T. (1977) Lunar highland melt rocks: Major element chemistry and comparisons with lunar highland glass groups (abstract). In *Lunar Science VIII,* p. 717–719. The Lunar Science Institute, Houston.

Nava D. F. (1974) Chemical compositions of some soils and rock types from the Apollo 15, 16, and 17 lunar sites. *Proc. Lunar Sci. Conf. 5th,* p. 1087–1097.

Nielsen R. L. and Drake M. J. (1979) Pyroxene-melt equilibria. *Geochim. Cosmochim. Acta* **43,** 1259–1272.

Nunes P. D., Tatsumoto M., Knight R. J., Unruh D. M., and Doe B. R. (1973) U-Th-Pb systematics of some Apollo 16 lunar samples. *Proc. Lunar Sci. Conf. 4th,* p. 1797–1822.

Nyquist L. E., Bansal B. M., and Wiesmann H. (1975) Rb–Sr ages and initial $^{87}Sr/^{86}Sr$ for Apollo 17 basalts and KREEP basalt 15386. *Proc. Lunar Sci. Conf. 6th,* p. 1445–1465.

Nyquist L. E., Bansal B. M., Wiesmann H., and Jahn B.-M. (1974) Taurus-Littrow chronology: Some constraints on early lunar crustal development. *Proc. Lunar Sci. Conf. 5th,* p. 1515–1539.

Nyquist L. E., Hubbard N. J., Gast P. W., Bansal B. M., Wiesmann H., and Jahn B.-M. (1973) Rb–

Sr systematics for chemically defined Apollo 15 and 16 materials. *Proc. Lunar Sci. Conf. 4th,* p. 1823–1846.

Nyquist L. E., Hubbard N. J., Gast P. W., Church S. E., Bansal B. M., and Wiesmann H. (1972) Rb–Sr systematics for chemically defined Apollo 14 breccias. *Proc. Lunar Sci. Conf. 3rd,* p. 1515–1530.

Oberli F., Huneke J. C., and Wasserburg G. J. (1979) U-Pb and K-Ar systematics of cataclysm and precataclysm lunar impactites (abstract). In *Lunar and Planetary Science X,* p. 940–942. Lunar and Planetary Institute, Houston.

O'Keefe J. D. and Ahrens T. J. (1975) Shock effects from a large impact on moon. *Proc. Lunar Sci. Conf. 6th,* p. 2831–2844.

Onorato P. I. K., Uhlmann D. R., and Simonds C. H. (1976) Heat flow in impact melts: Apollo 17 Station 6 boulder and some applications to other breccias and xenolith laden melts. *Proc. Lunar Sci. Conf. 7th,* p. 2449–2467.

Onorato P. I. K., Uhlmann D. R., and Simonds C. H. (1978) The thermal history of the Manicouagan impact melt sheet, Quebec. *J. Geophys. Res.* **83,** 2789–2798.

Papanastassiou D. A. and Wasserburg G. J. (1972a) Rb–Sr systematics of Luna 20 and Apollo 16 samples. *Earth Planet. Sci. Lett.* **17,** 52–63.

Papanastassiou D. A. and Wasserburg G. J. (1972b) The Rb–Sr age of a crystalline rock from Apollo 16. *Earth Planet. Sci. Lett.* **16,** 289–298.

Papike J. J., Hodges F. N., Bence A. E., Cameron M., and Rhodes J. M. (1976) Mare basalts: Crystal chemistry, mineralogy, and petrology. *Rev. Geophys. Space Phys.* **14,** 475–540.

Phinney W. C. and Simonds C. H. (1977) Dynamical implications of the petrology and distribution of impact melt rocks. In *Impact and Explosion Cratering* (D. J. Roddy, R. O. Pepin and R. B. Merrill, eds.), p. 771–790. Pergamon, N.Y.

Reid A. M. (1974) Rock types present in lunar highland soils. *The Moon* **9,** 141–146.

Reid A. M., Duncan A. R., and Richardson S. H. (1977) In search of LKFM. *Proc. Lunar Sci. Conf. 8th,* p. 2321–2338.

Reid A. M., Warner J., Ridley W. I., and Brown R. W. (1972) Major element composition of glasses in three Apollo 15 soils. *Meteoritics* **7,** 395–415.

Rhodes J. M. and Hubbard N. J. (1973) Chemistry, classification, and petrogenesis of Apollo 15 mare basalts. *Proc. Lunar Sci. Conf. 4th,* p. 1127–1148.

Rhodes J. M., Rodgers K. V., Shih C., Bansal B. M., Nyquist L. E., Wiesmann H., and Hubbard N. J. (1974) The relationships between geology and soil chemistry at the Apollo 17 landing site. *Proc. Lunar Sci. Conf. 6th,* p. 1097–1117.

Ridley W. I., Brett R., Williams R. J., Takeda H., and Brown R. W. (1972) Petrology of Fra Mauro basalt 14310. *Proc. Lunar Sci. Conf. 3th,* p. 159–170.

Ridley W. I., Reid A. M., Warner R. W., Brown R., Gooley R., and Donaldson C. (1973) Glass compositions in Apollo 16 soils 60501 and 61221. *Proc. Lunar Sci. Conf. 4th,* p. 309–321.

Roedder P. L. (1974) Activity of iron and olivine solubility in basaltic liquids. *Earth Planet. Sci. Lett.* **23,** 397–410.

Roedder P. L. and Emslie R. F. (1970) Olivine-liquid equilibrium. *Contrib. Mineral. Petrol.* **29,** 275–289.

Rose H. J. Jr., Cuttitta F., Annell C. S., Carron M. K., Christian R. P., Dwornik E. J., Greenland L. P., and Ligon D. T. Jr. (1972) Compositional data for twenty-one Fra Mauro lunar materials. *Proc. Lunar Sci. Conf. 3rd,* p. 1215–1229.

Rose H. J. Jr., Cuttitta F., Berman S., Carron M. K., Christian R. P., Dwornik E. J., Greenland L. P., and Ligon D. T. Jr. (1973) Compositional data for twenty-two Apollo 16 samples. *Proc. Lunar Sci. Conf. 4th,* p. 1149–1158.

Ryder G. (1976) Lunar sample 15405: Remnant of a KREEP basalt-granite differentiated pluton. *Earth Planet. Sci. Lett.* **29,** 255–268.

Ryder G. and Spudis P. (1979) Volcanism prior to the termination of the heavy bombardment: Evidence, characteristics, and concepts (abstract). In *Papers Presented to the Conference on the Lunar Highlands Crust,* p. 132–134. Lunar and Planetary Institute, Houston.

Ryder G., Stoeser D. B., and Wood J. A. (1977) Apollo 17 KREEPY basalt: A rock type intermediate between mare and KREEP basalts. *Earth Planet. Sci. Lett.* **35,** 1–13.

Ryder G. and Taylor G. J. (1976) Did mare-type volcanism commence early in lunar hisory? *Proc. Lunar Sci. Conf. 7th,* p. 1741–1755.

Schaeffer O. A., Husain L., and Schaeffer G. A. (1976) Ages of highland rocks: The chronology of lunar basin formation revisited. *Proc. Lunar Sci. Conf. 7th,* p. 2067–2092.

Schonfeld E. and Meyer C. Jr. (1973) The old Imbrium hypothesis. *Proc. Lunar Sci. Conf. 4th,* p. 125–138.

Simonds C. H. (1975) Thermal regimes in impact melts and the petrology of the Apollo 17 Station 6 boulder. *Proc. Lunar Sci. Conf. 6th,* p. 641–672.

Simonds C. H. (1979) Low speed (8 km/sec) impact-accretional heating and early fractionation of the moon (abstract). In *Papers Presented to the Conference on the Lunar Highlands Crust,* p. 148–150. Lunar and Planetary Institute, Houston.

Simonds C. H., Floran R. J., McGee P. E., Phinney W. C., and Warner J. L. (1978) Petrogenesis of melt rocks, Manicouagan impact structure, Quebec. *J. Geophys. Res.* **83,** 2773–2788.

Simonds C. H., Phinney W. C., Warner J. L., McGee P. E., Greeslin J., Brown R. W., and Rhodes J. M. (1977) Apollo 14 revisited, or breccias aren't so bad after all. *Proc. Lunar Sci. Conf. 8th,* p. 1869–1893.

Simonds C. H., Warner J. L., and Phinney W. C. (1973) Petrology of Apollo 16 poikilitic rocks. *Proc. Lunar Sci. Conf. 4th,* p. 613–632.

Simonds C. H., Warner J. L., Phinney W. C., and McGee P. E. (1976) Thermal model for impact breccia lithification: Manicouagan and the moon. *Proc. Lunar Sci. Conf. 7th,* p. 2509–2528.

Steele I. M. and Smith J. V. (1973) Mineralogy and petrology of some Apollo 16 rocks and fines: General petrologic model of moon, *Proc. Lunar Sci. Conf. 4th,* p. 519–536.

Stettler A., Eberhardt P., Geiss J., Grogler N., and Guggisberg S. (1975) Age sequence in the Apollo 17 station 7 boulder (abstract). In *Lunar Science VI,* p. 771–773. The Lunar Science Institute, Houston.

Stettler A., Eberhardt P., Geiss J., Grogler N., and Maurer P. (1973) Ar^{39}-Ar^{40} ages and Ar^{37}-Ar^{38} exposure ages of lunar rocks. *Proc. Lunar Sci. Conf. 4th,* p. 1865–1888.

Taylor G. J., Warner R. D., and Keil K. (1979) Lunar highland rocks with evolved compositions: Fractional crystallization or liquid immiscibility (abstract)? In *Papers Presented to the Conference on the Lunar Highlands Crust,* p. 166–168. Lunar and Planetary Institute, Houston.

Taylor L. A., Mao H. K., and Bell P. M. (1973) "Rust" in the Apollo 16 rocks. *Proc. Lunar Sci. Conf. 4th,* p. 829–839.

Taylor L. A., Mao H. K., and Bell P. M. (1974) -FeOOH, akaganeite, in lunar rocks. *Proc. Lunar Sci. Conf. 5th,* p. 743–748.

Turner G., Cadogan P. H., and Yonge C. J. (1973) Argon seleochronology. *Proc. Lunar Sci. Conf. 4th,* p. 1889–1914.

Turner G., Huneke J. C., Podosek F. A., and Wasserburg G. J. (1971) ^{40}Ar-^{39}Ar ages and cosmic ray exposure ages of Apollo samples. *Earth Planet.`Sci. Lett.* **12,** 19–35.

Turner G., Huneke J. C., Podosek F. A., and Wasserburg G. J. (1972) Ar^{40}-Ar^{39} systematics in rocks and separated minerals from Apollo 14. *Proc. Lunar Sci. Conf. 3rd,* p. 1589–1612.

Unruh D. M., Nakamura N., and Tatsumoto M. (1977) History of the Pasamonte achondrite: Relative susceptibility of the Sm–Nd, Rb–Sr, and U–Pb systems to metamorphic events. *Earth Planet. Sci. Lett.* **37,** 1–12.

Vaniman D. T., Lellis S. F., and Papike J. J., and Cameron K. L. (1976) The Apollo 16 drill core: Modal petrology and characterization of the mineral and lithic component. *Proc. Lunar Sci. Conf. 7th,* p. 199–239.

Vaniman D. T. and Papike J. J. (1978) The lunar highland melt-rock suite, *Geophys. Res. Lett.* **5,** 429–432.

Vaniman D. T. and Papike J. J. (1979) Silicate mineralogy of highland melt rocks (abstract). In *Papers Presented to the Conference on the Lunar Highlands Crust,* p. 178–180. Lunar and Planetary Institute, Houston.

Walker D., Longhi J., Grove T. L., Stolper E., and Hays J. F. (1973) Experimental petrology and origin of rocks from the Descartes Highlands. *Proc. Lunar Sci. Conf. 4th,* p. 1013–1032.

Walker D., Longhi J., and Hays J. (1972) Experimental petrology and the origin of Fra Mauro rocks and soil. *Proc. Lunar Sci. Conf. 3rd,* p. 797–817.

Wänke H., Baddenhausen H., Freibus G., Jagoutz E., Kruse H., Palme H., Spettel B., and Teschke F. (1973) Multielement analyses of Apollo 15, 16, and 17 samples and the bulk composition of the moon. *Proc. Lunar Sci. Conf. 4th,* p. 1461–1481.

Wänke H., Palme H., Kruse H., Baddenhausen H., Cendales M., Dreibus G., Hofmeister H., Jagoutz E., Palme C., Spettel B., and Thacker R. (1976) Chemistry of lunar highland rocks. A refined evaluation of the composition of the primary matter. *Proc. Lunar Sci. Conf. 7th,* p. 3479–3499.

Warner J. L. (1972) Metamorphism of Apollo 14 breccias. *Proc. Lunar Sci. Conf. 3rd,* p. 623–643.

Warner J. L., Phinney W. C., Bickel C. E., and Simonds C. H. (1977) Feldspathic granulitic impactites and pre-final bombardment lunar evolution. *Proc. Lunar Sci. Conf. 8th,* p. 2051–2066.

Warren P. H. and Wasson J. T. (1977) Pristine nonmare rocks and the nature of the lunar crust. *Proc. Lunar Sci. Conf. 8th,* p. 2215–2235.

Warren P. H. and Wasson J. T. (1978) Compositional-petrographic investigation of pristine nonmare rocks (abstract). *Lunar and Planetary Science IX,* p. 185–217. Lunar and Planetary Institute, Houston.

Wasson J. T., Warren P. H., Kallemeyn G. W., McEwing C. E., Mittlefehldt D. W., and Boynton W. V. (1977) SCCRV, a major component of highlands rocks. *Proc. Lunar Sci. Conf. 8th,* p. 2237–2252.

APPENDIX

Tables A-1 to A-4.

A-1: Olivine analyses. Where the analysis quoted is believed to represent the most Mg-rich olivine of the sample, the analysis is marked "Most Mg-rich".

A-2: Feldspar analyses. Where the analysis quoted is believed to represent the most Ca-rich feldspar of the sample, the analysis is marked "Most Ca-rich". Many of these may be xenocrysts, but only the texturally obvious xenocrysts are marked as such.

A-3: Pyroxene analyses. Specific characteristics of individual analyses are indicated (e.g., Most Mg-rich, Most Al-rich, etc.) In addition, each analyses is given a letter identification (A-E) indicating that part of the pyroxene quadrilateral which the analysis was chosen to represent. Where a hybrid identification is given (A/B), the analysis is from a position intermediate between two of the locations marked in this diagram:

A = Mg-rich opx
B = opx/cpx at Fe′ value for expansion of pyroxene into the augite field
C = Ca-poor, Mg-rich cpx
D = Ca-rich, Mg-rich cpx
E = Fe-rich cpx

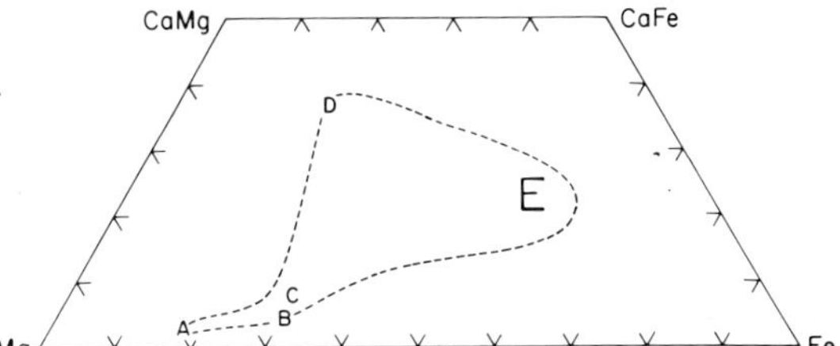

A-4: Selected oxide analyses.

* Analyses marked by an asterisk were kindly donated by A. Albee, B. Dymek, A. Gancarz and A. Chodos of California Institution of Technology.

Table A-1. Olivines.

	67559		68415*		68416		60335			66095		14310*	64455	
			Most Mg-rich				Most Mg-rich				Xenocryst	Fe-rich Olivine	Most Mg-rich	
SiO_2	36.5	36.5	38.2	37.4	36.9	36.1	39.3	38.9	38.8	38.0	38.8	31.9	39.0	39.1
Al_2O_3	0.11	0.09	0.06	0.02	0.18	0.22	0.06	0.10	0.11	0.13	0.12	0.44	0.16	0.10
TiO_2	0.02	0.02	0.05	0.06	0.06	0.06	0.02	0.02	0.03	0.08	0.02	0.21	0.03	0.00
FeO	34.1	34.3	25.4	28.8	28.1	33.3	15.8	18.0	20.2	22.2	18.7	58.3	17.4	20.1
MnO	0.28	0.25	0.33	0.28	0.31	0.36	0.18	0.21	0.27	0.20	0.20	0.55	0.14	0.19
MgO	30.5	30.6	37.6	34.5	34.5	30.4	44.6	42.0	41.2	39.2	42.5	8.4	43.0	41.4
CaO	0.40	0.37	0.16	0.16	0.31	0.26	0.12	0.13	0.17	0.22	0.16	0.30	0.15	0.16
Cr_2O_3	0.09	0.12	0.10	0.07	0.15	0.12	0.21	0.22	0.20	0.12	0.15	0.02	0.09	0.08
Σ	102.00	102.25	101.90	101.29	100.51	100.82	100.29	99.58	100.98	100.15	100.65	100.12	99.97	101.13
Si	0.986	0.985	0.989	0.991	0.985	0.986	0.989	0.995	0.990	0.987	0.986	1.004	0.992	0.994
Al	0.004	0.003	0.002	0.001	0.006	0.007	0.002	0.003	0.003	0.004	0.003	0.016	0.005	0.003
Ti	0.001	0.001	0.001	0.001	0.001	0.001	0.000	0.000	0.001	0.002	0.000	0.006	0.000	0.000
Fe	0.771	0.774	0.550	0.638	0.627	0.759	0.332	0.385	0.430	0.482	0.397	1.535	0.369	0.427
Mn	0.006	0.006	0.007	0.006	0.007	0.008	0.004	0.004	0.006	0.004	0.004	0.015	0.003	0.004
Mg	1.229	1.231	1.451	1.363	1.371	1.234	1.673	1.604	1.566	1.519	1.611	0.394	1.630	1.569
Ca	0.012	0.011	0.004	0.005	0.009	0.008	0.003	0.004	0.005	0.006	0.004	0.010	0.004	0.004
Cr	0.002	0.003	0.002	0.001	0.003	0.002	0.004	0.004	0.004	0.002	0.003	0.000	0.002	0.002
$\Sigma_{cations}$	3.011	3.014	3.006	3.006	3.009	3.005	3.007	2.999	3.005	3.006	3.008	2.980	3.005	3.003
Fo	61.5	61.4	72.5	68.1	68.4	61.6	83.3	80.5	78.2	75.7	80.0	20.4	81.4	78.4
Fa	38.5	38.6	27.5	31.9	31.6	38.4	16.7	19.5	21.8	24.3	20.0	79.6	18.6	21.6

Table A-1. *(Continued)*

	62295				60315		62235			65015*		77135	
	Groundmass		Xenocrysts										
	Most Mg-rich				Most Mg-rich		Most Mg-rich		"Ophitic"	Most Mg-rich		Most Mg-rich	
SiO_2	40.1	38.2	41.1	41.0	38.4	37.8	38.8	38.4	36.1	38.8	35.8	39.5	37.7
Al_2O_3	0.15	0.11	0.00	0.03	0.17	0.14	0.15	0.17	0.16	0.14	0.54	0.26	0.08
TiO_2	0.00	0.03	0.00	0.00	0.16	0.10	0.07	0.07	0.07	0.17	0.22	0.03	0.04
FeO	11.7	21.4	4.2	5.3	22.0	26.6	21.4	22.8	35.5	22.2	33.7	18.5	26.2
MnO	0.07	0.22	0.00	0.00	0.27	0.42	0.21	0.25	0.33	0.15	0.31	0.14	0.25
MgO	47.9	40.0	53.6	53.3	39.7	35.5	40.6	39.3	29.0	38.3	29.5	43.1	36.3
CaO	0.27	0.11	0.03	0.01	0.15	0.19	0.23	0.19	0.25	0.26	0.24	0.39	0.23
Cr_2O_3	0.03	0.03	0.08	0.12	0.10	0.06	0.21	0.20	0.12	0.07	0.07	0.20	0.04
Σ	100.22	100.10	99.01	99.76	100.95	100.81	101.67	101.38	101.53	100.09	100.38	102.12	100.84
Si	0.990	0.989	0.992	0.988	0.989	0.996	0.988	0.988	0.988	1.006	0.983	0.987	0.992
Al	0.004	0.003	0.000	0.001	0.005	0.004	0.004	0.005	0.005	0.004	0.017	0.008	0.003
Ti	0.000	0.000	0.000	0.000	0.003	0.002	0.001	0.001	0.001	0.004	0.006	0.001	0.001
Fe	0.241	0.463	0.085	0.106	0.472	0.586	0.456	0.490	0.812	0.481	0.774	0.387	0.576
Mn	0.001	0.004	0.000	0.000	0.006	0.009	0.004	0.005	0.007	0.003	0.007	0.003	0.005
Mg	1.762	1.543	1.928	1.912	1.522	1.395	1.540	1.506	1.182	1.480	1.207	1.607	1.421
Ca	0.007	0.003	0.001	0.000	0.004	0.005	0.006	0.005	0.007	0.007	0.007	0.010	0.006
Cr	0.001	0.001	0.002	0.002	0.002	0.001	0.004	0.004	0.002	0.001	0.002	0.004	0.001
$\Sigma_{cations}$	3.006	3.006	3.008	3.009	3.003	2.998	3.003	3.004	3.004	2.986	3.003	3.007	3.005
Fo	87.9	76.7	95.8	94.8	76.1	70.1	77.0	75.2	59.0	75.5	60.9	80.5	71.0
Fa	12.1	23.3	4.2	5.2	23.9	29.9	23.0	24.8	41.0	24.5	39.1	19.5	29.0

Table A-2. Feldspars.

	67559			68415*			68416			63549			65055		
	Most Ca-rich			Most Ca-rich			Most Ca-rich			Most Ca-rich			Most Ca-rich		
SiO_2	44.5	47.6	51.6	43.9	45.4	48.4	43.4	45.8	50.0	43.5	44.4	46.3	43.6	46.6	51.1
Al_2O_3	36.4	33.3	30.5	36.7	35.7	32.3	35.9	34.5	31.8	36.1	35.7	34.1	35.6	33.8	31.3
FeO	0.10	0.52	0.52	0.15	0.46	0.47	0.10	0.63	0.35	0.07	0.11	0.28	0.16	0.48	0.38
MgO	0.00	0.00	0.00	0.05	0.07	0.28	0.00	0.07	0.04	0.00	0.07	0.06	0.00	0.10	0.01
CaO	20.0	17.3	14.6	20.2	18.1	16.5	19.4	17.6	14.3	19.4	18.8	17.3	19.8	17.6	14.2
Na_2O	0.26	1.6	3.3	0.27	0.97	1.9	0.27	1.1	2.7	0.29	0.56	1.3	0.24	1.3	3.2
K_2O	0.01	0.10	0.33	0.01	0.09	0.14	0.01	0.06	0.24	0.00	0.01	0.08	0.01	0.07	0.43
Σ	101.27	100.42	100.85	101.28	100.79	99.99	99.08	99.76	99.43	99.46	99.65	99.42	99.41	99.95	100.62
Si	2.031	2.181	2.337	2.009	2.078	2.221	2.025	2.115	2.290	2.022	2.053	2.140	2.033	2.146	2.317
Al	1.962	1.796	1.630	1.980	1.926	1.747	1.974	1.882	1.716	1.981	1.949	1.859	1.954	1.837	1.673
Σ_{tet}	3.993	3.977	3.967	3.989	4.004	3.968	3.999	3.997	4.006	4.003	4.002	3.999	3.987	3.983	3.990
Fe	0.004	0.020	0.020	0.006	0.018	0.018	0.004	0.024	0.014	0.003	0.004	0.011	0.006	0.019	0.014
Mg	0.000	0.000	0.000	0.003	0.005	0.019	0.000	0.005	0.002	0.000	0.005	0.004	0.000	0.007	0.001
Ca	0.977	0.850	0.706	0.991	0.888	0.811	0.972	0.869	0.701	0.970	0.934	0.857	0.986	0.869	0.689
Na	0.023	0.139	0.292	0.024	0.086	0.169	0.024	0.096	0.240	0.026	0.050	0.114	0.021	0.114	0.278
K	0.001	0.006	0.019	0.001	0.005	0.008	0.000	0.004	0.014	0.000	0.001	0.005	0.001	0.004	0.025
$\Sigma_{x\text{-cations}}$	1.005	1.015	1.037	1.025	1.002	1.025	1.000	0.998	0.971	0.999	0.994	0.991	1.014	1.013	1.007
$\Sigma_{cations}$	4.998	4.992	5.004	5.014	5.006	4.993	4.999	4.995	4.977	5.002	4.996	4.990	5.001	4.996	4.997
Or	0.10	0.60	1.90	0.10	0.50	0.80	0.00	0.40	1.50	0.00	0.10	0.50	0.10	0.40	2.50
Ab	2.30	14.00	28.70	2.40	8.80	17.10	2.40	9.90	25.10	2.60	5.10	11.70	2.10	11.50	28.00
An	97.60	85.40	69.40	97.60	90.70	82.10	97.60	89.70	73.40	97.40	94.80	87.80	97.80	88.10	69.50
Fe/(Fe+Mg)	1.00	1.00	1.00	0.63	0.79	0.49	1.00	0.83	0.85	1.00	0.46	0.72	1.00	0.74	0.95

Table A-2. *(Continued)*

	60335			66095				14276*			14310*		
	Most Ca-rich			Xenolith				Most Ca-rich			Most Ca-rich		
SiO_2	44.6	45.8	48.5	44.1	43.8	45.5	46.9	44.7	47.4	52.7	43.8	46.4	50.5
Al_2O_3	35.9	34.3	32.5	36.1	36.1	35.4	34.5	34.7	33.4	29.2	36.1	34.0	31.9
FeO	0.12	0.18	0.20	0.12	0.15	0.30	0.33	0.21	0.39	0.43	0.15	0.06	0.79
MgO	0.01	0.07	0.13	0.00	0.00	0.15	0.15	0.12	0.14	0.08	0.19	0.29	0.06
CaO	19.3	17.9	16.4	19.8	19.7	18.7	18.1	19.5	17.2	12.9	19.2	17.1	14.7
Na_2O	0.40	0.92	1.9	0.27	0.39	0.71	1.0	0.49	1.6	3.5	0.51	1.4	2.6
K_2O	0.03	0.06	0.20	0.02	0.01	0.12	0.41	0.05	0.29	0.91	0.05	0.21	0.55
Σ	100.36	99.23	99.83	100.41	100.15	100.88	101.39	99.77	100.42	99.72	100.00	99.46	101.10
Si	2.051	2.123	2.226	2.030	2.024	2.082	2.131	2.073	2.173	2.405	2.026	2.143	2.286
Al	1.948	1.874	1.757	1.963	1.968	1.909	1.852	1.897	1.804	1.571	1.968	1.851	1.702
Σ_{tet}	3.999	3.997	3.983	3.993	3.992	3.991	3.983	3.970	3.977	3.976	3.994	3.994	3.988
Fe	0.005	0.007	0.008	0.005	0.006	0.012	0.013	0.008	0.015	0.016	0.006	0.002	0.030
Mg	0.001	0.005	0.009	0.000	0.000	0.010	0.010	0.008	0.010	0.005	0.013	0.020	0.004
Ca	0.953	0.888	0.809	0.979	0.976	0.916	0.882	0.969	0.845	0.631	0.952	0.846	0.713
Na	0.036	0.084	0.166	0.024	0.035	0.063	0.088	0.044	0.142	0.310	0.046	0.125	0.228
K	0.002	0.003	0.012	0.001	0.000	0.007	0.024	0.003	0.017	0.053	0.003	0.012	0.032
$\Sigma_{x\text{-cations}}$	0.997	0.987	1.004	1.009	1.017	1.008	1.017	1.032	1.029	1.015	1.020	1.005	1.007
$\Sigma_{cations}$	4.996	4.984	4.987	5.002	5.009	4.999	5.000	5.002	5.006	4.991	5.014	4.999	4.995
Or	0.10	0.30	1.20	0.10	0.0	0.70	2.40	0.30	1.70	5.30	0.30	1.30	3.30
Ab	3.60	8.50	16.80	2.40	3.50	6.40	8.80	4.30	14.20	31.20	4.60	12.70	23.50
An	96.20	91.20	82.00	97.50	96.50	92.90	88.80	95.40	84.10	63.50	95.10	86.00	73.20
Fe/(Fe+Mg)	0.90	0.58	0.46	1.00	1.00	0.53	0.55	0.50	0.61	0.75	0.31	0.10	0.88

Table A-2. *(Continued)*

	64455			62295			62235				65015*		
	Most Ca-rich			Most Ca-rich			Xenocryst	Most Ca-rich			Most Ca-rich		
SiO_2	45.3	45.4	46.4	45.2	45.9	46.7	45.0	44.1	44.9	45.8	44.1	45.5	50.7
Al_2O_3	34.1	36.3	32.6	34.1	32.7	33.1	35.3	35.3	34.5	34.3	35.2	35.4	31.1
FeO	0.08	0.06	0.16	0.20	0.26	0.39	0.19	0.22	0.27	0.22	0.25	0.27	0.44
MgO	0.20	0.36	0.24	0.28	0.37	0.42	0.05	0.00	0.00	0.00	0.03	0.11	0.06
CaO	19.1	18.5	18.6	19.4	19.0	18.7	19.1	19.2	19.0	18.7	20.0	18.6	16.6
Na_2O	0.67	0.69	0.91	0.56	0.78	0.99	0.55	0.34	0.66	0.88	0.41	0.91	1.0
K_2O	0.03	0.02	0.09	0.03	0.09	0.08	0.03	0.02	0.06	0.25	0.03	0.07	1.90
Σ	99.48	101.33	99.00	99.77	99.10	100.38	100.22	99.18	99.39	100.15	100.02	100.86	101.80
Si	2.103	2.062	2.160	2.095	2.142	2.147	2.072	2.054	2.087	2.112	2.045	2.082	2.294
Al	1.864	1.946	1.792	1.862	1.796	1.797	1.916	1.940	1.891	1.864	1.924	1.909	1.658
Σ_{tet}	3.967	4.008	3.952	3.957	3.938	3.944	3.988	3.994	3.978	3.976	3.969	3.991	3.952
Fe	0.003	0.002	0.006	0.008	0.010	0.015	0.008	0.009	0.011	0.009	0.010	0.010	0.017
Mg	0.014	0.024	0.017	0.019	0.026	0.029	0.004	0.000	0.000	0.000	0.002	0.008	0.004
Ca	0.951	0.899	0.925	0.964	0.949	0.929	0.944	0.956	0.946	0.924	0.994	0.912	0.805
Na	0.060	0.060	0.082	0.050	0.071	0.088	0.049	0.031	0.060	0.079	0.037	0.081	0.088
K	0.002	0.001	0.006	0.002	0.005	0.004	0.002	0.001	0.003	0.014	0.002	0.004	0.110
$\Sigma_{x\text{-cations}}$	1.030	0.986	1.036	1.043	1.061	1.065	1.007	0.997	1.020	1.026	1.045	1.015	1.024
$\Sigma_{cations}$	4.997	4.994	4.988	5.000	4.999	5.009	4.995	4.991	4.998	5.002	5.014	5.006	4.976
Or	0.20	0.10	0.50	0.20	0.50	0.40	0.20	0.10	0.30	1.40	0.20	0.40	10.90
Ab	5.90	6.30	8.10	4.90	6.90	8.70	4.90	3.10	5.90	7.80	3.50	8.10	8.80
An	93.90	93.60	91.40	94.90	92.60	90.90	94.90	96.80	93.80	90.80	96.30	91.50	80.30
Fe/(Fe+Mg)	0.19	0.09	0.26	0.29	0.28	0.34	0.66	1.00	1.00	1.00	0.82	0.58	0.80

Table A-2. *(Continued)*

	77135			15386				15434			
	Most Ca-rich			Most Ca-rich				Most Ca-rich			
SiO_2	44.2	45.8	48.0	47.3	47.8	49.5	51.0	47.1	48.6	49.3	51.3
Al_2O_3	35.9	35.3	34.4	34.0	31.9	30.8	30.0	32.5	32.3	31.3	30.8
FeO	0.18	0.26	0.30	0.19	0.20	0.38	0.48	0.16	0.26	0.34	0.48
MgO	0.00	0.00	0.00	0.19	0.16	0.10	0.08	0.17	0.16	0.12	0.08
CaO	19.7	18.7	17.0	17.1	16.4	15.9	14.1	17.1	16.7	15.7	14.7
Na_2O	0.28	0.98	1.7	1.4	1.7	2.2	2.4	1.5	1.7	2.0	2.4
K_2O	0.07	0.09	0.26	0.08	0.15	0.34	0.32	0.11	0.16	0.23	0.33
Σ	100.33	101.13	101.66	100.26	98.31	99.22	98.38	98.64	99.88	98.99	100.09
Si	2.039	2.090	2.166	2.163	2.228	2.285	2.356	2.191	2.228	2.275	2.334
Al	1.952	1.896	1.830	1.834	1.752	1.676	1.634	1.783	1.747	1.705	1.650
Σ_{tet}	3.991	3.986	3.996	3.997	3.980	3.961	3.990	3.974	3.975	3.980	3.984
Fe	0.007	0.010	0.012	0.007	0.008	0.015	0.019	0.006	0.010	0.013	0.018
Mg	0.000	0.000	0.000	0.013	0.011	0.007	0.005	0.012	0.011	0.008	0.006
Ca	0.971	0.914	0.821	0.840	0.818	0.787	0.696	0.855	0.822	0.777	0.719
Na	0.025	0.086	0.152	0.121	0.151	0.196	0.217	0.133	0.152	0.176	0.210
K	0.004	0.005	0.015	0.005	0.009	0.020	0.019	0.007	0.010	0.013	0.019
$\Sigma_{x\text{-cations}}$	1.007	1.015	1.000	0.986	0.997	1.025	0.956	1.013	1.005	0.987	0.972
$\Sigma_{cations}$	4.998	5.001	4.996	4.983	4.977	4.986	4.946	4.987	4.980	4.967	4.956
Or	0.40	0.50	1.50	0.50	0.90	2.00	2.00	0.70	1.00	1.40	2.00
Ab	2.50	8.60	15.40	12.50	15.40	19.60	23.30	13.30	15.50	18.20	22.10
An	97.10	90.90	83.10	87.00	83.70	78.40	74.70	86.00	83.50	80.40	75.90
Fe/(Fe+Mg)	1.00	1.00	1.00	0.35	0.42	0.68	0.78	0.34	0.48	0.62	0.76

Table A-3. Pyroxenes.

	67559					68415*						68416		
	(A) Most Al, Mg-rich	(B)	(C)	(D)	(E) Fe, Ca-rich Low Cr	(B) Cr, Al-poor	(B) Cr, Al-rich	(C)	(D)	(E) Fe, Ca-rich	(E) Fe-rich Ca-poor	(A/B)	(C)	(D)
SiO_2	54.3	53.4	52.8	51.8	49.8	54.0	52.9	53.1	50.6	49.0	50.6	53.2	53.1	52.0
Al_2O_3	4.1	1.0	1.6	1.4	0.9	0.77	2.2	1.4	1.8	1.4	0.86	1.1	1.6	2.1
TiO_2	0.32	0.28	0.50	0.78	1.1	0.28	0.82	0.66	1.7	2.0	1.2	0.23	0.46	0.69
FeO	9.8	15.6	14.9	14.9	26.4	16.3	15.2	14.3	13.8	19.4	26.4	13.2	15.2	11.3
MnO	0.09	0.16	0.21	0.16	0.38	0.32	0.32	0.24	0.27	0.31	0.50	0.19	0.26	0.18
MgO	30.3	25.7	21.7	15.4	9.9	25.0	22.0	20.2	14.8	11.5	15.9	27.3	22.0	18.1
CaO	2.1	2.5	8.2	15.9	11.6	3.8	6.4	9.4	16.1	14.3	5.3	2.2	7.2	14.7
Na_2O	0.00	0.00	0.00	0.00	0.00	0.00	0.00	0.10	0.10	0.10	0.10	0.00	0.00	0.10
Cr_2O_3	0.76	0.50	0.68	0.51	0.01	0.37	0.67	0.59	0.75	0.16	0.30	0.63	0.76	0.89
Σ	101.77	99.14	100.59	100.85	100.09	100.84	100.51	99.99	99.92	98.17	101.16	98.05	100.58	100.06
Si	1.882	1.956	1.932	1.939	1.958	1.958	1.928	1.954	1.902	1.916	1.930	1.952	1.943	1.924
^{IV}Al	0.118	0.044	0.068	0.060	0.042	0.033	0.072	0.046	0.080	0.065	0.039	0.047	0.057	0.076
Σ_{tet}	2.000	2.000	2.000	1.999	2.000	1.991	2.000	2.000	1.982	1.981	1.969	1.999	2.000	2.000
^{VI}Al	0.050	0.000	0.003	0.000	0.000	0.000	0.022	0.015	0.000	0.000	0.000	0.000	0.011	0.015
Ti	0.008	0.007	0.014	0.022	0.033	0.008	0.028	0.023	0.060	0.074	0.043	0.006	0.013	0.019
Fe	0.284	0.476	0.456	0.467	0.869	0.494	0.463	0.440	0.434	0.634	0.842	0.403	0.466	0.351
Mn	0.002	0.005	0.006	0.005	0.013	0.010	0.010	0.007	0.009	0.010	0.016	0.006	0.008	0.006
Mg	1.567	1.404	1.186	0.858	0.582	1.351	1.195	1.108	0.829	0.670	0.904	1.491	1.199	0.998
Ca	0.078	0.099	0.323	0.636	0.491	0.148	0.250	0.371	0.648	0.599	0.217	0.084	0.281	0.582
Na	0.000	0.000	0.000	0.000	0.000	0.000	0.000	0.007	0.007	0.008	0.007	0.000	0.000	0.008
Cr	0.020	0.014	0.020	0.015	0.000	0.011	0.019	0.017	0.022	0.005	0.009	0.018	0.022	0.026
Σ_{oct}	2.009	2.005	2.008	2.003	1.988	2.022	1.987	1.988	2.009	2.000	2.038	2.008	2.000	2.005
$\Sigma_{cations}$	4.009	4.005	4.008	4.002	3.988	4.013	3.987	3.988	3.991	3.981	4.007	4.007	4.000	4.005
Wo	4.0	5.0	16.4	32.3	25.1	7.4	13.1	19.3	33.9	31.5	11.0	4.3	14.4	30.1
En	81.1	70.7	60.2	43.7	29.8	67.8	62.6	57.7	43.4	35.2	46.1	75.1	61.3	51.5
Fs	14.9	24.3	23.4	24.0	45.1	24.8	24.3	23.0	22.7	33.3	42.9	20.6	24.3	18.4

Table A-3. *(Continued)*

	63549				65055					60335			66095	
	(A) Most Al, Mg-rich	(B)	(C)	(D/E)	(A) Most Al, Mg-rich	(B)	(C)	(D)	(E)	(A) Most Mg-rich	(B/C)	(D)	(C) Al-rich	(C) Al-poor
SiO_2	53.1	53.7	52.5	49.4	55.2	53.7	52.6	50.2	48.2	53.6	53.9	51.8	51.1	53.4
Al_2O_3	4.5	1.5	2.1	1.6	2.3	2.1	2.0	2.5	1.1	1.8	1.0	2.4	2.8	1.7
TiO_2	0.36	0.31	0.44	1.3	0.19	0.27	0.46	1.1	0.68	1.1	0.70	1.6	1.7	1.2
FeO	10.4	14.6	14.2	25.3	9.6	14.5	15.1	13.4	35.7	10.9	13.7	6.4	13.1	13.1
MnO	0.16	0.21	0.21	0.36	0.06	0.15	0.18	0.19	0.44	0.21	0.31	0.15	0.26	0.20
MgO	28.3	25.6	21.1	11.3	30.7	25.9	21.3	13.0	5.0	29.1	25.3	17.3	24.1	26.2
CaO	2.5	3.1	8.4	11.7	1.6	3.4	8.1	18.4	10.5	2.8	4.3	20.4	5.2	4.2
Na_2O	0.00	0.00	0.00	0.00	0.00	0.00	0.00	0.20	0.00	0.00	0.00	0.10	0.00	0.00
Cr_2O_3	0.61	0.47	0.52	0.26	0.54	0.53	0.56	0.50	0.03	0.87	0.63	0.87	0.81	0.54
Σ	99.93	99.49	99.47	101.22	100.19	100.55	100.30	99.49	101.65	100.38	99.84	101.02	99.07	100.54
Si	1.883	1.995	1.938	1.916	1.938	1.934	1.934	1.910	1.949	1.905	1.956	1.888	1.879	1.919
^{IV}Al	0.117	0.045	0.062	0.074	0.062	0.066	0.066	0.090	0.051	0.075	0.043	0.103	0.120	0.072
Σ_{tet}	2.000	2.000	2.000	1.990	2.000	2.000	2.000	2.000	2.000	1.980	1.999	1.991	1.999	1.991
^{VI}Al	0.070	0.018	0.030	0.000	0.034	0.022	0.021	0.021	0.002	0.000	0.000	0.000	0.000	0.000
Ti	0.009	0.008	0.012	0.037	0.005	0.007	0.013	0.030	0.021	0.030	0.019	0.044	0.047	0.033
Fe	0.309	0.444	0.439	0.819	0.281	0.437	0.464	0.426	1.208	0.324	0.414	0.194	0.402	0.394
Mn	0.005	0.006	0.007	0.012	0.002	0.004	0.006	0.006	0.015	0.006	0.010	0.005	0.008	0.006
Mg	1.498	1.386	1.159	0.655	1.605	1.388	1.164	0.738	0.301	1.541	1.369	0.945	1.317	1.405
Ca	0.095	0.122	0.334	0.487	0.060	0.132	0.317	0.753	0.456	0.107	0.167	0.797	0.205	0.160
Na	0.000	0.000	0.000	0.000	0.000	0.000	0.000	0.012	0.000	0.000	0.000	0.006	0.000	0.000
Cr	0.017	0.014	0.015	0.008	0.015	0.015	0.016	0.015	0.001	0.024	0.018	0.024	0.023	0.015
Σ_{oct}	2.003	1.998	1.996	2.018	2.002	2.005	2.001	2.001	2.004	2.032	1.997	2.015	2.002	2.013
$\Sigma_{cations}$	4.003	3.998	3.996	4.008	4.002	4.005	4.001	4.001	4.004	4.012	3.996	4.006	4.001	4.004
Wo	5.0	6.2	17.2	24.7	3.1	6.7	16.3	39.1	23.0	5.4	8.5	41.1	10.6	8.2
En	78.6	70.8	59.8	33.2	82.4	70.8	59.7	38.4	15.2	77.9	69.9	48.6	68.2	71.5
Fs	16.4	23.0	23.0	42.1	14.5	22.5	24.0	22.5	61.8	16.7	21.6	10.3	21.2	20.3

Table A-3. *(Continued)*

	14276*							14310*				64455	
	(A) Most Mg-rich	(A) Mg, Al-rich	(A/B) Most Al-rich	(B)	(C)	(D)	(E)	(A) Most Mg, Al-rich	(B)	(C)	(D)	(C) Al-rich	(C) Al-poor
SiO_2	54.6	54.6	52.1	52.4	53.5	50.3	45.2	53.2	53.6	52.2	49.3	53.3	54.8
Al_2O_3	2.0	3.1	3.5	1.2	0.81	2.4	1.3	3.7	0.85	1.7	2.6	2.5	0.89
TiO_2	0.34	0.48	0.66	0.69	0.40	1.30	1.50	0.74	0.37	0.89	2.50	0.86	0.36
FeO	11.2	10.9	14.6	19.2	18.0	12.2	41.2	11.4	17.1	17.9	15.9	11.1	11.9
MnO	0.28	0.21	0.17	0.31	0.30	0.26	0.45	0.23	0.28	0.26	0.28	0.17	0.19
MgO	30.0	29.0	26.3	23.4	22.3	14.2	4.4	28.2	23.3	19.8	11.6	27.7	28.8
CaO	1.6	1.8	2.0	2.3	5.0	17.5	4.6	2.6	3.2	6.7	17.5	2.9	2.6
Na_2O	0.00	0.00	0.00	0.10	0.10	0.10	0.10	0.00	0.00	0.10	0.10	0.00	0.00
Cr_2O_3	0.53	0.62	0.50	0.50	0.38	0.65	0.08	0.67	0.37	0.57	0.37	0.65	0.54
Σ	100.55	100.71	99.83	100.10	100.79	98.91	98.83	100.74	99.07	100.12	100.15	99.18	100.08
Si	1.927	1.918	1.884	1.934	1.962	1.905	1.911	1.881	1.979	1.937	1.872	1.916	1.955
^{IV}Al	0.073	0.082	0.116	0.052	0.035	0.095	0.065	0.119	0.021	0.063	0.116	0.084	0.037
Σ_{tet}	2.000	2.000	2.000	1.986	1.997	2.000	1.976	2.000	2.000	2.000	1.988	2.000	1.992
^{IV}Al	0.010	0.047	0.033	0.000	0.000	0.012	0.000	0.035	0.016	0.011	0.000	0.022	0.000
Ti	0.011	0.016	0.022	0.024	0.014	0.046	0.060	0.025	0.013	0.031	0.089	0.023	0.009
Fe	0.330	0.320	0.442	0.593	0.552	0.386	1.456	0.337	0.528	0.555	0.505	0.335	0.355
Mn	0.008	0.006	0.005	0.010	0.009	0.008	0.016	0.007	0.009	0.008	0.009	0.005	0.005
Mg	1.578	1.519	1.418	1.287	1.219	0.802	0.277	1.486	1.282	1.095	0.657	1.484	1.533
Ca	0.060	0.068	0.077	0.091	0.196	0.710	0.208	0.099	0.127	0.266	0.712	0.111	0.098
Na	0.000	0.000	0.000	0.007	0.007	0.006	0.005	0.000	0.000	0.007	0.007	0.000	0.000
Cr	0.015	0.017	0.014	0.015	0.011	0.019	0.003	0.019	0.011	0.017	0.011	0.019	0.015
Σ_{oct}	2.012	1.993	2.011	2.027	2.008	1.989	2.025	2.008	1.986	1.990	1.990	1.999	2.015
$\Sigma_{cations}$	4.012	3.993	4.011	4.013	4.005	3.989	4.001	4.008	3.986	3.990	3.978	3.999	4.007
Wo	3.1	3.6	4.0	4.6	10.0	37.4	10.7	5.1	6.5	13.9	38.0	5.7	4.9
En	80.1	79.6	73.2	65.3	62.0	42.2	14.3	77.3	66.2	57.1	35.0	76.7	77.0
Fs	16.8	16.8	22.8	30.1	28.0	20.4	75.0	17.5	27.3	29.0	27.0	17.6	18.1

Table A-3. *(Continued)*

	60315			62235					65015*			
	(A) Most Mg, Al-rich	(B)	(D)	(A) Most Mg, Al-rich	(B)	(C)	(D)	"Ophitic"	(A) Most Mg-rich	(B)	(C)	(D)
SiO_2	54.1	54.4	50.3	53.9	53.3	52.1	50.4	51.8	54.5	53.2	52.8	50.6
Al_2O_3	2.9	1.6	3.6	2.4	1.6	1.5	2.7	1.9	1.8	0.97	0.81	2.8
TiO_2	0.79	1.4	3.0	0.56	0.70	1.1	2.2	1.2	0.75	0.77	0.68	2.0
FeO	9.0	13.4	6.6	10.5	15.1	17.4	10.5	15.1	13.3	18.6	17.8	8.7
MnO	0.15	0.20	0.21	0.13	0.20	0.26	0.14	0.24	0.21	0.37	0.30	0.22
MgO	30.9	26.9	16.6	29.5	26.5	21.7	15.3	19.6	28.2	23.4	22.0	16.5
CaO	1.5	2.6	19.5	1.5	2.1	5.4	18.2	9.3	2.0	2.7	5.6	18.3
Na_2O	0.00	0.00	0.20	0.00	0.00	0.00	0.10	0.00	0.00	0.00	0.00	0.20
Cr_2O_3	0.83	0.36	0.72	0.64	0.44	0.38	0.48	0.47	0.58	0.29	0.13	0.75
Σ	100.17	100.86	100.73	99.13	99.94	99.84	100.02	99.61	101.34	100.30	100.12	100.07
Si	1.901	1.938	1.842	1.923	1.931	1.934	1.881	1.927	1.926	1.952	1.951	1.866
^{IV}Al	0.099	0.062	0.156	0.077	0.068	0.064	0.118	0.073	0.074	0.042	0.035	0.122
Σ_{tet}	2.000	2.000	1.998	2.000	1.999	1.998	1.999	2.000	2.000	1.994	1.986	1.988
^{IV}Al	0.019	0.004	0.000	0.025	0.000	0.000	0.000	0.011	0.001	0.000	0.000	0.000
Ti	0.021	0.037	0.082	0.015	0.019	0.030	0.062	0.033	0.025	0.027	0.024	0.069
Fe	0.265	0.399	0.202	0.314	0.456	0.539	0.327	0.468	0.393	0.571	0.550	0.268
Mn	0.004	0.006	0.006	0.004	0.006	0.008	0.004	0.008	0.006	0.011	0.009	0.007
Mg	1.616	1.429	0.906	1.567	1.432	1.200	0.854	1.086	1.486	1.280	1.212	0.907
Ca	0.058	0.101	0.766	0.057	0.083	0.213	0.729	0.371	0.076	0.106	0.222	0.723
Na	0.000	0.000	0.013	0.000	0.000	0.000	0.005	0.000	0.000	0.000	0.000	0.014
Cr	0.023	0.010	0.021	0.018	0.012	0.011	0.014	0.014	0.016	0.008	0.004	0.022
Σ_{oct}	2.006	1.986	1.996	2.000	2.008	2.001	1.995	1.991	2.003	2.003	2.021	2.010
$\Sigma_{cations}$	4.006	3.986	3.994	4.000	4.007	3.999	3.994	3.991	4.003	3.997	4.007	3.998
Wo	3.0	5.2	40.7	2.9	4.2	10.9	38.1	19.2	3.9	5.4	11.2	38.1
En	83.1	73.8	48.2	80.7	72.4	61.2	44.6	56.2	76.0	65.4	61.1	47.8
Fs	13.9	21.0	11.1	16.4	23.4	27.9	17.3	24.6	20.1	29.2	27.7	14.1

Table A-3. *(Continued)*

	77135				15386					15434						
	(A) Most Mg-rich	(C)	(C/D)	(D)	(A) Most Mg, Al-rich	(A/B)	(B)	(D)	(E)	(A/B)	(A/B)	(B)	(C)	(D)	(D)	(E)
SiO_2	55.1	53.0	52.6	50.7	53.4	52.9	51.8	50.1	49.5	53.5	53.2	52.4	51.5	49.6	49.2	48.3
Al_2O_3	1.8	2.1	2.2	4.9	3.0	2.3	1.9	2.0	1.1	1.6	1.9	1.3	1.5	2.4	2.6	1.4
TiO_2	0.54	1.2	1.3	1.8	0.60	0.72	0.85	1.6	0.82	0.51	0.66	0.77	0.97	1.6	2.0	0.91
FeO	10.9	13.3	12.1	14.7	11.7	15.7	19.0	16.9	30.8	14.1	14.8	20.4	20.4	17.7	13.4	32.3
MnO	0.16	0.23	0.23	0.20	0.09	0.20	0.25	0.25	0.49	0.20	0.24	0.32	0.34	0.33	0.24	0.52
MgO	30.2	23.8	22.4	14.4	28.9	26.3	23.0	12.3	11.0	26.9	26.3	22.2	19.6	12.7	13.0	8.8
CaO	1.7	6.7	9.1	14.4	1.40	1.7	2.2	17.1	6.2	1.7	1.7	2.4	5.1	15.2	17.9	8.9
Na_2O	0.00	0.00	0.00	0.20	0.00	0.00	0.00	0.00	0.00	0.00	0.00	0.00	0.00	0.00	0.00	0.00
Cr_2O_3	0.60	0.60	0.72	0.17	0.92	0.91	0.85	0.52	0.27	0.79	0.86	0.74	0.92	0.63	1.00	0.37
Σ	101.00	100.93	100.65	101.47	100.01	100.73	99.85	100.77	100.18	99.30	99.66	100.53	100.33	100.16	99.34	101.50
Si	1.932	1.912	1.910	1.871	1.901	1.910	1.920	1.904	1.958	1.941	1.929	1.940	1.930	1.897	1.878	1.920
^{IV}Al	0.068	0.088	0.090	0.129	0.099	0.090	0.081	0.091	0.042	0.059	0.071	0.055	0.066	0.103	0.116	0.066
Σ_{tet}	2.000	2.000	2.000	2.000	2.000	2.000	2.001	1.995	2.000	2.000	2.000	1.995	1.996	2.000	1.994	1.986
^{VI}Al	0.007	0.002	0.004	0.084	0.026	0.007	0.000	0.000	0.009	0.009	0.010	0.000	0.000	0.004	0.000	0.000
Ti	0.014	0.033	0.035	0.050	0.016	0.019	0.024	0.046	0.025	0.014	0.018	0.022	0.028	0.046	0.058	0.027
Fe	0.321	0.403	0.369	0.454	0.348	0.473	0.588	0.538	1.019	0.429	0.450	0.632	0.639	0.566	0.427	1.072
Mn	0.004	0.007	0.007	0.007	0.003	0.006	0.008	0.008	0.016	0.006	0.008	0.010	0.011	0.011	0.008	0.017
Mg	1.580	1.281	1.210	0.794	1.533	1.412	1.269	0.698	0.649	1.454	1.442	1.225	1.092	0.724	0.740	0.523
Ca	0.065	0.258	0.352	0.568	0.054	0.065	0.088	0.697	0.263	0.065	0.067	0.095	0.204	0.625	0.732	0.380
Na	0.000	0.000	0.000	0.016	0.000	0.000	0.000	0.000	0.000	0.000	0.000	0.000	0.000	0.000	0.000	0.000
Cr	0.017	0.017	0.021	0.005	0.026	0.026	0.025	0.016	0.008	0.023	0.025	0.022	0.027	0.019	0.030	0.011
Σ_{oct}	2.008	2.001	1.998	1.978	2.006	2.008	2.002	2.003	1.989	2.000	2.000	2.006	2.001	1.995	1.995	2.030
$\Sigma_{cations}$	4.008	4.001	3.998	3.978	4.006	4.008	4.003	3.998	3.989	4.000	4.000	4.001	3.997	3.995	3.989	4.016
Wo	3.3	13.3	18.2	31.2	2.8	3.3	4.5	35.9	13.5	3.3	3.4	4.9	10.5	32.4	38.4	19.1
En	80.2	65.7	62.4	43.6	79.1	72.2	65.0	36.0	33.3	74.4	73.0	62.4	56.1	37.6	38.8	26.2
Fs	16.5	21.0	19.4	25.2	18.1	24.5	30.5	28.1	53.2	22.3	23.5	32.7	33.4	30.0	22.8	54.7

Table A-4. Oxides.

	68416	65055	62295	62235	15386	15386	15434
	ilm (3 ox)	usp (4 ox)	sp (4 ox)	ilm (3 ox)	ilm lath (3 ox)	mesostasis ilm (3 ox)	ilm lath (3 ox)
Al_2O_3	0.17	2.5	67.8	0.09	0.19	0.16	0.33
TiO_2	54.2	33.4	0.00	52.6	52.1	51.3	52.9
FeO	40.6	64.4	8.7	42.0	45.7	47.2	45.7
MnO	0.33	0.30	0.00	0.31	0.31	0.33	0.43
MgO	4.2	0.20	21.7	3.4	0.67	0.04	1.2
Cr_2O_3	0.72	0.20	2.8	0.51	0.72	0.23	0.43
Σ	100.22	101.00	101.00	98.91	99.69	99.26	100.99
Al	0.005	0.108	1.963	0.003	0.006	0.005	0.009
Ti	0.998	0.905	0.000	0.980	0.984	0.981	0.990
Fe	0.830	1.940	0.179	0.873	0.959	0.999	0.949
Mn	0.007	0.010	0.000	0.006	0.007	0.007	0.009
Mg	0.150	0.112	0.795	0.126	0.025	0.001	0.044
Cr	0.014	0.006	0.054	0.010	0.014	0.005	0.008
Σ	2.004	3.081	2.991	1.998	1.995	1.998	2.009
Fe/(Fe+Mg)	0.84	0.99	0.18	0.87	0.97	1.00	0.95
(Al/2):(Cr/2):Ti	—	6:0:94	97:3:0	—	—	—	—

Papike, J.J. and Merrill, R.B., eds.
Proc. Conf. Lunar Highlands Crust (1980), p. 339-352
Printed in the United States of America

Silicate liquid immiscibility, evolved lunar rocks and the formation of KREEP

G. J. Taylor, R. D. Warner*, and K. Keil

Department of Geology and Institute of Meteoritics, University of New Mexico, Albuquerque, New Mexico 87131

M.-S. Ma and R. A. Schmitt

Department of Chemistry and the Radiation Center, Oregon State University, Corvallis, Oregon 97331

Abstract—Criteria exist to evaluate the role of silicate liquid immiscibility in the formation of highly fractionated lunar lithologies, such as high-Fe and high-Si in clasts in 77538, quartz monzodiorite (QMD) in 15405, and sodic ferrogabbro (SFG) in 67915. The clast assemblage in 77538 clearly represents a pair of melts formed by silicate liquid immiscibility; compared to the high-Si clasts, the high-Fe clasts are enriched in P_2O_5, FeO, TiO_2, and CaO and depleted in SiO_2, K_2O, Na_2O and Al_2O_3, precisely as expected for immiscible melt pairs. The evolution of the QMD in 15405 was not dominated by silicate liquid immiscibility. The rock has higher K_2O/P_2O_5 (4.8) than expected for fractional crysallization of a KREEP basalt magma ($K_2O/P_2O_5 \sim 1$), which suggests that liquid immiscibility took place, but other element ratios that ought to be much higher than in KREEP are only slightly so (K/La is 97 vs. 63 in KREEP) and some are lower (Ba/U is 170 vs. 240 in KREEP). A similar situation prevails for the SFG in 67915: K_2O/P_2O_5 is higher (5.0) than expected for fractionation of KREEP basalt, but other diagnostic element-ratios are only slightly greater than they are in KREEP (K/La is 63, Ba/U is 300). If silicate liquid immiscibility occurred during the evolution of the QMD or SFG, the high-Fe and high-Si melts produced did not separate from one another to a significant extent. The same arguments apply to the formation of KREEPy lunar crustal rocks: If silicate liquid immiscibility took place late in the crystallization of the lunar magma ocean, KREEPy rocks ought to have anomalies in their trace element abundances because of preferential partitioning into the one or the other immiscible melt. Because no extreme fractionations relative to KREEP are observed, we conclude that extensive silicate liquid immiscibility did not occur in the lunar magma ocean, or if it did occur, the two melts did not separate physically from one another.

INTRODUCTION

A small percentage of nonmare lunar samples (mostly clasts in breccias) have high Fe/Mg ratios, suggesting that they have undergone extensive differentiation. These evolved rocks, which range in composition from quartz monzodiorites to granites, are important because they contain information about the processes that

*Present address: Department of Chemistry and Geology, Clemson University, Clemson, South Carolina 29631.

operated during the last stages of fractionation in lunar magmas, possibly including the lunar magma ocean. Interpretations fall into two main categories: Some authors suggest that the petrogenesis of evolved lunar rocks involved extensive fractionational crystallization followed by silicate liquid immiscibility (e.g., Hess *et al.*, 1975, 1978; Rutherford *et al.*, 1976; Crawford and Hollister, 1977; Quick *et al.*, 1977; Warner *et al.*, 1978). Other investigators favor an origin of evolved rocks involving only fractional crystallization (e.g., Ryder, 1975, Ryder *et al.*, 1976; Lovering and Wark, 1975; Irving, 1977; Nyquist, 1977). We have studied clasts with evolved compositions in three lunar rocks: 77538, 15405, and 67915. We describe the results here and evaluate the roles that liquid immiscibility may have played in their genesis. We also speculate on the degree to which liquid immiscibility may have taken place late in the history of the lunar magma ocean. (The origin of rock 12013 may also have involved silicate liquid immiscibility, but because this complex breccia has been described and discussed in detail by Quick et al. (1977), we do not discuss it in this paper.)

SAMPLES AND EXPERIMENTAL PROCEDURES

We present new data on the sodic ferrogabbro from 67915 and the quartz monzodiorite from 15405. Two samples of the sodic ferrogabbro were analyzed: thin sections 67915,190 and ,191, cut from parent 67915,38, and sample 67915,163,01 (4.13 mg) separated from 67915,163, a sample allocated to a consortium headed by Dr. Kurt Marti. The sample of the quartz monzodiorite, 15405,152, weighed 14.7 mg and was made into a thin section after instrumental neutron activation analysis (INAA).

INAA was done at Oregon State University using procedures similar to those described by Laul and Schmitt (1973). Mineral compositions were determined with an automated ARL EMX-SM electron microprobe; data were corrected for differential matrix effects by the method developed by Bence and Albee (1968). Modal analyses were done with a petrographic microscope equipped with an automatic point-counting stage; two modal analyses were made of each thin section. Some bulk compositions were determined by combining modal abundances of minerals with the mineral composition determined by microprobe. In the case of silica and whitlockite, assumed mineral compositions, based on ideal chemical formulae, were used.

RESULTS

High-Fe and high-Si clasts in breccia 77538

Rake sample 77538 contains discrete high-Fe and high-Si clasts. These were described in detail by Warner *et al.* (1978), who interpreted them as immiscible melts. We briefly summarize Warner *et al.*'s (1978) observations here.

The high-Si clasts consist mostly of silica and K-feldspar ($An_{1-4}Ab_{5-4}OR_{88-93}$), frequently intergrown in a barred texture, with smaller amounts of sodic plagioclase ($\sim AN_{68}Ab_{30}OR_3$), fayalitic olivine (Fo_{4-13}), ferroaugite ($En_{6-9}Fs_{48-51}Wo_{40-44}$), ilmenite (Fe/Fe+Mg), 0.94–0.99), metal (nearly pure Fe with <0.1 wt.% Ni and ~0.2 wt.% Co), troilite, and a Ca-rich phosphate mineral. The high-Fe clasts are made mainly of ferroaugite ($En_{15-22}Fs_{42-48}Wo_{32-40}$) and ferropigeonite ($En_{30-36}Fs_{49-56}Wo_{14-15}$) that enclose blebs of silica and fayalitic olivine (Fo_{12-17}); troilite, metal (nearly pure Fe with <0.1 wt.% Ni and ~0.2 wt.% Co), and ilmenite

Table 1. Bulk compositions (wt.%) of lunar samples with evolved compositions.

	77538		15405 QMD[a]			15405 Gran	67915 SFG[b]				
	1	2	3	4	5	6	7	8	9	10	11
SiO_2	50.3	74.0	57.4	(55.4)		68.1	(57.3)	54.0	61.1	51.8	56.7
TiO_2	2.82	0.54	1.1	2.6		0.90	6.0	6.7	2.8	5.3	4.7
Al_2O_3	0.96	12.5	13.2	11.9		10.2	8.4	12.6	12.7	12.2	11.1
Cr_2O_3	0.17	<0.01	—	0.22	0.18	0.06	0.03	0.04	0.03	<0.05	0.03
FeO	31.3	2.21	11.2	14.1	15.1	7.0	13.6	11.5	8.4	16.5	12.8
MnO	—	—	—	0.18		0.12	0.20	0.2	0.1	0.2	0.2
MgO	4.3	0.08	3.4	3.8		1.5	3.8	3.0	2.9	2.2	3.0
CaO	9.8	1.86	9.0	8.9		4.9	8.9	9.0	9.0	8.8	8.9
Na_2O	0.12	0.90	1.0	0.81	0.87	0.79	1.35	1.6	1.6	0.4	1.1
K_2O	0.29	7.6	1.9	2.1	1.8	3.4	0.46	0.3	0.5	0.9	0.6
P_2O_5	0.28	0.07	0.4	—		0.52	—	0.2	0.1	<0.05	0.1
Total	100.34	99.76	98.60	100		97.98[c]	100	99.14	99.23	98.30	99.23

[c] Includes 0.45 wt.% BaO
[a] QMD = quartz monzodiorite clast
[b] SFG = sodic ferrogabbro

1) High-Fe clasts, broad-beam microprobe analysis (Warner *et al.*, 1978); 2) High-Si clasts, broad-beam microprobe analysis (Warner *et al.*, 1978); 3) 15405,51 modal recombination (Taylor, 1976); 4) 15405,152 INAA (SiO_2 by difference); 5) 15405,85, INAA (Shih *et al.*, unpubl. data, quoted by Ryder and Norman, 1978); 6) 15405,12, broad-beam microprobe analysis of granite (Ryder, 1976); 7) 67915,163,01 INAA (SiO_2 by difference); 8) 67915,163,01 modal recombination; 9) 67915,38, modal recombination; 10) 67915,45-1b and -1c, broad beam analyses 96 and 97 (Roedder and Weiblen, 1974); 11) Average of columns 6, 8 and 9.

(Fe/(Fe+Mg), 0.95–0.98) are also present. Occurrences of a high-Si area in a high-Fe clast, and vice versa, were noted by Warner *et al.* (1978).

Bulk compositions of the two types of clasts (Table 1), determined by broad-beam microprobe analysis, are strikingly different. Compared to the high-Si clasts, the high-Fe clasts are greatly enriched in FeO, MgO, CaO, and P_2O_5, and depleted in SiO_2, Al_2O_3, Na_2O, and K_2O. No trace-element data are available on separated clasts. An analysis of a bulk sample of 77538 (Laul and Schmitt, 1975), however, shows that it has a KREEP-like REE pattern with a La content of 174 times chondrites. This reflects an abundance of KREEP basalt material, consistent with the presence of pyroxene mineral clasts with compositions like those in KREEP basalts (see Warner *et al.*, 1978, Fig. 4).

Quartz-monzodiorite clasts in melt rock 15405

Ryder (1976) described a suite of KREEP basalt, quartz-monzodiorite (QMD), and granite clasts in melt rock 15405. The QMD lithology has a coarse-grained (pyroxene and plagioclase are up to 1 mm across) basaltic texture (Fig. 1A); its mode is presented in Table 2 (Taylor, 1976) and average mineral compositions appear in Table 3 (Taylor, 1976 and previously unpublished data). QMD consists of ~35% each of pyroxene and plagioclase, 10–15% each of silica and K-feldspar,

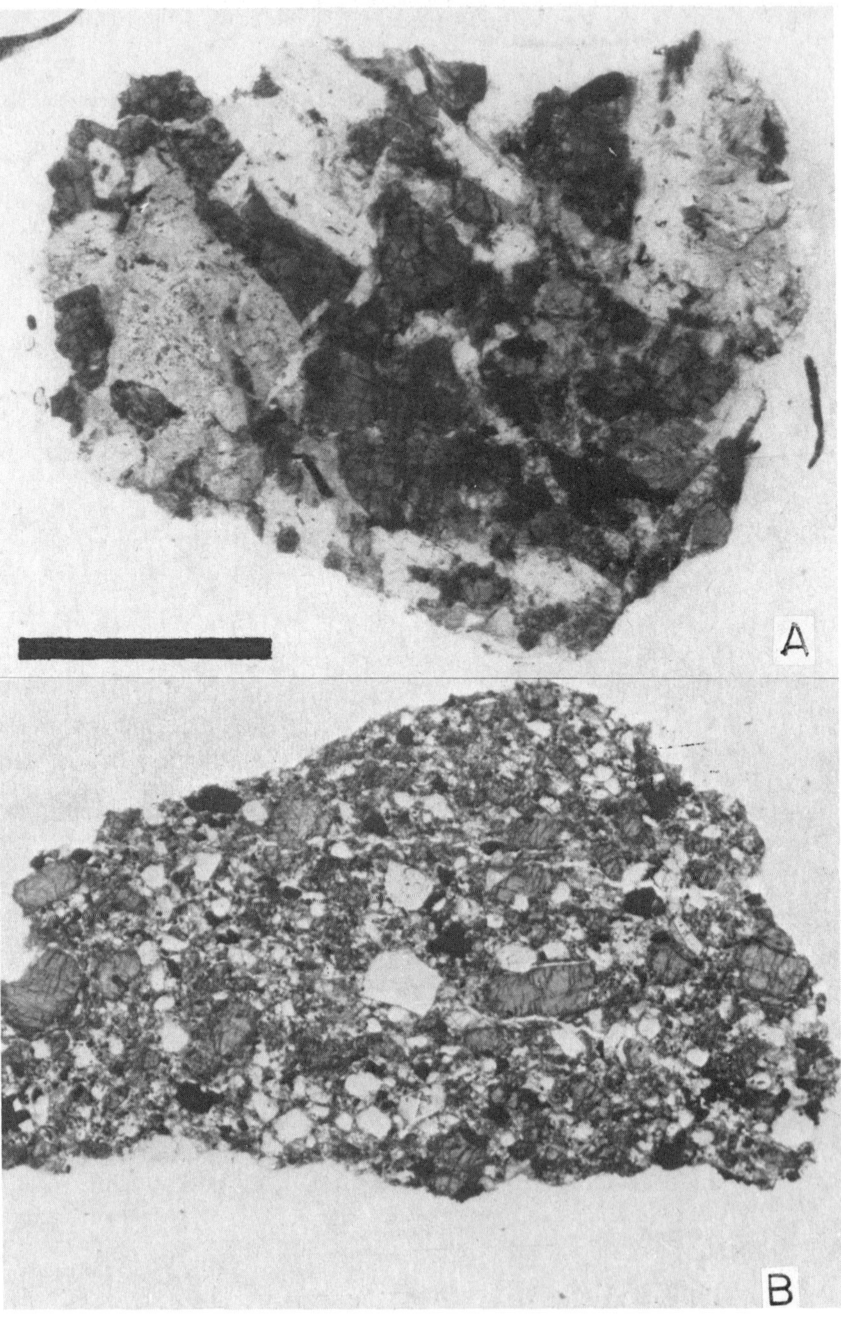

Fig. 1. Photomicrographs of evolved lunar rocks. A) Quartz-monzodiorite 15405,57; plagioclase (white), quartz and K-feldspar intergrowths (gray), pyroxene (dark gray) and ilmenite (black) intergrown in a basaltic texture. Scale bar is 1 mm. B) Sodic ferrogabbro 67915,190; plagioclase (white), pyroxene (gray) and ilmenite (black) occur in this cataclastic rock. (Silica is also present in small grains, but is indistinguishable from plagioclase in this photograph. The large white grains, however, are all plagioclase.) Scale bar is 1 mm.

Table 2. Modal compositions (vol. %) of samples of quartz monzodiorite (QMD) and sodic ferrogabbro (SFG)

	15405,51[b] QMD		67915,38[c] SFG		67915,163,01 SFG	
Augite[a]	18.0	36.0	23.0	27.8	22.6	27.4
Pigeonite[a]	18.0		4.8		4.8	
Plagioclase	34.8		42.2		42.3	
Ilmenite	1.6		4.8		12.3	
K-feldspar	11.0		1.2		0.9	
Silica	15.0		23.5		16.8	
Zircon	0.6		—		—	
Phosphates	0.8		0.2		0.3	
Metal	0.2		t		t	
Troilite	t		0.4		t	

[a] Modal analysis determined total pyroxene only; abundances of augite and pigeonite were determined from random microprobe survey of pyroxene compositions.
[b] One section, ~ 20 mm^2.
[c] Average of two sections (,190 and ,191) each ~ 20 mm^2.
[d] One section, ~ 4 mm^2.

Table 3. Average mineral compositions (wt.% for oxides, mole % for end members) in quartz monzodiorite (QMD) and sodic ferrogabbro (SFG); N is the number of analyses.

	15405,51 QMD				67915,38 SFG				
	1	2	3	4	5	6	7	8	9
N	6	5	22	5	19	4	20	2	11
SiO_2	49.3	49.3	50.1	64.1	51.2	50.3	53.6	63.6	0.07
TiO_2	0.93	0.55	—	—	0.97	0.61	—	—	52.8
Al_2O_3	1.39	0.78	31.0	18.5	1.02	0.52	28.9	19.2	0.09
Cr_2O_3	—	—	—	—	0.09	0.06	—	—	0.11
FeO	23.5	33.0	0.3	0.2	20.7	30.0	0.27	0.15	43.0
MnO	—	—	—	—	0.34	0.49	—	—	0.47
MgO	8.9	9.8	—	—	10.0	11.1	<0.02	—	1.93
CaO	14.9	6.4	13.7	0.3	15.2	5.8	12.3	1.59	—
Na_2O	—	—	2.82	0.6	—	—	3.7	0.25	—
K_2O	—	—	0.63	15.4	—	—	0.80	14.1	—
Total	98.92	99.83	98.55	99.10	99.52	98.88	99.57	98.89	98.47
En	27.2	29.8			30.9	34.6			
Fs	40.2	56.2			35.7	52.5			
Wo	32.6	14.0			33.4	12.9			
An			70.2	1.5			61.6	8.4	
Ab			26.1	5.5			33.6	2.4	
Or			3.7	93.0			4.8	89.2	

1) Augite macrocrystals; 2) pigeonite macrocrystals; 3) plagioclase; 4) K-feldspar; 5) augite macrocrystals; 6) pigeonite macrocrystals; 7) plagioclase; 8) K-feldspar; 9) ilmenite.

~0.5–1.5% each of ilmenite, zircon, and whitlockite, and minor chromite and metal (nearly pure Fe with <0.02 wt.% Ni and 0.2 wt.% Co). Two main pyroxenes are present, ferroaugite and ferropigeonite (Tables 2,3), both of which are unzoned and exsolved into low-and high-Ca pyroxenes (Ryder, 1976). Using average mineral compositions and the mode, Taylor (1976) determined the bulk composition of a sample of QMD (Table 1, analysis 3). Nyquist *et al.* (1977) and Shih *et al.* (unpublished data quoted by Ryder and Norman, 1978) determined trace element abundances in samples of QMD in 15405 (Table 4). We have now determined major, minor and trace elements by INAA on a sample of QMD separated from 15405 (Table 1, analysis 4 and Table 4). The results are generally consistent with one another. Slight differences in major element abundances between our sample and Taylor's (1976), particularly in TiO_2, Al_2O_3, and FeO, are almost certainly due to sampling errors, with the sample provided for INAA containing slightly more ilmenite and pyroxene, and slightly less plagioclase than the one upon which Taylor's (1976) estimate was based. Trace element contents in our sample are similar to those reported by Nyquist *et al.* (1977) (Table 4 and Fig. 2), although the REE abundances are about 20% lower in our sample. The REE and Ba are essentially unfractionated compared to KREEP; the slight Ta

Table 4. Trace element contents (PPM) of 15405 quartz monzodiorite (QMD) and 67915 sodic ferrogabbro (SFG).

	QMD			SFG
	1	2	3	4
Sc	29	—	30.7	34
V	33	—	—	<10
Co	8.0	—	7.8	6.6
Zr	1620	—	—	320
Ba	1900	1490	—	390
La	183	224	210	26.7
Ce	413	555	560	62
Nd	287	328	—	45
Sm	77.4	92.0	93	13.1
Eu	2.75	2.69	2.52	2.45
Tb	14.9	—	19.7	2.85
Dy	101	116	—	19
Yb	55.2	60.9	65	11.2
Lu	8.2	8.1	9.0	1.58
Hf	44.7	—	51	9.6
Ta	10.1	—	13	2.9
Th	39.4	—	43	4.7
U	11.1	—		1.3

1) 15405,152, INAA.

2) 15405,85, mass spectrometric analysis (Nyquist *et al.*, 1977).

3) 15405,85, INAA (Shih *et al.*, unpubl. data, quoted by Ryder and Norman, 1978).

4) 67915,163,01, INAA

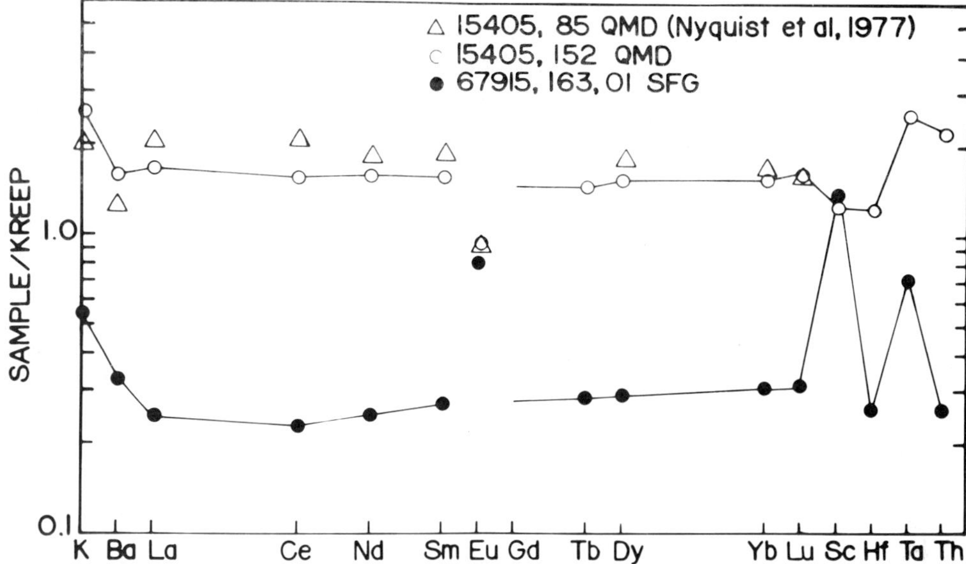

Fig. 2. Trace element abundances in two evolved lunar rocks, normalized to KREEP (Warren and Wasson, 1979).

anomaly (relative to KREEP) in our sample is consistent with the greater TiO_2 content compared with Taylor's (1976) estimate because Ta concentrates in ilmenite. The Zr/Hf ratio (36) is substantially lower than it is in KREEP (46; Warren and Wasson, 1979). Compared to the REE, both Zr and Hf are lower relative to KREEP (the modest Hf depletion is evident in Fig. 2), but Zr is much more depleted than Hf. In fact, rather than being enriched by a factor of ~1.6 relative to KREEP (as are the REE), Zr is lower than it is in KREEP (1620 ppm vs. 1700 in KREEP; Warren and Wasson, 1979).

The "granite" lithology in 15405 is enigmatic. It is as coarse-grained as the QMD, but does not have an igneous texture. It appears crushed and smeared out. Mineral compositions are the same as those in the QMD (Ryder, 1976). The one bulk analysis available (Table 1, analysis 6) was done by the broad-beam microprobe technique on one clast. Ryder (pers. comm.) now thinks that the granite clast may simply be a modally nonrepresentative, crushed QMD clast. We have examined section 15405,12, which contains the granite clast, and concur that the clast is likely to be a QMD.

Sodic ferrogabbro clasts in breccia 67915

Feldspathic breccia 67915 contains a variety of clasts types, one of which has been called a sodic ferrogabbro, or SFG (Weiblen and Roedder, 1973; Roedder and Weiblen, 1974). We have studied two clasts of SFG, one (67915,38) in thin section only and one (67915,163) by INAA and petrology. Our results are similar to those reported by Roedder and Weiblen (1974); modes of our two samples are given in Table 2 and average mineral compositions in Table 3. SFG clasts are generally coarse-grained (crystals of plagioclase, pyroxene, or ilmenite up to 1 mm), but cataclastic; no hint of the original (igneous?) texture remains (Fig. 1B). The clasts consist of plagioclase, ferroaugite and ferropigeonite, ilmenite, and silica, with minor amounts of K-feldspar, a phosphate mineral, troilite and metal (nearly pure Fe, with <0.04 wt.% Ni and ~ 0.4 wt.% Co). The silica mineral and K-feldspar are not intergrown as they frequently are in 77538 or the 15405 QMD; instead, they occur as separate, equant crystals.

The bulk composition of the SFG lithology varies from clast to clast. The sample (67915,163,01) we analyzed by INAA (Table 1, analysis 6) contains considerably more TiO_2 than the clast (67915,38) we studied by petrography and microprobe alone (Table 1, analysis 9). (This is not an artifact of the modal recombination method: our estimated bulk composition of sample 67915,163,01 agrees with that determined by INAA; compare Table 1, analyses 7 and 8.) The average of two broad-beam microprobe analyses done by Roedder and Weiblen (1974) on fine-grained sodic ferrogabbro clasts (Table 1, analysis 10) is also richer in TiO_2 than our estimated composition of 67915,38, but has a similar content of Al_2O_3. The average (Table 1, analysis 11) of these measurements is probably a reasonable estimate of the SFG's bulk composition. Compared to the QMD in 15405, the SFG contains more TiO_2 and considerably less K_2O and P_2O_5, but is otherwise broadly similar in composition. Not surprisingly, the low K_2O and P_2O_5 contents in the SFG are accompanied by lower REE contents (Table 4, Fig. 2): The SFG contains ~6x lower REE than the QMD and the trace element pattern is fractionated relative to KREEP (Fig. 2). The fractionation, however, is not severe, except for K, Sc and Ta. The Ta anomaly is probably due to the presence of more ilmenite in the sample analyzed by INAA. As in the QMD, Zr/Hf is lower in the SFG (33) than it is in KREEP (46).

SILICATE LIQUID IMMISCIBILITY IN THE FORMATION OF EVOLVED LUNAR ROCKS?

Both experimental data (Hess *et al,* 1975; Rutherford *et al.,* 1976) and observations of natural lunar rocks (Roedder and Weiblen, 1970, 1971, 1972a,b, 1973; Powell *et al.,* 1975; Crawford and Hollister, 1977) demonstrate unequivocally that extensive fractional crystallization can lead to silicate liquid immiscibility, producing one melt rich in FeO and one rich in SiO_2. All instances of liquid immiscibility in natural lunar materials, however, are restricted to the mesostases of basaltic rocks. Did it ever occur on a larger scale, such as late in the crystallization of the magma ocean, as speculated by Hess *et al.* (1975) and Roedder (1978)? The evolved lunar rocks described above, which are clearly the products of extensive fractionation, provide an opportunity to evaluate the role silicate liquid immiscibility may have played in the evolution of lunar magmas on a scale much larger than the mesostases of basalts.

Existing data, including experimental data on the partitioning of minor and trace elements between two silicate melts (Watson, 1976; Ryerson and Hess, 1978), provide criteria to identify immiscible melts. In practice, these criteria are not unambiguous, however, and the empirical and theoretical details of silicate liquid immiscibility are not yet firmly established. Nevertheless, the general relationships are clear enough to be useful. In general, elements with high charge densities (that is, elements with high ratios of valence to ionic radius) tend to concentrate in the high-Fe, silica-poor melt of an immiscible pair, whereas elements with low charge densities tend to reside in the high-Si melt (Hess and Rutherford, 1974). (Al is an exception; although it has a relatively high charge density, it concentrates in the high-Si melt because it copolymerizes with Si.) Thus, the elements P, Ti, and the REE concentrate in the high-Fe melt and the elements K, Al, Na, Rb and Cs concentrate in the high-Si melt. The degree to which the elements concentrate in one melt or the other varies, of course, with some strongly concentrated (e.g., P in the high-Fe, K in the high-Si) and others only weakly so (e.g., Ba) (Hess *et al.*, 1975; Watson, 1976). The REE although strongly concentrated into the high-Fe melt, do not fractionate significantly from one another in either melt (Ryerson and Hess, 1977; Watson, 1976). Armed with these criteria, which are summarized in Table 5, we evaluate the role that silicate liquid immiscibility has played in forming the evolved rocks described above.

77538 high-Fe and high-Si clasts

The collection of millimeter-sized clasts of high-Fe and high-Si composition in 77538 probably represents the best example that silicate liquid immiscibility took place on a scale larger than the glassy mesostasis in lunar basalts. The high-Fe clasts contain much higher FeO, MgO, TiO_2, CaO and P_2O_5, and much lower SiO_2, K_2O, Na_2O and Al_2O_3 than do the high-Si clasts (Table 1), precisely as expected for an immiscible-melt pair. The only negative evidence is the relatively high K_2O/P_2O_5 in the high-Fe clasts (1.0); this ratio, if the assemblage in 77538 represents fractionation of a KREEP magma (as the bulk rock composition suggests), ought to be <0.4 (Table 5). Perhaps the high-Fe clasts do not precisely reflect the composition of the high-Fe melt. They may be pyroxene-rich concentrates from a high-Fe immiscible melt. The low bulk Al_2O_3 and P_2O_5 in the high-Fe clasts support this interpretation. Or possibly the parent liquid that existed

Table 5. Elemental ratios expected for silicate liquid immiscibility in a KREEP magma and ratios observed in evolved lunar rocks.

	77538		15405	15405	67915		Predicted for KREEP magma[2]	
	High-Fe	High-Si	QMD	Granite	SFG	KREEP[1]	High-Fe	High-Si
K_2O/P_2O_5	1.0	109	4.8	6.5	5.0	1.1	<0.4	>4
K/La	—	—	97	—	140	63	$\ll 63$	$\gg 63$
K/U	—	—	1600	—	2900	1400	$\ll 1400$	$\gg 1400$
Ba/U	—	—	170	—	300	240	<240	>240

[1] KREEP Component (Warren and Wasson, 1979).

[2] Inferred from experimental data and compositions of natural immiscible melts (see text for references) and the composition of KREEP (Warren and Wasson, 1979).

before immiscibility took place had a composition near the Fe-rich end of the immiscibility field; if so, then only the high-Si melt would show the perturbed element ratios. Another possibility is that the complementary high-Fe melt is unsampled in 77538 and the Fe clasts represent pyroxene concentrates from the high-Si melt. If so, the high-Si melt must have fractionated after it formed and ought to have higher Fe/Mg than the high-Fe clasts. Pyroxene compositions and bulk compositions support this interpretation: Ferroaugites in the high-Fe clasts have lower Fe/(Fe+Mg) (0.71) than they do in the high-Si clasts (0.87) and bulk Fe/(Fe+Mg) is lower in the high-Fe clasts (0.80) than in the high-Si clasts (0.94). In spite of these complexities, the bulk of the data, including K_2O/P_2O_5 in the high-Si clasts, strong points towards an origin involving silicate liquid immiscibility late in the crystallization of a KREEP basalt magma.

15405 QMD

Ryder (1976), Nyquist *et al.* (1977) and Irving (1977) suggest that the QMD lithology in 15405 formed by fractional crystallization of a KREEP basalt magma. Their arguments are based on major and trace element trends among Apollo 15 KREEP basalts. All trends, such as the progressive increase in SiO_2, REE, and Fe/Mg, appeared to terminate with the QMD. In contrast, Rutherford *et al.* (1976) suggested that silicate liquid immiscibility played an important part in the evolution of the QMD. These authors emphasized the importance of the granite lithology in 15405, suggesting that it formed by liquid immiscibility. In their view, the QMD could have formed by mixing (possibly by impact-melting) of the granite with plagioclase fractionated from the complementary (unsampled) high-Fe melt. A serious problem with this idea is that the granite may not really be a separate lithology. As discussed above, the one large clast analyzed (Ryder, 1976) is crushed and smeared out and may simply be a nonrepresentative sample of the QMD. Nevertheless, the idea merits careful consideration, particularly since trace element data on the QMD have become available.

The K_2O/P_2O_5 ratio in the QMD is 4.8. This is considerably greater than it is in KREEP (~1.1) and in the lower range of values measured in experimental or natural high-Si immiscible melts (Table 5). This suggests that liquid immiscibility was involved in the formation of the QMD. If so, then evidence for it ought to be apparent in the abundances of other elements. Three parameters (K/La, K/U, and Ba/U are listed in Table 5. If silicate liquid immiscibility took place and the QMD represents the high-Si melt (perhaps modified by subsequent fractionation), the K/La and K/U ratios should be much greater than they are in KREEP, and Ba/U should be somewhat greater. It turns out the K/La and K/U in the QMD are greater than in KREEP, but not by much, and Ba/U is less than it is in KREEP. These data rule out the possibility that liquid immiscibility played a dominant role in the formation of the QMD. Data on the compositions of residual liquids produced during crystallization experiments on KREEP basalts (Hess *et al.,* 1978) also suggest that the QMD found by fractional crystallization: although not mimicing the QMD composition precisely, residual liquids formed during crystallization of a melt with the composition of KREEP basalt 15382 resembled the QMD composition and eventually entered the two-liquid field. This suggests that the QMD represents a fractionated basalt that had not yet reached the immiscibility field.

If silicate liquid immiscibility was not significant, what process accounts for the high K_2O/P_2O_5 ratio in the QMD? The obvious choice is crystal fractionation: perhaps the QMD magma lost a phosphate phase or gained K-feldspar. Loss of a phosphate seems unlikely because the K_2O/P_2O_5 ratio implies loss of ~3 wt.% apatite or whitlockite, which would have affected the relative abundances of the REE, altering the KREEP pattern observed (Fig. 2). Gain of ~9 wt.% K-feldspar is a possibility, but this seems to be too large a fractionation of material in a silica-rich, hence relatively viscous, magma. Alternatively, the high K_2O/P_2O_5 ratio might be a sampling error. The only P_2O_5 measurement was made by the modal-recombination method on one thin section. Perhaps the P_2O_5 content is higher in other samples.This is, of course, plausible, but we note that the granite lithology in 15405, which we think is a nonrepresentative sample of QMD, also has a high K_2O/P_2O_5 ratio (Table 5).

In summary, the main evidence pointing towards a role for liquid immiscibility in the evolution of the QMD in 15405 is the high K_2O/P_2O_5, but other elemental ratios such as K/U and Ba/U do not support such an origin. The high K_2O/P_2O_5 cannot be easily explained by crystal fractionation or sampling errors, either. Perhaps the fractionated KREEP magma in which the QMD formed did begin to form immiscible melts, and high-Fe and high-Si blebs began to separate from one another. If crystallization was proceeding relatively rapidly at the same time, perhaps only a modest amount of fractionation of high-Si from high-Fe melt occurred, leaving the final assemblage only slightly affected by the episode of liquid immiscibility. The only evidence remaining is the high K_2O/P_2O_5, the ratio of the two elements most strongly fractionated during liquid immiscibility.

67915 SFG

The discussion about the QMD applies equally well to the SFG lithology found in breccia 67915. K_2O/P_2O_5 is high (5.0, Table 5) and, although K/La, K/U and Ba/U are all higher than they are in KREEP, they are not drastically so. As we speculated for the QMD, perhaps a period of liquid immiscibility was followed by or accompanied by crystallization so the high-Fe and high-Si melts were not physically separated by much, but the assemblage of melts from which the SFG crystallized contained a slight enrichment of high-Si over high-Fe melt.

There is, however, a critical distinction between the SFG and the QMD: SFG contains ~8 times less REE than does the QMD. (It also has a less fractionated REE pattern, with chrondrite-normalized La/Lu of ~1.7, versus 2.2–2.7 in the QMD. However, only one analysis of the SFG has been done and it is possible that sampling errors are the cause of the lower La/Lu.) Assuming the SFG represents a melt produced by extensive fractional crystallization, its lower REE contents imply that its parent magma was also low in REE. In fact, because the

REE contents of the SFG are about the same as those in LKFM (Taylor *et al*, 1973; Reid *et al*., 1977), the parent magma to the SFG must have been lower in REE than LKFM. We speculate that the SFG could have evolved in the lunar magma ocean and might represent the composition of the magma after ~95% crystallization (assuming an initial concentration of REE of 4 times chondritic). Before detailed quantitative modelling, however, the composition of the SFG must be determined more precisely and trace element abundances must be measured on a larger sample than the one analyzed so far.

SILICATE LIQUID IMMISCIBILITY IN THE MAGMA-OCEAN?

Hess *et al*. (1975) and Roedder (1978) speculate that silicate liquid immiscibility could have occurred late in the crystallization of the lunar magma ocean and may have had a significant effect in the evolution of the lunar crust. We acknowledge that liquid immiscibility probably occurred (indeed, it was probably unavoidable), but we suggest that it did not significantly affect the evolution of lunar crustal rocks. When, and if, silicate liquid immiscibility occurred in the magma ocean, the two melts would have drastically different compositions, with the high-Si melt characterized by very high K/Ba, K/U and Ba/U. Because the trace-element signature of KREEP rocks probably arose during crystallization of the magma ocean (Warren and Wasson, 1979), some KREEP rocks would have acquired a signature dominated by high-Si melt, whereas others would have acquired a signature from high-Fe melt. Consequently, the relative abundances of incompatible elements in KREEP rocks ought to reflect the relative amounts of high-Si and high-Fe melt they contained. They would reflect this by displaying anomalies in the abundances of incompatible elements. But as Warren and Wasson (1979) point out, KREEP rocks are characterized by nearly the same relative abundances of incompatible elements, including K/La, K/U and Ba/U. This seems to rule out a significant role for silicate liquid immiscibility in the formation of KREEP or related lunar crustal rocks. If *large-scale* liquid immiscibility occurred, it was not accompanied by extensive physical separation of the high-Fe and high-Si melts because such separation would have led to extreme element fractionations and these fractionations would be evident in the trace element contents of lunar crustal rocks. On the other hand, the rare samples of lunar granites suggest that some immiscibility did take place, possibly in the magma ocean (Rutherford *et al*., 1976). Furthermore, KREEP rocks do occasionally have modest anomalies in their K-contents (Warren and Wasson, 1979). Because K is one of the elements most strongly fractionated when immiscibility takes place, perhaps the K anomalies record an episode of liquid immiscibility in the lunar magma ocean during which high-Fe and high-Si melts physically separated from one another, but on a local scale only.

Finally, we draw attention to the low Zr/Hf ratios of the QMD (36) and the SFG (33) compared to this ratio in KREEP (46). If interlaboratory biases in the analyses or sampling errors are not the causes of the low ZR/Hf in the evolved rocks, it is difficult to understand the variations in Zr/Hf because these elements ought to exhibit nearly identical geochemical behavior, including during silicate liquid immiscibility. Consequently, we are initiating a detailed study of Zr/Hf fractionations in lunar samples.

Acknowledgments—This work was supported by NASA grants NGL 32-004-063 (K. Keil) and NGL 32-002-039 (Roman Schmitt).

REFERENCES

Bence A. E. and Albee A. L. (1968) Empirical correction factors for the electron microanalysis of silicates and oxides. *J. Geol.* **76,** 382–403.

Crawford M. L. and Hollister L. S. (1977) Evolution of KREEP: Further petrologic evidence. *Proc. Lunar Sci. Conf. 8th,* p. 2403–2417.

Hess P. C. and Rutherford M. J. (1974) Element fractionation between immiscible melts (abstract). In *Lunar Science V,* p. 328–330. The Lunar Science Institute, Houston.

Hess P. C., Rutherford M. J., and Campbell H. W. (1978) Ilmenite crystallization in nonmare basalt: Genesis of KREEP and high-Ti mare basalt. *Proc. Lunar Planet. Sci. Conf. 9th,* p. 705–724.

Hess P. C., Rutherford M. J., Guillemette R. N., Ryerson F. J., and Tuchfeld H. A. (1975) Residual products of fractional crystallization of lunar magmas: An experimental study. *Proc. Lunar Sci. Conf. 6th,* p. 895–909.

Irving A. J. (1977) Chemical variation and fractionation of KREEP basalt magmas. *Proc. Lunar Sci. Conf. 8th,* p. 2433–2448.

Laul J. C. and Schmitt R. A. (1973) Chemical composition of Luna 20 rocks and soil and Apollo 16 soils. *Geochim. Cosmochim. Acta* **37,** 927–942.

Laul J. C. and Schmitt R. A. (1975) Chemical composition of Apollo 17 samples: boulder breccias (2), rake breccias (8), and others (abstract). In *Lunar Science VI,* p. 489–491. The Lunar Science Institute, Houston.

Lovering J. F. and Wark D. A. (1975) The lunar crust-chemically defined rock groups and their potassium-uranium fractionation. *Proc. Lunar Sci. Conf. 6th,* p. 1203–1217.

Nyquist L. E., Wiesmann H., Shih C.-Y., and Bansal B. M. (1977) REE and Rb-Sr analysis of 15405 quartz-monzodiorite (super KREEP) (abstract). In *Lunar Science VIII,* p. 738–740. The Lunar Science Institute, Houston.

Powell B. N., Dungan M. A., and Weiblen P. W. (1975) Apollo 16 feldspathic melt rocks: Clues to the magmatic history of the lunar crust. *Proc. Lunar Sci. Conf. 6th,* p. 415–433.

Quick J. E., Albee A. L., Ma M.-S., Murali A. V., and Schmitt R. A. (1977) Chemical compositions and possible immiscibility of two silicate melts in 12013. *Proc. Lunar Sci. Conf. 8th,* p. 2153–2189.

Reid A. M., Duncan A. R., and Richardson S. H. (1977) In search of LKFM. *Proc. Lunar Sci. Conf. 8th,* p. 2321–2338.

Roedder E. (1978) Silicate liquid immiscibility in magmas and in the system K_2O-FeO-Al_2O_3-SiO_2: an example of serendipity. *Geochim. Cosmochim. Acta* **42,** 1597–1617.

Roedder E. and Weiblen P. W. (1970) Lunar petrology of silicate melt inclusions, Apollo 11 rocks. *Proc. Apollo 11 Lunar Sci. Conf.,* p. 801–837.

Roedder E. and Weiblen P. W. (1971) Petrology of silicate melt inclusions, Apollo 11 and Apollo 12 and terrestrial equivalents. *Proc. Lunar Sci. Conf. 2nd,* p. 507–528.

Roedder E. and Weiblen P. W. (1972a) Petrographic features and petrologic significance of melt inclusions in Apollo 14 and 15 rocks. *Proc. Lunar Sci. Conf. 3rd,* p. 251–279.

Roedder E. and Weiblen P. W. (1972b) Silicate melt inclusions and glasses in lunar soil fragments from the Luna 16 core sample. *Earth Planet. Sci. Lett.* **13,** 272–285.

Roedder E. and Weiblen P. W. (1974) Petrology of clasts in lunar breccia 67915. *Proc. Lunar Sci. Conf. 5th,* p. 303–318.

Rutherford M. J., Hess P. C., Ryerson F. J., Campbell H. W., and Dick P. A. (1976) The chemistry, origin and petrogenetic implications of lunar granite and monzonite. *Proc. Lunar Sci. Conf. 7th,* p. 1723–1740.

Ryder G. (1976) Lunar sample 15405: Remnant of a KREEP basalt-granite differentiated pluton. *Earth Planet. Sci. Lett.* **29,** 255–268.

Ryder G. and Norman M. (1978) *Catalog of Pristine Non-Mare Materials. Part 1.* Non-Anorthosites. JSC Publ. 14565. Johnson Space Center, Houston. 146 pp.

Ryder G., Stoesser D. B., Marvin U. B., and Bower J. F. (1975) Lunar granites with unique ternary feldspars. *Proc. Lunar Sci. Conf. 6th.* p. 435–449.

Ryerson F. J. and Hess P. C. (1978) Implications of liquid-liquid distribution coefficients to mineral-liquid partitioning. *Geochim. Cosmochim. Acta* **42,** 921–932.

Taylor G. J. (1976) Sample 15405: Further petrological studies of the KREEP-rich quartz monzodiorite. In *Interdisciplinary Studies by the Imbrium Consortium.* Vol. 1 (J. A. Wood, ed.), p. 94–95. LSI Contr. No. 267D. Center for Astrophysics, Cambridge.

Taylor S. R., Gorton M. P., Muir P., Nance W., Rudowski R., and Ware N. (1973) Lunar highlands composition: Apennine Front. *Proc. Lunar Sci. Conf. 4th* p. 1445–1459.

Warner R. D., Taylor G. J., Mansker W. L., and Keil K. (1978) Clast assemblages of possible deep-seated (77517) and immiscible melt (77538) origins in Apollo 17 breccias. *Proc. Lunar Planet. Sci. Conf. 9th,* p. 941–958.

Warren P. H. and Wasson J. T. (1979) The origin of KREEP. *Rev. Geophys. Space Phys.* **17,** 73–88.

Watson E. B. (1976) Two-liquid partition coefficients: Experimental data and geochemical implications. *Contrib. Mineral. Petrol.* **56,** 119–134.

Weiblen P. W. and Roedder E. (1973) Petrology of melt inclusions in Apollo samples 15598 and 62295, and of clasts in 67915 and several lunar soils. *Proc. Lunar Sci. Conf. 4th,* p. 681–703.

Papike, J.J. and Merrill, R.B., eds.
Proc. Conf. Lunar Highlands Crust (1980), p. 353-375
Printed in the United States of America

Volcanic rocks in the lunar highlands

Graham Ryder

Lunar Curatorial Laboratory, Northrop Services, Inc., P.O. Box 34416, Houston, Texas 77034

Paul Spudis

Department of Geology, Arizona State University, Tempe, Arizona 85281

Abstract—Volcanism is an important geological process because it provides information on the interior of a planet, provides constraints on its evolution and thermal history, and modifies the morphology and chemistry of the planet's surface. Volcanism prior to the extant mare volcanism potentially provides significant information for understanding the development of the Moon and its crust.

Clasts of basalt demonstrate that just after, during, and prior to the terminal lunar bombardment (3.9–4.0 b.y. ago) a wide variety of volcanic rocks, with differing sources and isotopic characteristics, was being erupted. At the Apollo 14 site alone, a contemporaneous, diverse suite was extruded. Chemical and site-geological studies support the concept of pre-bombardment KREEP volcanism at the Apollo 14 site and probably elsewhere, but the presence of KREEP-free feldspathic impactite breccias on the Moon suggests that KREEP volcanism was not extensive prior to ~4.2 b.y. ago. Mineralogical and chemical studies suggest that mare volcanism occurred at such ancient times but as yet such inferences are not definitive.

Although findings from the Apollo 16 mission destroyed the concept that the Cayley Plains at the Descartes site were volcanic, some light plains units apparently *are* volcanic. The best, but not the only example, the Apennine Bench, can be correlated with Apollo 15 KREEP volcanic clasts on the basis of remote chemical and morphological studies. Orientale ejecta and other light plains units may *blanket* volcanic rocks: they contain craters that excavated material of much lower albedo, possibly mare volcanics. Some post-bombardment non-mare volcanic domes in the highlands may be an indication of some kinds of volcanism more prevalent in ancient times.

For the most ancient, unsampled times, the distinction between volcanism and impact melting itself becomes obscure conceptually. A diverse suite of volcanic rocks was probably extruded, including KREEP and mare but also varieties not easily defined as either (as are the Apollo 17 KREEPy basalts). The term "non-mare" as a group name for ancient volcanics is unwarranted.

INTRODUCTION

The study of basaltic volcanism is important because basalts provide information on both the interior and thermal history of a planet. Further, volcanology is an important part of crustal studies because volcanic rocks modify both the morphology and the chemistry of the surface, masking the underlying units.

Potentially, the post-heavy bombardment volcanism (mainly mare) of the moon is fully explorable and many of its characteristics are already known (Papike *et al.*, 1976; Papike and Vaniman 1978; and others). However, the intense, probably cataclysmic, heavy bombardment of the Moon which terminated at ~3.95 b.y. obscures the record of earlier periods. It is commonly assumed that KREEP volcanism took place in this older period. It has also been shown that the possibility of ancient mare volcanism cannot be readily discarded (Shoemaker, 1972; Ryder and Taylor, 1976).

Evidence concerning ancient lunar volcanism can be sought directly and indirectly in both sample studies and in remote sensing (particularly photogeological) studies. This paper summarises such evidence. Most of it concerns volcanism near the time of the terminal bombardment; with increasing age, the record becomes increasingly obscure. Such volcanism, however, was important in the evolution of the surface and near-surface of the lunar highlands crust. Feldspathic (~25% Al_2O_3) basaltic-textured rocks enriched in meteoritic siderophiles, although considered volcanic by some authors, are most likely to be impact melts; they are not discussed in the present paper.

DIRECT EVIDENCE: BASALT SAMPLES

The Apollo collection contains several basalts dating from around the time of the final heavy bombardment i.e. 3.9–4.0 b.y. For several of these samples adequate siderophile and geochronological data are available to demonstrate that they are both "ancient" and volcanic using the criteria propounded by Anders (1978), Irving (1975) and Warren and Wasson (1977, 1978). In other cases either one or both of these characteristics is lacking and their interpretation rests on chemical, petrographic, and/or geological inferences. Tables 1–3 and Figures 1 and 2 summarize significant data for these samples.

Apollo 15 KREEP basalts

Numerous particles and several small fragments of xenolith-free, meteorite uncontaminated KREEP basalts exist among the soils and breccias of the Apollo 15 collection (e.g. Dowty *et al.*, 1976; Basu and Bower, 1976; Drake *et al.*, 1973; Dymek *et al.*, 1974; Ryder and Bower, 1976; Irving, 1977). These basalts are subophitic to intersertal (Fig. 1a) and consist of plagioclase, pigeonitic/augitic pyroxene with orthopyroxene cores, ilmenite, cristobalite, up to 20% glassy mesostasis and several accessory phases. Olivine is extremely rare. These aluminous, LIL-element enriched basalts show a chemical variation compatible with a series related by fractional crystallization (Irving, 1977). Geochronological data demonstrate that they were completely liquid at 3.9 b.y. (Table 3); a significant difference in the initial Sr isotopic ratio of one fragment indicates a real genetic distinction. Although impact remelting at 3.9 b.y. has frequently been invoked from model age arguments, it seems more likely that these liquids were produced

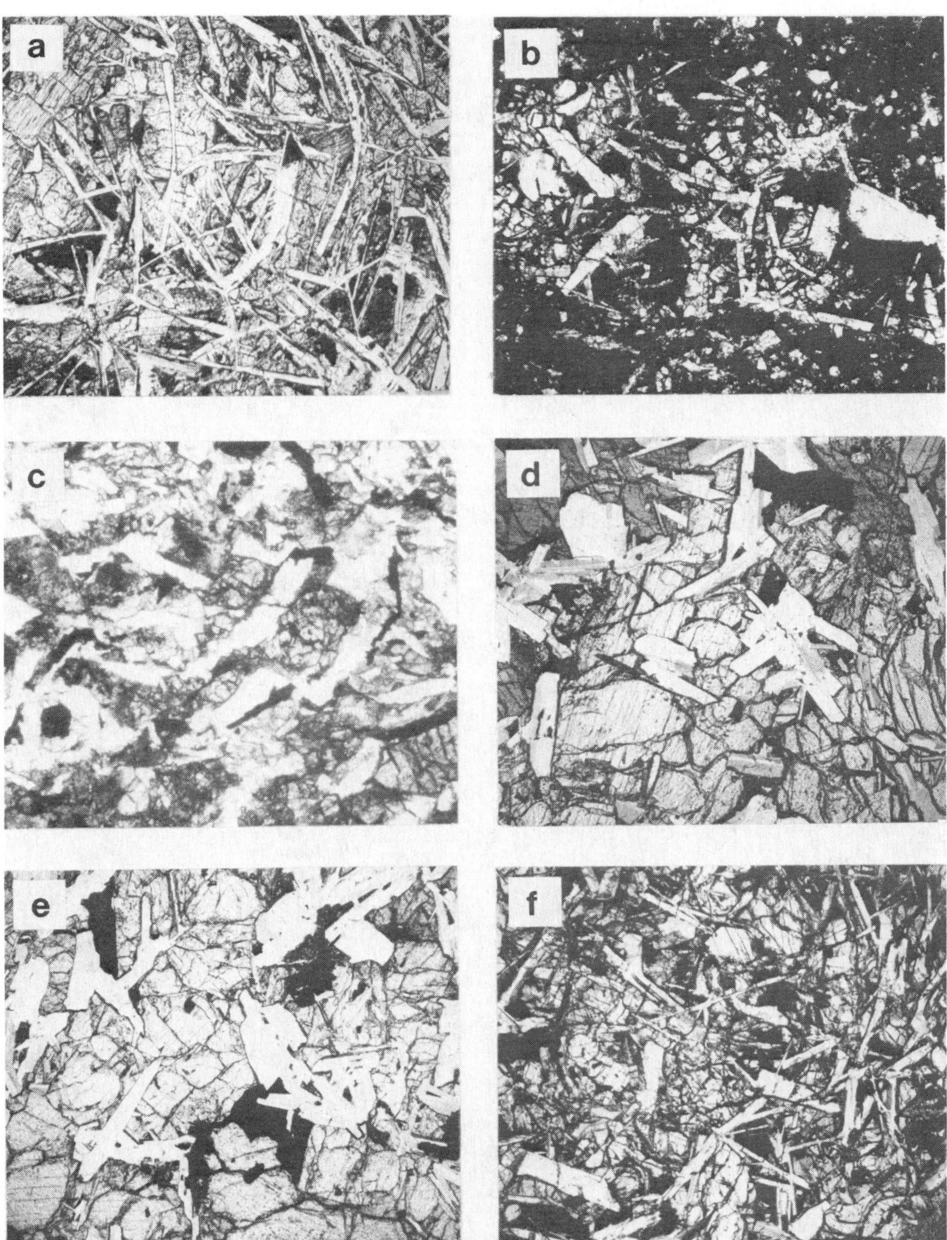

Fig. 1. Transmitted light photomicrographs of ancient basalts. All to same scale: width of views ~2 mm. a) 15382. Apollo 15 KREEP basalt. b) 72275, Apollo 17 KREEPy basalt clast in matrix. c) 67749, Apollo 16 KREEP basalt d) 14053, e) 14072, f) 14321, 1013.

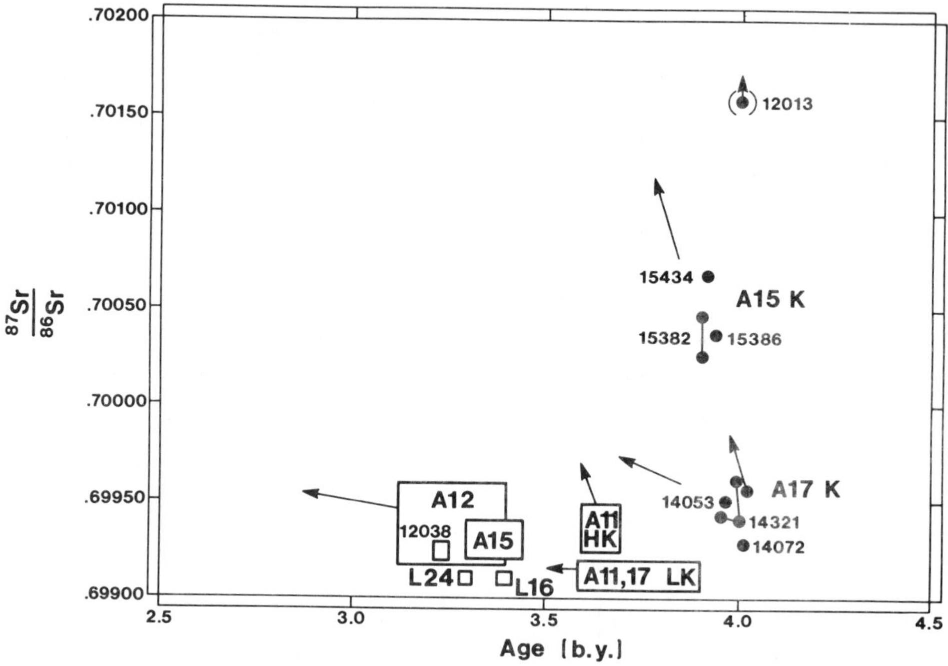

Fig. 2. $^{87}Sr/^{86}Sr$ evolution diagram for lunar basalts. Basalts are plotted at their crystallization age; arrows indicate locus of subsequent evolution for appropriate Rb/Sr. A17K = 72275 basalt.

by partial melting and fractional crystallization at 3.9 b.y. (Dowty *et al.* 1976). If so, their source was itself enriched in incompatible elements.

The Apennine Bench formation (Fig. 3) is a post-Imbrium light plains unit which has a high Th content (~11 ppm; Metzger *et al.*, 1979). Spudis (1978, 1979) has shown that the chemical and morphological features of the Apennine Bench are unlikely to represent Imbrium impact melt (or other Imbrium debris) but are compatible with being flows of Apollo 15 KREEP basalt. Thus, this KREEP volcanism occurred at or shortly after the formation of the Imbrium basin (it may even have been triggered by the Imbrium impact by the introduction of heat and a plumbing system; Papanastassiou and Wasserburg 1976; Spudis 1979). If this is correct, the Imbrium impact is older than, or equal to, 3.9 b.y.

Apollo 17 KREEPy basalts

72275 contains several clasts of a meteorite-free (Morgan *et al.*, 1975b) basalt. The subophitic basalt (Fig. 1b) contains plagioclase, clinopyroxene, ilmenite, and a complex glassy mesostasis (Stoeser *et al.*, 1974, Ryder *et al.*, 1977). Olivine is sparse, while chromite and Fe-metal are common. Chemically it is intermediate between KREEP and mare basalts and has only half the incompatible element content of Apollo 15 KREEP basalts (Blanchard *et al.*, 1975). Its Rb-Sr age of 4.01 b.y. (Compston *et al.*, 1975) shows the basalt to have been extruded significantly before the Apollo 15 KREEP basalts; its chemistry and its much lower initial $^{87}Sr/^{86}Sr$ ratio (Figure 2, Table 3) preclude any simple relationship with the latter.

Apollo 16 KREEP

A single basalt clast in 67749 (Fig. 1c) has a texture and mineralogy like the Apollo 15 KREEP basalts, but the pyroxenes are more Fe-rich (Steele and Smith, 1973). While the distinctive textures and mineralogy indicate that the fragment is from a distinct volcanic rock, an absence of chemical or geochronological data precludes further interpretation.

Apollo 14 high alumina mare basalts

Several fragments of high-alumina mare basalt have been identified in the Apollo 14 samples (Table 1). 14053 (251 g) and 14072 (45 g) were collected as individual rocks; the others are fragments in breccias and soils. 14321 contains a large

Table 1. Mineralogical Data for Ancient Basalt Clasts (for references see text).

Sample	Plagioclase Mol % An	Olivine Mol % Fo	Pyroxene Mol % En,Wo
14053	91–77	Phenocrysts 63 others ~12	55,10–0,40
14072	—	Phenocrysts 74–76 others to 35	59,11–15,31
14321 clasts			
hi-Al	96–72	76–54	70,8–5,25
14063 clasts	81–67	—	~50,27–34,34
14082 clast	87–65	—	50,20–40,35
14312 clasts			
(i)	89–83	67–61	65,10–30,40
(ii)	96–83	67–65	60,10–35,40
14318 clasts			
(i)	86–80	—	40,20–25,30
(ii)	96–83	—	66,10–22,40
(iii)	87–78	—	75,5
14321 clasts			
Mg-vitrophyres	95–79	89–80	—
A15 KREEP	91–76	—	82,3–25,25
A17 KREEPy	94–76	(69–54)	75,5–10,40
A17 B1 St. 2			
ol-clasts	98–90	83–51	75,5–30,25
A17 St. 6			
boulder clasts	93–66	—	65,10–40,40

Table 2. Summary Chemical Data for Ancient Basalt Clasts.

Sample	14053	14072	14321 hi-Al	14143,11A soil particle	14063	14082	14318
Reference	(1–12)	(10–16)	(2,4,5,15–25)	(26)	(27)	(28)	(29)
SiO_2	47.	45.	47.	—	46.	44.	45.
TiO_2	2.7	2.6	2.2	2.3	7.3	6.0	2.6
Al_2O_3	13.	11.1	12.5	13.2	14.4	11.4	12.2
Cr_2O_3	0.41	~0.5	~0.4	0.45	0.2	0.2	0.3
FeO	16.8	17.8	16.	17.4	10.7	11.8	18.1
MgO	8.5	12.2	~8.5	10.0	6.7	8.2	8.3
CaO	11.1	9.8	10.5	12.7	11.3	13.2	10.2
Na_2O	0.44	0.32	0.55	0.65	1.0	0.7	0.40
K_2O	0.11	0.08	0.15	0.23	0.1	0.1	0.72
P_2O_5	0.11	0.10	—	—	—	0.01	0.85
Sc	50–90	47	55	59	—	—	—
V	135	—	~100	135	—	—	—
Sr	100	80–100	110	—	—	—	—
La	13	8	24	18.7	—	—	—
Lu	0.90	0.61	1.2	0.94	—	—	—
Zr	215–310	165	440	—	—	—	—
U	0.60	0.25	0.60	0.7	—	—	—
Th	2.2	1.0	2.8	—	—	—	—
Rb	2.2	1.4	2–6	—	—	—	—
Ba	146–190	130	100–250	160	—	—	—
Ni	14	31	35	<190	—	—	—
Co	25–48	32	30	33	—	—	—
Ir (ppb)	0.017	0.13	0.04–1.1	—	—	—	—
Re (ppb)	0.0066	0.11	0.005	—	—	—	—
Au (ppb)	0.11	0.089	0.4	—	—	—	—

REFERENCES

1. Keith *et al.* (1972)
2. Papanastassiou and Wasserburg (1971)
3. LSPET (1971)
4. Ehmann *et al.* (1972)
5. Janghorbani *et al.* (1973)
6. Church *et al.* (1972)
7. Morgan *et al.* (1972)
8. Nyquist *et al.* (1972)
9. Willis *et al.* (1972)
10. Hubbard *et al.* (1972)
11. Helmke *et al.* (1972)
12. Tatsumoto *et al.* (1972)
13. Longhi *et al.* (1972)
14. Hughes *et al.* (1973)
15. Taylor *et al.* (1972)
16. Compston *et al.* (1972)
17. Wänke *et al.* (1972)
18. Duncan *et al.* (1975)
19. Lindstrom *et al.* (1972)
20. Grieve *et al.* (1975)
21. Morgan *et al.* (1975a)
22. Mark *et al.* (1974)
23. Baedecker *et al.* (1972)
24. Mark *et al.* (1973)
25. Mark *et al.* (1975)
26. Warren *et al.* (1978)
27. Ridley (1975)
28. Ryder and Bower (1976)
29. Kurat *et al.* (1974)

Table 2. *(Continued)*

Sample	14321 Mg	Al5 KREEP (26,)	A17 KREEP	B1 St. 2 ol-norm	73255		
Reference	(30)	(31–37)	(38–42)	(43–45)		(46)	
SiO_2	46.	51.	48.	46.	—	—	—
TiO_2	1.03	2.1	1.4	0.5	2.1	1.54	.34
Al_2O_3	10.3	15.5	13.5	11.	14.2	13.8	14.2
Cr_2O_3	0.2	0.3	0.46	0.6	.51	.46	.72
FeO	10.8	10.	15.	16.	16.6	17.1	15.4
MgO	24.3	9.	10.	13.	11.0	9.5	10.1
CaO	6.3	9.5	10.5	10.	11.5	11.0	5.9
Na_2O	0.78	0.8	0.29	0.16	0.23	0.26	0.18
K_2O	0.46	0.6	0.25	0.10	—	—	—
P_2O_5	—	0.6	0.5	0.01	—	—	—
Sc	—	22	61	—	52.9	63.4	51.5
V	—	60	—	—	120	160	220
Sr	—	200	90		—	50	95
La	—	~70	48	—	1.82	27.5	0.92
Lu	—	~3	1.75	—	0.30	0.36	0.16
Zr	—	1000	625	—	—	—	—
U	—	3.	1.6	—	—	—	—
Th	—	10.	6.3	—	—	—	—
Rb	—	16	7	—	—	—	—
Ba	—	750	355	—	—	—	—
Ni	—	10–30	43	—	—	—	—
Co	—	20	37	—	26.5	22	36
Ir (ppb)	—	.01–.06	0.023	—	—	—	—
Re (ppb)	—	0.0089	0.0066	—	—	—	—
Au (ppb)	—	0.0033–0.22	0.045	—	—	—	—

oxides in wt %, all others in ppm except as noted

REFERENCES

30. Allen *et al.* (1973)
31. Hubbard *et al.* (1973)
32. Murali *et al.* (1977)
33. Schonfeld *et al.* (1972)
34. Dowty *et al.* (1976)
35. Gros *et al.* (1976)
36. Hubbard *et al.* (1974)
37. Rhodes and Hubbard (1973)
38. Leich *et al.* (1975)
39. Morgan *et al.* (1975b)
40. Compston *et al.* (1975)
41. Nunes and Tatsumoto (1975)
42. Blanchard *et al.* (1975)
43. Stoeser *et al.* (1974)
44. Ryder *et al.* (1975)
45. Taylor *et al.* (1978)
46. Blanchard and Budahn (1979)

Table 3. Geochronological Data for Ancient Basalt Clasts.

Sample	Ar-Ar*** Age (b.y.)	Sm-Nd Age (b.y.)	Rb-Sr** Age (b.y.)	Initial $^{87}Sr/^{86}Sr$*	T_{BABI}	Reference
Apollo 15						
KREEP basalts						
			3.90±.02	0.70024±12		Papanastassiou & Wasserburg (1976)
15382				(0.70034)	4.29	Nyquist *et al.* (1973)
	(3.9)					Turner *et al.* (1973)
	(3.9)					Stettler *et al.* (1973)
			3.94±.01	0.70024±3	4.25	Nyquist *et al.* (1973)
15386		3.85±.08				Carlson and Lugmair (1979)
15434			3.91±.04	0.70052±10	4.27	Nyquist *et al.* (1973)
Apollo 17						
KREEPy basalt						
72275			4.01±.04	0.69935±14	4.15	Compston *et al.* (1975)
			3.96±.04	0.69948±6	4.60	Papanastassiou & Wasserburg (1971)
				(0.69971)	4.72	Nyquist *et al.* (1972)
14053	3.92±.08					Husain *et al.* (1972)
	3.93±.06					Stettler *et al.* (1973)
	3.94±.04					Stettler *et al.* (1973)
	3.94±.05					Turner *et al.* (1971)
14072			3.99±.09	0.69917±7	4.48	Compston *et al.* (1972)
	4.04±.05					York *et al.* (1972)
			4.05±.08	0.69919±13		Compston *et al.* (1972)
			4.08±.10	0.69928±18		Compston *et al.* (1971,72)
			4.01±.12	0.69957±13	4.53	Mark *et al.* (1973)
			3.99±.14	0.69928±12		Mark *et al.* (1975)
14321: hi-Al clasts			3.95±.04	0.69942±4	4.24	Papanastassiou & Wasserburg (1971)
				(0.69930)		Mark *et al.* (1974)
	3.91					York *et al.* (1972)
	3.99					York *et al.* (1972)
	3.94					York *et al.* (1972)
	3.91					Turner *et al.* (1971)

*** Plateau age except values in parentheses.

** $\lambda = 1.39 \times 10^{-11}$ yr.

* Corrected where necessary to conform with CalTech data. Values in parentheses indicate extrapolation from whole-rock data to age determined by other work.

number of these mare basalts, not all of which are the same but some of which may be from a single flow or related flows. 14053, 14072, and 14321 clasts have siderophile and internal Rb-Sr isochron data, whereas all the other clasts lack these data. Nonetheless, all can be presumed to be pre-Imbrium whether the original interpretation of the breccias as direct Imbrium ejecta (e.g. Sutton *et al.*, 1972) or the more recent interpretation that the Fra Mauro is dominated by locally derived, pre-Imbrium material (Head and Hawke 1975 and others) is correct.

14053 (Fig. 1d) is an ophitic basalt consisting mainly of clinopyroxene (50%), plagioclase (40%) and ilmenite (2%) with some olivine and cristobalite and several minor phases and interstitial glass. Gancarz *et al.* 1971, LSPET 1971, Kushiro *et al.* 1972, Bence and Papike (1972), Roedder and Weiblen (1972), Haggerty (1972), and El Goresy *et al.* (1972) have extensively analyzed and described the phases. Some of the olivine forms cores to pyroxene crystals. Chemically the sample is mare-like, with the distinctive high FeO, Sc, Cr, and V of mare samples (Table 2). The low siderophile contents (Morgan *et al.*, 1972) are clearly indicative of a volcanic origin; this is supported by the low Ni content (most less than 2%) of the Fe-metal grains (El Goresy *et al.*, 1972). Age determination by Ar-^{40} ^{39}Ar (Husain *et al.*, 1972; Stettler *et al.*, 1973; Turner *et al.*, 1971) and Rb-Sr (Papanstassiou and Wasserburg, 1971) defines crystallization close to 3.94 b.y. ago.

14072 (Fig. 1e) is a subophitic to ophitic basalt, rather like 14053 but for its olivine phenocrysts. It is slightly more fine-grained than 14053 and differs in that much of its pyroxene is twinned. It consists of clinopyroxene (50%), plagioclase (38%), olivine (2.5%), opaques (7.7%) with interstitial minerals, including cristobalite, and glass (Longhi *et al.*, 1972). The opaque phases have been analyzed and described by Longhi *et al.* (1972) and El Goresy *et al.* (1972). The sample is less fractionated than 14053 (Table 1) but is clearly a mare-like basalt. The siderophile content is low (Hughes *et al.*, 1973) and the metal has low Ni as in 14053 (El Goresy *et al.*, 1972). While it is possible that 14053 is related to 14072 by olivine fractionation, some features appear to be inconsistent with this conclusion; in particular 14072 may be slightly older and its initial $^{87}Sr/^{86}Sr$ is significantly lower (Table 3, Fig. 2). High-pressure experimental studies by Walker *et al.* (1972) show a multiply saturated liquidus of olivine plus aluminous low-Ca pyroxene at about 10 kbar (~200 kms depth); however 14072 may have accumulated some olivine (Longhi *et al.*, 1972; Walker *et al.*, 1972), thus may not represent a liquid.

14321 hi-Al clasts Numerous clasts of basalt have been observed in 14321 (see Meyer and King, 1979) and range from glassy and vitrophyric to medium-grained ophitic (Fig. 1f) (Grieve *et al.*, 1975). The latter are the most common. Olivine phenocryst contents vary from conspicuous to absent. Opaque minerals are less abundant, and plagioclase tends to be more acicular than in 14053 or 14072; otherwise the mineralogies and textures of all are roughly similar. Analyses of pyroxenes by Grieve *et al.*, (1975) and Ryder (unpublished) for 14321, by Bence and Papike (1972) for 14053, and some limited data by Longhi *et al.* (1972) for 14072 show similar major element zoning trends. However, the 14321 pyroxenes evolve to about 2% TiO_2 whereas 14053/14072 pyroxenes have fairly constant TiO_2 values of about 1%. The 14321 hi-Al mare clasts differ from 14053 and 14072 in including metal grains with up to 17% Ni in early olivine or pyroxene; Co contents are frequently higher than the meteoritic range (Grieve *et al.*, 1975). Rb-Sr determinations give ages around 4.0 b.y., contrasting with Ar-Ar ages which are generally slightly younger (Table 3). Mark *et al.* (1975) suggest that real differences between basalt clasts, manifested by initial Sr isotopic variations, exist.

14063 contains several fragments of hi-Al, intermediate TiO_2, mare basalts, analyzed and described by Ridley (1975). No siderophile or geochronological data exist. The clasts contain lathy to acicular plagioclase, intergranular pyroxene, ilmenite, and rare mesostasis glass. The pyroxenes show limited zoning and are TiO_2 (3–4%) and Al_2O_3 (3–5%) rich. Bulk Na_2O is high and the Cr_2O_3 and FeO are low as compared to normal mare basalts. The fragments are clearly distinct from other basalts except the clasts in 14082. Steele and Smith (1976) suggested that a second variety of mare basalt also contributed to the clastic material in 14063.

14082 contains rare very small clasts of basalt, similar to those in 14063, which were described and analyzed by Ryder and Bower (1976). No siderophile or geochronological data exist. The largest is an intermediate TiO_2, high Na_2O, low FeO basalt and chemically differs from the 14063 clast only in having lower Al_2O_3 and slightly higher MgO. Minor cristobalite, rather than mesostasis glass, is present.

14312 contains several basalt clasts, some of which were analyzed and described by Ryder and Bower (1976, 1977). No siderophile or geochronological data exist. The largest, 5 mm, clast is ophitic-subophitic containing olivine phenocrysts, zoned clinopyroxenes, plagioclase laths, ilmenite and other phases. Fe-metals contain less than 1% Ni and ~0.5% Co. The clast is similar to the 14321 basalts except for the limited Ni in the metal, a more limited zoning of pyroxenes, and differences in Cr-spinel compositions. Other basalt clasts in 14321 show somewhat different textures and mineral compositions, thus are possibly from distinct basalts.

14318 basalt clasts were noted by Kurat *et al.* (1974), Taylor (1976), and Ryder and Bower (1977). The clast analyzed by Kurat *et al.* (1974) is distinctly different from those analyzed by Ryder and Bower (1977) and a variety of basalts is probably present. 14318 is probably a lithified soil; the interpretation of these basalts as pre-Imbrian is less secure than for other Apollo 14 clasts because geochronological information is not available.

14321 Mg-vitrophyre clasts were analyzed and described by Allen *et al.* (1979). No siderophile or geochronological data is yet available. The vitrophyres contain 15–43% zoned olivine microphenocrysts in a brown glass matrix; olivine accumulation seems to have occurred. Some contain a few per cent of plagioclase laths and traces of Fe-metal are present. The chemistry is unusual, and seems to have highland affinities e.g. low CaO/Al_2O_3, low FeO, high MgO and high K_2O. They are possibly impact melts (Allen *et al.*, 1979) but at present only circumstantial evidence exists for this hypothesis.

Apollo 14 soil particles. Chemical analyses by Warren *et al.* (1978) and Ma *et al.* (1980) include aluminous mare basalts from Apollo 14 coarse fines. The particle analyzed by Warren *et al.* (1978) is similar to the 14321 fragments (Table 2). The four texturally diverse particles analyzed by Ma *et al.* (1980) are chemically diverse, including particles with much lower REE abundances. The geochemistry indicates that the four particles had origins distinct from each other *and* from other aluminous mare basalts. No ages for these basalts have been established.

Apollo 17 boulder clasts

Boulder 1 at Station 2 contains several small mare-like clasts which are olivine-normative (Stoeser *et al.*, 1974; Ryder *et al.*, 1975; Taylor *et al.*, 1978). Their petrography indicates that they are volcanic, but no siderophile data exists. They contain calcic plagioclase, olivine, clinopyroxene, and chromite, and are extremely low in titanium. They are constrained by ^{40}Ar-^{39}Ar age determinations on the matrix in which they are enclosed to be older than ~3.99 or 4.0 b.y. (Leich *et al.*, 1975) and are probably pre-Serenitatis basin.

The Station 6 boulders contain small ophitic/subophitic basalt clasts reported by Warner *et al.* (1976). They have 50% clinopyroxene and up to 6% ilmenite, indicating a mare affinity. No siderophile or direct geochronological data exist, but they are constrained by ^{40}Ar-^{39}Ar age determinations to be older than about 3.95 b.y. (Cadogan and Turner 1976). If these boulders indeed represent Serenitatis impact melt, then the basalts are pre-Serenitatis basin.

73255. Three samples of basalt clasts were analyzed by Blanchard and Budahn (1979). They appear to be pristine, hi-alumina mare basalts according to their FeO, Cr_2O_3, Sc, rare-earth, and Co contents. They appear to be generally similar in composition but differ in detail. The host breccia is ~3.87 b.y. old (Staudacher *et al.*, 1979.)

SIGNIFICANCE OF BASALT CLASTS

Despite the limited sampling, it is clear that there is a wide variety of basalt types with ages around 3.9–4.0, apparently just post-dating and pre-dating the Imbrium and Serenitatis impact events. These samples include KREEP basalts, mare basalts, and at least one type with some of the characteristics of both. The Apollo

14 site alone shows several different, roughly contemporaneous basalt types (analogous to the Apollo 12 mare collection in which several contemporaneous suites exist). All of these ancient basalts appear to be hi-Al types, whether KREEP or mare, whereas younger basalts include both hi-Al and low-Al types. KREEP basalts differ from the mare basalts not only in the elevated rare-earths but also in their higher Sr and lower Sc.

An unusual chemical feature of the ancient hi-Al basalts with mare affinities so far analyzed (14321, several analyses; 72275) is their high Ge content, greater than 500 ppb. This contrasts with Apollo 15 KREEP, less than 100 ppb Ge and common mare basalts, less then 40 ppb. The young hi-Al basalts at the Luna 16 site may also have high Ge, since the soil is enriched in Ge beyond that accountable by meteoritic contamination (Laul *et al.*, 1972; Vinogradov, 1971). However, 12038, a hi-alumina basalt, has very low Ge (Baedecker *et al.*, 1971). This enrichment of Ge in at least the ancient high-alumina mare basalts is so far unexplained but presumably has genetic significance.

The ancient (~3.9–4.0 b.y.) hi-Al mare basalts are distinct from other mare basalts on a Sr-evolution diagram (Fig. 2): they have more evolved sources, compatible with their higher Rb contents. *Younger* hi-Al mare basalts (Luna 16, Luna 24, 12038) have much less evolved Sr isotopic ratios than more ancient hi-Al group and must have distinct source types. The Apollo 17 KREEPy basalt is distinctly separated from the Apollo 15 KREEP basalts and falls with the hi-Al mare basalts in this figure. It must have a source distinct from that of the Apollo 15 KREEP basalts regardless of possible difference in partial melting, fractional crystallization, or assimilation (Ryder *et al.*, 1977; Irving, 1977). Plagioclase is probably required in the source of both KREEP types.

Ridley (1975) discussed the origin of hi-Al mare basalts and concluded that plagioclase is *not* required in the source materials, which may instead be fairly similar to "normal" mare basalt sources. Nonetheless the distinctly more evolved source region (for Sr isotopes) and the higher Mg/Fe of most of the ancient hi-Al mare basalts demands some significant differences. The question of plagioclase in the source is unresolved and will probably remain so until high-pressure experiments on *liquid* compositions (not yet established) and further chemical modelling provide constraints. Chemical and Sr-isotopic characteristics suggested to Nyquist *et al.* (1978) that plagioclase was present in the source of the Luna 24 aluminous basalts. (Indeed it may be present in some non-aluminous mare basalt sources; Papanastassiou *et al.*, 1977).

This ancient suite of basalts adds to the suite of better-sampled younger mare basalts and emphasizes the exceptional diversity of lunar volcanism. The relationships between individual types are extremely complex and are compounded by the fact that several types have many characters in common (e.g., young and old hi-Al basalts) but differ significantly in other characters. Source materials for this volcanism not only have a range of essential mineralogies, but a range of trace minerals and elements which are by no means dependent on the major minerals.

INDIRECT EVIDENCE: CHEMICAL AND MINERALOGICAL

Volcanic rock samples older than the final bombardment, i.e. older than 4.0 b.y., have not been identified. Since this lack may merely be a function of the intense bombardment itself (Ryder and Taylor, 1976) evidence for the existence and nature of older volcanism must come from less direct methods. Investigations of the chemistry and mineralogy of ancient breccias must necessarily be less definitive than studies of volcanic samples but can at least place some constraints on the types and the timing of more ancient volcanism.

Mineralogical studies

The compositions of mineral fragments can be used to indicate the presence or absence of volcanic lithologies in highland breccias. Plutonic rocks tend to have equilibrated pyroxenes, one hi-Ca and one low-Ca, a function of their slow cooling rates. Pyroxenes in most known impact melts display similar features; most also display very little Fe-enrichment. Conversely volcanics display non-equilibrated compositions and significant zoning. Such evidence has been used to infer the presence of mare debris in Apollo 16 light matrix breccias (Taylor *et al.*, 1973; Ryder and Taylor, 1976). More such work, with good data on trace elements in plagioclase as well as mafics, could provide better evidence for or against ancient volcanic products in dated breccias.

Chemical studies

KREEP has a distinct chemical signature and is present in most highlands breccias. This has been frequently taken to indicate that KREEP volcanism older than 4.0 b.y. occurred even though actual volcanic rocks have not been identified. The argument has been augmented by the model ages of ~4.3 b.y. for KREEP samples. However, neither model ages nor KREEP chemistry in themselves denote volcanism, because KREEP can reach the surface by impact excavation. It seems to have done so during the excavation of basins (Ryder and Wood, 1977; Charette *et al.*, 1977; and others). The best evidence for KREEP *volcanism* prior to 4.0 b.y. comes from an analysis of the Apollo 14 rocks and site by Hawke and Head (1977) which suggests that early KREEP volcanism occurred in topographic lows. This KREEP now forms the matrix of breccias. Because it is never found as pristine volcanics it must be presumed to be older, more degraded, than the hi-Al mare clasts which are up to 4.0 b.y. old. Thus KREEP volcanism both preceded (Apollo 14) and post-dated (Apollo 15) Imbrium.

Unlike KREEP, mare rocks do not have a distinct chemical signature. Several chemical mixing models of highlands breccias have postulated a mare component (e.g. Boynton *et al.*, 1976), but its requirement is dependent on the compositions chosen for other components and has not been shown to be necessary (Ryder 1979). However, Wänke *et al.* (1978) have noted that some rocks with low La/Mg ratios have higher Ti/Mg than can be accounted for with known highland rocks. Although they postulate that these compositions result from admixture of a pristine highland component, our own analysis shows that this high Ti/Mg component also has a high Fe/Mg and Ti/Sm and may well be low-Ti mare basalt. The presence or absence of a mare component in highlands breccias is still unresolved.

Feldspathic impactites

Warner *et al.* (1977) brought attention to a suite of highlands rocks, feldspathic impactites which are metamorphosed breccias virtually devoid of KREEP components. KREEP is currently pervasive at the lunar surface and it is difficult to envisage the formation of a lunar breccia even at ~4.0 b.y. which did not have a KREEP admixture. However, at the time of the construction of the feldspathic impactites KREEP was absent from the lunar surface or at least of far lesser extent than it was 4.0 b.y. ago. The formation age of these breccias appears to be about 4.2 b.y. (Warner *et al.,* 1977). This limits KREEP volcanism on a planetwide scale to be younger than ~4.2 b.y. (it also limits basin excavation of KREEP to younger than 4.2 b.y.).

EVIDENCE FOR REGIONAL DEPOSITS OF ANCIENT LUNAR VOLCANIC ROCKS

Depictions of regional lunar geologic units have been based primarily on photogeological techniques. These efforts have resulted in a general picture of mare volcanic episodes that postdate the youngest lunar basins and the cessation of final heavy bombardment. It was believed in pre-Apollo days that the Cayley Formation, a pre-mare light plains deposit in terra regions, represented an early episode of highland volcanism (Wilhelms, 1970). Results of the Apollo 16 missions demonstrated the probable impact origin of most of the Cayley Formation and this interpretation was extended to all lunar light plains (see Head, 1976).

It has become evident in the years since the Apollo missions that an impact origin for *all* lunar light plains is not likely. By utilizing remote sensing geochemical data obtained from lunar orbit (e.g. Adler and Trombka, 1977) and earth-based spectral information (Charette *et al.,* 1977) in combination with photogeology, some lunar light plains deposits can be related in chemistry and emplacement mechanism to returned lunar samples of known volcanic origin. Results from these studies indicate that regional deposits from ancient volcanism have been preserved on the moon for periods longer than 4 billion years.

The first lunar light plains unit shown to be of volcanic origin was the Apennine Bench Formation (Fig. 3). These light plains post-date the Imbrium basin but

Fig. 3. The Apennine Bench Formation (AB) embays terra associated with the Imbrium basin (arrows). The fractured deposit (M) is probably basin impact melt, distinctly different in surface morphology. (AS 17-2110).

pre-date all other geologic units in the region including the crater Archimedes and mare basalts (Hackman 1966). The chemical composition at these plains is that of medium-K Fra Mauro (KREEP) basalt (Spudis, 1978) and is essentially identical to the composition of volcanic KREEP basalt fragments found in the nearby Apollo 15 landing site soils and breccias. These plains embay highland terra in a morphological relationship similar to that seen at mare-terra contacts and do not mantle pre-existing topography in the manner of impact-melt deposits. On the basis of these data, the plains have been interpreted as pre-mare volcanic KREEP basalt flows (Spudis, 1978; Hawke and Head, 1978).

Since the identification of pre-mare volcanic light plains in the Apennine region, other light plains around the moon have been shown to have similar geologic relationships and compositions. Some of the light plains in the Fra-Mauro region mapped as Cayley Formation may in fact be isolated patches of Apennine Bench Formation. Other light plains, such as near the farside crater Van De Graff, also appear to be composed primarily of KREEP basalt and are possible remnants of pre-mare volcanic episodes (Hawke *et al.*, 1979).

Large expanses of light plains on the Moon occupy geologically old multi-ringed basins, particularly abundant on the eastern limb and farside. Many of these plains display small (5–15 km) impact craters that have dark haloes, *i.e.* low albedo ejecta (Fig. 4). Mafic anomalies have been observed where these dark halo craters have been overflown by orbital geochemical instruments. Schultz and Spudis (1979) proposed that plains containing numerous dark halo craters of this type are ancient basaltic surfaces that have been mantled by Imbrium age highland debris. Subsequent impact cratering has penetrated these debris blankets to expose these ancient basalts as crater ejecta. The age of the overlying highland debris is Nectarian and early Imbrium indicating a *minimum* age of these basalt deposits of ~3.9 b.y.; it is difficult to constrain the maximum age of the mare

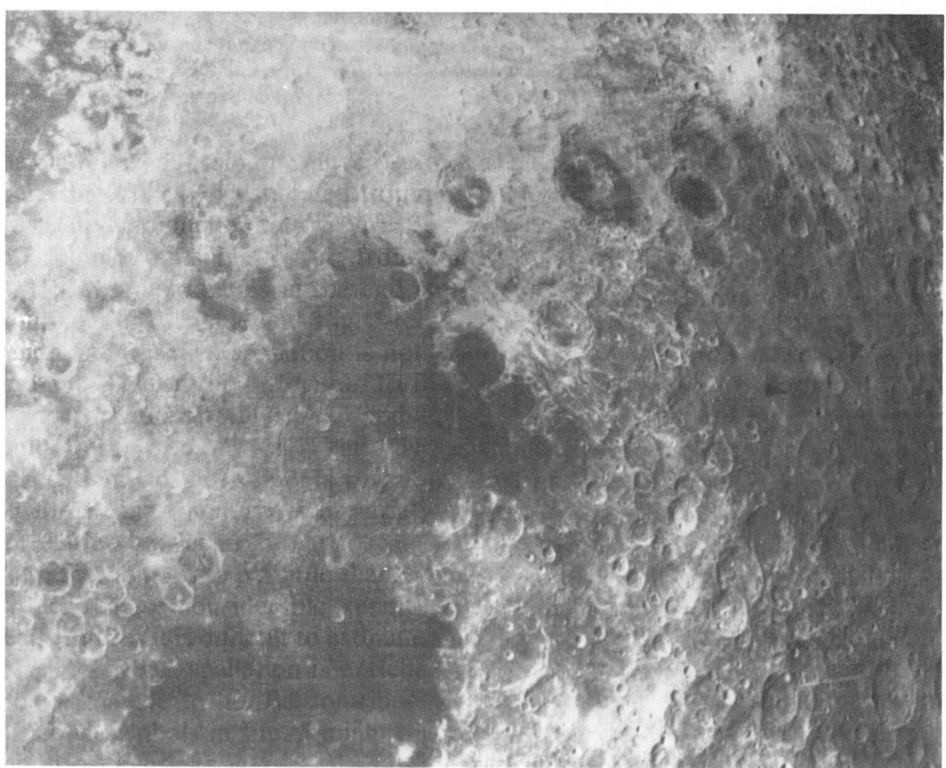

Fig. 4. Region near Mare Marginis (M) and Mare Smythii (S). Light plains on right display dark halo craters (arrows) indicating presence of buried ancient lava flows. (AS 17-152-23327).

deposits. The great extent of these light plains indicate that estimates of volumes of mare volcanics on the Moon may have to be revised to take into account these episodes of ancient mare volcanism (Schultz and Spudis, in preparation).

Numerous highland regions, particularly on the lunar western nearside, display an unusually strong red spectral reflectance (Whitaker, 1972). These red spots have been interpreted as pre-mare KREEP basalts (Malin, 1974) and studies of the morphology of some of these structures suggest extensive igneous dome-building processes (Head and McCord, 1978). The cause of the unusual reflectance characteristics is not known and it is not possible to relate these units compositionally to known lunar rock types. However, the strong red color has been shown to be an indication of low Fe content (Head and McCord, 1978) and may be related to a unique form of lunar rhyolitic volcanism, a composition favoring the formation of viscous extrusive landforms. The resolution of this question will depend upon higher quality remote sensing data and/or possible sample return from these unique highland units.

Photogeological/remote sensory data indicate that volcanism has been an important lunar surface process as early as pre-Imbrium times. KREEP and mare volcanic episodes have apparently developed in time and there is evidence for types of volcanic products that have not been sampled.

SUMMARY AND CONCLUSIONS

Volcanism at the time of the terminal bombardment was both varied and widespread according to sample and photogeological/remote sensing considerations. Volcanism at earlier periods is much more difficult to evaluate. It is possible that it did not even occur, but our understanding of planetary evolution suggests that it did. Ancient *plutonic* rocks (4.2–4.5 b.y. old) exist on the Moon and may well have had corresponding volcanic rocks. Thermal models of the Moon suggest a monotonic cooling most likely producing more volcanism in earlier than later epochs. Many achondritic meteorites have ages demonstrating that planets smaller than the Moon were capable of generating volcanic rocks as early as 4.5 b.y. ago; intuitively, one would expect at least the same for the Moon. Careful evaluation of the chemistry and mineralogy of breccias, and a more detailed search for basalt clasts, may improve our understanding of earlier volcanism.

Considerations of extremely ancient volcanism so far unsampled require that two concepts be addressed. *First,* the term "volcanism" generally implies the extrusion of silicate liquid originally created by internal partial melting. This is difficult to apply very precisely to early lunar history if our understanding that the crust formed by crystallization of a magma ocean produced by accretion is essentially correct. Such a magma ocean may have released liquids as flows but these would not be volcanic in a strict sense. More importantly, one model for the origin of KREEP is that it is a residual from the magma ocean episode. If so, it is not truly volcanic, since it crystallized from an impact melt, a total melt not a partial melt. (Neither then are the anorthosites, norites etc., produced in this episode strictly igneous). However, any residues would probably have re-equilibrated internally to some extent prior to extrusion. Clearly the definition of volcanism must be flexible in the context of earliest lunar history, even if it is in principal straightforward enough to distinguish volcanics from non-fractionating melts produced in single impacts later in lunar history (Simonds *et al.*, 1976). Some basin-sized impacts may have penetrated pre-existing magma chambers and excavated lava (Schultz and Spudis, 1979, Schultz and Mendenhall, 1979); although the magma is basin ejecta, it is not impact melt. The line distinguishing volcanic from impact melt may sometimes be thin, and the overlapping of volcanic and basin-forming epochs complicates the problem tremendously.

Second, ancient volcanism may have produced extremely varied rock types, and the term "non-mare" as a group name is unfounded if mare basalts are defined chemically. While we tend to consider volcanic rocks as either KREEP or mare, some volcanic rocks early in lunar history might not have been easily grouped with either groups e.g., there might have been rocks with KREEP major elements but much lower abundances or different patterns of incompatibles, rocks somewhat like rhyolites, or volcanics with high Mg/Fe ratios (however we are not stating that these particular volcanics did exist). Our understanding of the variety of ancient volcanism is far from complete.

Acknowledgments—Rand B. Schaal reviewed the manuscript. This work was supported by contract NAS 9-15425 between the NASA Johnson Space Center and Northrop Services, Inc. for the operation of the Lunar Curatorial Laboratory. The attendance of P.D.S. at the Highlands Conference was supported by Sigma Xi and NASA Grant NSG-7429.

REFERENCES

Adler I. and Trombka J. I. (1977) Orbital chemistry—lunar surface analysis from the x-ray and gamma-ray remote sensing experiments. *Phys. Chem. Earth* **10,** 17–43.

Allen F. M., Bence A. E., and Grove T. L. (1979) Olivine vitrophyres in Apollo 14 breccia 14321: Samples of the high-Mg component of the lunar highlands. *Proc. Lunar Planet. Sci. Conf. 10th,* p. 695–712.

Anders E. (1978) Procrustean science: Indigenous siderophiles in the lunar highlands, according to Delano and Ringwood. *Proc. Lunar Planet. Sci. Conf. 9th,* p. 161–184.

Baedecker P. A., Schandy R., Elzie J. L., Kimberlin J., and Wasson J. T. (1971) Trace element studies of rocks and soils from Oceanus Procellarum and Mare Tranquillitatis. *Proc. Lunar Sci. Conf. 2nd,* p. 1037–1061.

Baedecker P. A., Chou C.-L., and Wasson J. T. (1972) The extralunar component in lunar soils and breccias. *Proc. Lunar Sci. Conf. 3rd,* p. 1343–1359.

Basu A. and Bower J. F. (1976) Petrography of KREEP basalt fragments from Apollo 15 soils. *Proc. Lunar Sci. Conf. 7th,* p. 659–678.

Bence A. E. and Papike J. J. (1972) Pyroxenes as recorders of lunar basalt petrogenesis: Chemical trends due to crystal-liquid interaction. *Proc. Lunar Sci. Conf. 3rd,* p. 431–469.

Blanchard D. P. and Budhan J. R. (1979) Remnants from the ancient lunar crust: Clasts from consortium breccia 73255. *Proc. Lunar Planet. Sci. Conf. 10th,* p. 803–816.

Blanchard D. P., Haskin L. A., Jacobs J. W., Brannon J. C., and Korotev R. L. (1975) Major and trace element chemistry of Boulder 1 at Station 2, Apollo 17. *The Moon* **14,** 359–371.

Boynton W. V., Chou C.-L., Robinson K. L., Warren P. H., and Wasson J. T. (1976) Lithophiles, siderophiles and volatiles in Apollo 16 rocks and soils. *Proc. Lunar Sci. Conf. 7th,* p. 727–742.

Cadogan P. H. and Turner G. (1976) The chronology of the Apollo 17 Station 6 boulder. *Proc. Lunar Sci. Conf. 7th,* p. 2267–2285.

Carlson R. W. and Lugmair G. W. (1979) Sm-Nd constraints on early lunar differentiation and the evolution of KREEP. *Earth Planet. Sci. Lett.* **45,** 123–132.

Charette M. P., Taylor S. R., Adams J. B., and McCord T. B. (1977) The detection of soils of Fra Mauro basalt and anorthositic gabbro in the lunar highlands by remote spectral reflectance techniques. *Proc. Lunar Sci. Conf. 8th,* p. 1049–1061.

Church S. E., Bansal B. M., and Wiesmann H. (1972) The distribution of K, Ti, Zr, U, and Hf in Apollo 14 and 15 materials. In *The Apollo 15 Lunar Samples* (J. W. Chamberlain and C. Watkins, eds.) p. 210–213. Lunar Science Institute, Houston.

Compston W., Vernon M. J., Berry H., and Rudowski R. (1971) The age of the Fra Mauro formation: A radiometric older limit. *Earth Planet. Sci. Lett.* **12,** 55–58.

Compston W., Vernon M. J., Berry H., Rudowski R., Gray C. M., Ware N., Chappell B. W., and Kaye M. (1972) Apollo 14 mineral ages and the thermal history of the Fra Mauro formation. *Proc. Lunar Sci. Conf. 3rd,* p. 1487–1501.

Compston W., Foster J. J., and Gray C. M. (1975) Rb-Sr ages of clasts from within Boulder 1, Station 2, Apollo 17. *The Moon* **14,** 445–462.

Dowty E., Keil K., Prinz M., Gros J., and Takahashi H. (1976) Meteorite-free Apollo 15 crystalline KREEP. *Proc. Lunar Sci. Conf. 7th,* p. 1833–1844.

Drake M. J., Stoeser J. W., and Goles G. G. (1973) A unified approach to a fragmental problem: Petrological and geochemical studies of lithic fragments from Apollo 15 soils. *Earth Planet. Sci. Lett.* **20,** 425–439.

Duncan A. R., McKay S. M., Stoeser J. W., Lindstrom M. M., Lindstrom D. J., Fruchter J. S., and Goles G. G. (1975) Lunar polymict breccia 14321: A compositional study of its principal components. *Geochim. Cosmochim. Acta* **39,** 247–260.

Dymek R. F., Albee A. L., and Chodos A. A. (1974) Glass-coated soil breccia 15205: Selenologic history and petrologic constraints on the nature of its source region. *Proc. Lunar Sci. Conf. 5th,* p. 235–260.

Ehmann W. D., Gillum D. E., and Morgan J. W. (1972) Oxygen and bulk element composition of Apollo 14 and other lunar rocks and soils. *Proc. Lunar Sci. Conf. 3rd,* p. 1149–1160.

El Goresy A., Taylor L. A., and Ramdohr P. (1972) Fra Mauro crystalline rocks: Mineralogy, geochemistry, and subsolidus reduction of the opaque minerals. *Proc. Lunar Sci. Conf. 3rd,* p. 333–349.

Gancarz A. J., Albee A. L., and Chodos A. (1971) Petrologic and mineralogic investigation of some crystalline rocks returned by the Apollo 14 mission. *Earth Planet. Sci. Lett.* **12,** 1–18.

Greive R. A., McKay G. A., Smith H. D., and Weill D. F. (1975) Lunar polymict breccia 14321: a petrographic study. *Geochim. Cosmochim. Acta* **39,** 229–245.

Gros J., Takahashi H., Hertogen J., Morgan J. W., and Anders E. (1976) Composition of the projectiles that bombarded the lunar highlands. *Proc. Lunar Sci. Conf. 7th,* p. 2403–2425.

Hackman R. J. (1966) Geologic map of the Montes Apenninus region of the Moon. *U. S. Geol. Survey,* Misc. Investigation Series Map I–463.

Haggerty S. E. (1972) Apollo 14: subsolidus reduction and compositional variations of spinels. *Proc. Lunar Sci. Conf. 3rd,* p. 305–332.

Hawke B. R. and Head J. W. (1977) Pre-Imbrian history of the Fra Mauro region and Apollo 14 sample provinces. *Proc. Lunar Sci. Conf. 8th,* p. 2741–2761.

Hawke B. R. and Head J. W. (1978) Lunar KREEP volcanism: geologic evidence for history and mode of emplacement. *Proc. Lunar Planet. Sci. Conf. 9th,* p. 3285–3309.

Hawke B. R., Spudis P. D., and Clark P. E. (1979) Geochemical anomalies on the Lunar eastern limb and farside (abstract). *In Papers Presented to the Conference on the Lunar Highlands Crust,* p. 56–58. Lunar and Planetary Institute, Houston.

Head J. W. (1976) Lunar volcanism in space and time. *Rev. Geophys. Space Phys.* **14,** 265–300.

Head J. W. and Hawke B. R. (1975) Geology of the Apollo 14 region (Fra Mauro): Stratigraphic history and sample provenance. *Proc. Lunar Sci. Conf. 6th,* p. 2483–2501.

Head J. W. and McCord T. B. (1978) Imbrian-age highland volcanism on the Moon: the Gruitheisen and Marian domes. *Science* **199,** 1433–1436.

Helmke P. A., Haskin L. A., Korotev R. L., and Ziege K. E. (1972) Rare earths and other trace elements on Apollo 14 samples. *Proc. Lunar Sci. Conf. 3rd,* p. 1275–1292.

Hubbard N. J., Gast P. W., Rhodes J. M., Bansal B. M., Wiesmann H., and Church S. E. (1972). Non-mare basalts: Part II. *Proc. Lunar Sci. Conf. 3rd,* p. 1161–1179.

Hubbard N. J., Rhodes J. M., Gast P. W., Bansal B. M., Shih C.-Y., Wiesmann H., and Nyquist L. E. (1973) Lunar rock types: the role of plagioclase in non-mare and highland rock types. *Proc. Lunar Sci. Conf. 4th,* p. 1297–1312.

Hubbard N. J., Rhodes M. J., Wiesmann H., Shih C.-Y., and Bansal B. M. (1974) The chemical definition and interpretation of rock types returned from the non-mare regions of the Moon. *Proc. Lunar Sci. Conf. 5th,* p. 1227–1246.

Hughes T. C., Keays R. R., and Lovering J. F. (1973) Siderophile and volatile trace elements in Apollo 14,15 and 16 rocks and fines: Evidence for extralunar component and Tl-, Au-, and Ag-enriched rocks in the ancient lunar crust (abstract). In *Lunar Science IV,* p. 400–402. The Lunar Science Institute, Houston.

Husain L., Schaeffer O. A., Kunkhouser J., and Sutter J. (1972) The ages of lunar material from Fra Mauro, Hadley Rille and Spur Crater. *Proc. Lunar Sci. Conf. 3rd,* p. 1557–1567.

Irving A. J. (1975) Chemical, mineral and textural systematics of non-mare melt rocks: implications for lunar impact and volcanic process. *Proc. Lunar Sci. Conf. 6th,* p. 363–394.

Irving A. J. (1977) Chemical variation and fractionation of KREEP basalt magmas. *Proc. Lunar Sci. Conf. 8th,* p. 2433–2448.

Janghorbani M., Miller M. D., Ma M., Lindgren L. C., and Ehmann W. D. (1973) Oxygen and other elemental abundance data for Apollo 14,15,16, and 17 samples. *Proc. Lunar Sci. Conf. 4th,* p. 1115–1126.

Keith J. E., Clark R. S., and Richardson K. A. (1972) Gamma-ray measurements of Apollo 12,14, and 15 lunar samples. *Proc. Lunar Sci. Conf. 3rd,* p. 1671–1680.

Kurat G., Keil K., and Prinz M. (1974) Rock 14318: a polymict lunar breccia with chondritic texture. *Geochim. Cosmochim. Acta* **38,** 1133–1146.

Kushiro I., Ikeda Y., and Nakamura Y. (1972) Petrology of Apollo 14 high-alumina basalt. Proc. *Lunar Sci. Conf. 3rd,* p. 115–129.

Leich D. A., Kahl S. B., Kirschbaum A. R., Niemyer S., and Phinney D. (1975) Rare gas constraints on the history of Boulder 1, Station 2, Apollo 17. *The Moon* **14,** 407–444.

Lindstrom M. M., Duncan A. R., Fruchter J. S., McKay S. M., Stoeser, J. W., Goles G. G., and Lindstrom D. J. (1972) Compositional characteristics of some Apollo 14 clastic materials. *Proc. Lunar Sci. Conf. 3rd,* p. 1201–1214.

Longhi J., Walker D., and Hays J. F. (1972) Petrography and crystallization history of basalts 14310 and 14072. *Proc. Lunar Sci. Conf. 3rd,* p. 131–139.

LSPET (Lunar Sample Preliminary Examination Team) (1971) Preliminary examination of lunar samples from Apollo 14. *Science* **173,** 681–693.

Laul J. C., Wakita H., Showalter D. L., Boynton W. V., and Schmitt R. A. (1972) Bulk, rare earth, and other trace elements in Apollo 14 and 15 and Luna 16 samples. *Proc. Lunar Sci. Conf. 3rd,* p. 1181–1200.

Ma M.-S., Schmitt R. A., Warner R. D., Taylor G. J., Barker S., and Keil K. (1980) Aluminous mare basalts and basaltic-textured KREEPy rocks from Apollo 14 coarse fines (abstract). In *Lunar and Planetary Science XI,* p. 652–654. Lunar and Planetary Science Institute, Houston.

Malin M. C. (1974) Lunar red spots: possible pre-mare materials. *Earth Planet Sci. Lett.* **21,** 331–341.

Mark R. K., Cliff R. A., Lee-Hu C., and Wetherill G. W. (1973) Rb-Sr studies of lunar breccias and soils. *Proc. Lunar Sci. Conf. 4th,* p. 1785–1795.

Mark R. K., Lee-Hu C., and Wetherill G. W. (1974) Equilibration and ages: Rb-Sr studies of breccias 14321 and 15265. *Proc. Lunar Sci. Conf. 5th,* p. 1477–1485.

Mark R. K., Lee-Hu C., and Wetherill G. W. (1975) More on Rb-Sr in lunar breccia 14321. *Proc. Lunar Sci. Conf. 6th,* p. 1501–1507.

Metzger A. E., Haines E. L., Etchegaray-Ramirez M. I., and Hawke B. R. (1979) Thorium concentrations in the Lunar Surface III. Deconvolution of the Apenninus region. *Proc. Lunar Planet. Sci. Conf. 10th,* p. 1701–1718.

Meyer C. Jr. and King C. D. (1979) *Breccia Guidebook No. 1 14321,* JSC 14753, NASA Johnson Space Center, Houston. 56 pp.

Morgan J. W., Ganapathy R., and Krähenbühl U. (1975a) Meteoritic trace elements in lunar rock 14321, 184. *Geochim. Cosmochim. Acta* **39,** 261–264.

Morgan J. W., Higuchi H., and Anders E. (1975b) Meteoritic material in a boulder from the Apollo 17 site: implications for its origin. *The Moon* **14,** 373–383.

Morgan J. W., Krähenbühl U., Ganapathy R., and Anders E. (1972) Trace elements in Apollo 15 samples: Implications for meteorite influx and volatile depletion on the moon. *Proc. Lunar Sci. Conf. 3rd,* p. 1361–1376.

Murali A. V., Ma M.-S., Laul J. C., and Schmitt R. A. (1977). Chemical composition of breccias, feldspathic basalt and anorthosites from Apollo 15 (15308, 15359, 15382 and 15362), Apollo 16 (60618 and 65785), Apollo 17 (72435, 72538, 72559, 72735, 72738, 78526, and 78527) and Luna 20 (22012 and 22013) (abstract). In *Lunar Science VIII,* p. 700–702. The Lunar Science Institute, Houston.

Nunes P. D. and Tatsumoto M. (1975) U-Th-Pb systematics of selected samples from Apollo 17, Boulder 1, Station 2. *The Moon* **14,** 463–471.

Nyquist L. E., Hubbard N. J., Gast P. W., Bansal B. M., Wiesmann H., and Jahn B. (1973) Rb-Sr systematics for chemically defined Apollo 15 and 16 materials. *Proc. Lunar Sci. Conf. 4th,* p. 1823–1846.

Nyquist L. E., Hubbard N. J., Gast P. W., Church S. E., Bansal B. M., and Wiesmann H. (1972) Rb-Sr systematics for chemically defined Apollo 14 breccias. *Proc. Lunar Sci. Conf. 3rd,* p. 1515–1530.

Nyquist L. E., Wiesmann H., Bansal B., Wooden J., and McKay G. (1978) Chemical and Sr-isotopic characteristics of Luna 24 samples. In *Mare Crisium: The View from Luna 24* (R. B. Merrill and J. J. Papike, eds.), p. 631–655. Pergamon, N.Y.

Papanastassiou D. A., De Paulo D. J., and Wasserburg G. J. (1977) Rb-Sr and Sm-Nd chronology and geneology of mare basalts from the Sea of Tranquility. *Proc. Lunar Sci. Conf. 8th,* p. 1639–1672.

Papanastassiou D. A. and Wasserburg G. J. (1971) Rb-Sr ages of igneous rocks from the Apollo 14 mission and the age of the Fra Mauro Formation. *Earth Planet. Sci. Lett.* **12,** 36–48.

Papanastassiou D. A. and Wasserburg G. J. (1976) Rb-Sr age of troctolite 76535. *Proc. Lunar Sci. Conf. 7th,* p. 2035–2054.

Papike J. J., Hodges F. N., Bence A. E., Cameron M., and Rhodes J. M. (1976) Mare basalts: crystal chemistry, mineralogy, and petrology. *Rev. Geophys. Space Phys.* **14,** 475–540.

Papike J. J. and Vaniman D. T. (1978) Luna 24 ferrobasalts and the mare basalt suite: comparative chemistry, mineralogy, and petrology. In *Mare Crisium: The View From Luna 24* (R. B. Merrill and J. J. Papike, eds.) p. 371–401. Pergamon, N.Y.

Rhodes J. M. and Hubbard N. J. (1973) Chemistry, classification and petrogenesis of Apollo 15 mare basalts. *Proc. Lunar Sci. Conf. 4th,* p. 1127–1148.

Ridley W. I. (1975) On high-alumina mare basalts. *Proc. Lunar Sci. Conf. 6th,* p. 131–145.

Roedder E. and Weiblen P. W. (1972) Petrographic features and petrologic significance of melt inclusions in Apollo 14 and 15 rocks. *Proc. Lunar Sci. Conf. 3rd,* p. 251–279.

Ryder G. (1979) The chemical components of highlands breccias. *Proc. Lunar Planet. Sci. Conf. 10th,* p. 561–581.

Ryder G. and Bower J. F. (1976) Sample 14082 petrology. *In Interdisciplinary Studies by the Imbrium Consortium,* Vol. 1 (J. A. Wood, ed). LSI Contr. 267D, p. 41–50. Center for Astrophysics, Cambridge.

Ryder G. and Bower J. F. (1977) Petrology of 14312. In *Interdisciplinary Studies by the Imbrium Consortium,* Vol. 2 (J. A. Wood, ed.). LSI Contr. No. 268D, p. 13–15. Center for Astrophysics, Cambridge.

Ryder G., Stoeser D. B., Marvin U. B., Bower J. F., and Wood J. A. (1975) Boulder 1, Station 2, Apollo 17: Petrology and petrogenesis. *The Moon* **14,** 327–357.

Ryder G., Stoeser D. B., and Wood J. A. (1977) Apollo 17 KREEPy basalt: A rock type intermediate between mare and KREEP basalts. *Earth Planet. Sci. Lett.* **35,** 1–13.

Ryder G. and Taylor G. J. (1976) Did mare volcanism commence early in lunar history? *Proc. Lunar Sci. Conf. 7th,* p. 1741–1755.

Ryder G. and Wood J. A. (1977) Serenitatis and Imbrium impact melts: implications for large-scale layering in the lunar crust. *Proc. Lunar Sci. Conf. 8th,* p. 655–668.

Schonfeld E., O'Kelley G. D., Eldridge J. S., and Northcutt K. J. (1972) K, U, and Th concentrations in rake sample 15382 by non-destructive gamma-ray spectroscopy. In *The Apollo 15 Lunar Samples* (J. W. Chamberlain and C. Watkins, eds.) p. 253–254. The Lunar Science Institute, Houston.

Schultz P. H. and Mendenhall M. H. (1979) On the formation of basin secondary craters by ejecta complexes (abstract). In *Lunar and Planetary Science X,* p. 1078–1080. Lunar and Planetary Institute, Houston.

Schultz P. H. and Spudis P. D. (1979). Evidence for ancient mare volcanism. *Proc. Lunar Planet. Sci. Conf. 10th,* p. 2899–2918.

Shoemaker E. M. (1972) Cratering history and the early evolution of the moon (abstract). In *Lunar Science III,* p. 696–698. The Lunar Science Institute, Houston.

Simonds C. H., Warner J. L., Phinney W. C., and McGee P. E. (1976) Thermal model for impact breccia lithification: Manicouagan and the moon. *Proc. Lunar Sci. Conf. 7th,* p. 2509–2528.

Spudis P. D. (1978) Composition and origin of the Apennine Bench Formation. *Proc. Lunar Planet. Sci. Conf. 9th,* p. 3379–3394.

Spudis P. D. (1979) The extent and duration of KREEP volcanism (abstract). In *Papers Presented to the Conference on the Lunar Highlands Crust,* p. 157–159. Lunar and Planetary Institute, Houston.

Staudacher T., Jessberger E. K., Flohs I., and Kirsten T. (1979) ^{40}Ar-^{39}Ar age systematics of consortium breccia 73255. *Proc. Lunar Planet. Sci. Conf. 10th,* p. 745–762.

Steele I. M. and Smith J. V. (1973) Mineralogy and petrology of some Apollo 16 rocks and fines: general petrologic model of the moon. *Proc. Lunar Sci. Conf. 4th,* p. 519–536.

Steele I. M. and Smith J. V. (1976) Mineralogy and petrology of complex breccia 14063,14. *Proc. Lunar Sci. Conf. 7th,* p. 1949–1964.

Stettler A., Eberhardt P., Geiss J., Grögler N. and Maurer P. (1973) Ar^{39}-Ar^{40} ages and Ar^{37}-Ar^{38} exposure ages of lunar rocks. *Proc. Lunar Sci. Conf. 4th,* p. 1865–1888.

Stoeser D. B., Wolfe R. W., Wood J. A., and Bower J. F. (1974) Petrology and petrogenesis of Boulder 1. In *Interdisciplinary Studies of Samples* from *Boulder 1, Station 2, Apollo 17,* Vol. 1 (J. A. Wood, ed.). LSI Contr. No. 210D, p. 35–109. Smithsonian Astrophysical Observatory, Cambridge.

Sutton R. L., Hait M. H., and Swann G. A. (1972) Geology of the Apollo 14 landing site. *Proc. Lunar Sci. Conf. 3rd,* p. 27–38.

Tatsumoto M., Hedge C. E., Doe B. R., and Unruh D. M. (1972) U-Th-Pb and Rb-Sr measurements on some Apollo 14 lunar samples. *Proc. Lunar Sci. Conf. 3rd,* p. 1531–1555.

Taylor G. J. (1976) Sample 14318 petrology and previous studies. In *Interdisciplinary Studies by the Imbrium Consortium,* Vol. 1 (J. A. Wood, ed.). LSI Contr. No. 268D, p. 67–73. Center for Astrophysics, Cambridge.

Taylor G. J., Drake M. J., Hallam M. E., Marvin U. B., and Wood J. A. (1973) Apollo 16 Stratigraphy: The ANT hills, the Cayley Plains and a pre-Imbrian regolith. *Proc. Lunar Sci. Conf. 4th,* p. 553–568.

Taylor G. J., Warner R. D., and Keil K. (1978) VLT mare basalts: impact mixing, parent magma types, and petrogenesis. In *Mare Crisium: the View from Luna 24* (R. B. Merrill and J. J. Papike, eds.), p. 357–370. Pergamon, N.Y.

Taylor S. R., Kaye M., Muir P., Nance W., Rudowski R., and Ware N. (1972) Composition of the lunar highlands: chemistry of Apollo 14 samples from Fra Mauro. *Proc. Lunar Sci. Conf. 3rd,* p. 1231–1249.

Turner G., Cadogan P. H., and Yonge C. J. (1973) Argon selenochronology. *Proc. Lunar Sci. Conf. 4th,* p. 1889–1914.

Turner G., Huneke J. C., Podosek F. A., and Wasserburg G. J. (1971). ^{40}Ar-^{39}Ar ages and cosmic ray exposure ages of Apollo 14 samples. *Earth Planet. Sci. Lett.* **12,** 19–35.

Walker D., Longhi J., and Hays J. F. (1972) Experimental petrology and origin of Fra Mauro rocks and soils. *Proc. Lunar Sci. Conf. 3rd,* p. 797–817.

Wänke H., Baddenhausen H., Balacescu A., Teschke F., Spettel B., Dreibus G., Palme H., Quijano-Rico M., Kruse H., Wlotzka F., and Begemann F. (1972) Multielement analyses of lunar samples and some implications of the results. *Proc. Lunar Sci. Conf. 3rd,* p. 1251–1268.

Wänke H., Dreibus G., and Palme H. (1978) Primary matter in the lunar highlands: the case of the siderophile elements. *Proc. Lunar Planet. Sci. Conf. 9th,* p. 83–110.

Warner J. L., Phinney W. C., Bickel C. E., and Simonds C. H. (1977). Feldspathic granulitic impactites and pre-final bombardment lunar evolution. *Proc. Lunar Sci. Conf. 8th,* p. 2051–2066.

Warner J. L., Simonds C. H., and Phinney W. C. (1976) Apollo 17, Station 6 boulder sample 76255: Absolute petrology of breccia matrix and igneous clasts. *Proc. Lunar Sci. Conf. 7th,* p. 2233–2250.

Warren P. H., Afiattalab F., and Wasson J. T. (1978) Investigation of unusual KREEPy samples: Pristine rock 15386, Cone Crater soil fragments 14143, and 12023, a *typical* Apollo 12 soil. *Proc. Lunar Planet. Sci. Conf. 9th,* p. 653–660.

Warren P. H. and Wasson J. T. (1977) Pristine non-mare rocks and the nature of the lunar crust. *Proc. Lunar Sci. Conf. 8th,* p. 2215–2255.

Warren P. H. and Wasson J. T. (1978) Compositional-petrographic investigation of pristine non-mare rocks. *Proc. Lunar Planet. Sci. Conf. 9th,* p. 185–217.

Whitaker E. A. (1972) Lunar color boundaries and their relationship to topographic features. *The Moon* **4,** 348–355.

Wilhelms D. E. (1970) Summary of lunar stratigraphy—telescopic observation. *U.S. Geol. Survey Prof. Paper 599-F,* p. S1–S47.

Willis J. P., Erlank A. J., Gurney J. J., Theil R. H., and Ahrens L. H. (1972). Major, minor, and trace element data for some Apollo 11,12,14 and 15 samples. *Proc. Lunar Sci. Conf. 3rd,* p. 1269–1273.

Vinogradov A. P. (1971) Preliminary data on lunar ground brought to earth by automatic probe "Luna-16". *Proc. Lunar Sci. Conf. 2nd,* p. 1–16.

York D., Kenyon W. J., and Doyle R. J. (1972) ^{40}Ar-^{39}Ar ages of Apollo 14 and 15 samples. *Proc. Lunar Sci. Conf. 3rd,* p. 1613–1622.

Papike, J.J. and Merrill, R.B., eds.
Proc. Conf. Lunar Highlands Crust (1980), p. 377-394
Printed in the United States of America

Petrology of 60035: Evolution of a polymict ANT breccia

Richard D. Warner*, G. Jeffrey Taylor and Klaus Keil

Department of Geology and Institute of Meteoritics, University of New Mexico, Albuquerque, New Mexico 87131

Abstract—Sample 60035 is a polymict ANT breccia partly coated with glass. The breccia contains abundant clasts which have troctolitic/noritic anorthosite compositions. The troctolitic anorthosite clasts typically have a granoblastic texture, and the noritic anorthosite clasts a poikiloblastic texture. Both are texturally and mineralogically akin to granulitic impactites. 60035 also contains cataclastic and recrystallized ferroan anorthosite clasts, and an Mg-rich troctolite clast, all of which possibly represent original crustal rocks that have been texturally but not compositionally modified. The Mg-rich troctolite clast contains high-Ni (36–51 wt.%) metal grains that we interpret to be indigenous. Metal elsewhere in the rock is of low-Ni (~5–10 wt. %) composition, and likely is ultimately of meteoritic origin. At least two episodes of crushing and mixing were involved in the petrogenesis of 60035. Prior to the last of these, most clasts were thermally metamorphosed. Annealing and mineral equilibration since formation of the breccia has not been extensive. Both the glass coating and thin glass veins within the breccia formed after the rock was consolidated, and may have been generated in separate events: the glass veins by *in situ* shock melting or injection of superheated liquid, and the glass coating by mere splashing of liquid onto the breccia's surface.

INTRODUCTION

Most lunar highland rocks are impact breccias formed as a result of the intense meteorite bombardment of the original crustal highland rocks. The impact process has altered the textures and compositions of the primary crustal rocks, and the resulting impact breccias are therefore of only limited value when attempting to unravel the nature of the original rocks that made up the lunar highland crust. Nevertheless, some highland breccias contain clasts of apparently pristine character. Such clasts may reveal information concerning the early geologic evolution of the lunar crust.

Highland breccia 60035 is a 1,052 g sample that was collected at the LM/ALSEP area of the Apollo 16 landing site. It was set aside as a posterity sample, and has only recently been made available for study. Our interest in 60035 stems from the *Apollo 16 Lunar Sample Information Catalog* (Lunar Receiving Laboratory, 1972), which described the rock as an anorthositic gabbro. In order to assess the possibility of whether 60035 contains material of pristine crustal origin,

* Present address: Department of Chemistry and Geology, Clemson University, Clemson, South Carolina 29631.

we requested polished thin sections from widely separated portions of the rock for mineralogic-petrologic study. Our rationale was that if this preliminary examination proved the rock to be interesting, then a consortium could be organized to obtain supporting chemical and age data. Our investigation has revealed that 60035 contains several clasts (ferroan anorthosites and a Mg-rich troctolite) which may represent mineralogically "pristine", texturally altered crustal material. Also present are abundant clasts that are texturally and mineralogically akin to feldspathic granulitic impactites, which are interpreted to have formed before the final, ~4.0 AE lunar bombardment (Warner *et al.*, 1977). Thus, 60035 appears to contain significant pre-final bombardment material, and by unraveling the breccia's history we may gain insight into the evolution of the ancient lunar crust.

MATERIALS AND METHODS

We were allocated six polished thin sections from widely separated portions of 60035. These sampled the following materials: 60035,4 and 60035,6—a number of small chips which had broken off from the rock; 60035,5—glass surface coating and adjacent portion of breccia; 60035,17; 60035,19; 60035,21—various areas of the breccia. The polished thin sections were studied by optical microscopy and mineral and glass compositions were determined by electron microprobe using methods described in Keil (1967). Silicate and oxide mineral analyses and glass analyses were corrected for differential matrix effects following the procedures of Bence and Albee (1968) and Albee and Ray (1970). ZAF corrections were applied to analyses of metal grains. Interference of the Co K_α peak by the tailing edge of the Fe K_β peak was calibrated by measuring "Co" in pure Fe metal and in magnetite and deriving a correction factor (=0.0024). Co in the metal analyses was then corrected by multiplying this factor times the wt.% Fe, and subtracting the result from the observed Co value.

DESCRIPTION OF SAMPLE

General

Macroscopically, 60035 is a whitish rock that is partly glass-coated (Fig. 1). The glass coating is dark gray to black and covers about half the surface area. Light-colored, angular to subangular clasts are dispersed throughout the glass coating. In addition, numerous thin (<0.1 mm across), dark glass veins are visible in non-coated portions of the sample. The whitish rock is coherent and appears mostly to be fine-grained, although a few larger crystals of plagioclase up to ~1 mm can be seen. Small black specks are distributed uniformly throughout the rock, and some bright white clasts are discernible. Most of the rock is covered with patina.

Thin sections show that 60035 is a fine-grained, clast-rich, polymict breccia. In

Fig. 1. Photograph of bottom surface of 60035, showing macroscopic appearance of the rock. Note glass veins in upper right and right portions of photograph. Scale: cube side = 1 cm. NASA photograph #S-72-38300.

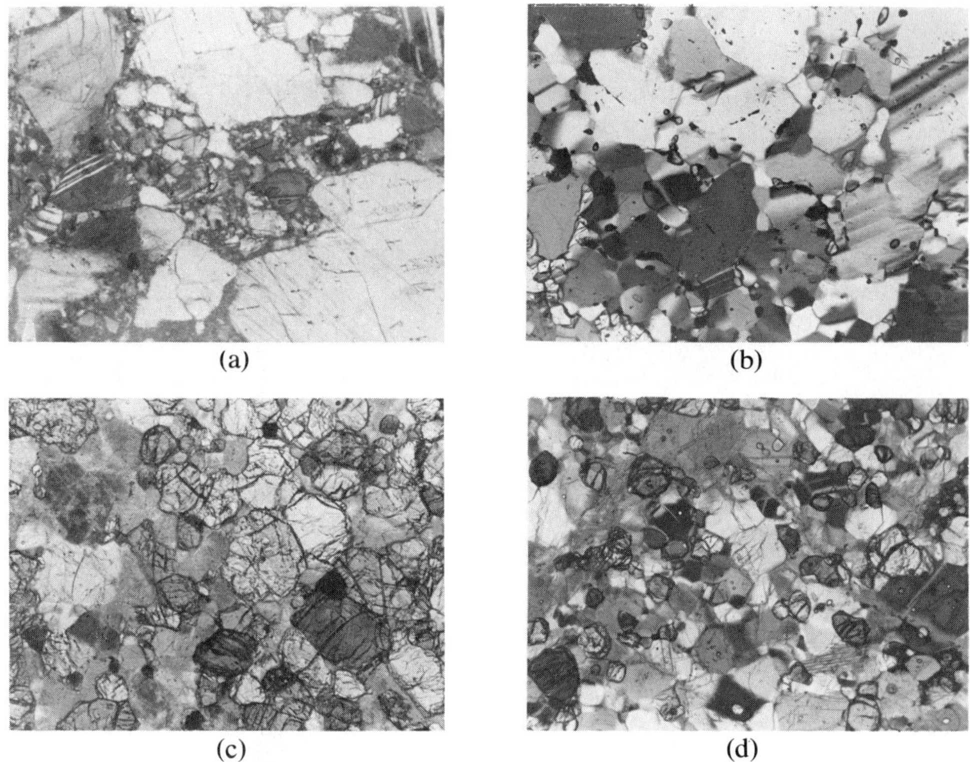

Fig. 2. Transmitted light photomicrographs (horizontal dimension equals ~1 mm) illustrating textures of major lithologies in 60035. (A) Cataclastic (ferroan) anorthosite chip (60035,6). Chip consists of larger, subangular or subrounded plagioclase crystals in a matrix of finely crushed plagioclase. Partly X nichols. (B) Recrystallized (ferroan) anorthosite chip (60035,6). Chip consists of partly mosaicized plagioclase crystals. Note development of some ~120° triple junctions. Minor interstitial pyroxene (high relief) and glassy patches (dark areas). Partly X nichols. (C) Troctolite clast (60035,21). Clast composed of equigranular olivine and plagioclase. Orthopyroxene (not shown) inhomogenously distributed. Partly X nichols. (D) Troctolitic anorthosite clast (60035,17). Note fine-grained, granoblastic texture, with abundant ~120° triple junctions. Olivine (high relief) principal mafic mineral. To see how typical of granulitic impactites this texture is, compare with Fig. 1 of Warner *et al.* (1977). Partly X nichols.

some thin section areas, clasts can be fairly readily distinguished, while in others clast-matrix boundaries are obscure. Lithic clasts are rounded to subrounded and range in size from <0.5 mm to ~1 cm. Mineral clasts are also rounded to subrounded and occasionally exceed 1 mm in size. The rock is traversed by irregular, thin, dark glass veins. In the following paragraphs we describe the major lithologies identifiable in thin sections of the rock. The mineralogy of the lithologies is summarized in Table 1, and textures are illustrated in Fig. 2.

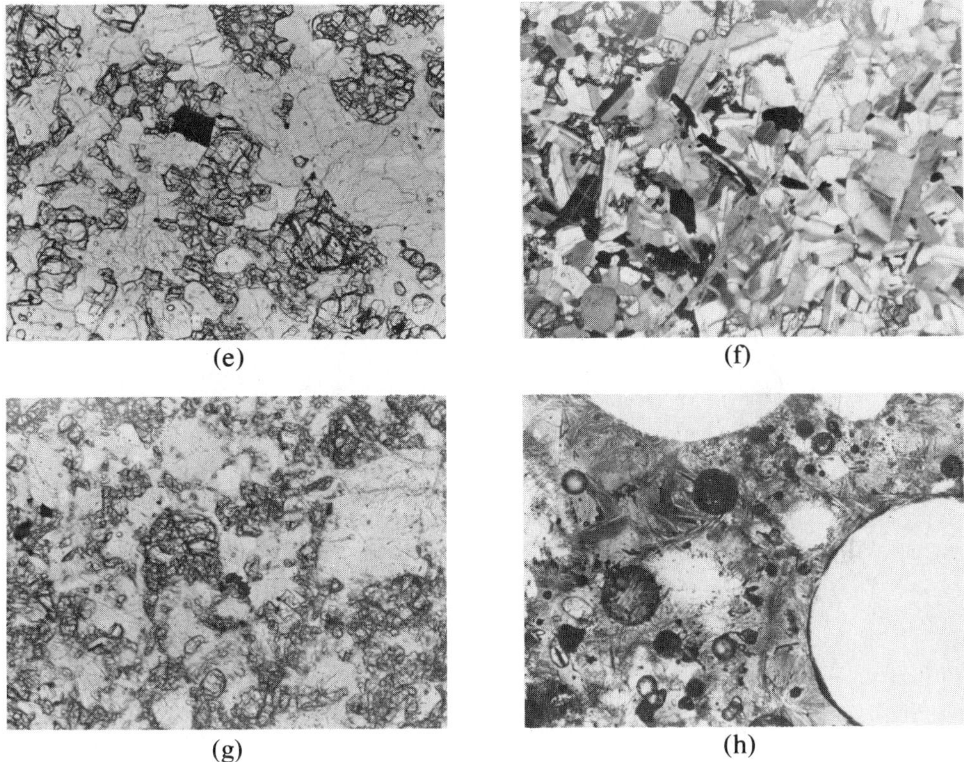

(E) Noritic anorthosite clast (60035,17). Spongy, poikiloblastic crystals with high relief are orthopyroxene. Minor, subequant olivine (also high relief). Plane polarized light. (F) Plagioclase-rich melt area (60035,19). Note lathlike habit commonly exhibited by plagioclase. Opaque grains are ilmenite. Partly X nichols. (G) Matrix (60035,21). Note abundance of partly granulated, sieve-textured, orthopyroxene crystals. Plane polarized light. (H) Glass surface coating (60035,5). Note subparallel feldspar crystallites and spherical vesicles (top and lower right). Plane polarized light.

Cataclastic ferroan anorthosite

This lithology is represented by an isolated chip (~1.25 × 2.5 mm) that had broken away from the sample. The chip is ~99% plagioclase and consists of subrounded to subangular, shocked crystals up to ~0.8 mm long surrounded by zones of very finely crushed material (Fig. 2A). Plagioclase is An_{95-96} in composition. Trace amounts of mafic minerals are present: these include Fe-rich pyroxenes, both orthopyroxene and high-Ca pyroxene (Fig. 3), and Fe-rich olivine (~Fo_{35}). The chip is thus texturally and mineralogically similar to the ferroan anorthosites described by Dowty *et al.* (1974). One edge of the chip is rimmed by glass.

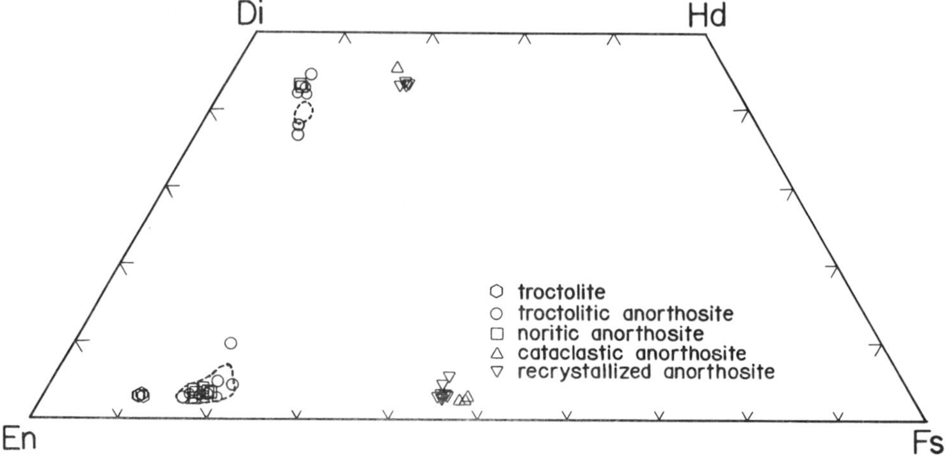

Fig. 3. Composition of pyroxenes (projected onto pyroxene quadrilateral) in 60035. Areas enclosed by dashed lines indicate ranges in composition of matrix grains.

Recrystallized ferroan anorthosite

This, too, is a chip (~1.75 × 2.5 mm) that had broken away from the sample. It has a recrystallized texture (Fig. 2B) that approaches granoblastic (e.g., some ~120° triple junctions are present). Plagioclase crystals (An_{97-98}) comprise ~95% of the chip and vary from <0.05 to ~0.7 mm in size. The larger grains are partly mosaicized. Minor interstitial mafics are present, mostly low-Ca pyroxene and some high-Ca pyroxene. Both are Fe-rich, similar to pyroxenes in the cataclastic ferroan anorthosite (Fig. 3). Trace amounts of ilmenite [Fe/(Fe+Mg), 0.92–0.94] are also present. Small patches of interstitial glassy mesostasis are scattered throughout the anorthosite.

Troctolite

A troctolite clast at the edge of thin section 60035,21 is at least 4 × 10 mm in size. Its mode and grain size are variable. One side of the clast consists of subequal amounts of equigranular olivine and plagioclase grains 0.05–0.25 mm across (Fig. 2C). The middle portion of the clast has more plagioclase (>60%), and includes some coarser grains (plagioclase up to ~0.6 mm long and olivine up to ~0.8 mm); minor poikiloblastic orthopyroxene (~1 mm long crystal) is also present. The other side contains considerable orthopyroxene and is finer-grained (e.g., most orthopyroxene grains are <0.05 mm). Overall, plagioclase (An_{95-97}) constitutes ~57% of the troctolite. Mafics are very Mg-rich, with tight compositional ranges: olivine is $Fo_{88\pm0.5}$ and orthopyroxene is $En_{86}Fs_{11}Wo_3$ with <1 mol % variation (Fig. 3). The troctolite also contains trace amounts of high-Ni (36–

51 wt.%) metal (Fig. 4), armalcolite [Fe/(Fe+Mg), 0.27–0.34], and aluminian chromite [Fe/(Fe+Mg), 0.37–0.61].

A second troctolite clast, in 60035,17, also occurs at the edge of the thin section, and is at least 1.75 × 3.5 mm. It is slightly more feldspathic (~60% plagioclase) and less magnesian (olivine: $Fo_{84\pm0.5}$) than the troctolite in 60035,21.

Troctolitic anorthosite and noritic anorthosite

These are by far the most abundant clast lithologies in 60035. The clasts are typically 1–2 mm across, but some are as large as ~5 mm. They are equivalent to feldspathic granulitic impactites (hereafter referred to simply as granulitic impactites), as defined by Warner *et al.* (1977). They contain ~80% plagioclase (Table 1), and have fine-grained, metamorphic textures: either granoblastic if olivine is the dominant mafic (Fig. 2D), or poikiloblastic if low-Ca pyroxene is the dominant mafic (Fig. 2E). Plagioclase occurs as equant to subequant crystals

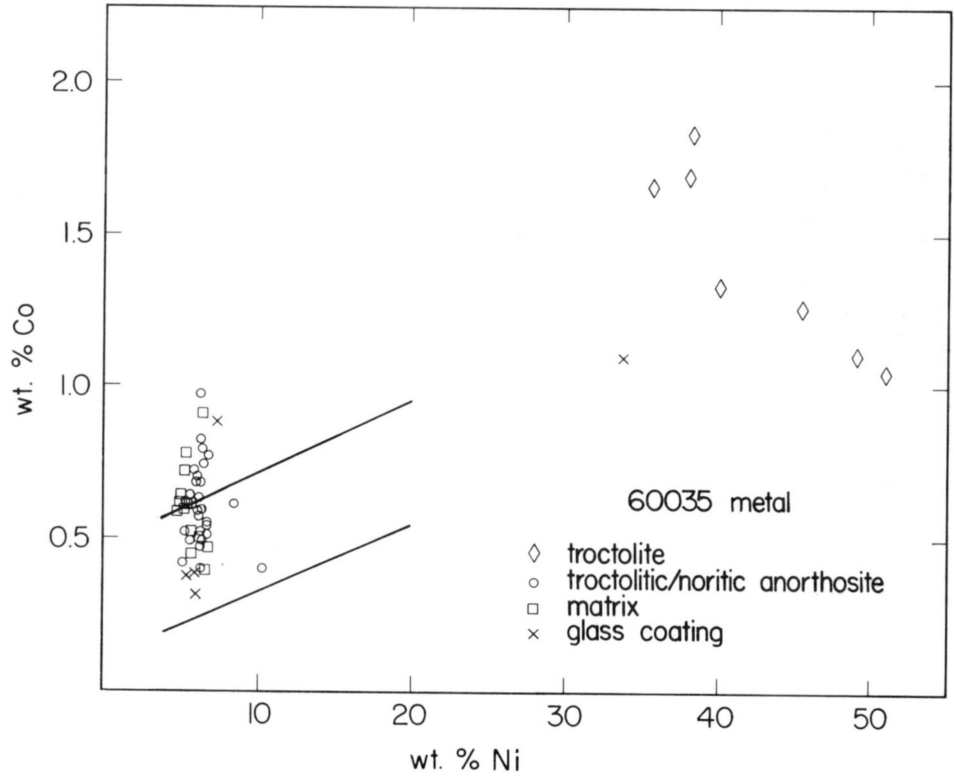

Fig. 4. Wt. % Ni vs. wt. % Co in 60035 metal grains. Area between solid lines denotes preferred compositional range of metal from sampled meteorites (from Goldstein and Yakowitz, 1971).

Table 1. Summary of 60035 lithologies.

Lithology	% Plagioclase	Average mineral composition (MOL %) Plagioclase	Olivine	Orthopyroxene
Cataclastic ferroan anorthosite	99	An_{96}	Fo_{35}	En_{50}
Recrystallized ferroan anorthosite	95	An_{97}	—	En_{52}
Troctolite	57	An_{96}	Fo_{88}	En_{86}
Troctolitic anorthosite	84(82–86)	An_{94-96}	Fo_{76-82}	En_{75-81}
Noritic anorthosite	81(71–87)	An_{95}	Fo_{76-81}	En_{77-80}
Matrix	75(72–79)	An_{95-96}	Fo_{77-81}	En_{77-79}

~0.05–0.2 mm across (some are larger), or as small rounded grains enclosed by pyroxene. Olivine crystals are also subequant to equant and are mostly 0.05–0.1 mm in size. Pyroxene (mostly low-Ca) tends to occur as irregular, spongy crystals that enclose numerous small plagioclase grains. Typically, a number of grains are shocked. The larger plagioclases, especially, have been cataclasized or exhibit undulose extinction.

Mineral compositions tend to be of very restricted range within a given clast, although there is some variation between clasts. For example, olivine and low-Ca pyroxene grains within individual clasts tend to show ≤ 1 mol. % variation in end members, whereas between clasts the range in average mineral compositions is ~6 mol% (Fo_{76-82}, olivine; En_{75-81}, low-Ca pyroxene: Table 1). Plagioclase compositions vary by several mol % within individual clasts, but show less variation in terms of average composition (An_{94-96}: Table 1) between clasts than mafic minerals. Accessory minerals include high-Ca pyroxene (Fig. 3), ilmenite (overall range in Fe/(Fe+Mg), 0.63–0.77), chromite [Fe/(Fe+Mg), 0.68–0.69], Cr-Zr armalcolite, and metal. The latter contain 5–10 wt.% Ni and 0.4–1.0 wt.% Co (Fig. 4).

Melt areas

Several areas in the rock consist predominantly of fine-grained, lathy plagioclase. One ~0.6×0.8 mm area in 60035,19 consists of intersecting plagioclase laths with minor interstitial mafic and ilmenite grains (Fig. 2F). A second, ~3 mm long area in the same thin section contains sieve-textured low-Ca pyroxene grains (~$En_{77}Fs_{19}Wo_4$), in addition to very fine-grained plagioclase laths. There is no association of these areas with the glass coating or glass veins. We interpret them as being clasts of plagioclase-rich melt which were incorporated into the breccia and subsequently partly recrystallized.

Matrix

In general, the matrix is finer-grained, less feldspathic, and more pyroxene-rich than the clast fraction of 60035. Crystal habits of plagioclase, olivine and pyroxene are similar to those elsewhere in the rock, except that the grains are more commonly granulated (Fig. 2G). Zones of very finely crushed mafic grains are often present. Some textural variation can be seen between different portions of the rock. For example, the matrix is slightly coarser-grained and somewhat more annealed in 60035,19 compared to 60035,17 and ,21. Matrix mineral compositions are similar to those observed in the troctolitic and noritic anorthosite lithologies (Table 1; Figs. 3 and 4). Mafic grains in 60035,19 tend to be several mol % more Fe-rich than in 60035,17 and ,21.

Glass coating and glass veins

The glass surface coating is highly vesicular and consists of subparallel bundles of feldspar crystallites separated by cryptocrystalline mesostasis (Fig. 2H). Locally, the texture is intersertal. The glass coating contains isolated mineral relics (mostly plagioclase). The transition between the glass and the underlying breccia occurs over a distance of <0.5 mm. The dark glass veins that cut 60035 are irregular and often branching. Individual veins vary in width from less than a μm to nearly 50 μm in thicker lenses. One larger glass vein connects directly to the glass surface coating. It varies in thickness from ~500 μm near the glass coating to 100–200 μm away from the coating. We have analyzed portions of the glass surface coating and glass veins in 60035 using the electron microprobe (Table 2). Although broadly similar, there are some significant differences in composition between the glass coating and glass veins. For example, FeO is lower in the glass

Table 2. Electron microprobe analyses (wt.%) of 60035 glass coating and glass veins.

	glass coating[1]	glass vein[1] (100–200 μm thick)	glass vein[3] (<1–50 μm thick)
SiO_2	44.1 (43.7–44.7)	44.3 (42.6–45.5)	45.6 (45.2–46.0)
TiO_2	0.29 (0.18–0.40)	0.16 (0.09–0.23)	0.12 (0.08–0.19)
Al_2O_3	29.0 (27.5–29.7)	29.3 (27.5–31.1)	31.3 (31.2–31.4)
Cr_2O_3	0.13 (0.11–0.14)	0.09 (0.01–0.15)	0.07 (0.04–0.09)
FeO	5.1 (4.8–5.7)	3.7 (2.4–4.8)	2.2 (2.1–2.3)
MnO	0.04 (0.01–0.06)	0.04 (0.01–0.09)	<0.01 (0–0.1)
MgO	6.0 (5.4–6.7)	6.5 (3.6–8.3)	4.5 (4.3–4.8)
CaO	15.7 (15.3–16.0)	16.0 (15.3–17.5)	16.8 (16.2–17.3)
Na_2O	0.26 (0.24–0.26)	0.43 (0.36–0.52)	0.39 (0.34–0.44)
K_2O	0.06 (0.06)	0.09 (0.07–0.11)	0.08 (0.06–0.10)
P_2O_5	0.02 (0–0.02)	0.04 (0.01–0.09)	0.03 (0–0.06)
Total	100.70	100.65	101.09
Mg/(Mg+Fe)	0.68	0.76	0.78

[1]Average and range of 4 analyses of ~25 μm size areas.

[2]Average and range of 25 analyses (from 5 different areas) obtained with ~5 μm size beam.

[3]Average and range of 3 analyses (from separate areas) obtained with ~5 μm size beam.

veins, particularly the thin veins in 60035,17, while Al_2O_3, CaO and Na_2O are higher. Thus, the glass veins are more feldspathic in composition than the glass coating.

DISCUSSION

Provenance of 60035 lithologies

In Fig. 5 we compare the data from the major lithologies observed in 60035 with those for pristine highland rocks on a plot of mean An in plagioclase versus mean *mg* [molar Mg/(Mg+Fe)] in mafic minerals. On such a plot, pristine highland rocks clearly divide into two distinct groups, ferroan anorthosites and Mg-rich plutonic rocks (Warner *et al.,* 1976; Warren and Wasson, 1977; Warner and Bickel, 1978).

The orthopyroxene data for the cataclastic anorthosite and the recrystallized anorthosite clasts both plot within the ferroan anorthosite field (Fig. 5). The olivine data point for the cataclastic anorthosite plots considerably below this field, but lies on the extension of the trend. This correlation does not necessarily imply that the clasts represent compositionally pristine highland material: in the absence of analyzable metal grains (see following section) or siderophile element data it is not possible to make such a judgment. The textures exhibited by the two ferroan anorthosite clasts (the one, cataclastic; the other, recrystallized) are clear indications that the original rocks have been texturally modified. The largest surviving grains in both clasts are <1 mm in size, and there is thus no evidence that the original rocks were ever of very coarse grain size, such as is assumed to be the case for pristine plutonic rocks (Warren and Wasson, 1977).

The data points for the troctolite clast in 60035,21 plot near the upper magnesian limit of the Mg-rich plutonic rock field (Fig. 5). Again, this does not *prove* that the troctolite is compositionally pristine. The original texture of the rock has been obliterated by metamorphic recrystallization. The largest grains are <1 mm in size. The inhomogeneous distribution of orthopyroxene, most of which is concentrated in a ~ 2 mm wide band near one side of the clast, suggests the possibility that the rock was originally coarser-grained and was subsequently granulated, more or less in place. The extreme homogenization of silicate mineral compositions, although possibly the product of metamorphic recrystallization, could also have inherited from a coarse-grained cumulate rock (see Ashwal, 1975). Hence, we believe that the ultimate precursor of the clast could well have been a plutonic troctolite.

We have also plotted (Fig. 5) data points for several representative troctolitic and noritic anorthosite clasts from 60035. For the most part they are within, but close to the edge of, the Mg-rich plutonic rock field. One data point falls below the aforementioned field. We have previously noted the textural and mineralogical similarities between the troctolitic/noritic anorthosite clasts in 60035 and the granulitic impactite suite of Warner *et al.* (1977). In a survey of lunar plutonic and

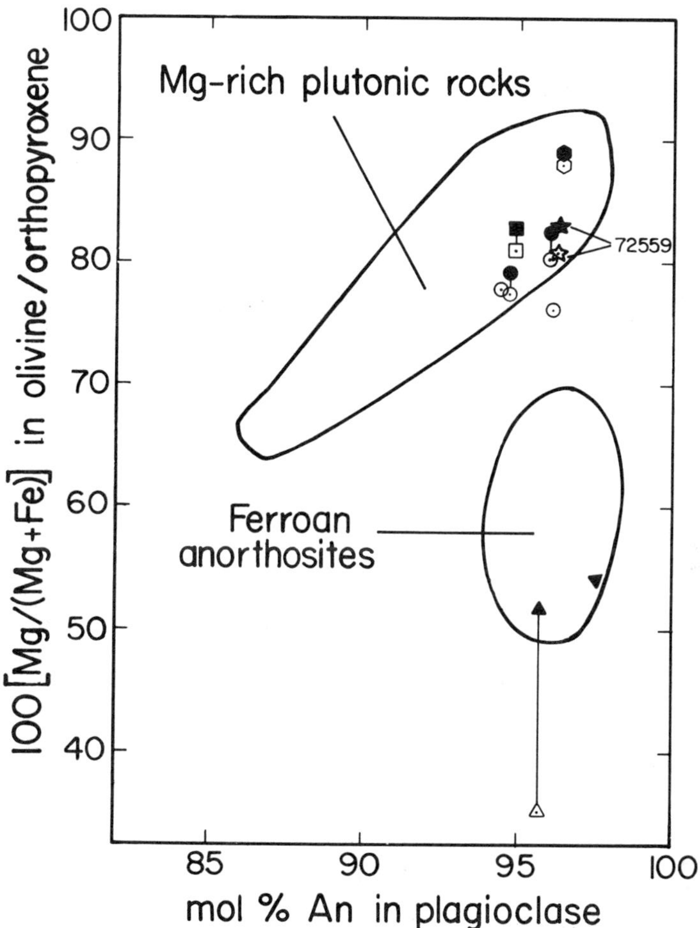

Fig. 5. Mol % An in plagioclase vs. 100 [Mg/(Mg+Fe)] in olivine (open symbols) and orthopyroxene (filled symbols) for various 60035 lithologies. Fields for Mg-rich plutonic rocks and ferroan anorthosites taken from Warren and Wasson (1977). Data for anorthositic troctolite 72559, a granulitic impactite, shown also for comparison. 60035 data points: triangles, cataclastic ferroan anorthosite; inverted triangle, recrystallized ferroan anorthosite; hexagons, troctolite; circles, troctolitic anorthosites; squares, noritic anorthosite.

granulitic lithic fragments, Bickel and Warner (1978) found that the majority of granulitic impactites cluster between the ferroan anorthosite and Mg-rich plutonic rocks fields in an An vs. *mg* diagram, but that many plot within the Mg-rich plutonic rocks field. The troctolitic/noritic anorthosites in 60035 all fall within the range that Bickel and Warner (1978) delineated for granulitic impactites. One set of data points (Fig. 5) is nearly coincident with that for anorthositic troctolite 72559, which is interpreted by Nehru *et al*. (1978) and Warren and Wasson (1978) as belonging to the granulitic impactite suite.

We note that the troctolitic/noritic anorthosite data points in Fig. 5 plot more or less in between those for the Mg-rich troctolite clast and the ferroan anorthosites. This suggests that the troctolitic/noritic anorthosites may be mixtures of precursors belonging to both primary trends. Bickel and Warner (1978) reached a similar conclusion for most granulitic impactites. Since ferroan anorthosites rarely exceed 10% mafic minerals by volume (Warren, 1979), it is to be expected that the *mg* value in such mixtures would be dominated by the mafics contributed by the Mg-rich plutonic rocks. The relation in Fig. 5, whereby the troctolitic/noritic anorthosite data points lie mostly with the Mg-rich plutonic rocks field, but at somewhat lower *mg* values than for the Mg-rich troctolite, are consistent with this expectation.

Origin of metal

Metal grains in 60035 cluster into two distinct compositional groups: one is high in Ni (36–51 wt.%) and is comprised of grains from the Mg-rich troctolite clast in 60035,21, while the other is of low-Ni (5–10 wt.%) composition and is comprised of grains found elsewhere in the rock (Fig. 4). In Fig. 6 we compare the fields corresponding to the high- and low-Ni metal groups in 60035 with those for

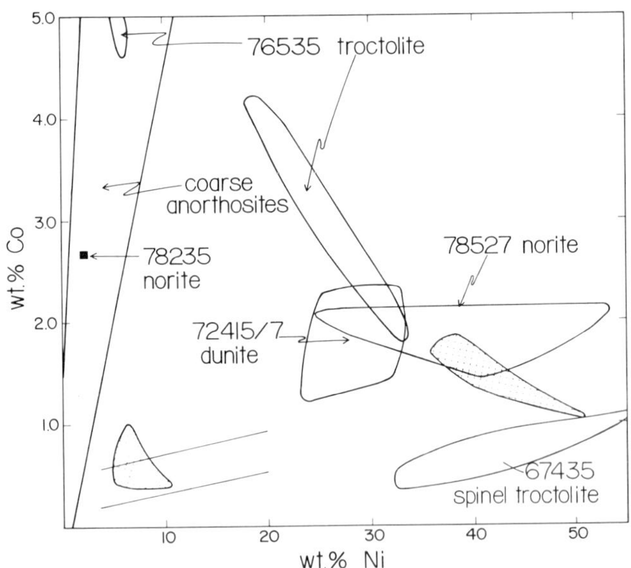

Fig. 6. Wt. % Ni vs. wt. % Co in 60035 metal (cross hatched areas) compared with ranges observed for metal grains in a number of representative monomict highland rocks. "Meteoritic" range given by parallel solid lines. Sources of data: coarse anorthosites, Hewins and Goldstein (1975); 67435, our unpublished data; 72415/7, Dymek *et al.* (1975); 76535, Gooley *et al.* (1974); 78235, McCallum and Mathez (1975); 78527, Nehru *et al.* (1978) and our unpublished data.

several representative monomict highland rocks. An interesting feature emerges from this plot: there is a tendency for a separation of metal compositions into two distinct groups similar to that encountered in an An vs. *mg* diagram. Coarse-grained (ferroan) anorthosites usually contain metal of low (≤10 wt.%) Ni and variable Co composition (Hewins and Goldstein, 1975). In contrast, Mg-rich plutonic rocks such as the spinel troctolite clast in 67435 (Prinz *et al.*, 1973), norite 78527 (Nehru *et al.*, 1978), and dunite 72415/7 (Dymek *et al.*, 1975) tend to be characterized by metal of high-Ni (>20 wt.%) composition. A major exception is norite 78235, which contains metal of low-Ni composition (McCallum and Mathez, 1975), and plots within the coarse anorthosite range of Hewins and Goldstein (1975). Troctolite 76535 contains both high- and low-Ni metal, but these are exsolution products of an initially homogeneous metal phase with moderate Ni (Gooley *et al.*, 1974).

The high-Ni metal of the Mg-rich troctolite clast is 60035,21 is of a composition broadly comparable to that of the high-Ni metal found in the Mg-rich plutonic rocks mentioned above (Fig. 6). Dunite 72415 and troctolite 76535 are known to contain low siderophile element abundances and are thought to lack a meteoritic component (Morgan *et al.*, 1974; Higuchi and Morgan, 1975). There is thus a strong possibility that the high-Ni metal in the Mg-rich troctolite clast in 60035 is of indigenous, non-meteoritic origin. The high-Ni content of the metal itself, which is far higher than that generally accepted as being within the iron meteorite range (Moore *et al.*, 1969), also appears to make a meteoritic origin unlikely.

As noted previously, metal grains elsewhere in 60035 contain ~5–10 wt.% Ni. This includes metal in troctolitic/noritic anorthosite clasts (= granulitic impactites) and in the matrix. Compared to metal grains in coarse-grained anorthosites, the low-Ni metal in 60035 is somewhat higher in Ni (Fig. 6). Much of the low-Ni metal in 60035 falls outside the preferred compositional range for meteoritic metal (Goldstein and Yakowitz, 1971), being higher in Co relative to the latter (Figs. 4, 6). The composition of metal in other granulitic impactites varies considerably from sample to sample, but for the most part is higher in Co relative to the meteoritic range and also in Ni relative to coarse-grained anorthosites, thus suggesting a common origin. For example, analyzed metal grains in granulitic impactite 79215 are high in Co relative to the meteoritic range and are high in Ni compared to coarse-grained anorthosites (Bickel *et al.*, 1976). Metal in 72559 is compositionally identical to the low-Ni metal in 60035, except for a few grains which have high-Ni (22–34 wt.%) areas (Nehru *et al.*, 1978). On the other hand, metal grains in the anorthositic gabbro clasts in 73215 (interpreted by Warner *et al.* (1977) as members of the granulitic impactite suite), have compositions that mostly reside in the meteoritic metal field (James and Hammarstrom, 1977).

The ultimate origin of the low-Ni metal in 60035 (and in many other granulitic impactites) is problematical. One possibility is that most or all of the metal is of meteoritic origin and originally had compositions within the meteoritic trend, but that re-equilibration during thermal metamorphism has increased Co in the metal, causing many grains to have compositions outside the meteorite range. Taylor *et al.* (1976) have experimentally determined that subsolidus annealing can substantially modify the Ni-Co contents of metal grains in lunar rocks. However, the general re-equilibration trend involves depletion of both Ni and Co in the metal. Metamorphic re-equilibration would therefore seem inadequate to account for the observed Co enrichment of the 60035 low-Ni metal compared to metal of meteoritic origin (Fig. 6). A second possibility is that the low-Ni metal is of meteoritic origin, but that the meteoritic component is an ancient one characterized by metal of different composition than found in sampled meteorites. A similar suggestion was offered by Bickel *et al.* (1976) as a possible explanation for the origin of the metal in 79215. In this case, little or no re-equilibration of metal compositions is required. A third possibility is that the low-Ni metal is indigenous and not of meteoritic origin, and that originally there were present both high-Ni grains from Mg-rich plutonic rock precursors and very low-Ni grains from ferroan anorthosite precursors. The present metal compositions could therefore represent an average composition achieved by subsolidus re-equilibration. This possibility seems very unlikely, however, because granulitic impactites for which trace siderophile element data exist, consistently show an enriched spectrum (Warner *et al.,* 1977), suggestive of the presence of a meteoritic component. Although no chemical data are available for 60035, we deem it likely that the troctolitic/noritic anorthosite clasts and the breccia matrix are also enriched in siderophile elements. A fourth possibility is that some of the metal is indigenous and some (perhaps most) is of meteoritic origin, and that the observed compositions are the result of homogenization achieved by subsolidus re-equilibration.

Geologic history

Our petrologic study of 60035 indicates that the rock is a polymict breccia composed wholly of ANT-derived material. In this section we consider how 60035, and polymict ANT breccias like it, evolved.

In their study of pristine nonmare rocks, Warren and Wasson (1977) concluded that anorthosites, troctolites, and norites are the most important constituents of the primary lunar crust. Assuming these as ultimate precursors, we illustrate in Fig. 7 how the lithologies observed in 60035 may have evolved. The first stage presumably involved an episode of brecciation whereby the anorthosite, troctolite, and norite parent rocks were crushed. A certain amount of mixing would undoubtedly have occurred, giving rise to a series of clastic rocks of intermediate compositions (e.g., troctolitic/noritic anorthosites; anorthositic troctolites/norites; etc.). Following this episode of crushing ± mixing, most of the material was thermally metamorphosed, perhaps within a thick ejecta blanket. This gave rise to rocks with granoblastic and/or poikiloblastic matrix textures, such as are observed in the recrystallized anorthosite, Mg-rich troctolite, and numerous troctolitic/noritic anorthosite clasts in 60035. The brecciation and metamorphism could well have been the result of a single large impact event. Warner *et al.* (1977) have suggested that a widespread, granulitic metamorphism may have occurred prior to the 3.9–4.0 AE final lunar bombardment. The metamorphism evidenced

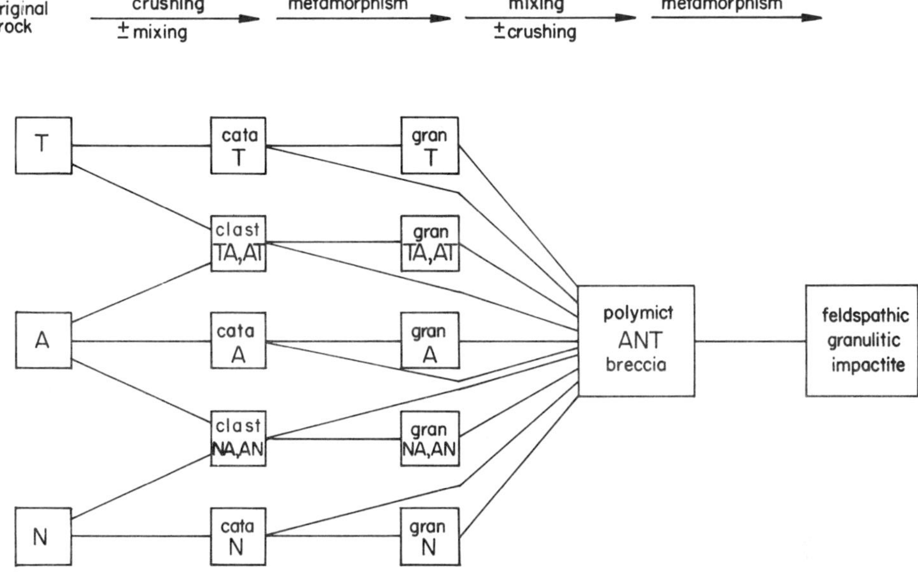

Fig. 7. Schematic diagram illustrating how polymict ANT breccia 60035 may have evolved. For explanation see text. T=troctolite; A=anorthosite; N=norite; TA=troctolitic anorthosite; etc.

in the various 60035 lithologies possibly could be equated with the lunar granulite metamorphism envisaged by Warner *et al.* (1977). Subsequently, another episode of mixing occurred whereby the various lithologies were incorporated into the present rock. The presence of the plagioclase-rich melt areas suggests that a minor amount of material of impact melt origin was added to the rock, and that subsequent to its consolidation into a polymict breccia, 60035 cannot have undergone prolonged annealing. Further, the distinctive mafic mineral compositions shown by the ferroan anorthosite and Mg-rich troctolite clasts attest to the fact that significant post-consolidation equilibration of mineral compositions has not occurred.

At some later time 60035 acquired its glass surface coating and thin, irregular glass veins. It is not clear to us whether the surface coating and the glass veins were produced in the same or separate events. The two differ somewhat in composition, especially in FeO (Table 2). The Mg/(Mg+Fe) ratios in the glass veins analyzed are 0.76 and 0.78, which are very close to the average Mg/(Mg+Fe) in the bulk rock, since olivine and orthopyroxene in the dominant lithologies range Fo_{76-82} and En_{75-81}, respectively (Table 1). In contrast, Mg/(Mg+Fe) of the glass coating is 0.68, which is considerably lower. Thus, given the limited compositional data that are available, it appears as if the glass veins are closer in composition to the host ANT breccia than is the glass coating. The glass coating probably represents an extraneous liquid which was splashed onto the exposed surface of the rock. The thin glass veins perhaps formed *in situ* via shock-induced local melting (cf. Roedder and Weiblen, 1977), or they may have been injected into the host rock under pressure. Whatever the origin of the glass veins, it seems probable that a shock event was involved (Roedder and Weiblen, 1977). This could have produced the shock features prevalent in the rock.

The sequence of events outlined above is the *simplest* history that 60035 could have undergone. One or more additional episodes of brecciation, mixing, melting, thermal metamorphism, etc., may have taken place. Further, not all clasts of similar mineralogy and texture need have undergone identical histories. For example, if the entire rock had undergone still another period of intense thermal metamorphism, the product (Fig. 7) would be a granulitic impactite that could be very difficult to distinguish from the troctolitic/noritic anorthosite clasts now present in 60035.

CONCLUSIONS

Based on our observations of the texture and mineralogy of six polished thin sections, we conclude that 60035 is a polymict ANT breccia. Most of the clasts in the rock are troctolitic or noritic anorthosites in composition and represent thermally metamorphosed mixtures of anorthositic, troctolitic, and noritic parent rocks. We equate them with the granulitic impactite suite of Warner *et al.* (1977). Two ferroan anorthosite clasts (one cataclastic, the other with a recrystallized texture), and a Mg-rich troctolite clast are present in the sections examined. These have had their original textures obliterated, but might be chemically pristine and may have retained their original mineral compositions. The breccia matrix appears to consist dominantly of ANT-derived material, and on the whole is less feldspathic and more orthopyroxene-rich, than the clast component. The rock is partly coated by dark glass, which appears to be more Fe-rich than the host rock and presumably was splashed onto the exposed surface of 60035. The breccia is cut by a number of thin, dark glass veins, whose composition is somewhat different from the glass surface coating, particularly in FeO. They may have formed in a separate event, either by *in situ* shock induced local melting or by injection into the rock of superheated liquid.

A more detailed interpretation of the petrogenesis of 60035 requires extensive chemical and age data on the various clasts and matrix. For example, siderophile data on selected clasts could clearly indicate whether or not any of the clasts are compositionally pristine. Such data could also bear on the origin of the high-Ni and low-Ni metal in 60035; however, because of the nondescript macroscopic appearance of the rock, it may be difficult to identify lithologies for separation from the hand specimen.

Acknowledgments—We thank George Conrad for maintaining the electron microprobe facility in excellent operating condition, Julie Dean for assistance in photomicrography, Judy Salas for drafting the line drawings, and Mary Fillmon for typing the manuscript. We especially thank Graham Ryder for calling our attention to this interesting rock. The manuscript has benefitted by thoughtful criticisms provided by Marc Norman, Graham Ryder, and William Phinney. This work was supported by the National Aeronautics and Space Administration, grant NGL 32-004-063 (Klaus Keil, Principal Investigator.)

REFERENCES

Albee A. L. and Ray L. (1970) Correction factors for electron probe microanalyses of silicates, oxides, carbonates, phosphates and sulfates. *Anal. Chem.* **42,** 1408–1414.

Ashwal L. D. (1975) Petrologic evidence for a plutonic igneous origin of anorthositic norite clasts in 67955 and 77017. *Proc. Lunar Sci. Conf. 6th,* p. 221–230.

Bence A. E. and Albee A. L. (1968) Empirical correction factors for the electron probe microanalysis of silicates and oxides. *J. Geol.* **76,** 382–403.

Bickel C. E. and Warner J. L. (1978) Survey of lunar plutonic and granulitic lithic fragments. *Proc. Lunar Planet. Sci. Conf. 9th,* p. 629–652.

Bickel C. E., Warner J. L., and Phinney W. C. (1976) Petrology of 79215: Brecciation of a lunar cumulate. *Proc. Lunar Sci. Conf. 7th,* p. 1793–1819.

Dowty E., Prinz M., and Keil K. (1974) Ferroan anorthosite: A widespread and distinctive lunar rock type. *Earth Planet. Sci. Lett.* **24,** 15–25.

Dymek R. F., Albee A. L., and Chodos A. A. (1975) Comparative petrology of lunar cumulate rocks of possible primary origin: Dunite 72415, troctolite 76535, norite 78235, and anorthosite 62237. *Proc. Lunar Sci. Conf. 6th,* p. 301–341.

Goldstein J. I. and Yakowitz H. (1971) Metallic inclusions and metal particles in the Apollo 12 lunar soil. *Proc. Lunar Sci. Conf. 2nd,* p. 177–191.

Gooley R., Brett R., Warner J., and Smyth J. R. (1974) A lunar rock of deep crustal origin: sample 76535. *Geochim. Cosmochim. Acta* **38,** 1329–1340.

Hewins R. H. and Goldstein J. I. (1975) The provenance of metal in anorthositic rocks. *Proc. Lunar Sci. Conf. 6th,* p. 343–362.

Higuchi H. and Morgan J. W. (1975) Ancient meteoritic component in Apollo 17 boulders. *Proc. Lunar Sci. Conf. 6th,* p. 1625–1651.

James O. B. and Hammarstrom J. G. (1977) Petrology of four clasts from consortium breccia 73215. *Proc. Lunar Sci. Conf. 8th,* p. 2459–2494.

Keil K. (1967) The electron microprobe X-ray analyzer and its application in mineralogy. *Fortschr. Mineral.* **44,** 4–66.

Lunar Receiving Laboratory (1972) *Apollo 16 Lunar Sample Information Catalog.* MSC-03210, NASA Johnson Space Center, Houston. 372 pp.

McCallum I. S. and Mathez E. A. (1975) Petrology of noritic cumulates and a partial melting model for the genesis of Fra Mauro basalts. *Proc. Lunar Sci. Conf. 6th,* p. 395–414.

Moore C. B., Lewis C. F., and Nava D. (1969) Superior analyses of iron meteorites. In *Meteorite Research* (P. M. Millman, ed.), p. 738–748. Springer-Verlag, N.Y.

Morgan J. W., Ganapathy R., Higuchi H., Krähenbühl U., and Anders E. (1974) Lunar basins: Tentative characterization of projectiles, from meteoritic elements in Apollo 17 boulders. *Proc. Lunar Sci. Conf. 4th,* p. 1703–1736.

Nehru C. E., Warner R. D., Keil K., and Taylor G. J. (1978) Metamorphism of brecciated ANT rocks: Anorthositic troctolite 72559 and norite 78527. *Proc. Lunar Planet. Sci. Conf. 9th,* p. 773–788.

Prinz M., Dowty E., Keil K., and Bunch T. E. (1973) Spinel troctolite and anorthosite in Apollo 16 samples. *Science* **179,** 74–76.

Roedder E. and Weiblen P. (1977) Shock glass veins in some lunar and meteoritic samples—Their nature and possible origin. *Proc. Lunar Sci. Conf. 8th,* p. 2593–2615.

Taylor L. A., Misra K. C., and Walker B. M. (1976) Subsolidus requilibration, grain growth, and compositional changes of native FeNi metal in lunar rocks. *Proc. Lunar Sci. Conf. 7th,* p. 837–856.

Warner J. L. and Bickel C. E. (1978) Lunar plutonic rocks: a suite of materials depleted in trace siderophile elements. *Amer. Mineral.* **63,** 1010–1015.

Warner J. L., Phinney W. C., Bickel C. E., and Simonds C. H. (1977) Feldspathic granulitic impactites and pre-final bombardment lunar evolution. *Proc. Lunar Sci. Conf. 8th,* p. 2051–2066.

Warner J. L., Simonds C. H., and Phinney W. C. (1976) Genetic distinction between anorthosites and Mg-rich plutonic rocks: New data from 76255 (abstract). In *Lunar Science VII,* p. 915–917. The Lunar Science Institute, Houston.

Warren P. H. (1979) *Certain* pristine nonmare rocks formed as cumulates from the magma ocean (but many others did *not*) (abstract). In *Papers Presented to the Conference on the Lunar Highlands Crust,* p. 192–194. Lunar and Planetary Institute, Houston.

Warren P. H. and Wasson J. T. (1977) Pristine nonmare rocks and the nature of the lunar crust. *Proc. Lunar Sci. Conf. 8th,* p. 2215–2235.

Warren P. H. and Wasson J. T. (1978) Compositional-petrographic investigation of pristine nonmare rocks. *Proc. Lunar Planet. Sci. Conf. 9th,* p. 185–217.

Papike, J.J. and Merrill, R.B., eds.
Proc. Conf. Lunar Highlands Crust (1980), p. 395-417
Printed in the United States of America

Genesis of highland basalt breccias: A view from 66095

James R. Garrison, Jr. and Lawrence A. Taylor

Department of Geological Sciences, The University of Tennessee, Knoxville, Tennessee 37916

Abstract—Highland breccia 66095, the "Rusty Rock", consists of a fine-grained subophitic matrix (78% of the rock) containing a variety of mineral and lithic clasts: anorthosite-troctolitic anorthosite, both intergranular and cataclastic (32% of clasts); plagioclase, both shocked and unshocked (22% of clasts); fine-grained, shocked, melt-rock clasts (i.e., basalts) with phenocrystic plagioclase (46% of clasts). The matrix is not chemically homogeneous, as it is for many breccias. Considerations of the chemistries of the matrix and clasts provides a basis for a qualitative 3-component mixing model: (1) an "ANT" plutonic complex, (2) Fra Mauro basalt, and (3) minor meteoritic material. With the inherent heterogeneities of the primitive highland crust, it is not possible to be more specific without adding several complications. Knowledge of the proportions of the "ANT" members, as well as Fra Mauro clasts, will necessitate an increased data base such as obtained by detailed "pull-apart" studies of highland melt breccias. It is the Fra Mauro component which is believed to be the source of the KREEP, abundant volatiles, and incompatible elements. Metamorphism of this component effectively mobilizes the volatiles. If high volatile fugacities are present during the melt stage of the breccia matrix, the volatiles will be incorporated into crystallizing phases (e.g., Cl_2 and P in chlorapatite). If abundant volatiles are present at subsolidus temperatures, they will deposit along cracks and grain boundaries where they slowly diffuse inward (e.g., Zn). If Cl_2 is present at low temperatures, it will react with ubiquitous native Fe to form $FeCl_2$, lawrencite, the precursor of rust. Subsequent metamorphism of these breccias can remobilize the volatiles—e.g., chlorpatite may release Cl_2 which can form $FeCl_2$. With such a scheme, the amount of leachable chlorine in a breccia may simply reflect its metamorphic history.

INTRODUCTION

The lunar highlands, forming over three-fourths of the near-side surface of the moon, are thought to represent primary crustal material subsequently modified by meteorite impacts through such processes as brecciation, cataclastic deformation, and partial to complete melting. The dominant rock type returned from the lunar highlands is breccia—impact produced melt rock containing clasts of primary crustal material, mineral fragments, minor meteoritic material, and fragments of earlier impact-derived materials. Many of these melt rocks cooled slowly and exhibit igneous textures making them difficult to distinguish from possible primary volcanic material. The complexities inherent in these highland breccias have complicated attempts to understand the nature of the primary lunar crust and the existence and nature of highland volcanic material.

Although a large data base of major and trace elements exists for these breccias, their polymict nature makes petrologic interpretation extremely difficult. Nevertheless, they provide us with an excellent opportunity to examine a representative sampling of the primitive lunar crust. Some attempts have been made to understand impact processes and the nature of the breccia components by detailed mineralogical studies of the breccia matrix material and individual clasts (e.g., J. Warner *et al.*, 1976; R. Warner *et al.*, 1976; James and Blanchard, 1976). Most recent attempts to characterize primitive or "pristine" crustal materials (e.g., Warren and Wasson, 1977; 1978; 1979) have centered around detailed chemical and isotopic studies of exotic clasts removed from highland breccias (e.g., norite, dunite, gabbro, QMD), although generally, the largest population of pristine nonmare material is considered to be anorthosite and troctolite (Ryder and Norman, 1978, 1979). Warren and Wasson (1977; 1979) have concluded that the lunar crust consists primarily of anorthosite, troctolite, and norite (i.e., ANT suite); anorthositic gabbro and low-K Fra Mauro represent non-pristine, 2nd generation rocks.

Highland breccia 66095, nicknamed "Rusty Rock" because of the abundance of FeOOH, is the most volatile-rich lunar rock thus far analyzed. It is unusually enriched in the halogens, Cl and Br, as well as other volatiles (e.g., Zn, Ge, Cd, Sb, Tl, Bi) (Jovanovic and Reed, 1975; Krähenbühl *et al.*, 1973). Many highland breccias contain the same volatiles in the same approximate proportions, but in lesser abundances. Therefore, 66095 can be viewed as a paradigm, *a representative of highland rocks in general*. Additional support for this can be found in the typical polymict nature of this rock, containing an assortment of lithic fragments (i.e., anorthosite, troctolitic anorthosite, and melt rocks). The understanding of the events surrounding the formation of this breccia will provide valuable insight into the problem of the origin and movement of volatile components in highland rocks.

In this study of highland breccia 66095, we characterize the mineralogy and petrology of the individual components of this polymict breccia and place them in a petrologic framework leading to a better understanding of breccia formation and impact processes. By characterizing the components, we should be able to look back through the impact event(s) and make some general statements about the nature of the primary crustal components.

ANALYTICAL METHODS

Mineral and glass analyses were obtained with an automated MAC 400S electron microprobe using the correction procedures of Bence and Albee (1968) with data from Albee and Ray (1970). The matrix and clast compositions of 66095 were determined by defocused beam analysis (DBA) using a spot diameter of approximately 50 μm. Corrections for heterogeneous target materials were applied according to the normative iteration procedure described by Albee *et al.* (1977). Replicate analyses of glass, matrix, and clast material indicate a precision of $\pm 2\%$ and $\pm 5\%$ (1σ) of the quantity present for major and minor elements, respectively.

We obtained a quantitative cumulative mode for 66095 based on modal analysis of all available petrographic thin sections, assuming the thin sections were made of random subsamples. Fine-grained basalt clasts, including Fra Mauro basalt, were subdivided into: (1) Fra Mauro basalt, (2) highland basalt (melt rock), and (3) gabbroic anorthosite (melt rock) based on textural similarities to selected clasts analyzed by DBA.

DESCRIPTIVE PETROLOGY

Lunar rock 66095 is a 1185 gm. medium gray, friable, clast-rich melt rock consisting of a fine-grained matrix (78% of the rock) containing a variety of mineral and lithic clasts (Bass, 1972; Garrison and Taylor, 1979a, 1979b; Fig. 1). The matrix of 66095 consists of a subophitic to ophitic intergrowth of 0.01 to 0.07 mm randomly oriented plagioclase laths (50 to 60% of the matrix), 0.1 to 0.3 mm subhedral to poikilitic pigeonite (up to 30% of the matrix), and 0.1 to 0.3 mm subhedral olivine (up to 10% of the matrix) with accessory amounts of FeNi metal, troilite, sphalerite, schreibersite, cohenite, and ilmenite in a fine mesostasis. A cumulative mode for 66095 is presented in Table 1; the separation of clasts into the Primary and 2nd Generation groups will be discussed in a subsequent section.

The most abundant clast type consists of a variety of fine-grained, shocked "basalt" clasts. These are comprised of clasts of poikiloblastic Fra Mauro basalt (9.5 modal %; Fig. 1b), as well as various melt-rock clasts identified by DBA as highland basalt (30.3 modal %) and gabbroic anorthosite (6.3 modal %; Fig. 1c, d). These basalt clasts differ mainly in abundance of phenocrystic plagioclase, as well as in the color of their mesostasis (gray to brown to black). In general, these clasts have sharp boundaries, although gradational, resorbed boundaries are not uncommon. Two of the melt-rock clasts contain small sphere-shaped, chondrule-like inclusions (Fig. 1e). These are similar in size and texture to chondrules described from lunar samples by Kurat *et al*. (1972).

The second most abundant clast population consists of anorthosite (3.6 modal %), cataclastic anorthosite (17.6 modal %), and troctolitic anorthosite (10.4 modal %) ranging in texture from intergranular to granoblastic aggregates of plagioclase (0.1 to 0.15 mm), with olivine occurring as subhedral to anhedral interlocking grains (0.01 to 0.07 mm), these rocks can range into porphyroclastic impactites (Fig. 1a, b). The anorthosites and troctolitic anorthosites contain rare FeNi metal (Fig. 2). Mineral clasts (up to 22 modal % of the clasts) are almost entirely plagioclase (0.1 to 0.5 mm), although rare olivine and pink spinel clasts do occur. The monomineralic plagioclase clasts show various degrees of shock, from slightly fractured to total conversion of the plagioclase into maskelynite.

Breccia 66095 contains numerous planar to slightly irregular fractures and lenticular rootless veins filled with brown to black shock-produced glass (Fig. 1f). Bass (1972) suggested that many of the veins of glass formed *in situ*. These glass veinlets contain abundant small spherules of FeNi metal and commonly exhibit flow banding. However, there is a lack of vesicles within the veins.

A peculiar feature of 66095 is the abundance of "rusty" stains and grains of akaganéite (β-FeOOH) surrounding the FeNi metal grains within the matrix;

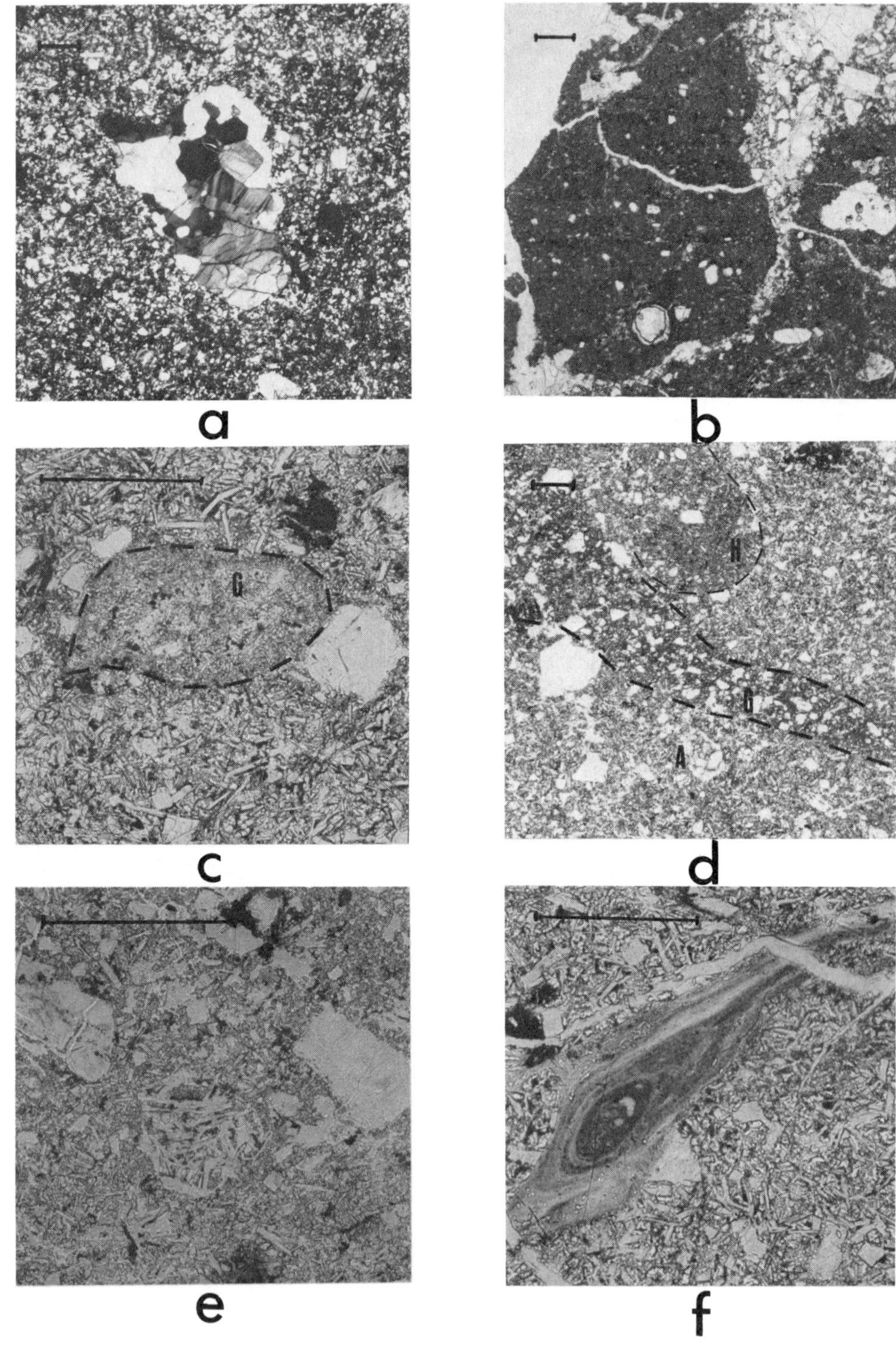
a
b
G
H
G
A
c
d
e
f

Table 1. Cumulative mode for 66095.

	% of rock		% of clast population	
PRIMARY CLASTS				
Anorthosite	0.8%	10.1%	3.6%	45.7%
Troctolitic anorthosite	2.3%		10.4%	
Fra Mauro basalt	2.1%		9.5%	
Plagioclase clasts	4.9%		22.2%	
Olivine clasts	tr		tr	
Pink spinel clasts	tr		tr	
2ND GENERATION CLASTS				
Cataclastic anorthosite	3.9%	12.0%	17.6%	54.2%
Gabbroic anorthosite	1.4%		6.3%	
Highland basalt	6.7%		30.3%	
MATRIX	77.9%		—	
	100.0%		99.9%	

some of the metal within the clasts has also been rusted. These stains occur both on the surface of this rock and deep within the chipped parent. Taylor *et al.* (1974) suggested that the oxhydration of lunar $FeCl_2$, lawrencite, by contaminating terrestrial water produced the akaganéite.

PREVIOUS CHEMISTRY

Bulk chemical analyses of 66095 subsamples are presented in Table 2. These analyses reflect different mixtures of impact-produced melt and included clast material; some actually approach the bulk composition of average Apollo 16 regolith; however, little is reported about the min/pet of these samples. It is evident that no one such analysis is representative of the breccia as a whole. In such cases, *the concept of whole-rock chemistry is meaningless*.

The volatiles in 66095 have received considerable attention (e.g., Krähenbühl *et al.*, 1973; El Goresy *et al.*, 1973; Brunfelt *et al.*, 1973; Taylor *et al.*, 1973; Jovanovic and Reed, 1974; 1975). In particular, the leachable Cl is very high, supposedly a reflection of loosely-held Cl along cracks and grain boundaries; P_2O_5 is also high, a contribution from the relatively abundant schreibersite. Although the absolute abundances of many volatile elements are high in this rock, the elemental ratios are similar to a large number of highland rocks with lesser amounts.

Fig. 1. Representative clasts and components of 66095. (a) anorthosite clast; (b) Fra Mauro basalt clast containing small amorthosite clasts. Upper right corner is an adjacent cataclastic anorthosite clast; (c) shocked gabbroic anorthosite clast (G); (d) highland basalt clast (H), gabbroic anorthosite clast (G), and cataclastic anorthosite clast (A); (e) chondrule; (f) foliated glass veinlet cutting ophitic matrix. Scale bar = 200μm.

Table 2. Chemical analyses of breccia 66095 and average Apollo 16 regolith (<1 mm fraction).

	66095,37[1]	66095,47[2]	66095,48[3]	66095,?[4]	Apollo 16 regolith[5]
SiO_2	44.07	44.92	—	44.47	44.94
TiO_2	0.18	0.72	0.50	0.71	0.58
Al_2O_3	30.02	23.66	25.89	23.55	26.71
Cr_2O_3	0.084	—	0.126	0.148	0.12
FeO	3.02	7.46	5.53	7.16	5.49
MnO	0.05	0.08	0.071	0.08	0.07
MgO	4.72	8.92	9.29	8.75	5.96
CaO	16.65	13.52	12.88	13.69	15.57
Na_2O	0.35	0.45	0.46	0.42	0.48
K_2O	0.08	0.158	0.088	0.15	0.13
P_2O_5	0.06	0.251	—	0.24	0.12
S	0.14	0.092	—	0.12	—
	99.434	100.231		99.488	100.17

[1] Hubbard *et al.* (1974) [2] Duncan *et al.* (1973) [3] Brunfelt *et al.* (1973)
[4] Lunar Sample Information Catalog Apollo 16 (1972) [5] Taylor (1975)

Table 3. Representative microprobe analyses of matrix minerals.

	CPX 79-M1*	CPX 79-M2	OLIVINE 83-M2 (6)**	OLIVINE 79-M1 (3)	PLAGIOCLASE 83-M2 (11)	PLAGIOCLASE 79-M1 (9)
SiO_2	51.8	53.0	37.7 (2)***	37.7 (2)	45.1 (2)	44.8 (4)
TiO_2	1.69	1.30	n.a.	n.a.	n.a.	n.a.
Al_2O_3	2.07	1.68	0.30(5)	0.24(2)	35.4 (2)	34.6 (4)
Cr_2O_3	0.46	0.48	0.11(1)	n.d.	n.a.	n.a.
FeO	11.5	12.9	21.6 (4)	21.3 (4)	0.24(7)	0.32(7)
MnO	0.23	0.27	0.31(11)	0.21(3)	n.a.	n.a.
MgO	23.1	25.9	40.3 (5)	40.9 (2)	0.27(4)	0.30(5)
CaO	8.17	3.74	0.28(5)	0.28(1)	17.9 (2)	19.0 (4)
Na_2O	n.d.	n.d.	n.a.	n.a.	0.61(9)	0.57(15)
K_20	n.d.	n.d.	n.a.	n.a.	0.11(4)	0.09(5)
	99.02	99.27	100.70	100.63	99.63	99.68
Si	1.903	1.926	0.975	0.972	2.082	2.078
Al	0.089	0.071	0.009	0.006	1.928	1.890
Ti	0.046	0.035	—	—	—	—
Cr	0.013	0.013	0.002	0.000	—	—
Mg	1.261	1.400	1.550	1.570	0.018	0.020
Fe	0.353	0.390	0.466	0.459	0.008	0.012
Mn	0.006	0.007	0.006	0.004	—	—
Ca	0.321	0.145	0.007	0.007	0.885	0.942
Na	0.000	0.000	—	—	0.054	0.050
K	0.000	0.000	—	—	0.005	0.004
	3.992	3.987	3.015	3.018	4.980	4.996

* number refers to 66095 subsample (e.g., 79-M1 = 66095,79 matrix area 1). ** number of analyses averaged. *** Units in ( ) represent standard deviation of replicate analyses in terms of the least units cited; n.a. = not analyzed; n.d. = not detected = <0.04%.

BRECCIA MATRIX AND GLASS

Average mineral analyses for matrix areas are presented in Table 3. Olivines have a uniform composition (Fo_{77}) throughout the rock. Plagioclase compositions are uniform on a local scale but vary in different areas of the rock (An_{89-94}). Pigeonite compositions are highly variable in all parts of the rock ranging from $Wo_7En_{72}Fs_{20}$ to $Wo_{17}En_{65}Fs_{18}$ with Al/Ti from 1.9 to 2.0. Taylor *et al.* (1973) found uniform FeNi metal compositions throughout the matrix (Fig. 2). For representative analyses of accessory schreibersite, cohenite, sphalerite, and akaganéite, the reader is referred to Taylor *et al.* (1973).

Defocused beam analyses of matrix and vein glass are presented in Tables 4 and 5. Figure 3 shows a plot of Al_2O_3 versus FeO + MgO on which 66095 matrix

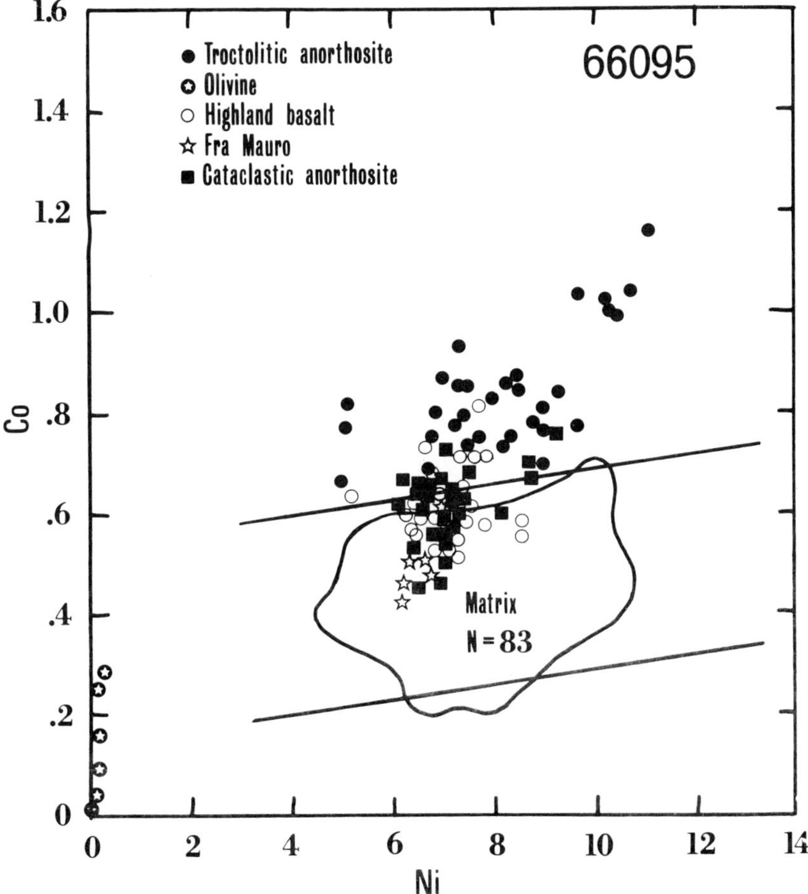

Fig. 2. Ni and Co compositions of native FeNi grains in 66095. The outlined area was taken from Taylor *et al.* (1973) and represents the metal grains in the matrix. The parallel lines are taken from Goldstein and Yakowitz (1973) and are used for reference.

Table 4. Defocussed beam analyses (DBA) of rock 66095.

	83-1* clast	83-2 clast	83-3 clast	83-4 matrix	83-5 matrix	83-6 matrix	83-7 chondrule	79-1 clast	79-2 matrix	79-3 matrix
SiO_2	44.5	45.2	45.7	45.7	45.0	46.2	44.7	46.2	46.5	46.7
TiO_2	0.53	0.68	0.67	1.19	1.20	0.57	0.62	0.61	0.75	0.74
Al_2O_3	29.8	27.5	25.6	23.4	21.6	25.0	31.5	19.7	24.3	22.6
Cr_2O_3	0.10	0.12	0.15	0.11	0.11	0.16	n.d.*	0.28	0.07	0.08
FeO	2.83	3.40	3.72	4.70	5.41	3.50	1.46	6.25	3.62	4.43
MnO	0.63	0.31	0.38	0.41	0.25	0.36	0.80	0.30	0.33	0.26
MgO	3.57	4.43	6.13	7.47	9.18	6.20	1.82	11.3	6.18	7.99
CaO	15.6	14.7	14.0	13.9	13.2	13.7	17.4	12.7	14.9	13.9
Na_2O	0.48	0.52	0.52	0.59	0.55	0.62	0.52	0.49	0.67	0.63
K_2O	0.13	0.28	0.20	0.44	0.36	0.22	0.35	0.11	0.34	0.24
Σ	98.17	97.14	97.07	97.91	96.86	96.53	99.17	97.94	97.66	97.57

	86-1a clast	86-1b clast	82-2 clast	86-3 clast	86-4 clast	86-5 clast	86-6 matrix	86-7 matrix	86-8 chondrule
SiO_2	46.6	46.5	46.7	44.9	45.9	46.1	46.4	46.6	45.9
TiO_2	0.51	0.52	0.61	0.51	0.41	0.50	0.79	0.50	0.67
Al_2O_3	21.9	22.5	21.4	27.9	24.5	27.5	24.4	22.7	29.5
Cr_2O_3	0.13	0.19	0.16	0.08	0.13	0.02	0.08	0.18	n.d.
FeO	5.67	5.34	5.61	3.10	4.39	3.69	3.77	5.10	1.64
MnO	0.14	0.14	0.09	0.22	0.22	0.61	0.39	0.13	0.69
MgO	9.87	9.15	10.1	5.14	7.20	4.80	7.30	9.26	2.47
CaO	13.0	13.7	13.0	15.7	14.3	16.0	14.4	12.7	17.4
Na_2O	0.50	0.46	0.42	0.47	0.48	0.48	0.57	0.56	0.55
K_2O	0.11	0.15	0.16	0.09	0.09	0.21	0.20	0.11	0.59
Σ	98.43	98.65	98.25	98.11	97.62	99.92	98.30	97.84	99.41

* The first two numbers refer to the subsample number for 66095. ** not detected; less than 0.03.

Table 5. Defocussed beam analyses (DBA) of shock-produced glass in 66095.

	79-1*	79-2a	79-2b	83-1	83-2	83-3a	83-3b	86-1	86-2
SiO_2	44.6	42.7	43.1	44.5	44.6	44.6	44.5	44.7	44.8
TiO_2	0.77	0.70	0.71	0.71	0.68	0.55	0.57	0.80	0.75
Al_2O_3	23.7	23.0	22.6	22.9	23.4	25.8	25.7	22.1	22.8
Cr_2O_3	0.17	0.16	0.05	0.13	0.11	0.09	0.05	0.19	0.19
FeO	5.36	8.98	8.28	7.27	6.92	4.90	5.15	8.50	7.60
MnO	0.04	0.10	0.12	0.07	0.09	0.16	0.36	0.10	0.13
MgO	9.12	9.19	9.29	9.65	9.29	7.24	7.41	10.1	9.82
CaO	13.9	12.7	13.3	12.9	13.0	14.0	14.7	12.8	13.1
Na_2O	0.61	0.63	0.60	0.52	0.49	0.44	0.45	0.48	0.50
K_2O	0.16	0.12	0.13	0.21	0.22	0.21	0.23	0.16	0.19
S	98.43	98.28	98.18	98.86	98.80	97.99	99.12	99.93	99.88

* The first two numbers refer to the subsample number for 66095.

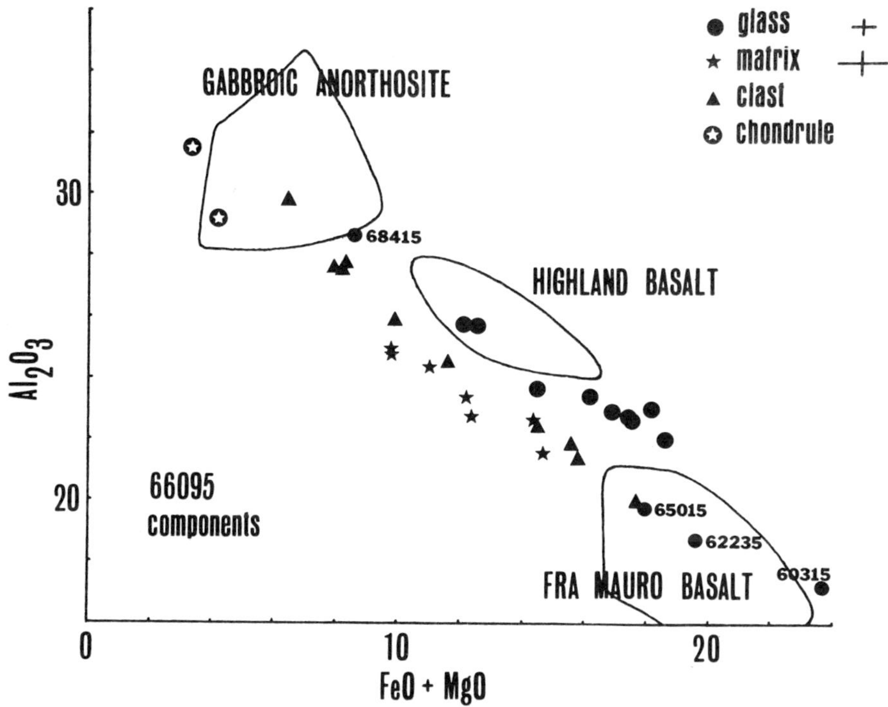

Fig. 3. Plot of Al_2O_3 versus MgO + FeO showing the compositions of 66095 matrix, vein glass, and clasts relative to the compositions of rocks and glasses of Fra Mauro basalts, highland basalts, and gabbroic anorthosites as discussed in the text. Crosses are 1σ error bars.

and glass DBA are shown relative to rocks and glasses of gabbroic anorthosite, highland basalt, and Fra Mauro basalt compositions (Reid *et al.*, 1972; Dence and Plant, 1972; Taylor, 1975; Ryder and Norman, 1978a; Reid, 1974; 1977). With the exception of K_2O and Na_2O, the majority of 66095 matrix (21.6–25.0 wt.% Al_2O_3) and glass (22.1–25.8 wt.% Al_2O_3) are similar to the "highland basalt" compositions defined by Reid *et al.* (1972) and Reid (1974) (i.e., 43.9–45.4 wt.% SiO_2, 0.3–0.9 wt.% TiO_2, 24.4–27.6 wt.% Al_2O_3, 0.07–0.11 wt.% Cr_2O_3, 4.9–6.6 wt.% FeO, 6.1–8.9 wt.% MgO, 14.5–15.6 wt.% CaO, 0.1–0.4 wt.% Na_2O, and 0.1 wt.% K_2O). The K_2O (0.12–0.23 wt.%) and Na_2O (0.44–0.63 wt.%) of the 66095 glass are also similar to Reid's highland basalt within our analytical precision, but the matrix K_2O (0.11–0.44 wt.%) and Na_2O (0.55–0.67 wt.%) deviate quite significantly from Reid's compositions. Realizing the inherent difficulties in DBA of heterogeneous targets, we feel that this deviation is not grounds for invalidating the classification of the matrix as "highland basalt". It is noteworthy that some of the low-Al_2O_3 compositions (e.g., 83–5 and 86–7 in Table 4 and 86–1 in Table 5) actually approach the composition of Fra Mauro basalt

(Reid, 1974; Reid *et al.*, 1972; 1977; Dence and Plant, 1972) (i.e., 46.1–51.6 wt.% SiO_2, 0.4–2.5 wt.% TiO_2, 14.5–21.0 wt.% Al_2O_3, 0.1–0.3 wt.% Cr_2O_3, 7.1–14.3 wt.% FeO, 4.7–11.0 wt.% MgO, 10.1–13.3 wt.% CaO, 0.3–1.1 wt.% Na_2O, 0.1–0.9 wt.% K_2O).

The matrix and vein glass compositions are quite heterogeneous from one part of the breccia to another. This may reflect compositional changes of the impact melt caused by local digestion of clasts of a variety of bulk compositions. It is generally considered that melt rocks are quite homogeneous—a consequence of melt sheet homogenization (Simonds *et. al.*, 1976, 1978), but clearly, highland breccia 66095 *does not* display this characteristic. Simonds *et al.* (1976) ascribed this homogenization to violent mixing such that the melt was homogenized on a submillimeter scale. Breccia 66095 may represent a more distal portion of a melt sheet such that mixing was not efficient.

From Fig. 3 it appears that the compositions of 66095 matrix and glass can be represented as mixtures of Fra Mauro basalt (see composition above) and anorthosite or gabbroic anorthosite (e.g., 44.5 wt.% SiO_2, 0.35 wt.% TiO_2, 31.0 wt.% Al_2O_3, 0.04 wt.% Cr_2O_3, 3.5 wt.% FeO, 3.4 wt.% MgO, 17.3 wt.% CaO, 0.1 wt.% Na_2O, 0.01 wt.% K_2O. This mixing model is also manifest in the abundance of the other six oxides as well.

To a first approximation, *66095 is a mixture of Fra Mauro basalt (KREEP) and* a more anorthositic component such as *gabbroic anorthosite, anorthosite, or troctolitic anorthosite*. From these observations, it is clear that melt rock 66095 contains a significant amount of KREEP-like component. Because of the inherent uncertainties in choosing the exact compositions of the end-members, particularly the KREEP component, any attempt to quantify this mixing model (e.g., with a least-squares treatment) would bc unrealistic.

Inspection of Fig. 3 reveals an apparent relative displacement of compositions of the matrix and glass. Examination of Table 4 and 5 shows that this is due mainly to FeO contents—matrix: 3.50–5.41%; glass: 4.90–8.98%. Although the normative correction procedure of Albee *et al.* (1977) was used for the DBA, there remains the possibility of a systematic analytical error connected to the correction scheme of the matrix and/or glass.

These glass veinlets could have two possible origins: (1) *in situ* melting as suggested by Bass (1972), or (2) injection of the host breccia by superheated melt as suggested by Roedder and Weiblen (1977). Bass (1972) found many of the 66095 glass veins to be *isolated* and *rootless* and stated that they did not appear to be injected, although injection could not be discounted for some of the larger black glass veins that appeared to occupy fractures. Many of the veins have sharp contacts with the matrix material, although gradational boundaries are not uncommon. The veins commonly contain mineral "fragments" but do not cut mineral or lithic clasts suggesting that fracture or vein formation was restricted to the melt rock matrix. Although injection of molten material into 66095 along fractures cannot be completely discounted, the lenticular, isolated and rootless nature of these veins is highly suggestive of *in situ* melt formation.

MELT-ROCK CLASTS

Highland breccia 66095 contains a varied suite of fine-grained clasts ranging in texture from porphyritic to poikiloblastic to intergranular (Fig. 1). Some of these clasts themselves contain small clasts of anorthosite (Fig. 1b). These melt-rock clasts consist of microphenocrysts, of plagioclase (An_{92-94}), olivine (Fo_{74-76}), and pigeonite in a very fine-grained, shocked mesostasis (e.g., Fig. 1c, d). Minor FeNi metal grains occur within the mesostasis, invariably associated with rust, akaganeite (Fig. 2).

The compositions of nine clasts and two chondrules were determined by DBA (Table 4). When plotted on the Al_2O_3 versus Fe + MgO diagram (Fig. 3) and compared to the compositions of Fra Mauro basalt, highland basalt, and gabbroic anorthosite as discussed above, a spectrum of compositions are realized: four of the clasts, 83-1, 83-2, 86-3, and 86-5, with 27.5–29.8 wt.% Al_2O_3 resemble gabbroic anorthosite; four clasts, 83-3, 86-1, 86-2, and 86-4, with 21.9–25.6 wt.% Al_2O_3 are similar to 66095 matrix and to highland basalt; and one clast, 79-1, (19.7 wt.% Al_2O_3) is similar to Fra Mauro basalt.

In general, these various melt-rock clasts are *not* similar to noritic or troctolitic compositions such as anorthositic norite (e.g., 15455) and troctolitic anorthosite (e.g., 76335) because of their much higher TiO_2 (0.41–0.68 wt.%); noritic and troctolitic rocks generally have less than 0.2 wt.% TiO_2, commonly less than 0.05 wt.% (Ryder and Norman, 1978).

The clast compositions are similar to compositions of other larger samples returned from the Apollo 16 site. Clast 79-1 is chemically identical to and petrographically similar to low-K Fra Mauro poikiloblastic rock 65015. The gabbroic anorthosite clasts (83-1, 83-2, 86-3, and 86-5) are chemically similar to 68415. The four highland basalt melt-rock clasts are chemically similar to any number of highland basalt melt rocks, even bulk 66095 sub-samples (Table 2).

MINERAL, ANORTHOSITE, AND TROCTOLITIC ANORTHOSITE CLASTS

Clasts of anorthosite (<5% olivine) cataclastic anorthosite, and troctolitic anorthosite (5–20% olivine) form up to 32% of all clasts included in melt-rock 66095. This clast group can be divided into an intergranular-texture class and a cataclastic class. The intergranular anorthosites (3.6% of the clasts) have plagioclase ranging from An_{92} to An_{96} and olivine ranging from Fo_{76} to Fo_{80}; there are two types of troctolitic anorthosites (10.4% of the clasts)—An_{96-98} with Fo_{58-69} and An_{92-96} with Fo_{76-82}. The cataclastic anorthosites (17.6% of the clasts) have plagioclase ranging from An_{92} to An_{94} and olivine ranging from Fo_{72} to Fo_{75}. Figure 4 shows a plot of mol % An in plagioclase versus the 100 × Mg/Mg+Fe in coexisting olivine for these three clast types. Both of the petrochemical plutonic rock types considered by Warren and Wasson (1978) to make up "pristine" lunar

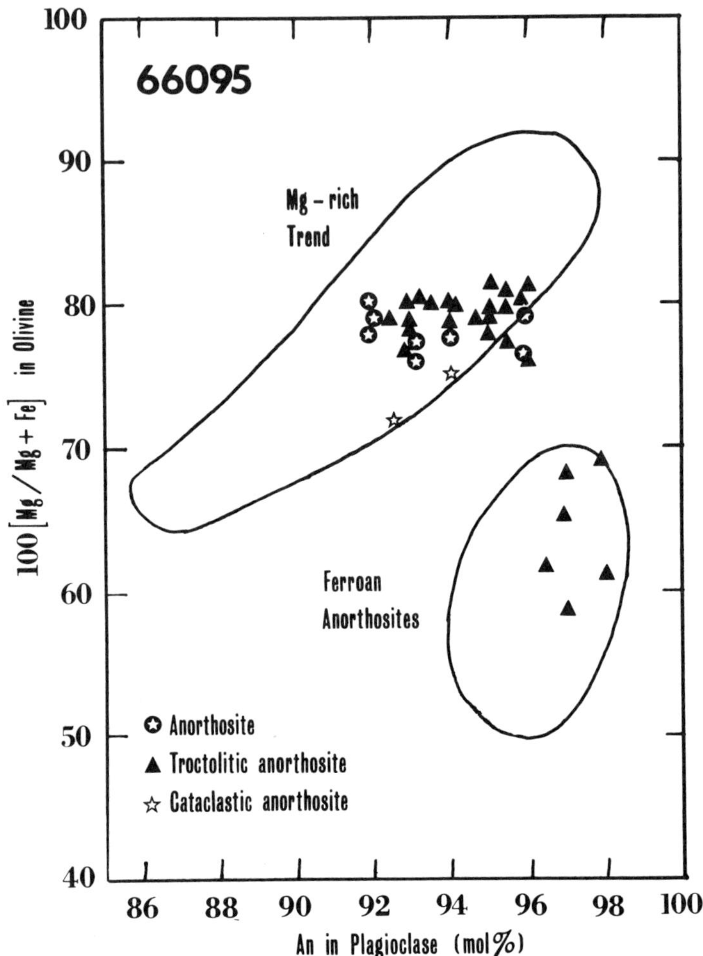

Fig. 4. Olivine versus plagioclase in several different clasts in 66095. The outlined areas and their names are taken from Warren and Wasson (1978).

plutonic crust are represented by the clasts of 66095. Anorthosites lie wholly within the Mg-rich plutonic trend; troctolitic anorthosites lie dominantly in the Mg-rich trend but several lie in the field of ferroan anorthosites.

Figure 2 shows the composition of FeNi metal grains from several clasts of troctolitic anorthosite and cataclastic anorthosite, as well as clasts of highland basalt and Fra Mauro basalt and the matrix of 66095. The troctolitic anorthosite metals are distinctly above the range of matrix metals and the reference boundary lines of Goldstein and Yakowitz (1971). Cataclastic anorthosite metals are intermediate between matrix and troctolitic anorthosite, as are those of the highland basalt melt-rock clasts.

Mineral clasts, dominated by plagioclase, form over 22% of the included material. The plagioclase clasts range in composition from An_{93} to An_{97}, distinctly more calcic than either matrix plagioclase or plagioclase microphenocrysts in the melt-rock clasts. Olivine clasts range in composition from Fo_{76} to Fo_{86}, but generally cluster near Fo_{76-80}. The compositional ranges of the plagioclase and olivine clasts are identical to those of the anorthosite and troctolitic-anorthosite clasts that fall within the Mg-rich plutonic trend suggesting that these mineral clasts represent disaggregated anorthositic and troctolitic material. Rare pink spinel clasts occur in subsamples 66095,11 and 66095,79. The spinel is a Cr-pleonaste; 66095,11 spinel has 3.25 wt.% Cr_2O_3, 12.4 wt.% FeO, and 18.4 wt.% MgO and 66095,79 spinel has 2.16 wt.% Cr_2O_3, 6.58 wt.% FeO, and 23.5 wt.% MgO. Some of the olivine clasts contain minute FeNi metal inclusions; their compositions are shown in Fig. 2.

The suite of clasts included in highland breccia 66095 contains every highland rock type, except norite, which suggests that 66095 probably provides us with a representative sampling of the lunar highland crust.

MODELING

The compositions of 66095 matrix and glass appear to represent a mixing line between low-K Fra Mauro basalt and a suite of plutonic rocks which includes anorthosite, troctolitic anorthosite, and gabbroic anorthosite. This suggests that the matrix of 66095, as well as other highland basalt or anorthositic gabbro melt rocks, were formed by impact-melting of strata consisting of Fra Mauro (KREEP) basalt and a chemically complex "ANT" plutonic suite.

Recent attempts to model highland rocks (e.g., Ryder, 1979; Wasson *et al.*, (1977) have involved the selection of "pristine" rocks as end-members. However, this is certainly not feasible for a breccia such as 66095 which consists of clasts of 2nd or higher-derivative remelts. In considering 66095, as well as other impact-melt breccias, we prefer a simple model in which a multiple-layer stratigraphy is subjected to a basin-forming impact. This impact generates large volumes of "highland basalt" or "anorthositic gabbro" melt containing a varied suite of clasts—a representative sampling of the pre-impact stratigraphy. Breccia 66095 contains clasts of anorthosite and troctolitic anorthosite (and gabbroic anorthosite) which represent one stratigraphic unit—a lithologically and chemically complex (ANT") plutonic complex. Fra Mauro clast 79-1 represents the Low-K Fra Mauro (LKFM) stratum. Since over 50% of the clasts in 66095 are 2nd generation cataclastic and melt-rock clasts, it is probable that to greater or lesser extents, the stratigraphy contained ejecta and/or regolith deposits (Fig. 5).

The general model for breccia formation presented in Figure 5 allows some latitude in the degree of "pristinity" (Warren and Wasson, 1977, 1978) required of each crustal end-member. If the LKFM (KREEP) component is indeed a highland extrusive unit (Hess *et al.*, 1978; Reid *et al.*, 1977), brecciation and regolith formation would be expected to modify the upper portions of such

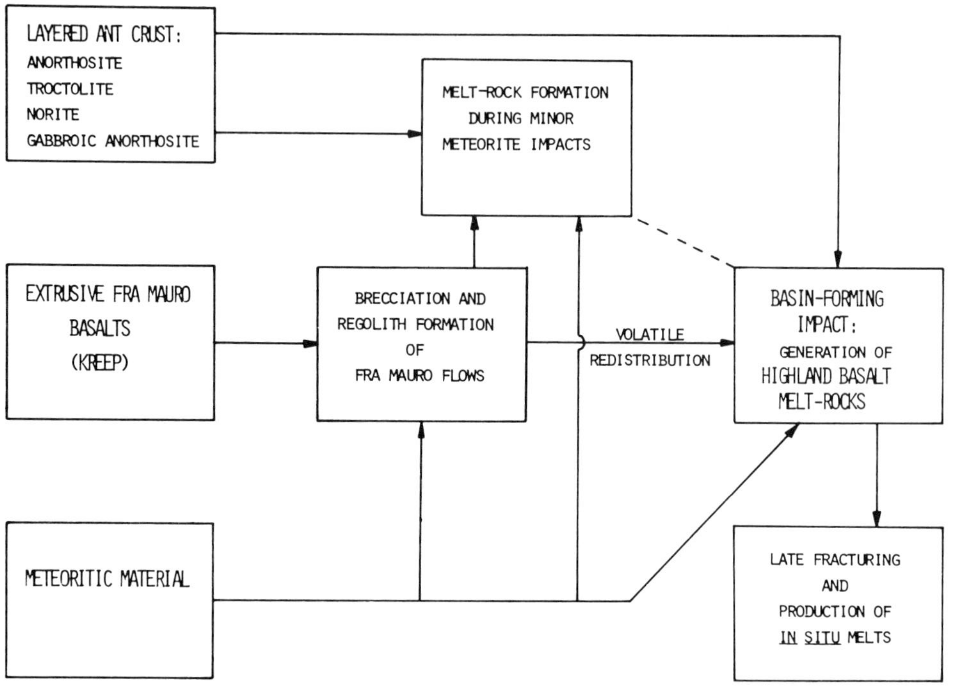

Fig. 5. Flow chart for the highland-basalt breccia mixing model.

"flows"; this leads to addition of a meteoritic component and subsequent reduction in "pristinity", with the overall bulk composition left relatively unchanged. It is highly unlikely that fragments of pristine Fra Mauro basalt flows could have survived the intense highland cratering. The most probable samples of "modified" Fra Mauro (KREEP) basalt are the low-K Fra Mauro poikiloblastic rocks returned from the Apollo 16 site (e.g., 65015, 62235, and 60315). Thus, rocks such as LKFM basalt, by virtue of their extrusive nature (i.e., exposure on the lunar surface), should not be expected to be void of at least some meteoritic material. Therefore, many such lunar rocks are not "pristine" according to siderophile criteria, but are the products of lunar volcanic or plutonic processes—a principle constituent of the *primary* lunar highland crust. Likewise, exposed "ANT" group plutonic strata would also have experienced local brecciation events (i.e., producing impactites such as 67955) and/or melting events (i.e., producing melt rocks such as 68415) during the highland cratering events. As in the above case, the bulk composition remains essentially unchanged except for a small addition of meteoritic material.

Available trace element and isotopic data allow us to quantitatively evaluate this general model. High abundances of U and Th in some of the Apollo 16 breccias (e.g., 66095) led Nunes *et al.* (1973) to conclude that these rocks contain a large amount of KREEP-like material. Breccia 66095 has an abundance of incompatible elements intermediate between the two proposed "end-members". Reported REE data for 66095 (Brunfelt *et al.*, 1973; Hubbard *et al.*, 1974) also support the mixing model (Figure 6). Sub-sample 66095,36 has a REE pattern similar to LKFM (e.g., 62235,57). Sub-sample 66095,37 has REE abundances much like gabbroic anorthosite or troctolite, while sub-sample 66095,48 shows intermediate REE abundances clearly reflecting a mixture.

Quantitative modeling—chasing one's own tail?

This general model discussed above (Fig. 5) is in reality a pseudo-ternary mixing model; its relationship to more finite end-member models is diagramatically displayed in Figure 7. The "ANT" plutonic component is multi-dimensional since it contains a complete spectrum of rock types realized in a differentiating plutonic complex. Given the lateral and vertical heterogeneities implicit in the primitive

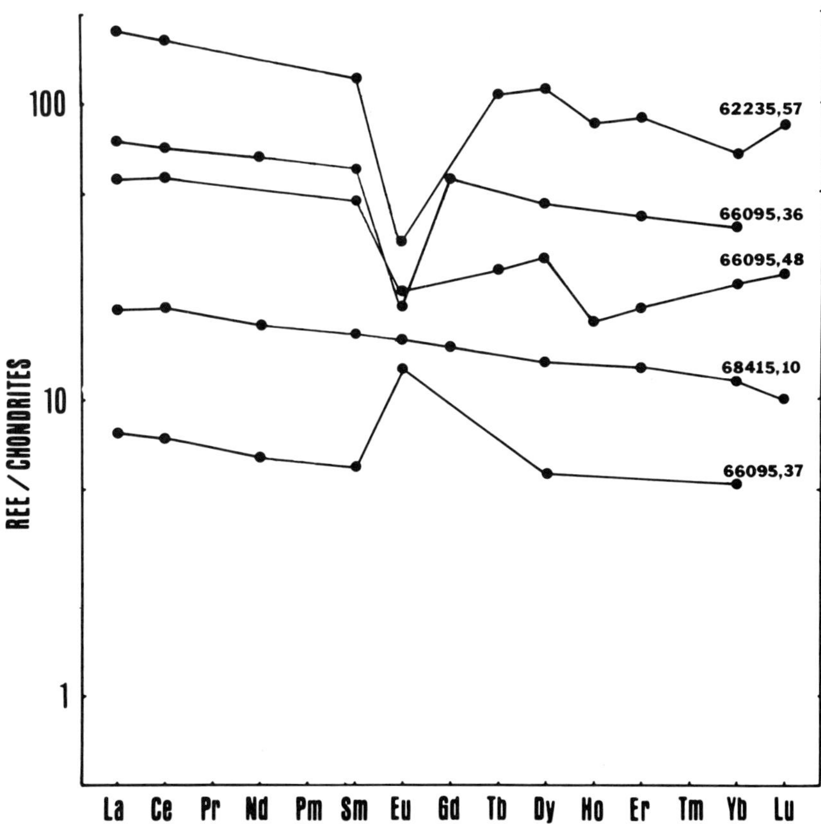

Fig. 6. The compositions of REE in subsamples of various highland rocks. Note the contrasting analyses for 66095. See text for further discussion.

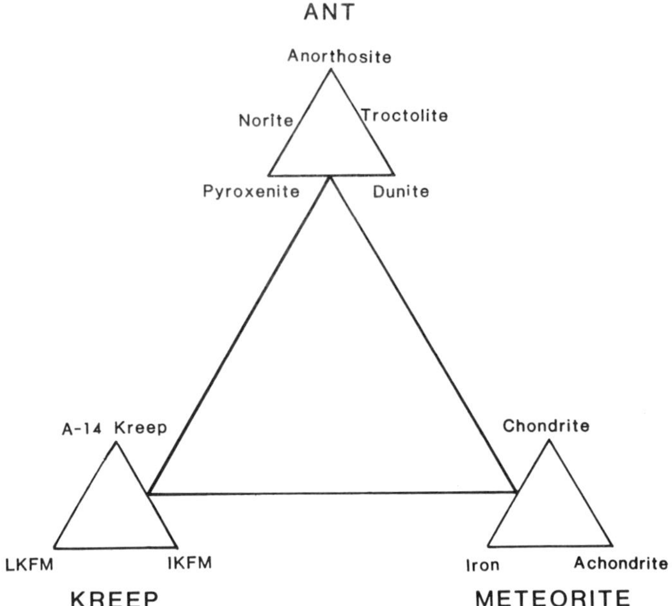

Fig. 7. The relative positions of the various components used to model the highland rocks.

highlands crust and the proportions of the lunar crust involved during major basin-forming impacts, use of this composite "ANT" plutonic component is more realistic than the very small plutonic subsystems commonly chosen as end-members (e.g., dunite, anorthosite, norite, etc.). Even the meteoritic component ELC (extra-lunar component) is in fact multi-dimensional, consisting of a suite of meteorite types currently realized.

Many of the shortcomings of the quantitative models with finite end-members are their failures to explain certain trace element systematics. These failures have led to the invention of elusive n+1 components such as SCCRV (Wasson *et al.*, 1977). Many of these shortcomings are manifestations of the *failures of investigators to correctly define the exact chemical compositions of the finite end-members*. Ryder (1979) discussed the arguments surrounding the exact compositions of a high-Mg component and the KREEP component. In his quantitative modeling, he avoids the complications surrounding the high-Mg component by using another finite component—norite. Examination of his quantitative calculations using KREEP, norite, anorthosite, dunite, and ELC reveals several interesting observations: (1) many of the elements contributed to the mixture by norite are easily contributed by the KREEP component as well, (2) dunite, an extremely rare lunar rock type, is included when troctolite is probably more realistic, and (3) the fits provided by Ryder's end-members are not much better than those of Wasson *et al.* (1977) using SCCRV. It appears that the sole function of norite in

Ryder's modeling is to overcome the imprecision introduced by using an arbitrarily defined KREEP end-member. The use of dunite as an end-member is unwarranted because of its rarity; troctolite is more realistic, although we realize that even dunite *is* more realistic than SCCRV.

We feel that many of the inconsistencies of these quantitative models are the result of the *non-representative nature* of the trace element systematics displayed by these small finite plutonic subsystems (e.g., anorthosite, dunite, and norite) or by the "KREEP". How can small sub-systems be representative of the trace element content of a large plutonic system subject to fractionation-controlled trace-element systematics? For example, can the compositions of one or two coarse-grained rocks from the Bushveld completely represent this complex? To seek to define the individual components of the plutonic suite is an ambitious endeavor, but when applying these components to breccia modeling, it must be realized that additional limitations exist because of the non-representative nature of the samples (i.e., size) and the limited data base. Recent modelings of lunar highland rocks and soils have fallen prey to the *over-interpretation* of chemical data.

We feel that the simplistic, although realistic, qualitative model outlined above removes some of the ambiguity associated with attempts to quantitatively model the lunar crust and subsequent breccias with small exotic, non-representative, finite magmatic subsystems. Our model *cannot be quantified,* but serves effectively as a working model in which to examine the primitive lunar crust and the subsequent impact processes. In fact, we do not feel that any model can be totally quantified and realistically explain the highland breccias.

The nature and role of the LKFM (KREEP) stratum

In the model above, we imply that the LKFM (KREEP) stratum is a volcanic unit extruded onto the surface of the primitive lunar crust (Hess *et al.,* 1978; Reid *et al.,* 1977). In our model, there are no constraints that the Fra Mauro basalts be uniformly distributed across the lunar surface or contemporaneously erupted. The occurrence of a KREEP component in all melt rocks, breccias, and soils suggests that the LKFM stratum was quite widespread. The absence of "pristine" LKFM samples (according to siderophile element abundances) suggests exposure to meteorite bombardment for a substantial length of time. Ryder and Spudis (1979) discussed the indirect evidence for the existence of highland KREEP volcanism prior to the end of major impact events near 4.0 b.y. ago. The occurrence of KREEPy clasts within melt rocks (e.g., 66095) with 3.9–4.0 b.y. ages implies KREEP (LKFM?) accumulations must predate these events. They suggested that the existence of KREEP-free impactites with 4.2–4.3 b.y. ages limits the abundance and distribution of KREEP volcanics. This possibility is not inconsistent with our model, although if we remove the constraint that LKFM be uniformly distributed across the lunar surface, LKFM could have occurred prior to 4.2–4.3 b.y. as suggested by Schonfeld and Meyer (1973). The high abundance of cataclastic anorthosite clasts in 66095 (17.6% of the clasts) also suggests that the primitive plutonic crust was exposed to meteorite bombardment for a substantial length of time and/or was never completely covered by LKFM.

Nunes and Tatsumoto (1973) found 66095 to contain an abnormally high abundance of Pb (13.75 to 16.26 ppm). They considered 85% of this Pb to be excess radiogenic Pb; an order of magnitude more than could be produced by radioactive decay in 4.5 b.y. This excess Pb is isotopically homogeneous which suggests Pb introduction at a discrete time. The $^{207}Pb/^{206}Pb$ age of 4.01 ± 0.06 b.y. and an internal U-Pb isochron age of 3.82 ± 0.28 b.y. suggest Pb introduction by some lunar event at about 4.0 b.y. This event was also noted by Nyquist *et al.* (1973) who reported a Rb-Sr BAB1 age of 3.9 ± 0.3 b.y. for a subsample of 66095. Using a three-stage model for Pb evolution, Nunes and Tatsumoto (1973) calculated an early crustal differentiation age of 4.47 b.y. This suggests that 66095 either contains primary crustal material or the Pb systematics retain a memory of primary crustal Pb.

The LKFM basalts were intensely metamorphosed and recrystallized as nearly all of their Pb was expelled, hence their high $^{207}Pb/^{206}Pb$ ratios (>10,000) (Nunes *et al.*, 1973). Many highland melt rocks, including 66095, and the Apollo 16 soils contain excess radiogenic Pb. Nunes and Tatsumoto (1973) found the uniform isotopic composition of the leachable Pb in 66095 to be consistent with Pb introduction at a discrete time. It is reasonable that this excess radiogenic Pb in the melt rocks and soils is that driven out of the intensely metamorphosed LKFM stratum.

Volatiles in highland breccias: effects of impact-induced metamorphism

It is likely that the volatilization and redistribution of Pb discussed above was accompanied by the redistribution of other volatile elements (e.g., Cl, Br, Zn, Rb, Ag, and Tl). Thus, it may be possible to explain the extreme enrichment of volatiles in melt-rock breccias such as 66095 and 61016 as the result of their containing substantial KREEP or LKFM component *and* the *addition of volatiles driven out of the volatile-rich LKFM stratum* during impact-induced metamorphism. Since the elemental ratios of the extremely volatile-rich highland breccia 66095 and 61016 mirror those for other rocks, it appears that highland rocks differ only in the absolute amount of volatile elements taken up during the impact events.

Volatile element mobilization is a direct consequence of the involvement of KREEP (LKFM?) material during basin-forming impacts. We can only speculate about the exact nature of this volatilization event, but some insight can be obtained by examining the exact location where these volatiles reside within the highland breccias. The presence of easily leachable Pb, Cl, and Zn (e.g., Jovanovic and Reed, 1974) suggests that many of the volatiles reside along cracks and grain boundaries. These volatiles (e.g., Cl) appear to have been introduced late in the formation of these rocks, although the exact sequence of events is not clear.

The role of Cl is illustrative of the complexities caused by impact-induced melting and metamorphism. If abundant volatiles were present at the time of melt formation (i.e., high f_{Cl_2}, f_P, etc.), it is likely that these fugacities would affect the crystallizing phases. Under these conditions, it is probable that chlorapatite would form and much of the Cl and P would reside *in* a mineral phase(s). If, however, the volatile content was low (i.e., low f_{Cl_2}, f_P, etc.) whitlockite or schreibersite might form instead of chlorapatite, with the additional effect of a slow rise the f_{Cl_2} resulting in formation of a *late-stage, volatile-rich vapor phase*. However, if fugacities were low because the volatiles were introduced after a major portion of crystallization was complete, *the volatile-rich vapor could permeate the rock and deposit along cracks and grain boundaries*. Low initial fugacities could also be the result of loss of volatiles from areas in the upper portions of melt sheets.

Native FeNi metal is very unstable at even moderate f_{Cl_2}, and $FeCl_2$ (lawrencite) readily forms. Other Pb-Zn salts may also form between grains. Oxhydrated $FeCl_2$, in the form of "rust" (FeOOH) is an indicator of the abundance of leachable Cl_2 in these rocks. It is the precursor $FeCl_2$ which is showing us the importance of post-crystallization permeation of volatiles in ejecta deposits. Rocks such as 66095 with high leachable and low residual Cl reflect low f_{Cl_2} during the melt stage and the dominance of late-stage residual and/or introduced Cl. Rocks such as 61016 with low leachable and high residual Cl reflect the effects of high initial f_{Cl_2}. Therefore, the abundance of leachable Cl may be a direct result of the f_{Cl_2} prevalent during the melt stage. Of course, some rocks contain high leachable Cl and high residual, a function of high f_{Cl_2} during crystallization and subsequent introduced Cl.

Cirlin and Housley (1980) have recently performed an experiment that is critical to our volatilization model. Starting with a synthetic lunar rock, they mechanically mixed in some chlorapatite, pelletized this powder, placed the pellet in Fe foil, and sealed this in an evacuated silica tube. At 850°C in only a few days, they noticed an apparent breakdown of the chlorapatite and the formation of $FeCl_2$ on the Fe foil. The native Fe acts as a getter for available chlorine. It is this deliquescent $FeCl_2$ which undergoes oxhydration to β-FeOOH (akaganéite) resulting in the notorious rust (Taylor *et al.,* 1973, 1974). If such chlorine mobilization is possible at such low subsolidus temperatures, it would seem that *the amount of "leachable chlorine"* (Jovanovic and Reed, 1974, 1975) in a lunar rock or soil *may simply be a reflection of the amount of Cl released during impact-induced metamorphism*.

This volatilization model suggests that many clasts included in these breccias may be expected to have anomalous volatile trace element abundances, although their bulk major and minor element chemical systems have effectively remained closed. Therefore, again we stress that we must be overly critical of trace element systematics of small isolated clasts which are interpreted to be representative of larger uncontaminated igneous systems. This clast "volatile contamination" is evidenced by the widespread occurrence of akaganéite surrounding the FeNi metal grains in the majority of the clasts in 66095.

The primitive pre-impact lunar crust

The highland melt rocks contain suites of clasts which provide a representative sampling of the pre-impact stratigraphy of the highland crust. In the case of large basin- forming impacts, this sampling represents a *significant* portion of the upper lunar crust. By careful examination of clast abundances and petrochemistries of the clast suites contained in *large* samples of lunar highland breccias, such as 66095, we should be able to characterize the nature of the upper lunar crust.

To illustrate this approach, we make use of the cumulative mode presented in Table 1. We divided the clasts into two classes: (a) a primary class composed of clasts formed by lunar plutonic or volcanic processes, and (b) a 2nd generation class consisting of clasts formed by impact processes (i.e., brecciation, cataclasis, and/or melting). As stated earlier, we consider mineral clasts to represent disaggregated primary anorthosite; therefore, they are primary themselves. All troctolitic anorthosites and anorthosites are considered primary because (1) mineral compositions (Fig. 4) are consistent with those of the pristine plutonic rock types of Warren and Wasson (1978), and (2) no unequivocal textural evidence of annealing or recrystallization was found. Their FeNi metal compositions are also suggestive of being primary (i.e., igneous fractionation). Fra Mauro basalt is considered a primary crustal constituent for reasons discussed above.

Clasts are sub-equally divided between these two classes (Table 1). Highland basalt and cataclastic anorthosite (impactite?) dominate the 2nd generation class, while plagioclase clasts dominate the primary class. The two most abundant primary rock types identified are Fra Mauro basalt (9.5% of the clasts) and troctolitic anorthosite (10.4% of the clasts).

Based on the above discussions, we can summarize with the following general statements about the stratigraphy of the upper lunar crust sampled by the basin-forming impact that produced highland breccia 66095:

1. Plutonic rocks were covered by one or more thick ejecta deposits.
2. The plutonic rocks were exposed to bombardment for a long period of time prior to being covered by LKFM, perhaps never completely covered.
3. Fra Mauro basalt formed a *significant* portion of the stratigraphic section and probably the primitive lunar crust as well.
4. The plutonic complex was dominated by anorthosite with subordinate troctolitic anorthosite.

SUMMARY

Polymict breccia 66095, the most volatile-rich of all lunar rocks, is the product of a complex history. The matrix is heterogeneous in composition, contrary to expectations. A characterization of the clasts and matrix of this rock has permitted a reconstruction of its genesis. *Inasmuch as 66095 is representative of a large suite of highland rocks rich in volatiles,* a model for the formation of highland breccias, in general, is proposed.

The model consists of a pseudo-ternary system with end-members of (1) an "ANT" plutonic complex, (2) Fra Mauro basalt (LKFM), and (3) minor meteoritic material. The plutonic component represents all rock types of a differentiating plutonic complex and obviously is multi-dimensional. Likewise the meteoritic component (ELC) is also multi-dimensional. The source of KREEP, volatiles, and incompatible elements in the highlands may be a LKFM component which is envisaged as an extensive unit once covering the plutonic rocks to greater and lesser extents. The stratigraphic section also contains one or more ejecta and/or regolith deposits. Thus, the model entails a multi-layer stratigraphy which was subjected to major impacting events. Huge vc'umes of "highland basalt" melt containing clasts of the various strata were produced. The resulting highlands rocks would also contain evidence of the Fra Mauro component in the overall enrichment in volatiles which were driven out of the initial volatile-rich LKFM rocks by impact-induced metamorphism.

The primitive highlands crust would display, in reality, both lateral and vertical heterogeneities of rock types. The basin-forming impacts have excavated and melted various quantities of this crust. Therefore, our use of a composite plutonic component for our model is justified. *The complications involved in using small poorly-defined subsystems (e.g., norite, SCCRV, and dunite) as end-members are a rudiment of components which are too specific and not entirely realistic.* The precise proportions of plutonic members will only be realized by an increase in the data base as a result of detailed "pull apart" studies of many polymict melt breccias.

The majority of the volatile elements are envisaged as coming from thermal metamorphism of the Fra Mauro basalt. These volatiles were effectively distilled from these rocks and migrated to favorable sites. The impact-generated melts may incorporate significant volatiles manifest in chlorapatite and resulting in low leachable Cl (i.e., high Cl_r; Jovanovic and Reed, 1974; 1975) as seems the case with 61016. If the volatiles are introduced after the major crystalization is complete, they will occur along cracks and grain boundaries (i.e., high leachable Cl, 66095; low Cl_r) and form $FeCl_2$ with the FeNi grains. The measured parameter of Cl_r may be an expression of the timing of Cl introduction, indirectly a result of metamorphism.

Acknowledgments—We wish to thank Drs. J. W. Head, P. Hess, R. Housely, O. B. James, J. J. Papike, G. J. Taylor, J. V. Smith and J. Warner for helpful discussions and constructive criticisms of our ideas expressed in this paper. Mr. R. Meyer assisted with electron microprobe analyses, and Dr. G. J. Taylor provided valuable discussions and assistance with DBA data reduction. This research was made possible by NASA Grant NGL 43-001-140.

REFERENCES

Albee A. L., Quick J. E., and Chodos A. A. (1977) Source and magnitude of errors in "broad-beam analysis" (DBA) with the electron probe (abstract). In *Lunar Science VIII*, p. 7–9. The Lunar Science Institute, Houston.

Albee A. L. and Ray L. (1970) Correction factors for electron probe microanalysis of silicates, oxides, carbonates, and sulfates. *Anal. Chem.* **42,** 1408–1414.

Apollo 16 Lunar Sample Information Catalog (1972) MSC-03210, NASA Johnson Space Center, Houston. 372 pp.

Bass M. (1972) Description of 66095. In *Apollo 16 Lunar Sample Information Catalog*, p. 268–274. MSC-03210, NASA Johnson Space Center, Houston.

Bence A. E. and Albee A. L. (1968) Empirical correction factors for the electron microanalysis of silicates and oxides. *J. Geol.* **76,** 382–403.

Brunfelt A. O., Heier K. S., Nilssen B., and Sundvoll B. (1973) Geochemistry of Apollo 15 and 16 materials. *Proc. Lunar Sci. Conf. 4th*, p. 1209–1218.

Cirlin E. H. and Housley R. M. (1980) Volatile transfer during lunar metamorphism: (1) Basalt-apatite system (abstract). In *Lunar and Planetary Science XI*, p. 146–148. Lunar and Planetary Institute, Houston.

Dence M. R. and Plant A. G. (1972) Analysis of Fra Mauro samples and the origin of the Imbrium Basin. *Proc. Lunar Sci. Conf. 3rd*, p. 379–399.

Duncan A. R., Erlank A. J., Willis J. P., and Ahrens L. H. (1973) Composition and inter-relationships of some Apollo 16 samples. *Proc. Lunar Sci. Conf. 4th*, p. 1097–1113.

El Goresy A., Ramdohr P., Pavicevic M., Medenbach O., Muller O., and Gentner W. (1973) Zinc, lead, chlorine and FeOOH-bearing assemblages in the Apollo 16 sample 66095: Origin by impact of a comet or a carbonaceous chondrite, *Earth Planet. Sci. Lett.* **18,** 411–419.

Garrison J. R. Jr. and Taylor L. A. (1979) Breccia Guidebook #2: 66095 "Rusty Rock." Houston Curatorial Branch Johnson Space Center, Report #JSC 16198, 27 pp.

Garrison J. R. Jr. and Taylor L. A. (1979b) Petrology of lunar rock 66095: Implications for the genesis of highland basalt and the stratigraphy of the Apollo 16 landing site (abstract). In *Papers Presented to the Conference on the Lunar Highlands Crust*, p. 18–20. Lunar and Planetary Institute, Houston.

Goldstein J. I. and Yakowitz H. (1971) Metallic inclusions and metal particles in the Apollo 12 lunar soil. *Proc. Lunar Sci. Conf. 2nd*, p. 177–191.

Hess P. C., Rutherford M. J., and Campbell H. W. (1978) Ilmenite crystallization in nomare basalt: Genesis of KREEP and high-Ti mare basalt. *Proc. Lunar Planet. Sci. Conf. 9th*, p. 705–724.

Hubbard N. J., Rhodes J. M., Wiesmann H., Shih C.-Y., and Bansal B. M. (1974) The chemical definition and interpretation of rock types returned from the non-mare regions of the moon. *Proc. Lunar Sci. Conf. 5th*, p. 1227–1246.

James O. B. and Blanchard D. P. (1976) Consortium studies of light-gray breccia 73215: Introduction, subsample distribution data, and summary of results. *Proc. Lunar Sci. Conf. 7th*, p. 2131–2143.

Jovanovic S. and Reed G. W. Jr. (1974) Labile and nonlabile element relationships among Apollo 17 samples. *Proc. Lunar Sci. Conf. 5th*, p. 1685–1701.

Jovanovic S. and Reed G. W. Jr. (1975) Soil breccia relationships and vapor deposits on the moon *Proc. Lunar Sci. Conf. 6th*, p. 1753–1759.

Krähenbühl U., Ganapathy R., Morgan J. W., and Anders E. (1973) Volatile elements in Apollo 16 samples: Implications for highland volcanism and accretion history of the moon. *Proc. Lunar Sci. Conf. 4th*, p. 1325–1348.

Kurat G., Keil K., Prinz M., and Nehru C. E. (1972) Chondrules of lunar origin. *Proc. Lunar Sci. Conf. 3rd*, p. 707–721.

Nunes P. D. and Tatsumoto M. (1973) Excess lead in "rusty rock" 66095 and implications for an early lunar differentiation. *Science* **182,** 916–919.

Nunes P. D. and Tatsumoto M. (1973) Excess lead in "rusty rock" 66095 and implications for an early lunar differentiation. *Science* **182,** 916–919.

Nunes P. D., Tatsumoto M., Knight R. J., Unruh D. M., and Doe B. R. (1973) U-Th-Pb systematics of some Apollo 16 lunar samples. *Proc. Lunar Sci. Conf. 4th,* p. 1797–1822.

Nyquist L. E., Hubbard N. J., Gast P. W., Bansal B. M., Wiesmann H., and Jahn B. (1973) Rb-Sr systematics for chemically defined Apollo 15 and 16 materials. *Proc. Lunar Sci. Conf. 4th,* p. 1823–1846.

Reid A. M. (1974) Rock types present in lunar highland soils. *The Moon* **9,** 141–146.

Reid A. M., Duncan A. R. and Richardson S. H. (1977) In search of LKFM. *Proc. Lunar Sci. Conf. 8th,* p. 2321–2338.

Reid A. M., Warner J., Ridley W. I., Johnson D. A., Harmon R. S., Jakes P., and Brown R. W. (1972) The major element compositions of lunar rocks as inferred from glass compositions in the lunar soils. *Proc. Lunar Sci. Conf. 3rd,* p. 363–378.

Roedder E. and Weiblen P. (1977) Shock glass veins in some lunar and meteoritic samples— Their nature and possible origin. *Proc. Lunar Sci. Conf. 8th,* p. 2593–2615.

Ryder G. (1979) Mixing the lunar crust: The components of highland rocks (abstract). In *Lunar and Planetary Science,* p. 1046–1048. Lunar and Planetary Institute, Houston.

Ryder G. and Norman M. (1978) Catalog of pristine non-mare materials. Part 1. Non-anorthosites, JSC Publ. 15465. Johnson Space Center, Houston. 146 pp.

Ryder G. and Norman M. (1979) Catalog of pristine non-mare materials. Part 2. Anorthosites [Revised]. JSC Publ. 14603. Johnson Space Center, Houston. 86 pp.

Ryder G. and Spudis P. (1979) Volcanism prior to the termination of the heavy bombardment: Evidence, characteristics, and concepts (abstract). In *Papers Presented to the Conference on the Lunar Highlands Crust,* p. 132–134. Lunar and Planetary Institute, Houston.

Schonfeld E. and Meyer C. Jr. (1973) The old Imbrium hypothesis. *Proc. Lunar Sci. Conf. 4th,* p. 125–138.

Simonds C. H., Floran R. J., McGee P. E., Phinney W. C. and Warner J. L. (1978) Petrogenesis of melt rocks, Manicouagan impact structure, Quebec. *J. Geophys. Res.* **83,** 2773–2788.

Simonds C. H., Warner J. L., Phinney W. C. and McGee P. E. (1976) Thermal model for impact breccia lithification: Manicouagan and the moon. *Proc. Lunar Sci. Conf. 7th,* p. 2509–2528.

Taylor L. A., Mao H. K., and Bell P. M. (1973) Rust in the Apollo 16 rocks. *Proc. Lunar Sci. Conf. 4th,* p. 829–839.

Taylor, L. A., Mao H. K., and Bell P. M. (1974) β-FeOOH, akaganéite, in lunar rocks. *Proc. Lunar Sci. Conf. 5th,* p. 743–748.

Taylor S. R. (1975) *Lunar Science: A Post-Apollo View.* Pergamon, N.Y. 372 pp.

Warner J., Simonds C. H., and Phinney W. C. (1976) Apollo 17, station 6 boulder sample 76255: Absolute petrology of breccia matrix and igneous clasts. *Proc. Lunar Sci. Conf. 7th,* p. 2233–2250.

Warner R. D., Planner H. N., Keil K., Murali A. V., Ma M.-S., Schmitt R. A., Ehmann W. D., James W. D. Jr., Clayton R. N., and Mayeda T. K. (1976) Consortium investigation of breccia 67435. *Proc. Lunar Sci. Conf. 7th,* p. 2379–2402.

Warren P. H. and Wasson J. T. (1977) Pristine nomare rocks and the nature of the lunar crust. *Proc. Lunar Sci. Conf. 8th,* p. 2215–2235.

Warren P. H. and Wasson J. T. (1978) Compositional-petrographic investigation of pristine non-mare rocks. *Proc. Lunar Planet. Sci. Conf. 9th,* p. 185–217.

Warren, P. H. and Wasson J. T. (1979) The compositional-petrographic search for pristine non-mare rocks: Third foray. *Proc. Lunar Planet. Sci. Conf. 10th,* p. 583–610.

Warren P. H. and Wasson J. T. (1979) The origin of KREEP. *Rev. Geophys. Space Phys.* **17,** 73–88.

Wasson J. T., Warren P. H., Kallemeyn G. W., McEwing C. E., Mittlefehldt D. W., and Boynton W. V. (1977) SCCRV, a major component of highlands rocks. *Proc. Lunar Sci. Conf. 8th,* p. 2237–2252.

Papike, J.J. and Merrill, R.B., eds.
Proc. Conf. Lunar Highlands Crust (1980), p. 419-439
Printed in the United States of America

Fission xenon in troctolite 76535

C. M. Hohenberg, B. Hudson, B. M. Kennedy and F. A. Podosek

Department of Physics and McDonnell Center for the Space Sciences, Washington University, St. Louis, Missouri 63130

Abstract—Xenon released in stepwise heating of troctolite 76535 shows remarkable similarity to that from the gas-rich breccias. The low temperature fractions are highly enriched in fissiogenic xenon from ^{244}Pu. Although the quantity of fission xenon is not inconsistent with reported age data, release patterns demonstrate that this component is not due to *in situ* decay. However, in contrast with the gas-rich breccias, the parentless radionuclear component in 76535 is not accompanied by solar wind xenon. Mechanisms for the emplacement of this component in 76535 cannot, therefore, involve exposure on the lunar surface but most likely involve adsorption along intergranular boundaries or thermally driven redistribution. The fact that xenon released from a gas-poor igneous cumulate can mimic that from gas-rich breccias, except for the presence of solar wind xenon in the latter, suggests that direct exposure to space is not required for the incorporation of these weakly bound, surface-correlated, parentless radionuclear components in lunar samples.

INTRODUCTION

Amounts of fission xenon in excess of that which could be produced by *in situ* decay have been observed in a majority of the gas rich breccias from the Apollo 14 and 16 sites. These effects have been widely investigated (e.g., Behrmann *et al.*, 1973; Bernatowicz *et al.*, 1978; Bernatowicz *et al.*, 1979), and some basic characterizations have begun to emerge. In certain instances where it is possible to determine the isotopic composition of the excess xenon, it conclusively matches the yield spectrum of ^{244}Pu spontaneous fission, sometimes with and sometimes without an accompanying monoisotopic contribution at ^{129}Xe, presumably from the decay of extinct ^{129}I. Stepwise heating experiments on gas-rich breccias indicate that excess radionuclear components are enriched relative to the solar wind in the low temperature fractions indicating, at least in these fractions, that it is more weakly bound than solar wind xenon. Recent results on size fractions from 14301 (Bernatowicz *et al.*, 1979) indicate that the excess component is preferentially sited along grain surfaces. However, differences in the release characteristics of excess monoisotopic ^{129}Xe from ^{129}I decay and excess fission ^{244}Pu decay suggest that even though they are similarly sited, significant differences exist in the respective activation energies for diffusive loss. Implications include differences in times or manner of incorporation of these two effects.

In this work we report data from 76535 which show remarkable similarity with

Table 1a. Xenon from Stepwise Heating of 76535,75 (^{132}Xe = 100)

Temp. °C	[132] × 10^{-15}ccSTP/g	124	126	128	129	130	131	134	136
500	10.7	0.81	0.39	8.85	81.8	13.17	77.1	51.21	43.34
		24	24	72	3.8	70	2.6	2.10	1.94
800	116.1	0.604	0.82	4.94	46.73	7.78	55.0	79.98	82.68
(blank)	(2.1)	85	12	.38	91	38	7	1.49	1.33
900	35.4	1.65	2.60	8.92	61.6	10.95	71.5	70.48	71.57
		26	31	46	1.6	95	2.3	1.67	1.65
1000	49.1	3.58	6.32	15.71	84.7	17.09	89.2	49.01	50.15
		34	30	58	2.1	66	2.9	2.03	1.20
1100	248.4	2.96	4.09	13.86	101.4	18.41	92.3	41.14	35.26
		13	16	32	1.2	51	1.1	75	68
1200	210.5	5.04	8.17	18.81	103.58	21.34	105.6	38.46	33.03
(blank)	(1.9)	17	16	33	72	35	1.0	51	46
1300	153.2	14.28	22.92	40.55	122.1	33.93	162.6	32.94	26.90
		37	52	65	1.5	69	1.5	70	50
1400	283.0	13.87	23.92	41.50	121.49	34.02	162.1	33.56	26.94
		20	31	34	75	37	1.3	40	26
1600	4505.3	36.56	66.01	103.60	143.13	79.36	347.0	20.34	13.86
(blanks)	(2.1)	12	18	22	26	13	6	10	6
1700	1661.7	41.98	76.62	118.97	149.57	90.81	395.1	15.44	8.42
		22	32	40	37	35	1.1	10	9
1800	2530.1	42.06	76.48	119.84	149.14	91.95	397.2	15.21	8.19
		14	20	29	36	21	9	9	7
2000	165.6	36.88	66.43	105.78	144.8	80.93	354.4	18.33	11.45
		60	81	82	1.7	94	3.1	46	38
Total	9969.0	35.630	64.471	101.54	141.18	78.31	342.42	20.71	14.20
		76	110	14	17	10	40	7	5

the Apollo 14 and 16 gas rich breccias in the low temperature release of plutonium fission xenon. 76535 is, however, neither a breccia nor is it rich in solar wind gases; it is a troctolite cumulate derived by igneous processes. In contrast with some of the gas rich breccias, it does not contain quantities of fission xenon that are incompatible, in principle, with *in situ* decay. However, the low temperature release of a large fraction of the fission xenon, large variations of the effect among three different samples analysed, and the absence of association with volume-correlated components suggest that the effect cannot be explained simply by *in situ* decay of an undisturbed system. Rather, like the gas-rich breccias, the extinct isotope effects in 76535, at least in the low temperature fractions, is a parentless component in the sense that decay took place elsewhere and only the daughter products now reside at the weakly retentive sites in 76535. The decoupling of parentless extinct isotope effects from solar wind xenon in 76535 provides some

Table 1b. Krypton from Stepwise Heating of 76535,75 ($^{84}Kr = 100$)

Temp. °C	[84] $\times 10^{-14}$ccSTP/g	78	80	81	82	83	86
500	2.8	3.58	4.83	−0.34	20.59	20.28	31.04
		.72 + 6.4	35	.34 + .84	58	61	82
800	25.9	3.53	13.40	−0.11	32.38	37.68	26.21
(blank)	(0.9)	.65 + .58	49	.27 + .01	75	1.01	76
900	47.3	7.89	26.59	0.22	53.75	68.02	20.17
		.43 + .82	73	.18 + .01	92	1.26	53
1000	133.8	8.63	29.36	−0.05	55.96	70.01	19.89
		.36 + .29	51	.11 + 0	79	78	38
1100	461.8	5.72	19.97	0.00	42.95	52.46	23.92
		.14 + .08	22	.04 + 0	32	30	22
1200	333.1	9.55	32.27	0.065	60.12	76.32	19.83
(blank)	(0.7)	.13 + .15	38	.053 + .002	49	59	24
1300	246.2	15.84	52.67	0.074	88.55	117.74	11.57
		.27 + .16	38	.062 + .002	72	71	24
1400	262.0	14.44	48.78	0.149	83.09	109.30	13.25
		.23 + .16	42	.047 + .009	60	87	27
1600	2414.4	21.35	68.79	0.136	110.23	148.34	6.68
(blank)	(1.2)	.08 + .01	13	.009 + .005	21	29	6
1700	874.0	21.78	72.57	0.126	115.42	156.23	3.78
		.13 + .03	32	.018 + .008	39	57	8
1800	1308.4	21.11	69.95	0.111	111.91	152.33	4.36
		.09 + .02	20	.014 + .006	27	38	5
2000	179.0	7.41	26.25	0.101	51.14	63.08	21.19
		.24 + .23	43	.083 + .003	61	56	37
Total	6286.7	18.24	59.96	—	98.12	131.40	9.10
		.05 + .00	9		13	17	4

new insights into mechanisms responsible for the redistribution and trapping of these effects in lunar material.

PROCEDURE AND RESULTS

A 0.7251 gram sample of 76535,75 was wrapped in .0717 grams of nickel foil and loaded into the sample arm of the low-blank extraction bottle of our ion-counting mass spectrometer system, described by Hohenberg (1980) and designated MS3 in this laboratory. The sample was heated overnight at approximately 125°C under vacuum prior to analysis. The tungsten heating coil was degassed at 2200°C for several hours and appropriate procedural blanks which included similar amounts of the nickel foil were run and are reported in Table 1.

Gases were liberated by stepwise extraction in the resistively heated tungsten coil, with the temperature at each step maintained for one hour for extractions at 1000°C and less, and 30 minutes for temperatures in excess of 1000°C. Temperatures are only approximate in that they were obtained by using an optical pyrometer prior to sample analysis to calibrate the coil temperature as a function of input power. Melting of the nickel foil blank, which occurred on schedule at 1453°C, serves as confirmation of the pyrometer calibration. Because of the large sample, which extended to the top of the coil, and the open spacing of the coil, actual sample temperatures are appreciably less than the nominal values reported.

The gases were exposed to a freshly depositied titanium film (flash getter) during the extraction. After heating the gases were admitted to a manifold volume containing a second flash getter and an incandescent filament. Ten minutes later they were exposed to a third flash getter and the heavy gases adsorbed on activated charcoal, maintained at −95°C, for 40 minutes. After adsorption the light gases were pumped away. The krypton and xenon were then desorbed by heating the charcoal to 150°C and physically partitioned for separate analysis in the mass spectrometer. Approximately 20% of the released gases were admitted into the spectrometer for krypton analysis, the remaining 80% admitted for xenon analysis. Instrumental sensitivity and isotopic discrimination were determined by similar treatments of pipetted aliquots of air (Hohenberg, 1980) and the mass spectrometer was operated statically in the ion-counting mode. No spectrometer background was observed in the xenon region and xenon blanks were atmospheric within errors. In the analysis of large quantities of xenon, however, we have noted the presence of small signals at 133, 135, and 137, in proportion to the peaks at 132, 134, and 136 respectively. These are either due to xenon hydride or simply ghost peaks generated by optical abberration or scattered ions. In general the effects are small ($^{i}XeH/^{i}Xe \simeq 2.7 \times 10^{-4}$) and, with the relatively low gas amounts for 76535, they are essentially negligible. The largest effect occurs at ^{130}Xe where the correction is approximately 0.1 percent, small in relation to other uncertainties, and consequently no such corrections were made for this sample. Krypton blanks are atmospheric with the exception of small hydrocarbon peaks at ^{78}Kr and ^{81}Kr at 2.7×10^{-14} and 1.1×10^{-14} equivalent ccSTP respectively. Corrections to ^{78}Kr were made on the basis of the observed signal at mass 77; corrections to ^{81}Kr were made on the basis of the observed signal at mass 79. Signals at mass 77 and 79 were 0.7×10^{-14} and 1.1×10^{-14} equivalent ccSTP respectively for both blanks and the sample runs.

Xenon and krypton data from stepwise heating of 76535,75 are given in Table 1. Included in the table are nickel foil procedural blanks whose gas amounts are indicated in parentheses following the temperatures at which they were run. The second uncertainty indicated for each ^{78}Kr and ^{81}Kr ratio is the additional uncertainty due to the hydrocarbon background at these masses and is additive to the statistical uncertainty listed first. The hydrocarbon background is quite stable and an uncertainty of 25% was assigned to the correction. Except for hydrocarbon corrections for krypton no other blank corrections were made to the data in Table 1. Xenon data for the reconstructed total compositions of 76535,75, for 76535,15

Table 2. Xenon from Various Apollo 17 Samples (^{132}Xe = 100)

Sample	[132] × 10^{-13}ccSTP/g	124	126	128	129	130	131	134	136
75035,4[a]	188.	26.64	46.45	74.53	137.88	54.14	324.2	26.29	20.27
		15	44	52	88	16	1.8	22	11
76015,39[a]	181.	18.83	32.30	52.49	108.90	38.74	154.16	48.55	48.05
		15	21	39	30	25	58	22	15
76235,13[a]	34.	8.45	14.87	28.64	95.40	26.67	119.87	48.98	49.21
		21	25	30	88	37	1.09	58	30
76315,41[a]	211.	16.17	27.82	45.85	108.86	35.97	142.83	45.65	44.28
		15	13	23	35	12	39	20	17
76535,15[a]	148.	24.76	44.50	69.61	95.14	51.53	234.92	55.53	56.11
		16	35	23	38	26	77	20	23
76535,75[b]	99.7	35.63	64.47	101.54	141.18	78.31	342.42	20.71	14.20
		8	11	14	17	10	40	7	5
76535,64[c]	1885.0	2.10	3.49	11.78	101.0	18.20	91.42	37.82	32.03
		4	6	14	3	9	27	9	9
(–air)[d]	(100)	(34)	(62)	(98)	(150)	(75)	(321)	(20)	(14)
77135,83[a]	155.	17.19	30.63	52.60	117.06	42.35	205.78	35.99	31.67
		10	19	37	20	23	44	0	18
79215,9[a]	433.1	30.59	60.43	101.14	134.89	87.33	569.0	13.63	5.92
		15	29	21	11	17	1.5	9	5

[a] Morgan (1975); krypton data for these rocks reported by Crozaz et al. (1974).
[b] This work
[c] Lugmair and Marti (1976).
[d] Computed for 76535,64 by removal of contaminant xenon of atmospheric composition (see text).

(Morgan, 1975) and for 76535,64 (Lugmair *et al.*, 1976) are given in Table 2 along with data for six other Apollo 17 samples.

As can be seen the isotopic spectra and amounts of xenon in the three samples of 76535 differ substantially. We will first address the source of some of these differences and then go on to interpret information contained in the specific isotopic structure. There is good evidence that mechanical crushing can introduce substantial contamination from terrestrial atmospheric xenon (Niemeyer and Leich, 1976) and contamination, so introduced, cannot be separated from xenon indigenous to the sample. The La Jolla sample (76535,64) was crushed to less than 90μm and the data indeed suggest the presence of an atmospheric component. Partitioning of the observed ^{126}Xe/^{130}Xe for this samples between the atmospheric ratio and the ratio observed in either of the other two samples suggests that about 79% of the ^{130}Xe in 76535,64 may in fact be atmospheric contamination. Subtracting the corresponding atmospheric contributions from all of the isotopes of 76535,64 results in the corrected xenon spectrum shown in parentheses in

Table 2. Note that the spectrum, so corrected, closely resembles that obtained for 76535,75 both in amount and in composition. The quantity of ^{136}Xe in 76535,75 (and corrected 76535,64) differs substantially from that in 76535,15 (1.4×10^{-12} and 8.3×10^{-12} ccSTP/g respectively). In spite of the difference in the quantity of ^{136}Xe among these samples, the quantity of spallation-produced ^{126}Xe is essentially identical in each sample, including 76535,64 (6×10^{-12}ccSTP/g), and is compatible with the 211 m.y. ^{81}Kr-Kr exposure age for 76535 found by Morgan (1975) and a similar value computed from the data of Table 1b.

76535 (wt. 155.5 g) was part of the Apollo 17 station 6 rake sample collected at the base of the North Massif. It is a coarse-grained troctolitic granulite whose texture and mineralogy indicate formation as an igneous cumulate at depths between 10 and 30 km followed by an extended period of slow cooling. Modal abundances indicate it to be 58 vol. % plagioclase, 37 vol. % olivine, 4 vol. % pyroxene, and < 1 vol. % accessory minerals (Gooley *et al.*, 1974). Isotopic structures show varying degrees of disturbance as indicated by a disturbed U-Pb system (Tera and Wasserburg, 1974) and excess radiogenic ^{40}Ar (Bogard *et al.*, 1975; Huneke and Wasserburg, 1975). The Sm-Nd data of Lugmair *et al.* (1976) *do* show a good internal isochron and Rb-Sr data of Papanastassiou and Wasserburg (1975), largely dominated by inclusion-bearing olvivines, also provides a reasonably good internal isochron. The two internal isochrons, however, yield distinctly different apparent ages of 4.26 and 4.61 billion years, respectively. ^{39}Ar-^{40}Ar studies (Husain and Schaeffner, 1975; Huneke and Wasserburg, 1975) yield apparent ages that are generally compatible with the Sm-Nd internal isochron, 4.23 to 4.26 billion years. These data suggest that some of the isotopic systems in some phases of 76535 were reset at about 4.26 billion years ago; but that some isotopic systems in selected structures, namely the olivines and their associated inclusions, may have remained essentially closed systems from the time of original mineralization. The apparent ^{81}Kr-Kr exposure age, computed from data in Table 1b (1600, 1700, and 1800 fractions) is 209 m.y., essentially identical to the 210 m.y. age found by Morgan (1975).

Figure 1 shows a three isotope correlation plot for xenon released in stepwise heating of 76535,75 for a single stage melt of 76535,15, and for stepwise heating of 14301,57 and ,110 (Drozd *et al.*, 1972; Bernatowicz *et al.*, 1979). Numbers on the points refer to the filament temperature in hundreds of degrees Celsius for the various 76535,75 temperature fractions. Three distinct components are represented in this correlation plot: solar wind xenon, approximated by surface-correlated xenon in mare soil (SUCOR, Podosek *et al.*, 1971; BEOC-12, Eberhardt *et al.*, 1972), spallation xenon to the upper left, and fission xenon to the right. The cluster of points for the 1600 through 2000°C extractions approximate pure spallation xenon at these temperatures. Mixtures of such spallation xenon and trapped xenon would plot along the line indicated in the figure connecting SUCOR with this cluster of points. Additions of fission xenon causes departures to the right of the line in proportion to the relative amounts of fission-produced ^{136}Xe which provides a means of computing the fission contribution at each temperature fraction.

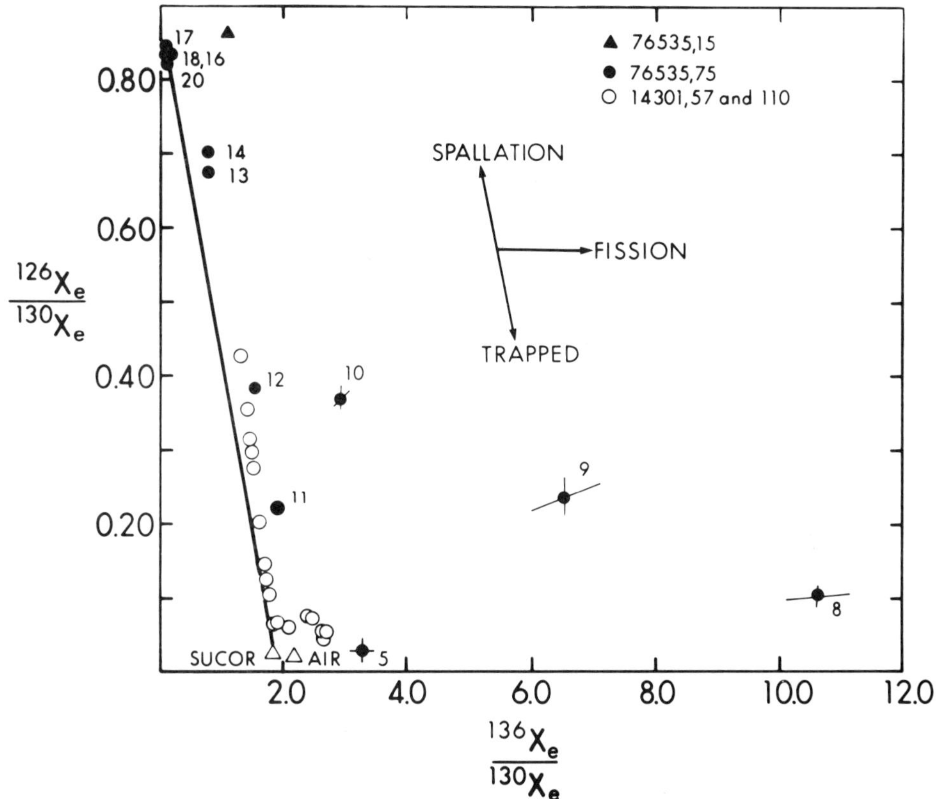

Fig. 1. Three-isotope correlation plot involving ^{126}Xe and ^{130}Xe, which have contributions from trapped xenon (SUCOR) and cosmogenic xenon, and ^{136}Xe, composed to trapped and fission xenon. Numbers on the 76535,75 points refer to the extraction temperature in hundreds of degrees Celsius. Addition of spallation xenon to a trapped component displaces the points toward the upper left; addition of fission xenon displaces the points to the right. The 800°C fraction is the most fissiogenic xenon ever seen in lunar samples. Approximately 84% of the ^{136}Xe is due to ^{244}Pu spontaneous fission. The small bump (open circles) above the "air" symbol, represents the low-temperature (fission-rich) fractions of the gas-rich breccia 14301 which, prior to the analyses of 76535, were the most enhanced in ^{244}Pu fission xenon. In spite of the difference in scale, the release pattern of fissiogenic xenon from 14301 is similar to that of 76535 as can be seen by comparing this figure with Fig. 4 of Drozd *et al.* (1972). 14301,57 data are from Drozd *et al.* (1972); 14301,110 are from Bernatowicz *et al.*, (1979); 76535,15 data from Morgan (1975).

Enhancement of the fission contribution in the low temperature fractions of 14301 can be seen in Fig. 1 (open circles) as a small protuberance to the right of the spallation-SUCOR mixing line, directly above the "air" point. Although 14301 exhibits the largest absolute fission effect at low temperatures of all the gas rich breccias (Drozd *et al.*, 1972), isotopically it is clearly dwarfed by the fission enhancement in the low temperature fractions of 76535, which represent the largest fission xenon effects seen in any lunar sample studied to date (for instance, 84% of the ^{136}Xe in the 800°C fraction is due to fission). In spite of the differences in magnitude the release pattern of the low temperature fission effects is quite similar to that of the gas rich breccias except that in the case of 76535 it is not accompanied by appreciable solar wind xenon.

Fission yield spectra can be obtained from the low temperature fission excesses by suitable component decomposition. The 800°C and other low temperature fractions are dominated by fission at the heavy xenon isotopes; the high temperature fractions, particularly the 1600°C through 2000°C points which cluster tightly in Fig. 1, are largely spallogenic. If the spallation component which dominates these temperature fractions is similar in composition to the (small) spallation contribution to the low temperature (fission-rich) fractions, then the fission yield spectra can be precisely determined by component superposition. A new component composite is constructed by adding sufficient quantities of solar wind xenon, whose composition is known, to a spallation-rich fraction so that it is made to match the spectra of the more fissiogenic fractions at the fission-shielded isotopes. The new composite, so constructed, will differ in composition from the fission-rich fraction only by the presence of more fission xenon in the latter. The fission-yield pattern can then be obtained simply by subtracting the two and normalizing the difference spectrum to ^{136}Xe. Generally good agreement with the known ^{244}Pu fission yield pattern (Table 3) is obtained for both the low temperature fractions of 76535,75 and 76535,15 (using whole rock 76535,75 as the spallation-rich component). This technique is relatively insensitive to the choice of the trapped component (SUCOR, BEOC-12, and terrestrial atmospheric compositions give similar results), but it does depend rather strongly upon the assumed similarity of the spallation component in the two fractions. The two most effective targets for the production of spallation xenon, barium and the light rare earth elements (REE), have quite different yields at ^{130}Xe and ^{132}Xe due primarily to shielding effects of ^{130}Ba and ^{132}Ba (Hohenberg *et al.*, 1978). Differences in the Ba to REE ratio among the temperature fractions involved will introduce spurious results in the apparent fission spectra, most pronounced at ^{129}Xe and ^{132}Xe.

Figure 2 is a three isotope correlation plot of ^{124}Xe, ^{126}Xe, and ^{130}Xe, all shielded in fission. Components present should include only trapped and cosmogenic xenon. If there were a single spallation component, individual temperature fractions would lie along a mixing line between these two end members. However, the spread of points on this figure indicates that the composition of spallation xenon differs among the various temperature fractions, reflecting variations in the ratio of barium to REE targets. Lines shown in the figure are the approximate mixing lines between spallation from the pure targets and solar wind xenon (Hohenberg *et al.*, 1978). Whereas the 1600°C–2000°C points seem to be dominated by Ba spallation, with correspondingly larger abundances of cosmogenic ^{130}Xe and ^{132}Xe, the 800°C and 900°C spallation component can contain substantial contributions from REE targets, depressing somewhat the ^{130}Xe and

^{132}Xe in the composite spallation spectrum. Therefore, mixtures of spallation-rich fractions and solar wind which match the fission-rich fraction at ^{124}Xe, ^{126}Xe, and ^{130}Xe will contain too little solar wind and the inferred apparent fission yield patterns will contain anomalously large quantities of ^{129}Xe and, to a lesser extent, ^{132}Xe. Anomalous values at ^{132}Xe are somewhat smaller than those at ^{129}Xe since more spallation produced ^{132}Xe, favored in Ba spallation, will be subtracted. As one can see from the data of Table 3, the closest fit to ^{244}Pu occurs in the fission yield spectrum derived for the 800°C fraction where the magnitude of the spallation correction is the smallest. Furthermore, it is apparent that 76535,15 contains a greater contribution from REE spallation than does 76535,75 as can readily be seen by close examination of Figs. 2, 3, and 4. REE spallation is characterized by larger $^{126}Xe/^{130}Xe$ and $^{129}Xe/^{130}Xe$ ratios ($\gtrsim$ 10). The measured $^{126}Xe/^{130}Xe$ ratio for the total rock point of 76535,15 is greater than those of the spallation-rich cluster of points for 76535,75 (1600°–2000°) which are essentially pure spallation indicating a higher $^{126}Xe/^{130}Xe$ ratio in cosmogenic xenon at 76535,15. Consequently, the $^{129}Xe/^{136}Xe$ inferred for the "fission" component in 76535,15 is overestimated due to failure of the assumption of identical spallation spectra. The same failure begins to produce noticeable anomalies in 900, 1000, and 1100°C fractions of 76535,75 (Table 3). In view of these problems, we believe the most accurate determination of the isotopic spectrum of the fission component in 76535 is given by decomposition of the 800°C fraction where the spallation contribution is very small and the fission contribution large. The yield spectrum from that point is consistent with ^{244}Pu with perhaps minor contributions from ^{238}U and there is no evidence for excess ^{129}Xe from decay of ^{129}I.

Table 3. Fission Yield Spectra

Sample (temp)	$^{F}136$	129	131	132	134	136
76535,15	.86	(.18)	.17	.861	.913	1.0
76535,75	.84	.01	.24	.812	.918	1.0
(800°C)		4	3	31	28	
(900°C)	.79	(.10)	.33	.845	.917	1.0
		12	9	41	41	
(1000°C)	.63	(.31)	.27	(1.01)	.812	1.0
		15	14	11	75	
(1100°C)	.26	(.64)	.35	(1.17)	.97	1.0
^{244}Pu		.05	.25	.88	.92	1.0
		5	2	3	4	
^{238}U			.076	.60	.83	1.0

The value for the apparent yields at ^{129}Xe and ^{132}Xe are overestimated in some cases (parentheses) due to differences in the spallation spectra (see text).

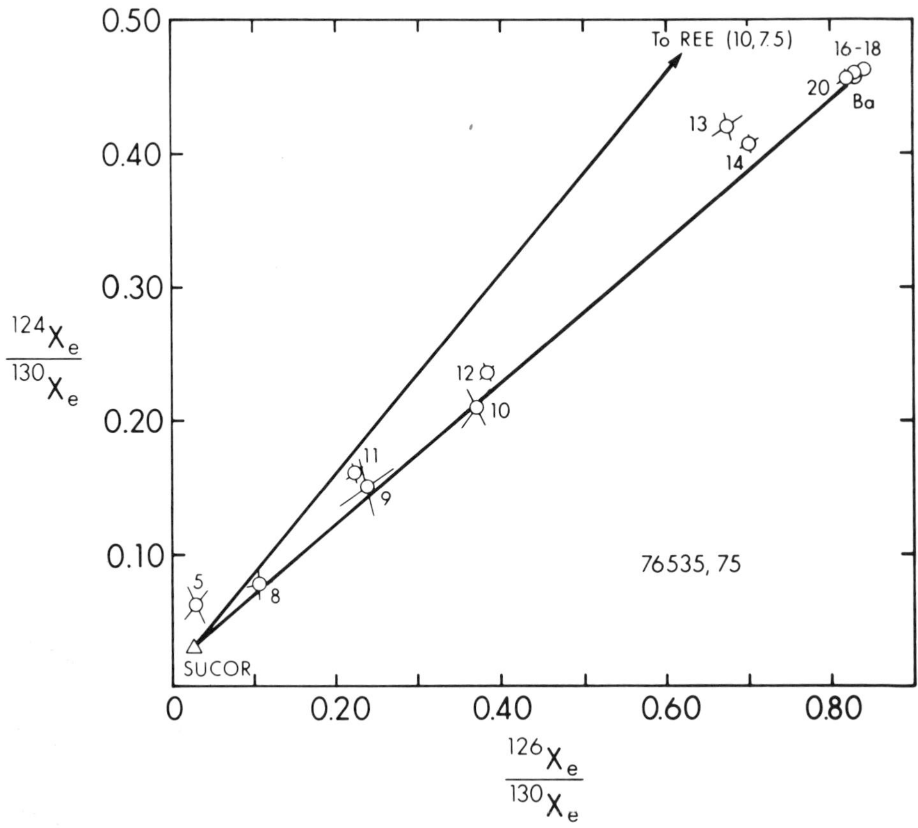

Fig. 2. Light isotope correlation plot for ^{124}Xe, ^{126}Xe and ^{130}Xe all shielded in fission. Mixtures of trapped xenon and a single spallation component should lie on a mixing line between the two. The temperature fractions of 76535,75 depart from a simple mixing line, suggesting that the spallation component is not uniform but reflects a distribution of Ba/REE ratios. The two indicated lines represent mixing between trapped and spallation from pure Ba and REE targets respectively (Hohenberg *et al.*, 1978).

Figure 3 is a three isotope correlation plot involving ^{126}Xe, ^{129}Xe and ^{130}Xe. Displacement of points to the right of a mixing line between SUCOR and the 1600°C–2000°C cluster is not to be interpreted as indicating excesses at ^{129}Xe, but, as was pointed out above, is symptomatic of variations in the spallation composition. Confirmation of this interpretation is found in Fig. 4 where ^{128}Xe replaces ^{130}Xe as the normalization isotope. These three isotopes are produced in roughly similar proportions in REE and Ba spallation, the composition of which is suggested by the cluster of points at the upper left. Note that now there is no evidence for displacement to the right of the line leading us to conclude that the displacements in Fig. 3 do not imply excess ^{129}Xe but reflect the variable ^{130}Xe yield in spallation. Since ^{128}Xe is also shielded in fission, it is also possible to infer a fission yield spectrum by using the ^{124}Xe, ^{126}Xe, and ^{128}Xe to generate a component composite from SUCOR and a spallation-rich fraction, the ^{128}Xe now replacing ^{130}Xe in the technique described above. This treatment has the advantage that the differences between REE and Ba spallation are less, but it has the disadvantage that some excess ^{128}Xe exists in the low temperature fractions, seen as departures *below* the line in Fig. 4. This excess, presumably due to ^{127}I (n, γ) reactions occurring on grain surfaces, has previously been observed in the Apollo 14 soils and gas-rich breccias (Basford *et al.*, 1973; Drozd *et al.*, 1975; Bernatowicz *et al.*, 1979). Although the fractional effect of this extra ^{128}Xe is small, it is of a similar magnitude as the errors in the inferred spallation spectra introduced by differences in the Ba/REE ratio between the composite component and the fission-rich fraction and it occurs in precisely those temperature fractions where we wish to obtain good fission yield spectra. Consequently, the subtraction

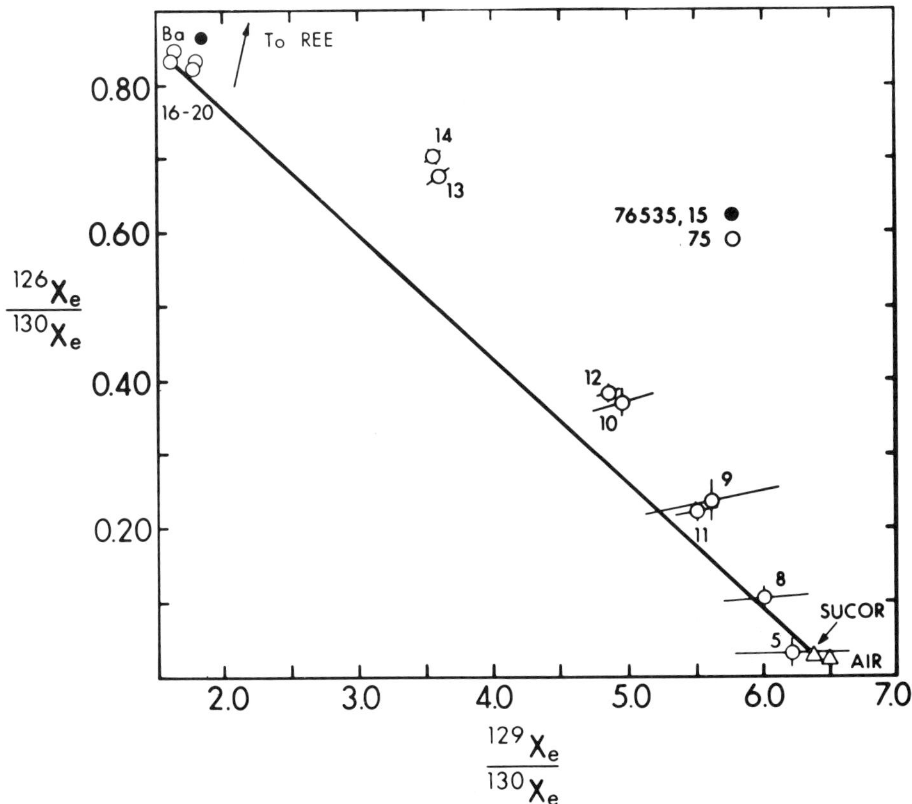

Fig. 3. Three-isotope correlation plot showing mixing between the trapped (SUCOR) and cosmogenic components. Differences in the composition of Ba and REE spallation are quite prominent in this plot. The pure Ba end member lies near the cluster of points at the upper left; the pure REE end member lies off the figure, in the direction of the arrow, at approximately (10,10). Although departures to the right of the indicated mixing line might be interpreted as indicating excess ^{129}Xe, they in fact represent differing mixtures of Ba and REE-derived cosmogenic xenon.

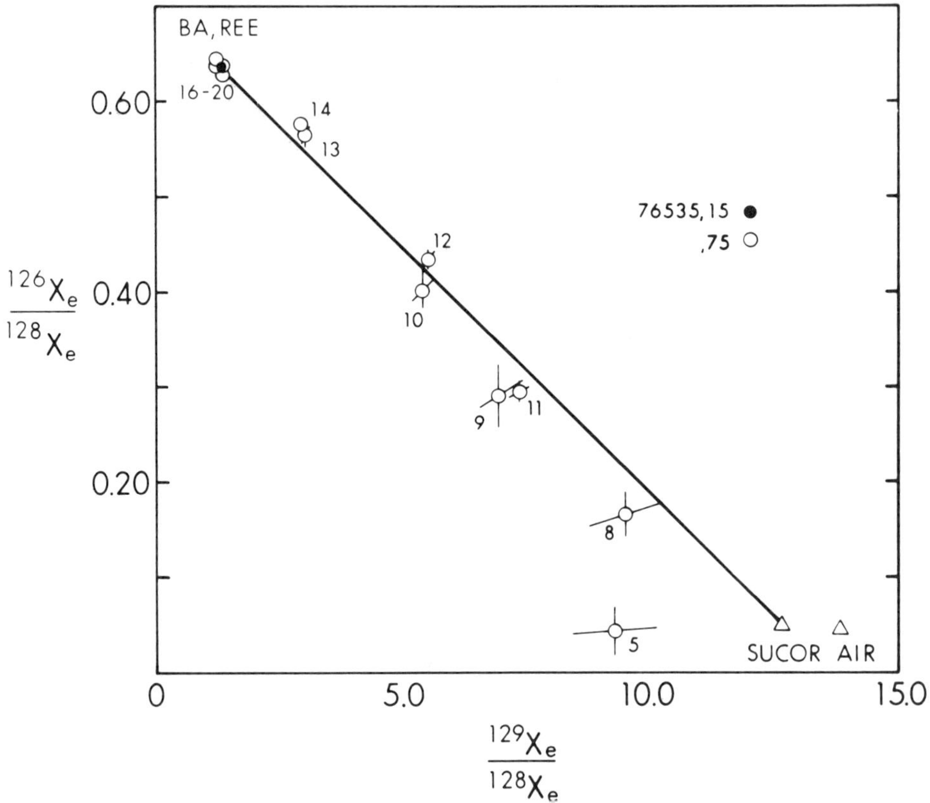

Fig. 4. This figure is similar to Fig. 3 except that the isotope of normalization is ^{128}Xe instead of ^{130}Xe. For the isotopes shown in this figure there is very little difference in the composition of REE and Ba spallation xenon. Both plot near the upper left end of the indicated mixing line. There is no indication for excess ^{129}Xe, which would be indicated by departures to the right of this line; small departures to the *left* of the mixing line may be due to minor contributions from the ^{127}I (n, $\gamma\beta$)^{128}Xe reaction (see text).

technique based upon ^{124}Xe, ^{126}Xe, and ^{128}Xe results in a fission yield spectrum somewhat more uncertain than those shown in Table 3 for the 800°C fraction. Two points, however, are worth noting. First, since variations in the spallation spectra generated by varying REE/Ba ratios are less at ^{124}Xe, ^{126}Xe, and ^{128}Xe, errors introduced by this method are not as strongly related to the spallation content of the fission-rich fraction. Second, spectra so derived are those of ^{244}Pu spontaneous fission and, though not as precise as that previously deduced for the 800°C fraction, they do not show the progressive apparent enrichments of ^{129}Xe and ^{132}Xe at 900, 1000, and 1100°C as the spallation contribution increases. We conclude again that the best measure of the fission-yield spectra for 76535 is that given by the 800°C fraction, decomposed by the first method (Table 3). Increasing apparent anomalies at ^{129}Xe and ^{132}Xe are introduced at the higher temperatures due to failure in the assumption of a uniform spallation component.

Radiometric ages for 76535 range from ^{39}Ar-^{40}Ar and Sm-Nd ages of around 4.3 b.y., to a Rb-Sr age of 4.6 b.y., whose spread is taken as evidence of varying degrees of isotopic disturbance (c.f. Papanastassiou and Wasserburg, 1976). The uranium content has been measured in two different samples of 76535 with values of 56 and 54 ppb (Haskin *et al.*, 1974; Keith *et al.*, 1974) providing about 1.7×10^{-13}ccSTP/g of ^{136}Xe from the spontaneous fission of ^{238}U. The observed quantities of fission-produced ^{136}Xe are 7.2×10^{-12}ccSTP/g for 76535,15 and 0.9×10^{-12} ccSTP/g for 76535,75. As expected from their spectra, the amount of fission xenon suggests that these samples are dominated by plutonium fission. If one were to attribute the fission xenon to *in situ* decay of plutonium, xenon closure times of about 4.4 and 4.2 b.y. for 76535,15 and 76535,75 (and 76535,64) respectively are suggested assuming an initial Pu/U ratio of 0.015 at 4.55 b.y. (Podosek, 1972). While this does not appear to be inconsistent with the range of apparent radiometric ages, there is evidence that the fission effects, at least in the lower temperature fractions, are not due to *in situ* decay but are most likely attributed to xenon redistribution either on a local or global scale.

Considering the whole rock data for 76535,15, 76535,75 and 76535,64 it is difficult to imagine conditions under which three bulk samples of the same small rock (separated by ~1 cm) could differ in xenon retention times by more than 200 m.y. or differ in their respective initial plutonium contents by a factor of 8, especially when there is no evidence for a strongly varying uranium abundance. Additional evidence for origin other than by *in situ* decay comes from the absence of an association between the fission xenon and the rare earth elements.

There is now substantial evidence that plutonium is at least approximately geochemically coherent with the rare earth elements (Lewis, 1975; Boynton, 1978; Hohenberg *et al.*, 1980[1]). Such an association is fortunate for xenon isotopic studies since ^{126}Xe from the spallation of rare earth targets is a natural tracer for xenon released from sites containing rare earth elements. If reasonable chemical coherence between REE and plutonium can be assumed and if the fission xenon were the result of *in situ* decay of plutonium one should observe a reasonably constant ratio between fission-produced ^{136}Xe and ^{126}Xe from rare earth spallation. Such a correlation has in fact been demonstrated in St. Severin whitlockite (Lewis, 1975) and in Angra dos Reis (Lugmair and Marti, 1977; Hohenberg *et al.*, 1980[1]) and both of these yield approximately the same ratio of $^{136}Xe_f$ to $^{126}Xe_{REE}$ of about 8 when normalized to the ^{81}Kr-Kr exposure age of Angra dos Reis, 55 m.y. (Lugmair and Marti, 1977). Adjusting for the difference in cosmic produced ^{126}Xe due to the difference in exposure ages, one would predict a ratio of about 4 for an object 4.55 billion years old with the same exposure history as 76535 and a 2π exposure geometry. Note that (Table 4) the whole rock data of 76535,15 yields a ratio of more than 9 on the basis of total composition, suggesting that it would *predate* ADOR by about 100 million years if the fission effects in 76535 were *in situ* produced. However, Table 4 also shows the total amounts of spallation-produced ^{126}Xe, the estimated contributions from the rare earths, and

---

[1] Hohenberg C. M., Hudson B., Kennedy B. M., and Podosek F. A. (1980) Xenon spallation systematics in Angra dos Reis. Submitted to *Geochim. Cosmochim. Acta*.

Table 4.

	($\times 10^{-15}$ cc STP/g)			
76535,75	$^{126}Xe_{sp}$	$^{126}Xe_{REE}$	$^{136}Xe_f$	$\frac{^{136}Xe_f}{^{126}Xe_{REE}}$
500	0.006	.0007	2.1	3000
800	0.74	0.09	81.	900
900	0.84	0.10	20.	200
1000	3.0	0.36	15.6	43
1100	9.2	1.1	23.6	21
1200	16.4	2.0	22.9	11
1300	34.6	4.2	21 ± 2	5.0 ± .4
1400	66.7	8.1	44 ± 4	5.4 ± .4
1600	2965	359	464 ± 160	1.3 ± .4
1700	1271	154	113 ± 69	0.7 ± .4
1800	1931	234	104 ± 104	0.4 ± .4
2000	110	13	9 ± 6	0.7 ± .4
Total Rock	6408	775	921 ± 344	1.2 ± .4
76535,15	6330	766	7200	9.4

Table 2. $^{126}Xe_{REE}$ is the amount of ^{126}Xe from spallation of REE, estimated using whole rock abundances (33ppm Ba and 7.6 ppm La + Ce + Nd (Haskin *et al.*, 1974) and a production rate of 0.60 ± .15 for REE (norm. to La + Ce + Nd) relative to Ba (Hohenberg *et al.*, 1980[1]).

the quantities of fission-produced ^{136}Xe for each temperature fraction of 76535,75. Here the $^{126}Xe_{REE}$ is estimated from the whole rock elemental composition using the relative production ratio from Hohenberg *et al.* (1980[1]). Although it is possible in principle to resolve $^{126}Xe_{Ba}$ from $^{126}Xe_{REE}$ using systematics of Hohenberg et al. (1978), uncertainties in the end member compositions preclude its use here. Column 4 of the table and Fig. 5 (closed symbols) show that the ratio of fission xenon to rare earth spallation is not even approximately constant (note the logarithmic scale of the ordinate). To the extent that Ba and REE do not occur in fixed proportion, as discussed above, we show in Fig. 5 (open symbols) lower limits to the $^{136}Xe_f/^{126}Xe_{REE}$ ratio by assigning *all* the cosmogenic ^{126}Xe to REE production. Even this limit is well above the ADOR line. Therefore, even allowing for the maximum range of variation in the target abundance ratio, we find no correlation between rare earth spallation and fission xenon in the lower temperature fractions. It is possible that the higher temperature fractions (>1200°C) do in fact release considerable xenon from the *in situ* fission of ^{244}Pu. At these temperatures it is difficult to estimate the amount of cosmogenic xenon from REE targets. If REE spallation is enriched at the lower temperatures, it must be depleted (relative to the Ba contribution) in the higher temperature fractions and the solid symbols in Fig. 5 might then understate the true $^{136}Xe_f/^{126}Xe_{REE}$ ratios for these fractions. Nonetheless, we can unambiguously observe that the low temperature fractions are dominated by fission and not accompanied by appre-

---

[1] Hohenberg C. M., Hudson B., Kennedy B. M., and Podosek F. A. (1980) Xenon spallation systematics in Angra dos Reis. Submitted to *Geochim. Cosmochim. Acta*.

ciable spallation xenon, a conclusion that is virtually independent of any assumptions. To be specific, more than 25% of the fission xenon and less than 2% of the REE spallation xenon is released in the low temperature fractions. On the other hand, the high temperature fractions are totally dominated by spallation, although we do not know precisely how much of this is due to REE targets. We conclude that fission effects in 76535 do not correlate with rare earth spallation effects or any other volume-correlated component. As was the case for the gas-rich breccias, the low temperature release of fission xenon in 76535 requires low

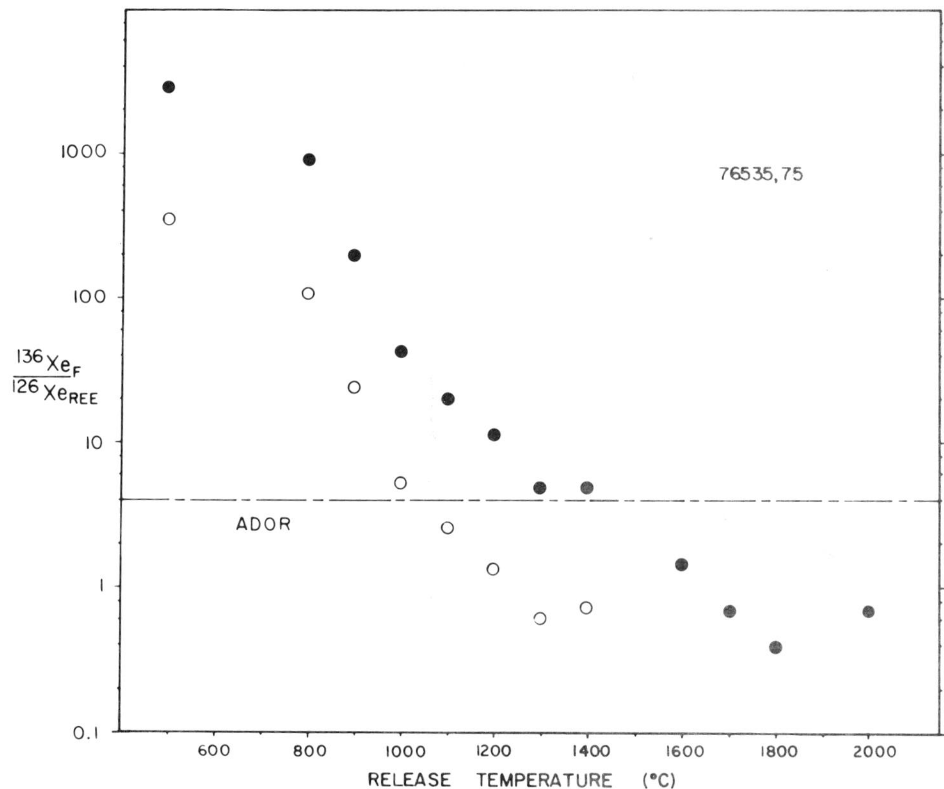

Fig. 5. Ratio of ^{136}Xe from fission to ^{126}Xe from the spallation of rare earth elements, plotted logarithmically (solid circles), as a function of the extraction temperature where the fraction of $^{126}Xe_{sp}$ from REE targets is estimated from whole rock elemental abundances. Open circles are striit lower limits to this ratio and are computed by assigning *all* the ^{126}Xe to REE (Table 4). The horizontal broken line represents the ratio found in Angra dos Reis (Hohenberg *et al.*, 1980[1]) normalized to the 209 million year exposure age of 76535 and corrected for the difference in exposure geometry. This plot clearly shows the absence of association between fission xenon and REE spallation in the lower temperature fractions.

[1] Hohenberg C. M., Hudson B., Kennedy B. M., and Podosek F. A. (1980) Xenon spallation systematics in Angra dos Reis. Submitted to *Geochim. Cosmochim. Acta.*

activation energies for diffusion, suggestive of similar siting conditions in both types of samples, most probably along intergranular boundaries.

Among the other Apollo 17 samples listed in Table 2, only the station 6 samples contain enrichments in fission xenon that can be easily estimated. Although these data are for single-stage melts, the amounts of fission-produced ^{136}Xe can be estimated from the three-isotope correlation plot shown in Fig. 6. Mixtures be-

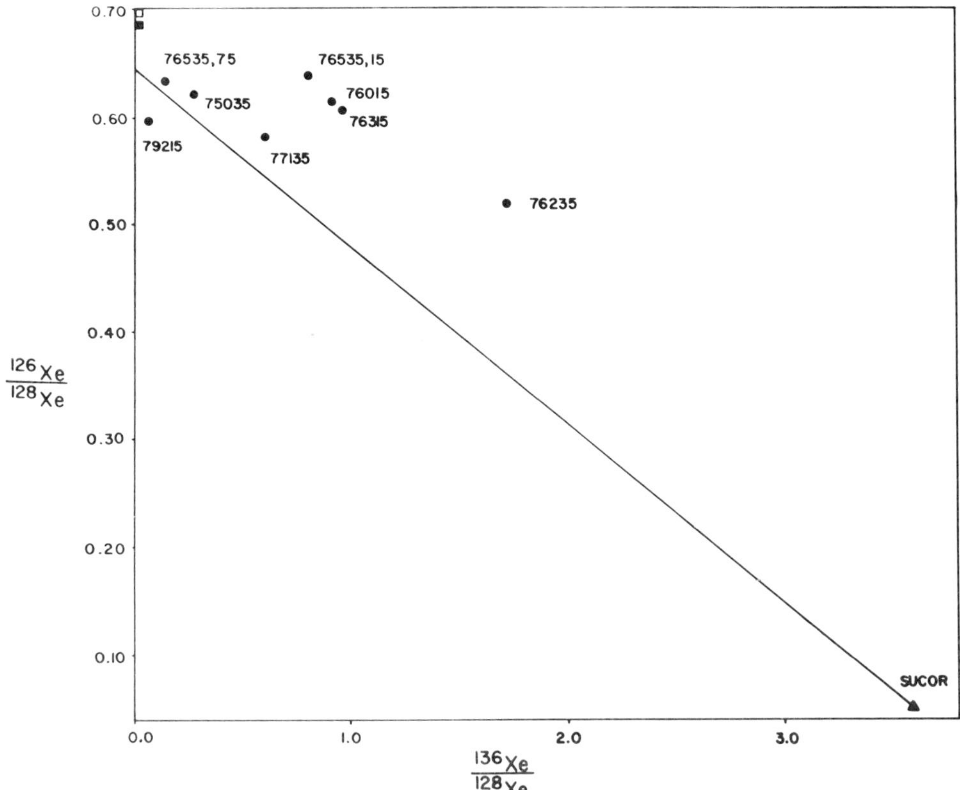

Fig. 6. Three isotope correlation plot involving ^{126}Xe, ^{128}Xe, and ^{136}Xe for the Apollo 17 (whole rock) data of Table 2. As in Fig. 1, cosmogenic contributions displace the points to the upper left, trapped (SUCOR) to the lower right, and fission to the right. Normalization to ^{128}Xe removes much of the uncertainty introduced by variations in Ba and REE; the line shown is a mixing line between SUCOR and spallation xenon characteristic of average (whole rock) spallation xenon in 76535,75. To the extent that the $^{126}Xe/^{128}Xe$ ratios are similar in the spallation components of these samples, mixtures of trapped and cosmogenic xenon should plot along the indicated mixing line. The filled square represents the composition of cosmogenic xenon derived from the systematics of Hohenberg *et al.* (1978) for the chemistry and exposure history of 76535; the open square is the derived cosmogenic xenon composition for 76315, whose Ba/REE ratio differs from that of 76535 by nearly a factor of two. While the systematics are not precise enough to provide an accurate $^{126}Xe/^{128}Xe$ yield, they do suggest that this ratio is not very sensitive to variations in target chemistry. Displacements to the right of the line indicate the presence of fissiogenic ^{136}Xe whose abundance can be accurately estimated only for 76535,15, 76015, 76315, and 76235 (see text).

tween SUCOR and a single spallation component should plot along a mixing line between the two. Since these samples were not analysed in stepwise heating, we have little direct knowledge of the isotopic composition of spallation xenon for each sample; the line indicated in the figure is a mixing line between SUCOR and the average spallation component for whole rock 76535,75. Samples 75035, 77135, and 79215 do not have fission contributions that can be accurately inferred since the uncertainty in the composition of spallation xenon in these samples implies a corresponding uncertainty in the location of the mixing line. The Station 6 samples are sufficiently displaced to the right of the indicated mixing line to allow us to estimate the quantity of fission-produced ^{136}Xe even though the exact placement of the line is not precisely known. On this basis one would estimate that the fraction of ^{136}Xe due to fission is approximately .82 for 76015, .78 for 76315, and .57 for 76235, leading to quantities of $^{136}Xe_f$ of 7.1×10^{-12}ccSTP/g, 7.3×10^{-12}ccSTP/g, and 0.95×10^{-12}ccSTP/g respectively. Note that the inferred fission xenon content of 76015 and 76315 are essentially identical to that found in 76535,15 and that the fission xenon inferred for 76235 is approximately the same as that found for 76535,75.

Since we do not know the spallation compositions of these samples very well the validity of the above estimations of the fission xenon content might be open to some question. We note, however, that the three samples in question, 76015, 76235, and 76315, were all collected from the Station 6 Boulder. The closed square in Fig. 6 represents the spallation xenon composition as computed for 76535 from the systematics of Hohenberg *et al.* (1978), using a Ba content of 33 ppm and a La content of 1.51 ppm (Haskin *et al.*, 1974), suitably correcting for surface erosion. The open square is the spallation composition for 76315, similarly computed, using a Ba content of 72.8 ppm and a La content of 5.41 ppm (Hubbard *et al.*, 1974). While the systematics are not refined to the point of providing precise isotopic ratios for the spallation component, they do suggest that the differences in the $^{128}Xe/^{126}Xe$ ratio in the spallation components are not very large. A qualitatively similar conclusion can be reached by comparing Figs. 3 and 4. As we previously noted, 76535,15 and 76535,75 differ in the ratio of Ba to REE. 76535,15 is displaced toward REE spallation in Fig. 3, but lies in the spallation-rich cluster of 76535,75 points in Fig. 4. The significance of this is that the $^{126}Xe/^{130}Xe$ and $^{129}Xe/^{130}Xe$ ratios are very different in the two spallation contributions, but the $^{126}Xe/^{128}Xe$ and $^{129}Xe/^{128}Xe$ ratios are not. Consequently, one would expect the $^{126}Xe/^{128}Xe$ ratio in the spallation component that is present in these rocks to be relatively insensitive to modest variations in the Ba to REE ratios. The position of the correlation line in Fig. 6 and the inferred fission xenon contents should therefore be correspondingly insensitive to these variations.

76315 contains 0.343 ppm uranium (Hubbard *et al.*, 1974), producing about 1.1×10^{-12}ccSTP/g of fission ^{136}Xe, which is roughly one seventh of the estimated amount of fission-produced ^{136}Xe in this sample. Turner and Cadogan (1975) obtained ^{39}Ar-^{40}Ar plateau ages for three clasts from this rock of 3.98×10^9 yrs, but one anorthositic clast appears to have retained radiogenic argon from times earlier than this. The observed quantity of fission xenon could be produced by *in situ* plutonium decay for a xenon closure time of about 4.25 b.y. and uncertainties in this quantity are rather modest since a factor of two error in estimating the quantity of fission xenon translates into an error of only 82 m.y. in the apparent xenon closure time. We, of course, have no assurance that the fission xenon in 76315, and the other Station 6 samples, have retained their fission xenon as an *in situ* decay product. Like extinct isotope xenon in 76535, this component may well be parentless in the sense that it is presently sited remotely from its source and, like 76535, the quantity of fission xenon may be highly variable from sample-to-sample.

SUMMARY AND DISCUSSION

Behrmann *et al.* (1972), Drozd *et al.* (1975), and Drozd *et al.* (1976) discuss mechanisms by which extinct isotope xenon may have been incorporated into the gas-rich breccias. Apparent association with solar wind xenon and the low activation energies for diffusive loss suggested to them that the component was preferentially located on grain surfaces. Recent work by Bernatowicz *et al.* (1979) on grain size separates of disaggregated 14301 have confirmed this suggestion. In addition, they showed that, on a whole rock basis, xenon from the various size fractions of 14301 are to a good approximation organized into only two functional components, one a volume-correlated spallation component, the other a surface-correlated component. Moreover, the surface-correlated component is a superposition largely of SUCOR and extinct isotope xenon with minor contributions from spallation-produced xenon. The two major surface-correlated contributions (solar and extinct isotope xenon) cannot be separated by size fraction analysis. That is, on a whole rock basis, extinct isotope xenon in 14301 resides in constant proportion to solar wind xenon. In stepwise heating extractions, extinct isotope effects are distinguishable from solar xenon by the preferential release of the former at lower temperatures.

Previously postulated mechanisms responsible for the incorporation of parentless extinct isotope xenon onto grain surfaces (Behrmann *et al.*, 1972; Drozd *et al.*, 1975; and Drozd *et al.*, 1976) fall into two general groups. In one model the radionuclear components are exhaled from the lunar interior into a transient lunar atmosphere from which the xenon is ionized (by photoionization or charge exchange reactions) and driven into grain surfaces by solar wind fields. The other model invokes simple adsorption of the radionuclear component on cold grain surfaces within the body of the regolith. The convincing association with solar wind xenon revealed by the 14301 grain-size study and the occurrence of extinct isotope effects among Apollo 16 samples (Bernatowicz *et al.*, 1978), which suggests a global rather than a local effect, tended at the time to support implantation from a transient lunar atmosphere, requiring actual residence on the lunar surface.

Results from stepwise heating of troctolite 76535 reported here indicate that parentless extinct isotopic xenon is also found in objects that have been derived by igneous processes and which contain very little solar wind xenon. Enhancement of the effect at low extraction temperatures and the general similarity of the release pattern with that of the gas-rich breccias (cf. Fig. 1 and Fig. 4 of Drozd *et al.*, 1972) suggest a similar siting, presumably along intergranular boundaries. Since solar wind xenon is practically absent from 76535 we conclude that the establishment of a parentless component with low activation energy for diffusive loss, presumably sited at grain surfaces, does *not* require exposure to space. Whether the mechanism for incorporation of parentless xenon components in 76535 is related to that responsible for similar effects in the gas-rich breccias is unclear, but the similarities are highly suggestive. It is important to remember that the quantity of fission xenon in 76535 is not inconsistent with *in situ* production; it is only the present siting that precludes such an origin. What is clear is that the present study tends to suggest grain surface siting. If 76535 did not experience the extensive regolith history of the gas-rich breccias it may have acquired its complement of parentless xenon in a local atmosphere during a period of extensive re-equilibration, diffusion, and homogenization (cf. Gooley *et al.*, 1974). On the other hand, if 76535 did experience an extended regolith history then adsorption on cold surfaces may play an important role in the placement of weakly bound xenon components. Although it remains to be seen which of these two mechanisms is responsible for the incorporation of the parentless components in 76535 (and perhaps the gas-rich breccias and other residents of the lunar regolith) it is apparent that the results of this study imply that electromagnetic implantation mechanisms are no longer required.

Adsorption on surfaces at thermal energies results in components that are too weakly bound to be permanently retained. As mentioned earlier, the fixing of such components has been all too readily demonstrated by mechanical crushing of lunar samples in the terrestrial atmosphere. Shock has also been shown to produce tenaciously bound components by Pepin *et al.* (1964) and Davis (1977). Experiments are currently underway in this laboratory to explore the modification of superficial components at low (70Kb) to modest (200Kb) shock pressures. Pending the results of these studies it does not appear that the initially superficial nature of surface adsorbed components presents any real problems for long-term retention. Surface adsorption (from either an open or closed atmosphere) is consequently the most favored process for the placement of parentless extinct isotope xenon on lunar grain surfaces. Adsorption within the regolith would seem to require such a fixing process; local redistribution at elevated temperatures and pressures, as may be the case for 76535, would not.

During the freeze-thaw disaggregation of 14301, it was noted that much of the more weakly bound parentless component was lost from those samples which were processed under water (Bernatowicz *et al.*, 1979). This was tentatively attributed to surface hydration reactions. It should be mentioned that 76535,15 was received from the lunar sample curator in July, 1973 and run shortly thereafter; 76535,75 was received from the curator in April, 1976 and run in the mass spectrometer in August, 1979. They differ by a fraction of 8 in the quantity of ^{244}Pu fission xenon, a large fraction of which is lightly bound. One disturbing possibility is that surface reactions, possibly hydration, may have contributed to the decrease in the observed amount of lightly-bound parentless xenon during the intervening 6 years. We are not suggesting that such surface reactions have actually occurred but simply pointing out that these effects cannot be excluded at this time and the possibility should be explored further. The other alternative is, of course, that extinct radionuclear effects are simply variable.

Differences in the composition of spallation xenon between the two fractions utilized in composite component subtraction techniques can lead to erroneously large estimations of ^{129}Xe in the inferred extinct isotope component. Of the previous two samples for which the isotopic composition of this component has been derived, 14301 and 14318 (Behrmann *et al.*, 1972), the former showed large enrichments in ^{129}Xe, the latter did not. The observed ^{129}Xe enrichment in 14301, however, cannot be a phantom effect generated by the failure in the assumptions of the method. That the reported enrichments of ^{129}Xe in 14301 are in fact real can be most easily seen in Fig. 11 of Bernatowicz *et al.* (1979) where the low temperature data contain ^{129}Xe/^{130}Xe that substantially exceed the corresponding value in SUCOR, atmospheric, or any viable trapped component. Moreover, Bernatowicz *et al.* (1979) demonstrated that this ^{129}Xe effect is separable from the parentless fission component and seems to suggest a time-ordered incorporation of these two extinct radionuclear effects with important implications for the chronology of these objects and the early lunar regolith in general.

Acknowledgments—The authors wish to thank C. White for manuscript preparation, and L. Ross and P. Suntharothok for help with the drafting. This work was supported in part by NASA grants NGL 26-008-065 and NSG-070016.

REFERENCES

Basford J. R., Dragon J. C., Pepin R. O., Coscio M. R., and Murthy V. R. (1973) Krypton and xenon in lunar fines. *Proc. Lunar Sci. Conf. 4th,* p. 1915–1955.

Behrmann C. J., Drozd R. J., and Hohenberg C. M. (1973) Extinct lunar radioactivities: xenon from ^{244}Pu and ^{129}I in Apollo 14 breccias. *Earth Planet. Sci. Lett.* **17,** 446–455.

Bernatowicz T. J., Hohenberg C. M., Hudson G. B., Kennedy B. M., and Podosek F. A. (1978) Excess fission xenon at Apollo 16. *Proc. Lunar Planet. Sci. Conf. 9th,* p. 1571–1597.

Bernatowicz T. J., Hohenberg C. M., and Podosek F. A. (1979) Xenon component organization in 14301, *Proc. Lunar Planet. Sci. Conf. 10th,* p. 1587–1616.

Bogard D. D., Nyquist L. E., Bansal B. M., Wiesmann H., and Shih C.-Y. (1975) 76535: An old lunar rock. *Earth Planet. Sci. Lett.* **26,** 69–80.

Boynton W. V. (1978) Fractionation in the solar nebula, II: Condensation of Th, U, Pu and Cm. *Earth Planet. Sci. Lett.* **40,** 63–70.

Crozaz G., Drozd R., Hohenberg C., Morgan C., Ralston C., Walker R., and Yuhas D. (1974) Lunar Surface Dynamics: Some general conclusions and new results from Apollo 16 and 17. *Proc. Lunar Sci. Conf. 5th,* p. 2475–2499.

Davis P. K. (1977) Effects of shock pressure on ^{40}Ar-^{39}Ar radiometric age determinations. *Geochim. Cosmochim. Acta* **41,** 195–206.

Drozd R., Hohenberg C. M., and Morgan C. J. (1975) Krypton and xenon in Apollo 14 samples: fission and neutron capture effects in gas-rich samples. *Proc. Lunar Sci. Conf. 6th,* p. 1857–1877.

Drozd R., Hohenberg C. M., and Ragan D. (1972) Fission xenon from extinct ^{244}Pu in 14301. *Earth Planet. Sci. Lett.* **15,** 338–346.

Drozd R. J., Kennedy B. M., Morgan C. J., Podosek F. A., and Taylor G. J. (1976) The excess fission xenon problem in lunar samples. *Proc. Lunar Sci. Conf. 7th,* p. 599–623.

Eberhardt P. J., Geiss J., Graf H., Grögler N., Mendia, Mörgeli M., Schwaller H., and Stettler A. (1972) Trapped solar wind gases in Apollo 12 lunar fines 12001 and Apollo 11 breccia 10046. *Proc. Lunar Sci. Conf. 3rd.* p. 1821–1856.

Gooley R., Brett R., and Warner J. (1974) A lunar rock of deep crustal origin: Sample 76535. *Geochim. Cosmochim. Acta* **38,** 1329–1339.

Haskin L. A., Shih C.-Y., Bansal B. M., Rhodes J. M., Weismann H., and Nyquist L. E. (1974) Chemical evidence for the origin of 76535 as a cumulate. *Proc. Lunar Sci. Conf. 5th,* p. 1213–1225.

Hohenberg C. M. (1980) High sensitivity ion-counting mass spectrometer system for noble gas analysis. *Rev. Sci. Inst.* In press.

Hohenberg C. M., Marti K., Podosek F. A., Reedy R. C., and Shirck J. (1978) Comparisons between observed and predicted cosmogenic noble gases in lunar samples. *Proc. Lunar Planet. Sci. Conf. 9th,* p. 2311–2344.

Hubbard N. J., Rhodes J. M., Weismann H., Shih C.-Y., and Bansal B. M. (1974) The chemical definition and interpretation of rock types returned from the non-mare regions of the moon. *Proc. Lunar Sci. Conf. 5th,* p. 1227–1246.

Huneke J. C. and Wasserburg G. J. (1975) Trapped ^{40}Ar in troctolite 76535 and evidence for enhanced ^{40}Ar-^{39}Ar age plateaus (abstract). In *Lunar Science VI,* 417–419. The Lunar Science Institute, Houston.

Husain L. and Schaeffer O. A. (1975) Lunar evolution: the first 600 million years. *Geophys. Res. Lett.* **2,** 29–32.

Keith J. E., Clark R. S., and Bennett L. J. (1974) Determination of natural and cosmic ray induced radionculides in Apollo 17 lunar samples. *Proc. Lunar Sci. Conf. 5th,* p. 2121–2138.

Lewis R. S. (1975) Rare gases in separated whitlockite from the St. Severin chondrite: xenon and krypton from fission of extinct ^{244}Pu. *Geochim. Cosmochim. Acta* **39,** 417–432.

Lugmair G. W. and Marti K. (1977) Sm-Nd-Pu timepieces in the Angra dos Reis meteorite. *Earth Planet. Sci. Lett.* **35,** 273–284.

Lugmair G. W., Marti K., Kurtz J. P., and Scheinen N. B. (1976) History and genesis of lunar troctolite 76535 or: How old is old? *Proc. Lunar Sci. Conf. 7th,* p. 2009–2033.

Morgan C. J. (1975) Exposure age dating of lunar features: Lunar heavy rare gases. Ph.D. thesis, Dept. of Physics, Washington Univ., St. Louis, Mo.

Niemeyer S. and Leich D. A. (1976) Atmospheric rare gas in lunar rock 60015, *Proc. Lunar Sci. Conf. 7th,* p. 587–597.

Papanastassiou D. A. and Wasserburg G. J. (1976) Rb-Sr age of troctolite 76535, *Proc. Lunar Sci. Conf. 7th,* p. 2035–2054.

Pepin R. O., Turner G., Reynolds J. H., DeCarli P., and Fredriksson K. (1964) Shock emplaces argon in a stony meteorite. *J. Geophys. Res.* **69,** 1403–1411.

Podosek F. A., Huneke J. C., Burnett D. S., and Wasserburg G. J. (1971) Isotopic composition of xenon and krypton in lunar soil and in the solar wind. *Earth Planet. Sci. Lett.* **10,** 188–216.

Podosek F. A. (1972) Gas retention chronology of Petersburg and other meteorites. *Geochim. Cosmochim. Acta* **36,** 755–772.

Tera F. and Wasserburg G. J. (1974) U-Th-Pb systematics on lunar rocks and inferences about lunar evolution and the age of the moon. *Proc. Lunar Sci. Conf. 5th,* p. 1571–1599.

Turner G. and Cadogan P. H. (1975) The history of lunar bombardment inferred from ^{40}Ar-^{39}Ar dating of highland rocks. *Proc. Lunar Sci. Conf. 5th,* p. 1509–1538.

Papike, J.J. and Merrill, R.B., eds.
Proc. Conf. Lunar Highlands Crust (1980), p. 441-456
Printed in the United States of America

Planetary crusts: A comparative review

J. V. Smith

Department of the Geophysical Sciences, University of Chicago, Chicago, Illinois 60637

Abstract—This paper brings up to date the references and ideas in the review "Mineralogy of the Planets: A Voyage in Space and Time", and selects properties of planetary crusts which have special significance for the origin of the lunar highlands crust. Particular emphasis is placed on the effects caused by absence of atmophile volatiles in lunar rocks, and on the use of lunar impact chronology to place constraints on development of crusts on other planets.

INTRODUCTION

Although each planetary crust is different, and although ideas cannot be transferred simply from one planet to another, intercomparison of properties leads to stimulating propositions. This review concentrates on those properties of planetary crusts which are particularly relevant to the highlands of the moon, and tries to relate them in terms of an overall scheme for growth and development of planets from a presumed solar nebula.

For brevity, readers are referred to an earlier review (Smith, 1979) for justification of the following basic assumptions:

(a) Condensates from the solar nebula varied from reduced, volatile-poor material near the Sun to oxidized, volatile-rich material outwards from the asteroid region.
(b) Accretion produced planetesimals up to some hundreds of kilometers diameter, whose development involved complex processes of physical and chemical differentiation, as illustrated by the diversity of meteorites.
(c) Each terrestrial planet began to grow from near-by slow planetesimals and ended with distant fast planetesimals.
(d) Only the Earth and Venus were big enough to retain all debris from late volatile-rich planetesimals deflected at high speed by the giant planets, and Mercury and the Moon lost much of the incoming volatiles; Mars was more successful than the latter two planets in retaining volatiles, because of the lower ratio of impact velocity to escape velocity.
(e) A powerful heat source (perhaps short-lived radioactivity) caused melting of at least some planetesimals, as demonstrated by crystal-liquid differentiation of many meteorites, and the large terrestrial planets began to frac-

tionate into crystalline and liquid components at an early stage of accretion; even the small planets underwent at least some differentiation by −4.3Ae, during at least the last one-sixth of accretion.

(f) The silicate mantle of the Earth must have been mostly crystalline at the end of accretion, with little reaction between a liquid metallic core and late accreted material; this rules out catastrophic core formation and a fission origin of the Moon during this era (but not earlier).

(g) The entire inner solar system was bombarded by projectiles large enough to form basins on the Moon up to −3.9Ae.

(h) Partial melting and volcanism was the major mechanism for heat loss of a planet, followed by outward radiation to space; such activity must have decreased with time.

(i) The major differences between chemical composition, pressure and temperature of the atmospheres of the inner planets provide a warning that caution is needed in attempting to model an early atmosphere and its chemical reaction with a crust.

(j) Because the Earth's atmosphere results from secondary outgassing from the interior, and not from primary accretion, some mechanism, probably a powerful solar wind, was responsible for removal of a presumed primary atmosphere; furthermore, preferential escape of H with respect to heavier gases from the secondary atmosphere might have caused oxidation of the early crust.

Since the literature review in Smith (1979), the following references to the general development of the solar system and planets have been noticed: Burns (1978), Chamberlain (1978), Consolmagno and Lewis (1978), Flynn *et al.* (1978), Gehrels (1978), Hartmann (1979), Herndon (1978), Lewis *et al.* (1979), Matsui (1978, 1979), McDonnell (1978), Phillips and Ivins (1979), Pollack (1979), Pollack *et al.* (1979), Reynolds and Cassen (1979), Schubert (1979), Schubert *et al.* (1979), Solomon (1979), Walker (1978) and Wetherill (1979).

Earth

Although intense disagreement remains about the properties of the deep mantle and core, there is increasing agreement about the present properties of the crust and uppermost mantle (up to 200 km depth) as geochemical constraints are coupled with detailed geological and petrographic observations. Because the development of the Earth's crust cannot be decoupled from that of the entire Earth, the following propositions about the development of the latter are presented. Readers should consult Ringwood (1979) for a different scheme. A new book on the origin, structure and evolution of the Earth, edited by McElhinney (1979), has been advertised, but not yet seen by me. Books edited by Tarling (1978) on the evolution of the Earth's crust, and by Barker (1979) on tonalitic and other intermediate rocks in the Earth's crust cover in detail much of the material summarized here: see also forthcoming papers by Lambert (1980a,b), Shaw (1980a,b) and Smith (1980).

The following observations and propositions are based mainly in interpretation of literature reviewed in Smith (1979), and only subsequent papers are referenced:

(a) *The Earth underwent major chemical differentiation* into a liquid Fe,Ni-rich core, a mantle rich in high-pressure silicates, an outer solid-plus-liquid region rich in magmatophile and volatile elements, a hydrosphere and an atmosphere. Intense volcanism should have taken care of the massive peak of heating near the end of accretion according to a Safronov-type model (Kaula, 1979a,c), thereby ruling out catastrophic volatilization. This is consistent with present-day emission of rare gases from inside the Earth (e.g. Matsuo *et al.*, 1978; Schwartzmann, 1978; Saito *et al.* 1978); these gases could have been dissolved in early liquid and trapped in early crystals. The low content of volatiles in the Moon's crust (Taylor, 1979) with respect to the Earth's crust requires detailed discussion. If volatile-rich material was hitting the Moon at the same time as the Earth, one possible answer is that the ratio of impact velocity/escape velocity for late volatile-rich planetesimals was too high for volatile-rich debris to be captured by the Moon. Accretion of the Moon in a volatile-poor region of the solar region followed by transfer to Earth orbit poses such severe dynamical and energetic problems that it is not advocated here.

(b) *The oldest surviving relics of terrestrial crust* date back only to −3.8Ae, and two simple explanations are possible: (i) the mantle was so cool after accretion that it did not heat up sufficiently to produce magmas until 700 million years had passed, (ii) crust was generated continuously and was destroyed continuously by recycling into the mantle until some crustal nuclei became stable. The first explanation is not accepted here because radiogenic heat would be 5 times greater ($\sim 1 \times 10^{14}$W) at −4.5Ae than at present, and simple calculations show that melting would occur with transfer of heat to the surface by volcanic activity. Indeed the presence of ultrabasic lavas in greenstone belts of ~−3Ae age testifies to the existence of deep-seated volcanism some 1.5×10^9 y after accretion, and rare ultrabasic lavas occur even in recent times. Consequently the second explanation is accepted. It is tempting to attribute destruction of early crust to bombardment by projectiles like those which formed the lunar basins, but self-destruction of the crust should also be considered.

It is difficult to estimate the number of basin-forming projectiles which hit the Earth because of uncertainty in the age distribution of basins on the Moon and in the velocities and orbits of the late planetesimals in the inner solar system. Non-definitive arguments suggest that at least 2,000 basins and more likely 20,000 basins were formed on Earth between −4.4 and −3.9Ae (Smith, 1980). The flux should decrease near-exponentially, and if the 40 well-defined basins on the Moon were formed in the period −4.3 to −3.9Ae, about 2,000 basins might be expected on Earth. Certainly the formation of a basin 200 km diameter and 20 km deep would trigger partial melting to some hundreds of kilometers depth in a mantle *already close to melting,* but would not trigger volcanism in a region some hundreds of Kelvin below solidus temperature. Mantle upwelling should occur under a basin and slumping should develop around the basin.

Although impacts must have destroyed crust, perhaps a more important process is self-destruction of crust. Volcanism would lead to huge piles of basic to ultrabasic volcanic rocks "floating" on a depleted peridotitic zone which would itself float on undepleted peridotite. Extrusive rocks would be quenched upon contact with the hydrosphere or atmosphere. Continued extrusion would result in forcible sinking of volcanic rocks. Complex devolatilization reactions and partial melting would occur at increasing pressure. Radiogenic heat production would be concentrated in the upper 30 km as K, U and Th were differentiated into silica-rich rocks. Depleted residues rich in pyroxene and especially garnet would founder and become mechanically mixed into the depleted perioditic zone thereby generating peridotite depleted strongly in alkalis but not so much in Ca and Al. Convection in the upper mantle would lead to incomplete mixing of the various types of peridotites. This qualitative scheme requires detailed examination before it can be taken seriously: however, it is worth emphasizing that even present-day continental plates undergo rifting (e.g., in China) and that partial destruction occurs at some margins (e.g., central zone of western coast of S. America). Earlier crustal nuclei would be smaller and less rigid than modern plates, and perhaps continents did not become stable until development of a granulite base, whose low content of H_2O and often of radiogenic elements inhibited melting and yielded greater mechanical strength. Flushing out of H_2O from amphibolites by CO_2-rich vapor from the mantle has been proposed as an important factor (Newton *et al.*, in preparation).

Whatever the fate of these arguments, which incorporate ideas in Lambert (1980a,b), Shaw (1980a,b) and Smith (1980), there seems no doubt that the presence of volatiles on Earth, and its larger size, led to a greater complexity of volcanic and metamorphic processes than on the dry Moon (c.f. Walker *et al.*, 1979).

Isotope systematics cannot place a definitive control on the temporal development of reservoirs but recent data and models are consistent with (i) transfer of a large fraction of magmatophile elements to within 600km of the Earth's surface during the first 100my (ii) retention of a significant amount in sub-crustal reservoirs for the first 4.5Ae. (iii) downward mixing of early crust into uppermost mantle for the first 500 my, (iv) irreversible growth of crust after −3.8Ae with particularly rapid growth at certain periods, and (v) melting inside the crust with development of a lower crust depleted in radiogenic elements. Some new references are: DePaolo and Wasserburg (1979). Hamilton *et al.* (1978), Hart *et al.* (1979), Hurst (1978), Jacobsen and Wasserburg (1979), Jordan (1978), Manhes *et al.* (1979), O'Nions *et al.* (1978; 1979a,b), Vidal and Dosso (1978). Further progress will require understanding of.the partition of radiogenic elements between crystals, liquids and vapors.

Whatever the future of thinking about the early crust of the Earth, it is certain that (i) volatiles played an important role, and (ii) much of the mantle was cool enough to store magmatophile elements for several Ae. Furthermore the concept of an Earth-covering magma ocean up to −3.8Ae is quite implausible when the rate of radiogenic heat production is compared with radiative cooling (c.f. Jakosky and Ahrens, 1979).

(c) *The nature of the early atmosphere and hydrosphere* is difficult to determine (Walker, 1977; Shimizu, 1979a), but new mineralogical evidence from Archaean sediments (Dimroth and Lichtblau, 1978) supports earlier evidence for an oxidizing state of early Archaean atmosphere and hydrosphere. At what stage oxygen-generating photosynthesis began is uncertain (Baur, 1978; Towe, 1978), but carbon isotopes in Archaean sediments are not inconsistent with biological activity from at least −3.2Ae. The theoretical expectation of lower luminosity of the Sun in pre-Archaean time makes it very difficult to obtain enough energy to produce a massive atmosphere at the expense of a hydrosphere, and indeed an enhanced greenhouse effect was proposed to avoid glaciation on the pre-Archaean Earth (Owen *et al.*, 1979; Shimizu, 1979a). Currently it seems easiest to assume that a hydrous ocean covered a substantial fraction of the Earth's surface in at least most of pre-Archaean time. The view popular some 10 years ago that the pre-Archaean atmosphere was rich in NH_3 and CH_4 seems to have become unpopular, and an atmosphere rich in N_2 and CO_x molecules ($x \leq 1$), e.g., Shimizu (1979a), would be easier to convert into the slightly oxidizing atmosphere of Archaean times than a highly reduced one; indeed it may be even easier to assume that much C was in the form of atmospheric CO_2 and combined into carbonate minerals in the crust and uppermost mantle. Janardhan *et al.* (1979) and Newton *et al.* (1980) assembled evidence in favor of a massive source of CO_2 in the mantle, whose release into the crust caused widespread granulite metamorphism at −3Ae. If H_2O was transported efficiently up to the ionosphere, H would be lost from the Earth and O made available for oxidation of atmospheric gases and crustal rocks. Although much H_2O remains in the oceans and crustal rocks, and a small amount remains as OH in upper-mantle rocks along with carbonate, it is not possible to rule out loss of (say) two-thirds of terrestrial hydrogen after accretion. Particularly important for the present comparison with the Moon is the simplicity of phase equilibria and mineralogy of low-volatile lunar rocks and the extraordinary complexity of phase equilibria and mineralogy for terrestrial rocks containing H_2O and CO_2 (Wyllie, 1979).

(d) *Igneous differentiation of the Earth is still active* today after many complex episodes, in contrast to igneous differentiation on the Moon which consisted of a major episode at ~4.5Ae involving most or all of the Moon, and a prolonged waning episode of mare volcanism probably terminating by about −2Ae. It is convenient to summarize the present activity on the Earth in plate-tectonic terms before looking backwards at Archaean events. Seismic activity is concentrated at the margins of six large plates which are fairly strong and coherent except at rifts. The nature of volcanic and plutonic activity depends on geographic position with respect to plate margins and boundaries between oceanic and continental crust. Volcanic rock types (to be described exhaustively by the Basaltic Volcanism Project) include (i) alkali basalts, produced by hot-spot melting of the mantle at ~100 km depth, (ii) mid-ocean-ridge basalts, probably produced by differentiation in a near-surface chamber fed continuously by upwelling magma, (iii) diverse varieties of island-arc basalts, probably produced by hybridism of partial melts from subducted oceanic basalts and the overlying mantle wedge, and (iv) tholeiitic flood basalts and rare carbonatitic, kimberlitic and alkali-rich rocks in continental regions produced by melting of underlying mantle ranging down to 200 km depth. Probably all magmas become contaminated during passage through overlying crust. Intrusive igneous rocks include massive bodies of granitoids, ranging from tonalitic-granodioritic batholiths near active plate margins to K-rich granite plutons mainly in plate interiors. The origin of all granitoids is complex, probably involving remelting of both igneous rocks originally derived from the mantle and metamorphosed sediments. Continental blocks are probably stabilized by high-grade metamorphism at 20–40 km depth. Although this broad-brush portrait of the present Earth is far too concise for most geologic purposes, it suffices to show how very much more complex is the geology of the Earth than the Moon.

(e) *Turning back to the Archaean era,* the geology is very difficult to interpret because all surviving rocks have undergone strong deformation + metamorphism, and chemical compositions of many rocks have been changed. Complex details for several regions are summarized in Windley (1976, 1977), Tarling (1978) and Barker (1979). There is little surviving evidence of impacts by large projectiles (Weiblen and Schulz, 1978), and even for the Sudbury basin an impact origin is disputed (see Grieve and Dence (1979) and Grieve and Robertson (1979) for discussion of the cratering record). Archaean rocks can be separated into two metamorphic suites. A high-grade suite consists mainly of tonalitic-granodioritic gneisses of calc-alkaline affinity, intercalated with layered igneous complexes and metamorphosed volcanic and sedimentary rocks which include marbles, graphite-bearing mica schists, and banded magnetite-quartzites. A low-grade suite in greenstone belts originally consisted of volcanic rocks of ultramafic, mafic and calc-alkaline affinity, and of sedimentary rocks which include cherts and banded iron formations. Anorthosite layers occur in layered complexes of overall basaltic composition, and meta-anorthosites are reported in greenstone belts (e.g., Naqvi and Hussain, 1979), but there is no evidence for an anorthositic crust of the type found in the Moon; indeed one is not expected because of the effect of H_2O and CO_2 on terrestrial volcanism. Whether plate-tectonic processes occurred in Archaean times is disputable, and certainly there is no reason to expect large stable plates in Archaean times. Perhaps there was a trend from erratic hot-spot tectonics in pre-Archaean times to unstable small-plate tectonics in Archaean times to large-plate tectonics in post-Archaean times (many relevant references listed in a paper by R. C. Newton *et al.,* in preparation.) Again it is necessary to emphasize the role of fluids in sedimentary, igneous and metamorphic rocks in the Earth's crust in contrast to the lunar rocks.

(f) *The role of the mantle* is quite different for development of crustal rocks on the Moon and Earth. For the Earth, the composition of spinel lherzolite xenoliths in alkali basalts and porphyroclastic garnet lherzolite nodules in kimberlites (many references including Jagoutz *et al.*, 1979) testifies to long-term survival in the uppermost mantle of some rocks undepleted in a basaltic component. Formation of the crust has led to complementary residues of depleted peridotites, harzburgites and dunites, all of which are represented by xenoliths from the upper mantle. Particularly important is the fate of subducted oceanic crust, for which phase-equilibrium data (Wyllie, 1979) are consistent with transformation to eclogite and garnetite in present-day tectonic models, but not necessarily for Archaean models. Anderson (1979a,b) pointed out that eclogite and garnetite should not be able to sink lower than 650 km because of density-pressure relations, and that there should be a massive region (450 km thick) of eclogite and garnetite in the upper mantle. This proposal is qualitatively consistent with the proposition by McKenzie and Weiss (1975) and Richter (1979) that the Earth's mantle has at least two convecting layers rather than a single convection system (e.g., Elsasser *et al.*, 1979: see also Sharpe and Peltier, 1979). Whereas the crystal-liquid relationships on the Moon are governed by low-pressure minerals, those on the Earth are affected also by high-pressure minerals, of which garnet and perovskite are particularly important (Ringwood, 1979; Liu, 1979; Watt and O'Connell, 1978). A final point is the importance of sulfide in both terrestrial and lunar magmas as a result of storage in the mantle—whereas the sulfide in lunar basalts is troilite (FeS), that in most terrestrial rocks is an Fe,Ni,Cu-sulfide which undergoes complex subsolidus processes.

To conclude this section, it is obvious that the crusts of the Earth and Moon have quite different properties, and simple analogies just do not exist. Impact chronology for the Moon provides an important control on speculations about the early history of the Earth's crust, while the detailed corpus of knowledge gained by centuries of study of terrestrial geology, geochemistry and geophysics provided the necessary basis for lunar studies.

Mercury

There is little to add to the review in Smith (1979). The surface morphology suggests that crystal-liquid differentiation was completed early enough to allow preservation of basins most easily attributed to the same population that formed the lunar basins. Global lineaments are attributed to despinning and contraction (Pechmann and Melosh, 1979). Reflectance spectra of Mercury's surface are like those of Apollo 16 soils, and suggest about 5.5 wt.% FeO (McCord and Clark, 1979).

Perhaps the main value of Mercury for thinking about the Moon is the firm evidence for its dipole magnetic field with the implication that even a small planet can have a liquid core: furthermore, melting of Mercury must have occurred well before −4Ae, with the implication that Mercury accreted from hot material (~1000K). This gives encouragement to thinking about the possibility of similar early melting on the Moon (see later).

Venus

Because the data on the crust of Venus are so pitifully few, comparison with the crust of the Moon is rather presumptuous!

The following propositions are based mainly on interpretation of literature reviewed in Smith (1979), and only subsequent papers are referenced:

(a) *Radar observations of the Venusian crust* have poor resolution from the geological viewpoint, and there is a danger of over-interpretation of features, as for early telescopic views of Mars. Current interpretation (Campbell *et al.*, 1979) repeats assignment of central-spotted rings of over 100 km diameter to impact craters with central peaks like those seen on Mercury and Mars. If the projectiles came from the population responsible for mare basins, some of the Venusian crust must have retained its shape since −4Ae. Origin of the rings from capture of a Venusian moon which fractured inside the Roche limit is very unlikely unless the moon were much smaller than the Earth's moon. If the slow rotation of Venus resulted from capture of a retrograde moon, the event would probably be so catastrophic that it must have occurred well before growth of stable crust. If linear features, including double curves 1000 km long and about 100 km apart (Campbell *et al.*, 1979, Fig. 3) and a complex "beaded" feature (Pettengill *et al.*, 1979) are rifts, Venus must have undergone convection, as is expected anyway from thermal considerations (e.g. Schubert, 1979). The tectonic-time history is speculative (Malin, 1979; McGill, 1979), and it will now be assumed that, although Venus and Earth have almost the same size and probably almost the same composition (next paragraph), at least parts of the Venusian crust stabilized faster than the terrestrial one.

(b) *The bulk composition of Venus* is probably close to that of Earth because of comparable K/U ratios of surface material and similar inventories of CO_2 in combined near-surface reservoirs (Anderson, 1980; Khodakovsky *et al.*, 1979; Warner, 1979). In addition, the amount of N_2 in the Venusian atmosphere (Hoffman *et al.*, 1979; Istomin *et al.*, 1979) is comparable to that in the Earth's atmosphere and crust. This implies that Venus should have accreted about the same mass of H_2O as CO_2 in any model in which Earth and Venus form in a similar way. Although it is theoretically possible to store ~500 ppm H_2O in mica and amphibole in Venus (Smith, 1979), this amount is much too high for several reasons, one of which is the need to stabilize the crust of Venus by −4Ae (Warner, 1979). Consequently, it is deduced that the Venusian crust lost all or most of its water to the atmosphere thereby forcing up the minimum temperature of melting. Loss of H by ionization in the ionosphere would have yielded a massive amount of O capable of oxidizing ~100 km thickness of basaltic rocks (Smith, 1979; Khodakovsky *et al.*, 1979).

(c) *The high surface temperature results in high-grade metamorphism* of the Venusian crust, and most minerals found on Earth are unstable on Venus. Because of the dense atmosphere (90bar CO_2 plus minor constituents), metamorphic assemblages for some bulk compositions change with elevation (Khodakovsky *et al.*, 1979). Ignoring details, all hydrous minerals are unstable on Venus, and carbonate and hydroxylated minerals have only narrow PTX ranges of stability. Because of the higher temperature expected for the Venusian than the terrestrial crust at a given depth, the transition basalt to eclogite may not occur on Venus (Anderson, 1979c); if so, basalt should not be subducted on Venus, and plate tectonics may not have developed.

(d) *Speculation on atmosphere-crust reaction* on Venus is in its infancy. Because of evidence for H_2SO_4-rich clouds and perhaps for a drizzle of sulfur droplets (Pollack *et al.*, 1978; Knollenberg and Hunten, 1979; Young, 1979; Prinn, 1979), the possibility of reaction with surface rocks and of volcanic emission must be considered. Such speculation should take into account feldspathoid minerals which can store various molecules including S and Cl compounds up to ~1200K. Finally, it is necessary to speculate on the reason for the 100-fold greater abundance of ^{36}Ar in the Venusian than the terrestrial atmosphere (Oyama *et al.*, 1979)—is it inherited from accretion (e.g., Shimizu, 1979b)?

Perhaps the main question raised for the highland crust by this discussion of Venus is whether gaseous volatiles (*not* magmatophile volatiles) could have been lost from the Moon into a transient atmosphere during an early high-temperature stage. If temperatures become higher than those needed (~1300–1400K) for stability of mica and amphibole (Smith, 1979, Fig. 8), H_2O should be expelled from basaltic magma. The meionite variety of scapolite is capable of storing CO_2 up to 1800K (Smith, 1979, Fig. 9) in appropriate rock compositions, and dolomitic carbonate is stable to high temperatures above about 20 kb in periodotitic compositions (Wyllie, 1979). Speculation is desirable on the possibility that Venus was heated catastrophically above stability conditions for H_2O- and CO_2-bearing minerals, perhaps because of capture of a Moon; if so, the crust developed from an essentially volatile-free magma, and the atmosphere lost its H and retained most of its CO_2.

Mars, Phobos and Deimos

That Mars, Earth and Venus contain substantial amounts of CO_2 testifies to the general availability of gaseous material in the inner solar system, thereby emphasizing the problem of explaining the lack of gaseous volatiles on the Moon.

Phobos and Deimos, the two small satellites of Mars, are quite different from the Moon, and may be asteroids ultimately captured by protoatmospheric drag (Hunten, 1979). The network of depressions and ridges on Phobos probably results from a large impact (Thomas, 1979), and there is no evidence of igneous activity as for many lunar features. Orbital evolution of Phobos and Deimos is quite different from the Moon (Lambeck, 1979).

Interpretation of the surface morphology of Mars is complicated because (i) it is difficult to distinguish volcanic from impact-debris deposits, and (ii) there may be problems in distinguishing between the erosional effects of flowing water and wind. Of course, hydrologic and aeolian processes do not occur on the Moon, but chemical identification of Martian dust clouds (Chýlek and Grams, 1978) is important because aeolian deposits may cover most of Mars (e.g. in the layered deposits around the north pole; Squyres 1979). Probably most dust results from disintegration of surface rocks, by analogy with terrestrial aeolian sediments (Smalley and Krinsley, 1979), but volcanic emission of sulfate-bearing aerosols was proposed by Settle (1979). If most of Mars is covered by aeolian sediments, reflectance spectra (McCord *et al.*, 1978) may be giving information merely on altered material, in contrast to lunar spectra which appear to correlate well with the chemistry of local rocks. Volatiles (Clark and Baird, 1979) can be expected to be trapped as permafrost and mineral-surface adsorbates in aeolian deposits. Because of the theoretical prediction of large excursions in the obliquity of Mars (Ward *et al.*, 1979), climatic changes should have occurred on Mars, and a complex evolution of H_2O and CO_2 in the sedimentary layer and the atmosphere (plus temporary hydrosphere?) can be expected (e.g. Levine, 1978; Owen, 1978).

The latest moment-of-inertia of Mars (0.365; Kaula, 1979b) requires a center-heavy mass distribution, but considerable tolerance exists in petrologic models. Prolonged volcanism from deep-seated hot spots is needed to explain the huge volcanic constructs, and a thick lithosphere (50% greater than on Earth?; Frey, 1979) is needed to support surface relief, including huge canyons.

Badly needed is an absolute chronology of the Martian surface; comparison with the Moon allows control only of features ascribed to the early heavy bombardment. If the chronology proposed by Hartmann (1978) is correct, igneous activity on Mars lasted longer than on the Moon, and probably than on Mercury. Was the higher content of H_2O and CO_2 on Mars than on the other two planets partly responsible for this? Certainly the beginning of melting is at a lower temperature when H_2O and CO_2 are present, but so many factors enter into planetary volcanism including planetary size (Walker *et al.*, 1979) that a simple answer cannot be expected.

The bulk composition of Mars is poorly constrained because of lack of knowledge of distribution of volatile elements between reservoirs, but the composition given by Morgan and Anders (1979) is a useful basis for discussion. Using data and arguments reviewed in Smith (1979), extrusive rocks on Mars may be rather rich in Fe, and a sedimentary layer may consist largely of clay and sulfate minerals resulting from alteration. Certainly there is a little analogy with lunar rocks, though it is amusing to recall the pre-Apollo period when aqueous sediments were proposed as fill for lunar maria!

Asteroids

Perhaps the most important feature of the asteroid belt from the viewpoint of the lunar highlands is the dominance of C-type asteroids in the outer part, and the similarity of infra-red spectra for C-type meteorites and the surface of the asteroid Ceres (Larson *et al.*, 1979). Many other types of reflectance spectra have been recorded for asteroids (reviewed in Smith, 1979), and the spectra for M-type asteroids have been matched with powdered metal probably of Fe,Ni-rich composition (Dollfus *et al.*, 1979). Because of the theoretical argument that the mass of asteroids must have been much greater in the past to give the present relative velocities, and because of the Kirkwood resonance gaps, it is presumed that a substantial mass was deflected into the inner solar system. Consequently it is assumed that volatile-rich material from C-type asteroids hit the lunar crust. The presence of E-type asteroids in the inner part of the asteroid belt was attributed to rare fortuitous transport outwards from the inner solar system (Smith, 1979). These non-definitive arguments are useful in building a scheme in which the inner planets, including the Moon, accreted mainly from volatile-poor, reduced material from the inner solar system, and were bombarded at the end by high-speed volatile-rich material from outside the orbit of Mars.

Jupiter and satellites

For a review of the mineralogy of the planets (Smith, 1979), I deliberately omitted discussion of the giant planets because so little was then known. The Voyager missions have produced spectacular results, especially for the satellites of Jupiter, but their relevance to the lunar highlands crust is not clear at this time; indeed, the proximity of these satellites to such a large planet as Jupiter may modify the accretion processes and dynamical history so much that attempts to draw inferences for the Moon would be tenuous at best and downright misleading at worst. For example, tidal heating appears to be a viable energy source for volcanism on Io (Peale *et al.*, 1979), but probably provides only a minor input of energy for lunar volcanism (Peale and Cassen, 1978). Although volcanism on Io is so exotic compared to terrestrial volcanism, there is no need to propose unusual condensates for the outer solar system. When account is taken of prolonged sputtering, the absence or near-absence of H_2O on Io is reasonable. It is not clear how the proposed SO_2 and S_n species involved in igneous activity on Io are stored below the surface, and feldspathoids should be investigated. Let it suffice that the densities of the satellites of the giant planets can be interpreted plausibly in terms of the same mineral phases as in the inner solar system, except for an abundance of ices for satellites of low density (e.g., Ganymede 1.93; Callisto, 1.79 g/cm^3). There is no need to propose any startling changes in current theory for condensation of the solar nebula. Relevant information is given in the June 1 and November 23, 1979 issues of *Science*, the August 30, 1979 issue of *Nature*, and in Cruikshank (1979).

Implications for the Moon

Perhaps the most important implication for the highland crust of the Moon is the presence of atmophile volatile elements in the crusts of the Earth and Mars, and the possible presence of such elements in an early crust of Venus. Absence in the Moon's crust requires an explanation. The suggestion by Binder (1978) that the Moon was fissioned from the Earth, and that volatile elements were "pumped" back into the Earth while refractory elements stayed on the Moon is intriguing, but quantification is needed of the physical and chemical processes. Because the Venusian crust has lost at least most of its CO_2 (and probably H_2O) to the atmosphere, the possibility of a primeval volatile-bearing crust on the Moon is worth further exploration, even though prospects do not appear promising. Certainly the possibility of ice in lunar polar regions (Arnold, 1979) should be explored. Currently, I prefer to explore the possibility that volatile-rich material impacted the Moon at too high a velocity to be retained in significant amounts, especially if such impacting took place when the Moon was near the Roche limit (Hodges, 1979). Such an approach effectively "decouples" this problem from the origin of most of the Earth and Moon, and allows other features to be explained by complex processes such as collision-induced fission or disintegrative capture (note that there may be semantic problems here).

Both horizontal and vertical tectonics are important on Earth, and vigorous mantle convection is almost certainly required for the horizontal component to be strongly developed. On the Moon, tectonics appear to be driven almost entirely by vertical movement (Solomon and Head, 1979), and the relative importance of rigidity of the crust and driving forces for mantle convection is worth exploring. Because of low pressures in the Moon, conversion of basalt to eclogite cannot be important, in contrast to the Earth.

To conclude, it is worth reiterating that, although planetary crusts are so different from each other that direct comparison is difficult, simultaneous study of all planetary crusts provides a tremendous stimulation to the imagination; indeed, only by assembling together data and ideas on the entire solar system can progress be made and incorrect ideas discarded.

Acknowledgments—I thank all those listed in Smith (1979) for their help in building up knowledge about the solar system, J. J. Papike for his encouragement to prepare this paper, and C. T. Herzberg, S. A. Morse and J. Warner for helpful criticism. Financial support came from NASA grant 14-001-171.

REFERENCES

Anderson D. L. (1979a) Chemical stratification of the mantle. *J. Geophys. Res.* **84,** 6297–6298.

Anderson D. L. (1979b) The upper mantle transition region: eclogite? *Geophys. Res. Lett.* **6,** 433–436.

Anderson D. L. (1980) Tetonics and composition of Venus. *J. Geophys. Res.* **7,** 101–102.

Arnold J. R. (1979) Ice in the lunar polar regions. *J. Geophys. Res.* **84,** 5659–5668.

Barker F. (ed.) (1979) *Trondhjemites, Dacites, and Related Rocks*. Elsevier, Amsterdam. 659 pp.

Baur M. E. (1978) Thermodynamics of heterogeneous iron-carbon systems: implications for the terrestrial primitive reducing atmosphere. *Chem. Geol.* **22,** 189–206.

Binder A. B. (1978) On fission and the devôlatilization of the moon of fission origin. *Earth Planet. Sci. Lett.* **41,** 381–385.

Burns J. A. (ed.) (1977) *Planetary Satellites*. Univ. Arizona Press, Tucson. 598 pp.

Campbell D. B., Burns B. A., and Boriakoff U. (1979) Venus; further evidence of impact cratering and tectonic activity from radar observations. *Science* **204,** 1424–1427.

Chamberlain J. W. (1978) *Theory of Planetary Atmospheres: an Introduction to their Physics and Chemistry*. Academic, N.Y. 330 pp.

Chýlek P. and Grams G. W. (1978) Scattering by nonspherical particles and optical properties of Martian dust. *Icarus* **36,** 198–203.

Clark B. C. and Baird A. K. (1979) Volatiles in the Martian regolith. *Geophys. Res. Lett.* **6,** 811–814.

Consolmagno G. J. and Lewis J. S. (1978) The evolution of icy satellite interiors and surfaces. *Icarus* **34,** 280–293.

Cruikshank D. P. (1979) The surfaces and interiors of Saturn's satellites. *Rev. Geophys. Space. Phys.* **17,** 165–176.

DePaolo D. J. and Wasserburg G. J. (1979) Petrogenetic mixing models and Nd-Sr isotopic patterns. *Geochim. Cosmochim. Acta* **43,** 615–628.

Dimroth E. and Lichtblau A. P. (1978) Oxygen in the Archean ocean: Comparison of ferric oxide crusts on Archean and Cainozoic pillow basalts. *Neues Jahrb. Mineral Abh.* **133,** 1–22.

Dollfus A., Mandeville J., and Duseaux M. (1979) The nature of the M-type asteroids from optical polarimetry. *Icarus* **37,** 124–132.

Elsasser W. M., Olson P., and Marsh B. D. (1979) The depth of mantle convection. *J. Geophys. Res,* **84,** 147–155.

Flynn G. J., Fraundorf P., Shirck J., and Walker R. M. (1978) Chemical and structural studies of "Brownlee" particles. *Proc. Lunar Planet. Sci. Conf. 9th,* p. 1187–1208.

Frey H. (1979) Martian canyons and African rifts: structural comparisons and implications. *Icarus* **37,** 142–155.

Gehrels E. (ed.) (1978) *Protostars and Planets. Studies of Star Formation and of the Origin of the Solar System.* Univ. Arizona Press, Tucson. 756 pp.

Grieve R. A. F. and Dence M. R. (1979) The terrestrial cratering record. II The crater production rate. *Icarus* **38,** 230–242.

Grieve R. A. F. and Robertson P. B. (1979) The terrestrial cratering record. I Current status of observations. *Icarus* **38,** 212–229.

Hamilton P. J., O'Nions R. K., Evensen N. M., Bridgwater D., and Allaart J. H. (1978) Samarium-neodymium isotopic investigations of Isua supracrustals and implications for mantle evolution. *Nature* **272,** 41–43.

Hart, R., Dymond J., and Hogan L. (1979) Preferential formation for the atmosphere-sialic crust system from the upper mantle. *Nature* **278,** 156–159.

Hartmann W. K. (1978) Martian cratering V: toward an empirical Martian chronology, and its implications. *Geophys. Res. Lett.* **5,** 450–452.

Hartmann W. K. (1979) A special class of planetary collisions: Theory and evidence. *Proc. Lunar Planet. Sci. Conf. 10th,* p. 1897–1916.

Herndon J. M. (1978) Re-evaporation of condensed matter during the formation of the solar system. *Proc. Roy. Soc. London.* **A363,** 283–288.

Hodges R. R. Jr. (1979) Effects of orbit recession of the moon on the escape of materials from the lunar surface. *Proc. Lunar Planet Sci. Conf. 10th,* p. 555–567.

Hoffman J. H., Hodges R. R. Jr., McElroy M. B., Donahue T. M., and Kolpin M. (1979) Composition and structure of the Venus atmosphere: results from Pioneer Venus. *Science* **205,** 49–52.

Hunten D. M. (1979) Capture of Phobos and Deimos by protoatmospheric drag. *Icarus* **37,** 113–123.

Hurst R. W. (1978) Strontium evolution in the West Greenland-Labrador craton: a model for early rubidium depletion in the mantle. *Geochim. Cosmochim. Acta* **42,** 39–44.

Istomin V. G., Grechnev K. V., and Kochnev V. A. (1979) Mass-spectrometric measurements of the composition of the Venus lower atmosphere. *Pis'ma Astron. Zh.* **5,** 211–216 (In Russian).

Jacobsen S. B. and Wasserburg G. J. (1979) The mean age of mantle and crustal reservoirs. *J. Geophys. Res.* **84,** 7411–7427.

Jagoutz E., Palme H., Baddenhausen H., Blum K., Cendales M., Dreibus G., Spettel B., Lorenz

V., and Wänke H. (1979) The abundances of major, minor, and trace elements in the earth's mantle as derived from primitive ultramafic nodules. *Proc. Lunar Planet. Sci. Conf. 10th,* 610–612.

Jakosky B. M. and Ahrens T. J. (1979) The history of an atmosphere of impact origin. *Proc. Lunar Planet. Sci. Conf. 10th,* p 2727–2739.

Janardhan A. S., Newton R. C., and Smith J. V. (1979) Ancient crustal metamorphism at low P_{H_2O}: charnockite formation at Kabbaldurga, South India. *Nature* **27,** 511–514.

Jordan T. H. (1978) Composition and development of the continental tectosphere. *Nature* **274,** 544–548.

Kaula W. M. (1979a) Thermal evolution of Earth and Moon growing by planetesimal impact. *J. Geophys. Res.* **84,** 999–1008.

Kaula W. M. (1979b) The moment of inertia of Mars. *Geophys. Res. Lett.* **6,** 194–196.

Kaula W. M. (1979c) The beginning of the Earth's thermal evolution. In *The Continental Crust and its Mineral Deposits* (D. Strangway, ed.), Proc. Wilson Conf. Geol. Soc. Can. In Press.

Khodakovsky I. L., Volkov V. P., Sidorov Yu, I., Borisov M. V., and Lomonosov M. V. (1979) Venus: preliminary prediction of the mineral composition of surface rocks. *Icarus* **39,** 352–363.

Knöllenberg R. G. and Hunten D. M. (1979) Clouds of Venus: particle size distribution measurements. *Science* **203,** 792–795.

Lambeck K. (1979) On the orbital evolution of the Martian satellites. *J. Geophys. Res.* **84,** 5651–5658.

Lambert R. St. J. (1980a) Earth tectonics and thermal history: review and a hot-spot model for the Archean. In *Precambrian Plate Tectonics* (A. Kröner, ed.), p. Elsevier, N.Y.

Lambert R. St. J. (1980b) The thermal history of the earth in the Archean. *Precambrian Res.* In press.

Larson H. P., Feierberg M. A., Fink U., and Smith H. A. (1979) Remote spectroscopic identification of carbonaceous chondrite mineralogies: applications to Ceres and Pallas. *Icarus* **39,** 257–271.

Levine J. S. (1978) The evolution of H_2O and CO_2 on earth and Mars. In *Comparative Planetology* (C. Ponnamperuma, ed.), p. 165–182. Academic, N.Y.

Lewis J. S., Barshay S. S., and Noyes B. (1979) Primordial retention of carbon by the terrestrial planets. *Icarus* **37,** 190–206.

Liu L. (1979) On the 650-km seismic discontinuity. *Earth Planet. Sci. Lett.* **42,** 202–208.

Malin M. C. (1979) Geology of Venus (abstract). In *Lunar and Planetary Science X,* p. 772–774. Lunar and Planetary Institute, Houston.

Manhes G., Allègre C. J. Dupré B., and Hamelin B. (1979) Lead-lead systematics, the "age of the Earth" and the chemical evolution of our planet in a new representation space. *Earth Planet. Sci. Lett.* **44,** 91–104.

Matsui T. (1978) Collisional evolution of mass-distribution spectrum of planetesimals. *Proc. Lunar Planet. Sci. Conf. 9th,* p. 1–13.

Matsui T. (1979) Collisional evolution of mass-distribution spectrum of planetesimals. II. *Proc. Lunar Planet. Sci. Conf. 10th,* p. 1881–1895.

Matsuo S., Suzuki M. and Mizutani Y. (1978) Nitrògen to argon ratio to volcanic gases. *Adv. Earth Planet. Sci.* **3,** 17–25.

McCord T. B. and Clark R. N. (1979) The Mercury soil: presence of Fe^{2+}. *J. Geophys. Res.* **84,** 7664–7668.

McCord T. B., Clark R. N. and Huguenin R. L. (1978) Mars: near-infrared spectral reflectance and compositional implications. *J. Geophys. Res.* **83,** 5433–5441.

McDonnell J. A. M. (ed.) (1978) *Cosmic Dust.* Wiley-Interscience, N.Y.

McElhinney M. W. (ed.) (1979) The Earth. Its Origin, Structure and Evolution. Academic, N.Y. 598 pp.

McGill G. E. (1979) Venus tectonics: another Earth or another Mars. *Geophys. Res. Lett.* **6,** 739–741.

McKenzie D. P. and Weiss N. (1975) Speculation on the thermal and tectonic history of the Earth. *Geophys. J. Roy. Astron. Soc.* **18,** 1–32.

Morgan J. W. and Anders E. (1979) Chemical composition of Mars. *Geochim. Cosmochim. Acta* **43,** 1601–1610.

Naqvi S. M. and Hussain S. M. (1979) Geochemistry of metaanorthosites from a greenstone belt in Karnataka, India. *Can. J. Earth Sci.* **16,** 1254–1264.

O'Nions R. K., Carter S. R., Evensen N. M., and Hamilton P. J. (1979b) Geochemical and cosmochemical applications of neodymium isotope analysis. *Ann. Rev. Earth Planet. Sci.* **7,** 11–38.

O'Nions R. K., Evensen N. M., and Hamilton P. J. (1979a) Geochemical modeling of mantle differentiation and crustal growth. *J. Geophys. Res.* **84,** 6091–6101.

O'Nions R. K., Evensen N. M., Hamilton P. J., and Carter S. R. (1978) Melting of the mantle past and present: isotope and trace element evidence. *Phil. Trans. Roy. Soc. London* **A288,** 547–559.

Owen T. (1978) The composition and history of the Martian atmosphere. *Adv. Earth Planet Sci.* **3,** 93–96.

Owen T., Cess R. D., and Ramanathan V. (1979) Enhanced carbon dioxide greenhouse to compensate for reduced solar luminosity on early Earth. *Nature* **277,** 640–642.

Oyama V. I., Carle G. C. Wreller F., and Pollack J. B. (1979) Venus lower atmosphere compositions: analysis by gas chromatography. *Science* **203,** 802–805.

Peale S. J. and Cassen P. (1978) Contribution of tidal dissipation to lunar thermal history. *Icarus* **36,** 245–269.

Peale S. J., Cassen P., and Reynolds R. T. (1979) Melting of Io by tidal dissipation. *Science* **203,** 892–894.

Pechmann J. B. and Melosh H. J. (1979) Global fracture patterns of a despun planet: Application to Mercury. *Icarus* **38,** 243–250.

Pettengill G. H., Ford P. G., Brown W. E., Kaula W. M., Keller C. H., Masursky H., and McGill G. E. (1979) Pioneer Venus radar mapper experiment. *Science* **203,** 806–807.

Phillips R. J. and Ivins E. R. (1979) Geophysical observations pertaining to solid-state convection in the terrestrial planets. *Phys. Earth. Planet. Inter.* **19,** 107–148.

Pollack J. B. (1979) Climatic change on the terrestrial planets. *Icarus* **37,** 479–553.

Pollack J. B., Burns J. A., and Tauber M. E. (1979) Gas drag in primordial circumplanetary envelopes: a nechamism for satellite capture. *Icarus* **37,** 587–611.

Pollack J. B., Strecker D. W., Witterborn F. C., Erickson E. F., and Baldwin B. J. (1978) Properties of the clouds of Venus, as inferred from airborne observations of its near-infrared reflectivity spectrum. *Icarus* **34,** 28–45.

Prinn R. G. (1979) On the possible roles of gaseous sulfur and sulfanes in the atmosphere of Venus. *Geophys. Res. Lett.* **6,** 807–810.

Reynolds R. T. and Cassen P. M. (1979) On the internal structure of the major satellites of the outer planets. *Geophys. Res. Lett.* **6,** 121–124.

Richter F. M. (1979) Focal mechanisms and seismic energy release of deep and intermediate earthquakes in the Tonga-Kermadec region and their bearing on the depth extent of mantle flow. *J. Geophys. Res.* **84,** 6783–6795.

Ringwood A. E. (1979) Origin of the Earth and Moon. Springer, N.Y. 350 pp.

Saito K., Basu A. R. and Alexander E. C., Jr. (1978) Planetary-type rare gases in the upper mantle-derived amphibole. *Earth Planet. Sci. Lett.* **39,** 274–280.

Schubert G. (1979) Subsolidus convection in the mantles of terrestrial planets. *Ann. Rev. Earth Planet. Sci.* **7,** 289–342.

Schubert G., Cassen P. and Young R. E. (1979) Subsolidus convective cooling histories of terrestrial planets. *Icarus* **38,** 192–211.

Schwartzman D. W. (1978) On the ambient mantle helium-4/argon-40 ratio and the coherent model of degassing of the earth. *Adv. Earth Planet. Sci.* **3,** 185–191.

Settle M. (1979a) Dispersion and deposition of volcanic aerosols on Mars. Production of volcanic sulfate aerosols on Mars (abstract). In *Lunar and Planetary Science X,* p. 1101–1103. Lunar and Planetary Institute, Houston.

Settle M. (1979b) Production of volcanic sulfate aerosols on Mars (abstract). In *Lunar and Planetary Science X,* p. 1110–1112. Lunar and Planetary Institute, Houston.

Sharpe H. N. and Peltier W. R. (1979) A thermal history model for the Earth with parameterized convection. *Geophys. J. Roy. Astron. Soc.* **59,** 171–203.

Shaw D. M. (1980a) Evolutionary tectonics of the Earth in the light of early crustal structure. In *The Continental Crust and its Mineral Deposits* (D. Strangway, ed.), *Proc. Wilson Conf. Geol. Soc. Can.* In press.

Shaw D. M. (1980b) Development of the early continental crust Part 3. Depletion of incompatible elements in the mantle. *Precambrian Res.*, **10,** 281–299.

Shimizu M. (1979a) An evolutionary model of the terrestrial atmosphere from a comparative planetological view. *Precambrian Res.* **9,** 311–324.

Shimizu M. (1979b) Implications of argon-36 excess on Venus. *Moon and Planets* **20,** 317–319.

Smalley I. J. and Krinsley D. H. (1979) Eolian sedimentation on Earth and Mars: some comparisons. *Icarus* **40,** 276–288.

Smith J. V. (1979) Mineralogy of the planets: a voyage in space and time. *Mineral. Mag.* **43,** 1–89.

Smith J. V. (1980) The first 800 million years of Earth's history. *Phil. Trans. Roy. Soc. London.* In press.

Solomon S. C. (1979) Formation, history and energetics of cores in the terrestrial planets. *Phys. Earth Planet. Inter.* **19,** 168–182.

Solomon S. C. and Head J. W. (1979) Vertical movement in mare basins: relation to mare emplacement, basin tectonics, and lunar thermal history. *J. Geophys. Res.* **84,** 1667–1682.

Squyres S. W. (1979) The evolution of dust deposits in the Martian north polar region. *Icarus* **40,** 244–261.

Tarling D. H. (ed.) (1978) Evolution of the Earth's Crust. Academic, London. 442 pp.

Taylor S. R. (1979) Structure and evolution of the Moon. *Nature* **281,** 105–110.

Thomas P. (1979) Surface features of Phobos and Deimos. *Icarus* **40,** 223–243.

Towe K. M. (1978) Early Precambrian oxygen: a case against photosynthesis. *Nature* **274,** 657–661.

Vidal P. and Dosso L. (1978) Core formation: catastrophic or continuous? Strontium and lead isotope geochemistry constraints. *Geophys. Res. Lett.* **5,** 169–172.

Walker D., Stolper E. M. and Hays J. F. (1979) Basaltic volcanism: the importance of planetary size. *Proc. Lunar Planet. Sci. Conf. 10th,* p. 1995–2015.

Walker J. C. G. (1977) *Evolution of the Atmosphere*. MacMillan, London. 318 pp.

Walker J. C. G. (1978) Atmospheric evolution on the inner planets. In *Comparative Planetology* (C. Ponnamperuma, ed.), p. 141–163. Academic, N.Y.

Ward W. R., Burns J. A. and Toon O. B. (1979) Past obliquity oscillations of Mars: the role of the Tharsis uplift. *J. Geophys. Res.* **84,** 243–259.

Warner J. L. (1979) A model for the lithosphere of Venus (abstract). In *Lunar and Planetary Science X,* 1295–1297. Lunar and Planetary Institute, Houston.

Watt J. P. and O'Connell R. J. (1978) Mixed-oxide and perovskite-structure model mantles from 700–1200 km. *Geophys. J. Roy. Astron. Soc.* **54,** 601–630.

Weiblen P. W. and Schulz K. J. (1978) Is there any record of meteorite impact in the Archean rocks of North America? *Proc. Lunar Planet. Sci. Conf. 9th,* p. 2749–2771.

Wetherill G. W. (1979) Steady state populations of Apollo-Amor objects. *Icarus* **37,** 96–112.

Windley B. F. (ed.) (1976) *The Early History of the Earth*. Wiley-Interscience, London. 618 pp.

Windley B. F. (1977) *The Evolving Continents*. Wiley, London. 385 pp.

Wyllie P. J. (1979) Magmas and volatile components. *Amer. Mineral.* **64,** 469–500.

Young A. T. (1979) Chemistry and thermodynamics of sulfur on Venus. *Geophys. Res. Lett.* **6,** 49–50.

Papike, J.J. and Merrill, R.B., eds.
Proc. Conf. Lunar Highlands Crust (1980), p. 457-466
Printed in the United States of America

Laboratory verification of the lunar orbital X-ray fluorescence experiment: Initial results

Norman Hubbard

NASA Johnson Space Center, Houston, Texas 77058

Bi-Shia King

U.S. Geological Survey, Menlo Park, California 94025

Abstract—The combined use of remote sensing and returned sample chemical data in lunar studies presumes a direct link between the two types of data. Laboratory experiments have demonstrated that within the precision of the lunar orbital X-ray fluorescence data there is a direct relationship between the fluorescence K-series X-radiation emitted from Mg, Al and Si in the surfaces of lunar soil samples and the concentrations of these elements in the bulk soil samples. This relationship encompasses both mare and terra soils. The degree of maturity is not an important variable for the existing orbital X-ray data. It is argued that these results are directly applicable to the lunar surface analysed by the Apollo orbital X-ray fluorescence experiment.

INTRODUCTION

Natural X-radiation emitted from the surfaces of Mercury, asteroids, comets and atmosphereless moons can be used to obtain the chemical composition of their surfaces, as was done by the orbital X-ray fluorescence experiment that was flown on Apollo 15 and Apollo 16 (Adler *et al.*, 1972). The accuracy of such analyses is ultimately dependent on the relationship between the average chemical composition of the surfaces of the regolith grains relative to that of the bulk regolith. This is especially critical because only a thin layer of the surface (approximately 10 microns) is sampled by the X-radiation from Mg, Al and Si. Thus the confidence that can be placed on these data for geological studies must ultimately be based on understanding the processes that produced the planetary regolith or on experiments that directly address the link between chemical data obtained by a surface analysis (remote sensing) and that obtained by a volume analysis (returned samples). In practice, planetary regoliths on atmosphereless bodies have only a few significantly different physical states. These are (1) agglutinated regoliths of Moon and Mercury, (2) pulverized regoliths of asteroids and (3) regoliths especially rich in ices composed of low Z materials (comets and some moons of Jupiter). In this research, agglutinated regoliths are studied using lunar soils, and pulverized regoliths are approximated using powdered terrestrial rocks. Icy regoliths are much more difficult to simulate and are not addressed in this report. Laboratory experiments have been initiated to determine (1) the relationship of concentration ratios for bulk lunar soils to the intensity ratios of the X-radiation from Mg, Al and Si for varying degrees of regolith maturity and (2) the degree of linearity between Al/Si and Mg/Si intensity ratios and Al_2O_3/SiO_2 and MgO/SiO_2 concentration ratios for a wide range of chemical compositions of lunar soils, and (3) the relationship between the results for pulverized terrestrial rocks and the results for lunar soils. Initial results are presented here.

BACKGROUND INFORMATION

One major uncertainty in using orbital X-ray fluorescence data for geochemical studies has been the errors associated with the conversion of Al/Si and Mg/Si intensity ratios to Al_2O_3 and MgO concentrations that can be directly compared with those from lunar sample analyses. Attempts to use lunar sample data to calibrate the orbital data have been made (Hubbard, 1979; Maxwell *et al.*, 1977; Bielefield, 1977) but are very unsatisfying because of the assumptions and extrapolations required. Theoretical calculations are no more satisfying, because modeling the physical conditions of the planetary surface and its effect on the intensity ratio measured is difficult and was not considered in the calculations (Adler *et al.*, 1972).

One facet of the above mentioned major uncertainty has been the lack of knowledge about the extent to which the measured intensity ratios are directly related to the concentration ratios for actual lunar soils. Lunar soils are the complex products of prolonged exposure to the space environment. This exposure has, to a limited extent, subjected the lunar soils to the sample preparation procedures that are routinely used with high quality X-ray fluorescence analyses made in the laboratory; in this case, the micrometeorite bombardment has partially homogenized the chemical composition and particle size of lunar soils and extensively vitrified them. However, lunar soils have a wide range of maturities and it can be presupposed that this range in maturities will be reflected as a scatter in the relationship between chemical analyses obtained by surface vs. volume methods of analyses. The original experimenters (Adler *et al.*, 1972) did the obvious thing and ratioed the Al and Mg signals to those from Si in order to partially obviate these and other undesirable effects. However, there has never been a demonstration of the extent to which the forming of intensity ratios does circumvent the problems. There is ample evidence that concentrations or concentration ratios can be obtained that are within about 30% or better (two sigma) of the correct values (Adler *et al.*, 1972; Bielefeld, 1977; Hubbard, 1979; Maxwell *et al.*, (1977). If better results can be obtained from the orbital X-ray data, then its potential usefulness in future lunar studies is enhanced. With this in mind we have performed laboratory experiments with actual lunar soils to learn the extent to which intensity ratios are directly related to concentration ratios over a wide range of soil maturities and chemical compositions.

For any future X-ray fluorescence flight experiments, we recommend that a base of experimental results be established for real or realistic materials which will identify and quantify the important variables and allow their incorporation into theoretical calculations. This will be especially necessary for other atmosphereless planetary bodies where the "ground truth" approach is expected to be even less viable because samples are not expected until well after the orbital chemical data have been obtained. This may be especially important for a cometary mission because the physical state and gross chemical composition of the material being analysed can be expected to vary as the comet approaches and then recedes from the sun.

THE EXPERIMENT

Loose soil samples (less than 1 mm and unground) are powdered terrestrial rocks (international standards for igneous rocks) were placed in a commercial X-ray fluorescence spectrometer (KEVEX 0700) and irradiated sequentially under vacuum with K-series X-rays of Ti, Ca and S and the continuum radiation from a Rh tube operated at 3.0 KV. These X-ray energies were chosen to simulate the complex solar X-ray flux over a wide range of solar activities because they have penetration depths similar to those of solar X-rays that do the effective excitation of Mg, Al and Si in lunar soils on the moon. Penetration depth is the presumed parameter of major importance if lunar soil grains have any surface conditions that are an important influence on the X-ray fluorescence of Mg, Al and Si from one lunar soil sample to another. The samples rested on a thin organic film (less than 4 microns) that allowed high transmission of Si, Al and Mg X-rays. Between runs the samples were shaken in their sample cups to expose new material to the X-ray beam. A few samples were run in duplicate and show no differences greater than those found for repeat analyses of the same sample. All samples had previously been chemically analysed by others using standard techniques as a part of the lunar sample program. These data and other information about these soils are available in the Catalog of Lunar Soils (Heiken, 1974).

PRESENTATION OF RESULTS

The preliminary data are shown in Figs. 1 through 5. X-rays from S have a penetration depth similar to that for the effective excitation energy of solar X-rays during periods of "quiet" sun, such as was common during Apollo 15 and Apollo 16 (SOLRAD 9, 1972). Figure 1 shows that for excitation by S X-rays a well-defined relationship exists between the intensity and concentration ratios for typical lunar soils and that the data for soils of basaltic and anorthositic compositions are colinear. Terrestrial rocks and lunar sample 74220 form a different line; 74220 is not a typical lunar soil because it is composed almost entirely of glass spheres instead of being extensively agglutinated. Increasing the energy of the incident X-radiation to Ti (Fig. 2) simulates a more "active" sun where the average excitation energy can increase substantially (SOLRAD 9, 1972) and results in a colinear plot for lunar soils and terrestrial rocks. It was also found that when X-rays of energies intermediate between S and Ti were used the plots of intensity ratio vs. concentration ratio remained linear and parallel for lunar soils. Thus, mixtures such as the complex solar X-ray spectra will also produce linear relationships between intensity and concentration ratios for lunar soils.

Figure 3 shows an error analysis of the data in Fig. 1 and the effects of variations in soil maturity on the data. For the error analysis in Fig. 3, the intercept in Fig. 1 is shifted to 0.0 by subtracting the intercept on the intensity ratios axis from the measured intensity ratios. The concentration ratio has been ratioed to the shifted intensity ratio and plotted against the best available measure of soil maturity (I_s/FeO) (Morris, 1977). The agglutinate content increases with increasing soil maturity, which is a measure of how long the lunar soil has been exposed on the very surface. A two sigma error bar of $\pm$ 4.0% is shown for multiple measurements on one sample. For all samples except 74220, which plots at about 2.7 on the vertical axis, the scatter of the concentration/intensity ratios has a two

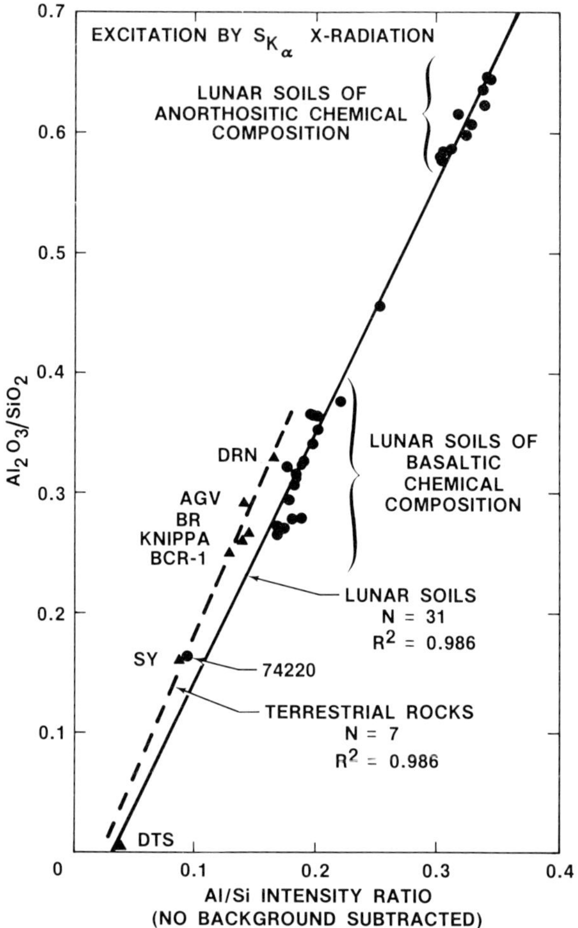

Fig. 1. The experimentally measured ratio of Al Kα to Si Kα X-radiation is plotted against the concentration ratio of these elements in the bulk sample. The lunar soils of basaltic chemical composition are from various Apollo landing sites in maria, and the lunar soils of anorthositic chemical composition are from the Apollo 16 site. The Al/Si intensity ratios were formed by ratioing the signals resulting from Al Kα and Si Kα, which were corrected for dead time, no background was subtracted, nor were there any corrections for detector efficiency. Background corrections were not made because scheduling conflicts prohibited delaying the initial experiments until the required equipment could be installed on the KEVEX 0700. This does not significantly degrade these results for the following reasons: (1) Si concentrations in these materials are nearly constant, so that variations in the already small correction for tailing of the signal from Si Kα under that from Al Kα are negligible. (2) The general background of scattered radiation under the signals from Si Kα and Al Kα has a variation from sample to sample that is negligible in these initial results and not expected to be important in more refined experiments.

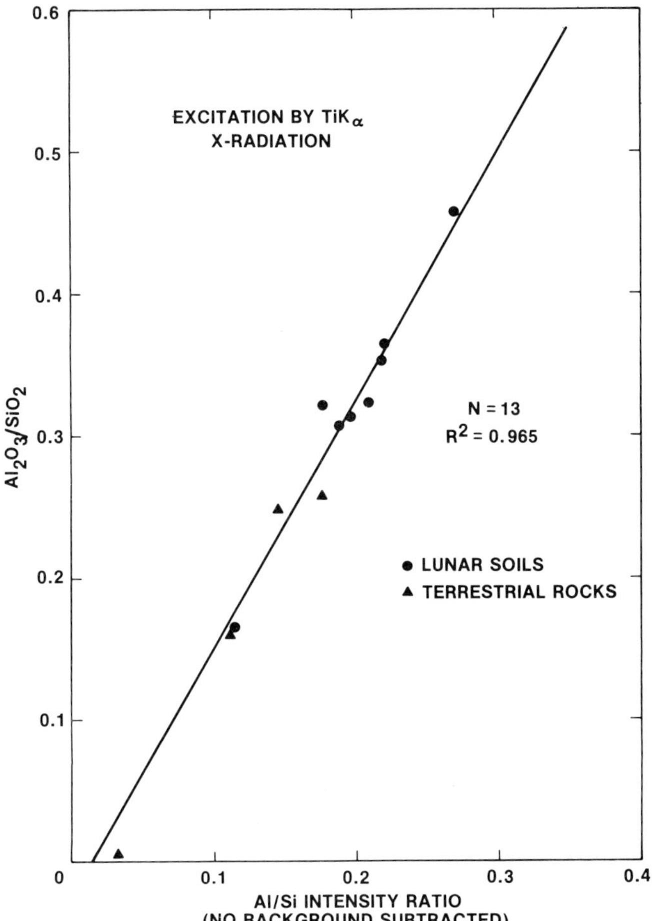

Fig. 2. The effects of excitation by X-rays from a more active sun is simulated by the use of Ti Kα X-rays. For the Al/Si ratio, the differences found between lunar soils and pulverized terrestrial rocks when S Kα X-rays are used disappears when the more energetic Ti Kα X-rays are used.

sigma value of ± 9.4%. Because the areas sampled by the Apollo orbital XRF experiment are very largely mature soils, we computed the scatter for samples that have I_s/FeO greater than 40, which gives a two sigma of ± 6.6%. These two sigma values are similar to or better than those implied by a recent error analysis of the orbital X-ray data (Hubbard, 1979). Therefore, even though these measurements can and will be refined, the basic answers are provided by these initial results.

Results for Mg are shown in Fig. 4, where it is seen that data for all lunar soils form a single line and that data for terrestrial rocks form a line with a different

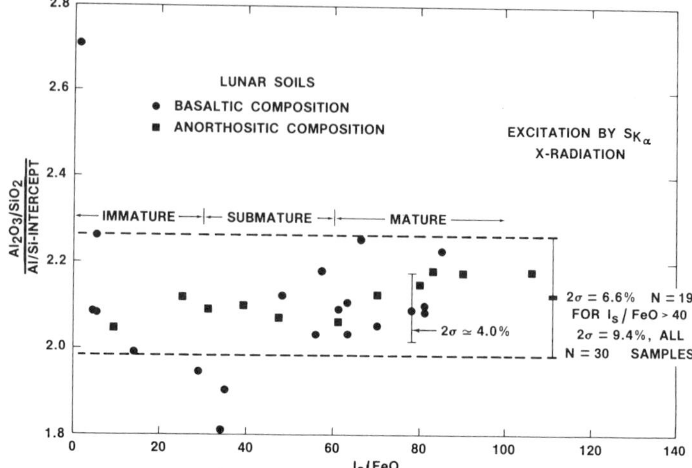

Fig. 3. The two axes of Fig. 1 are ratioed and plotted against the best available measure of soil maturity, I_s/FeO (Morris, 1977), to show the effects of variations in soil maturity on the relationship of Al/Si intensity ratios measured on the outer ~ 10 microns of the soil and the Al_2O_3/SiO_2 concentration ratios measured on the bulk soil. For this graph the intercept in Fig. 1 has been shifted by 0.0 by subtracting the intercept of the intensity ratio axis from the measured intensity ratios.

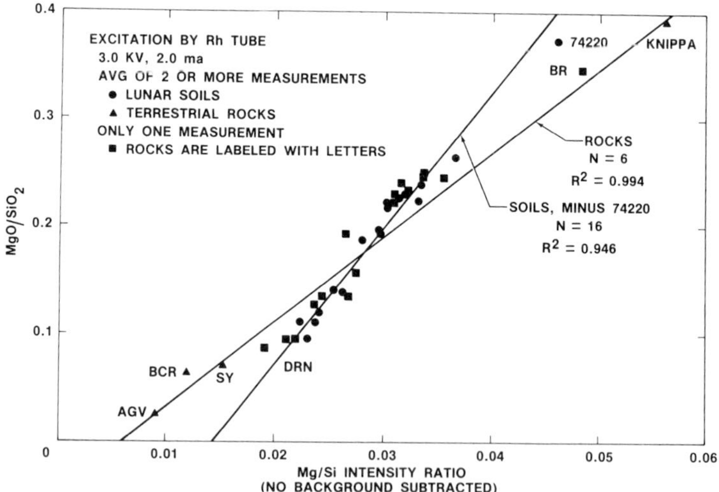

Fig. 4. As for Fig. 1, except that the relationship shown is between Mg/Si intensity ratios and MgO/SiO_2 concentration ratios and that the excitation is by the continuum X-radiation from a Rh target in a tube operated by 3.0 KV and 2.0 ma. This excitation is less energetic than that by S Kα and these results are shown here because they have less scatter produced by poor counting statistics for Mg Kα; in other words, this source excites Mg Kα better than does S Kα.

slope and intercept. Data for terrestrial rocks do not become colinear with those for lunar soils over the energy range of the incident radiation investigated (not shown). The intersection of the two lines shifts as the energy of the incident radiation changes. An error analysis of the lunar data is shown in Fig. 5 and gives a two sigma of 14.5% for the scatter of these data and no relationship to soil maturity is seen.

The difference between the lines for lunar soils versus terrestrial rocks may be due to differences in effective particle size, to the presence of a coating on the lunar soil grains, or to the vitreous nature of lunar soil grains. The possibilities will be investigated.

It has long been suggested by numerous authors that the finer grain size fractions of lunar soils are richer in Al than the coarser size fractions. Recently some data have been published (Laul *et al.*, 1979) showing that some soils of basaltic chemical composition have higher Al/Si ratios in the finer size fractions than do the bulk soils. Situations can be imagined where this would result in preferential sensing of the finer size fractions by the orbital X-ray experiment, with the result that erroneously higher Al_2O_3 values would be assigned to the bulk soil from the orbital data. However, the data in Fig. 1 show that either this is not the predominant situation, or that the effect is rather constant for both basaltic (mare) and anorthositic (terra) soils. Closer examination of the data in Figs. 1 and 3 leads to the suggestion that perhaps a few samples used in this study are exhibiting this effect. Specifically, the three basaltic soils with I_s/FeO = 30–40 (Fig. 3) differ from the other soils in a manner consistent with the preferential sensing of an Al-rich, fine-grained fraction. It is realistic to suggest that these soils may differ from the others in that they may be mixtures of finer grained (more mature) soils richer in Al and coarser grained (less mature) soils poorer in Al such that the two soils have not had their differences averaged by subsequent maturation.

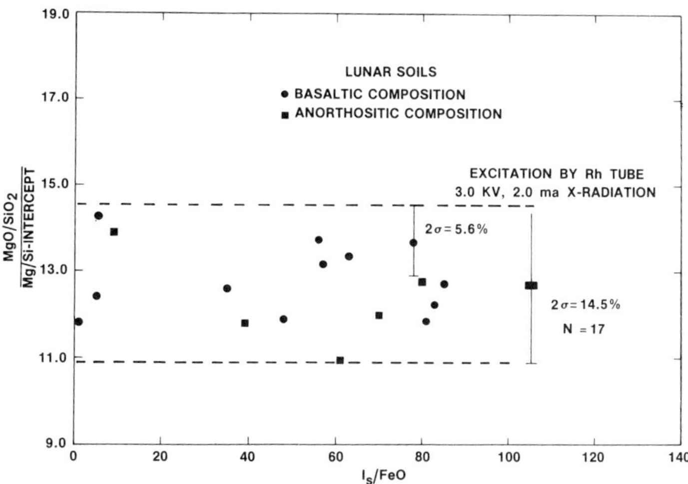

Fig. 5. Same as for Fig. 3, except that Mg/Si data are shown.

RELEVANCE TO ORBITAL DATA

The orbital data sample an approximately 10 micron layer of the lunar surface material which is controlled by the penetration depth of Mg, Si and Al X-rays in lunar materials. Consideration of the processes that have produced the lunar regolith (McKay *et al.*, 1974; Lindsay, 1972; Poupeau *et al.*, 1975; Gault *et al.*, 1974) suggests that the extensive reworking of the upper centimeters of the lunar surface will lead to a high degree of homogeneity between the average chemical composition of the outer approximately 10 microns of soil grains and the bulk soil. Likewise, over the large areas (greater than 10^3 km^2) sampled by the Apollo orbital experiment, impact gardening commonly will have been adequate to assure that the upper centimeters thick layer of regolith is representative of the upper meters. From the meter scale on up, assessment and analysis of the orbital X-ray data is in the realm of geology and beyond the scope of this paper.

The results presented in this paper show that the outer approximately 10 microns of lunar soil grains have average Al/Si and Mg/Si ratios that are directly proportional to those of the bulk soil as measured by conventional analyses. Thus, the required link between the orbital and lunar sample data is now known to exist.

It can also be argued that this experiment includes the critical item, i.e., the actual lunar surface, because consideration of the published literature on lunar soils leads to the conclusion that the physical state of the grains in the lunar soils collected by Apollo are typical of that for all lunar surface soils. All but a small percentage of the grains in mature lunar soils have been on the very lunar surface many times (Comstock, 1978; Gault *et al.*, 1974; Goswami and Lal, 1974). Thus, the experiment analysed the materials of a typical lunar surface, as seen by the orbital instruments. No attempt was made to reproduce the physical structure of the lunar surface in these experiments; it is argued that this is not important. First, a serious variation of Al/Si and Mg/Si concentration ratios with grain size is not expected because of the homogenization of chemical composition and particle size by agglutination during the maturation process (McKay *et al.*, 1974; Via and Taylor, 1976; Hu and Taylor, 1977). Second, the physical arrangement of grains has not been found to be important in this experiment. Specifically, shaking and reanalysis of the samples used in this experiment yielded no significant variation, as shown by the error bar in Figs. 3 and 5 for repeat analyses.

These results can be applied to all of the existing orbital X-ray data and very probably to all future such data for the following reasons. The soil maturation process is driven by fragmental debris that impacts the moon omnidirectionally, except perhaps in polar regions. All, or nearly all, of the lunar surface can be assumed to be composed of materials (plagioclase, pyroxene, olivine, glass) that respond the same way to high velocity impacts. Thus, the physical properties of lunar surface soils should be those known from the study of lunar samples. Chemical matrix effects can significantly alter the relationship between intensity ratio and concentration ratio if there are large variations in the concentrations of elements other than Mg, Al and Si. Most plausibly, this would have to be caused by variations in Ca, Ti and Fe that are well outside the wide limits known from the returned lunar soil samples and covered by the chemical compositions of the lunar soil samples used in this experiment. Soils having the composition of pure olivine or pure plagioclase are perhaps the most likely candidates to have serious chemical matrix effects relative to the other soil samples. The presence of both is readily deducible from the orbital data and this would constitute the primary result, with the precise MgO or Al_2O_3 concentration being the secondary result whose accuracy can be improved by standard corrections.

MAJOR CONCLUSIONS

A. Over the entire known range of chemical compositions of typical lunar soils, and to a precision similar to or better than that of the Apollo orbital X-ray data, there is a direct relationship between the Al/Si and Mg/Si intensity ratios measured on the outer approximately 10 microns of lunar soil samples and the Al/Si and Mg/Si concentration ratios of the bulk soil.

B. Soil maturity is not an important variable for the orbital X-ray fluorescence experiment when analysing the moon, and is not expected to be for Mercury. This conclusion must be reexamined if detectors having high spacial resolution are used because smaller areas of "fresh" soil may dominate the field of view.

C. There are physical differences between the finely ground terrestrial rocks used in this experiment and lunar soils that seriously affect the relationship of X-ray intensity to elemental concentration, even when data for elements of adjacent atomic number are ratioed. From this it is expected that the surfaces of asteroids (which are not agglutinated) will have to be well characterized and realistic laboratory experiments performed in order to obtain accurate chemical analyses of asteroids by remote X-ray fluorescence.

D. As the energy of the incident X-rays increases, corresponding to increasingly "active" solar conditions, the differences in response noted in Fig. 1 for lunar soils vs. terrestrial rocks becomes insignificant for Al/Si but remain for Mg/Si. This suggests that periods of active sun may be especially favorable for analyses of asteroids because uncertainties in the physical state of the surface will have a minimal effect on the Al/Si data and will increase the intensity of the measured fluorescent X-rays from the asteroids, thus improving the signal-to-noise ratio of the data.

REFERENCES

Adler I., Trombka J., Gerard J., Lowman P., Schadebeck R., Blodget H., Eller E., Yin L., and Lamothe R. (1972) The Apollo 15 X-ray fluorescence experiment. *Science* **175,** 436–440.

Bielfeld M. J. (1977) Chemistry of regions common to the orbital X-ray and gamma ray experiments. *Proc. Lunar Sci. Conf. 8th,* p. 1131–1147.

Comstock G. M. (1978) Miniregoliths I: Dusty lunar rocks and lunar soil layers. *Proc. Lunar Planet. Sci. Conf. 9th,* p. 2557–2577.

Gault D. E., Hörz F., Brownlee D. G., and Hartung J. B. (1974) Mixing of the lunar regolith. *Proc. Lunar Sci. Conf. 5th,* p. 2365–2386.

Goswami J. N. and Lal D. (1974) Cosmic ray irradiation pattern at the Apollo 17 site: Implications to lunar regolith dynamics. *Proc. Lunar Sci. Conf. 5th,* p. 2643–2662.

Heiken G. (1974) A catalog of lunar soils. Curator's Office, NASA Johnson Space Center, Houston. 221 pi.

Hu H-N. and Taylor L. A. (1977) Lack of chemical fractionation in major and minor elements during agglutinate formation. *Proc. Lunar Sci. Conf. 8th,* p. 3645–3656.

Hubbard N. (1979) Regional chemical variations in lunar basaltic lavas. *Proc. Lunar Planet. Sci. Conf. 10th,* p. 1753–1774.

Laul J. C., Lepel E. A., Vaniman D. T., and Papike J. J. (1979) The Apollo 17 drill core: Chemical systematics of grain size fractions. *Proc. Lunar Planet. Sci. Conf. 10th,* p. 1269–1298.

Lindsay J. F. (1972) Development of soil on the lunar surface. *J. Sed. Petrol.* **42,** 876–888.

Maxwell T. A., Strain D. L., and El-Baz F. (1977) Mare Crisium: Compositional inferences from low altitude X-ray fluorescence data. *Proc. Lunar Sci. Conf. 8th,* p. 933–944.

McKay D. S., Fruland R. M., and Heiken G. H. (1974) Grain size and the evolution of lunar soils. *Proc. Lunar Sci. Conf. 5th,* p. 887–908.

Morris R. V. (1977) Origin and evolution of the grain size dependence of the concentration of fine grained metal in lunar soils: The maturation of lunar soils to a steady state stage. *Proc. Lunar Sci. Conf. 8th,* p. 3719–3747.

Poupeau G., Walker R. M., Zinner E., and Morrison D. A. (1975) Surface exposure history of individual crystals in the lunar regolith. *Proc. Lunar Sci. Conf. 6th,* p. 3433–3448.

SOLRAD 9 (1972) Solar-Geophysical Data, U.S. Department of Commerce, Boulder, Colorado 80302.

Via W. N. and Taylor L. A. (1976) Chemical aspects of agglutinate formation: Relationships between agglutinate composition and the composition of the bulk soil. *Proc. Lunar Sci. Conf. 7th,* p. 393–403.

Papike, J.J. and Merrill, R.B., eds.
Proc. Conf. Lunar Highlands Crust (1980), p. 467-481
Printed in the United States of America

Geochemical anomalies on the eastern limb and farside of the moon

B. Ray Hawke

Planetary Sciences, Hawaii Institute of Geophysics, University of Hawaii, Honolulu, Hawaii 96822

Paul D. Spudis

Department of Geology, Arizona State University, Tempe, Arizona 85281

Abstract—A variety of orbital geochemistry and photogeologic data was used to investigate major geochemical anomalies on the east limb and farside of the Moon and to determine the processes responsible for their formation. These anomalies are located in the following regions: (1) near Langemak crater, (2) Mare Marginis, (3) Balmer crater, (4) terrain north of Taruntius crater, and (5) Van de Graaff. It was concluded that these anomalies are related to episodes of extrusive igneous activity which produced deposits with a variety of ages and compositions. Some of the associated volcanic deposits exhibit albedos and surface morphologies similar to those of the nearside maria, while others are now thinly covered by highland debris and are commonly represented by light plains units which exhibit clusters of dark-haloed craters. Little evidence was found for a genuine episode of highland volcanism. (e.g. Apollo 15 KREEP basalt, highland basalt, etc.) The geochemical anomalies appear to be related to basaltic deposits with either mare affinities or that are intermediate in composition between KREEP and mare basalts. The existence of such intermediate basalts on the eastern limb and farside supports previous suggestions that KREEP-like volcanism was not limited to a small area of the lunar nearside. Eastern limb and farside volcanic activity has been more extensive in both space and time and has produced a greater variety of compositions than has previously been thought.

Since thinly covered basaltic units can exert a strong influence on regional surface chemistry as determined from orbit, caution is urged in interpreting the surface compositions of highland terrains which exhibit a high density of dark-haloed impact craters.

INTRODUCTION

Analyses of the orbital geochemistry data have shown that some lunar regions have unusual abundances of certain elements relative to surrounding or adjacent areas, or have a surface chemistry unlike that which would be anticipated from the examination of local geologic relationships. Investigation of the formation of these geochemical anomalies can provide important clues to understanding impact and volcanic processes operative during the early phases of lunar evolution as well as the lateral and vertical composition of the highlands crust. For example, such studies have been of critical importance in outlining the history of KREEP volcanic activity in the western maria (Haines *et al.*, 1978; Spudis, 1978; Hawke and Head, 1978; Metzger *et al.*, 1979). The purpose of this study was to investigate the geology and geochemistry of selected anomalous highlands regions on the eastern limb and farside and attempt to understand the processes responsible for their formation.

GENERAL CAUSES OF GEOCHEMICAL ANOMALIES

There has been an increasing tendency in recent years to relate geochemical anomalies and other variations in lunar surface chemistry to lateral heterogeneities produced by the chemical differentiation of the global magma ocean. While this explanation is probably valid and may be particularly important in accounting for large-scale variations, it must be recognized that other processes are capable of producing geochemical anomalies in the lunar highlands. These include the following:

Basin-forming impacts

Large objects which impacted near the end of crustal formation may have produced localized melting and lateral chemical heterogeneities. More recent basin-forming impacts created major bodies of impact melt and excavated crustal material from a variety of depths which was deposited in a systematic fashion. Systematic changes in the composition of large basin ejecta deposits may represent changes in the vertical composition of the lunar crust at the target site. The composition of the associated body of impact melt may represent a homogenized section of the upper crust (Ryder and Wood, 1977) or may have differentiated to form a variety of rock types (Head, 1974).

Highland volcanism

Volcanic activity has also been an important process in determining the composition of the lunar surface. Although this is most obvious in the case of mare volcanism, premare or highland volcanism may have been very important. Very early (>4.0–4.1 b.y.) highland volcanic deposits were probably disrupted and largely destroyed during the terminal bombardment of the lunar surface, but may have left geochemical and mineralogic signatures which can be detected with existing remote sensing techniques. The heavily impact reworked KREEP-rich breccias collected at the Apollo 14 site are probably the products of such an episode of early highland volcanism (Hawke and Head, 1977; 1978).

More recent episodes of highland volcanism may have produced deposits which are both compositionally and morphologically distinct. Remote sensing and geologic evidence has recently been presented which strongly suggests that the Apennine Bench Formation is the surface expression of a post-Imbrium highland volcanic unit composed of KREEP basalt (Hawke and Head, 1978; Spudis,

1978; Metzger *et al.*, 1979). In addition, the Gruithuisen and Mairan Domes may represent morphologically and spectrally distinct non-mare extrusive volcanic constructs of Imbrian age (Head and McCord, 1978).

Mare volcanism

Mare volcanic deposits emplaced considerably before the end of the terminal bombardment would have been thoroughly reworked and mixed with highland material (see Ryder and Taylor, 1976). While little morphologic evidence of these very early basalts would remain, they could still exert an influence on the remotely-sensed surface composition of the area in which they were emplaced. Later mare basalts, extruded close to the end of the terminal bombardment, may have been thinly buried by layers of highland debris. These ancient mare basalts are represented on the lunar surface by certain light plains which exhibit dark-haloed craters (see Schultz and Spudis, 1979). These covered basaltic deposits, though present in the subsurface, could still exert a powerful influence on regional chemistry since some mare basalt would have been mixed with the highland material during the emplacement of the debris unit. In addition, variable amounts of vertical mixing of mare basalt with overlying material has occurred during the interval since the emplacement of the highlands debris.

Pyroclastic deposits in regions of explosive volcanic activity are capable of producing areas of mare-like chemistry in the lunar highlands. A prime example is the Sulpicius Gallus Formation which occurs in a portion of the Haemus Mountains southwest of Mare Serenitatis (see Schonfeld and Bielefeld, 1978). Since pyroclastic activity appears to have been concentrated in the vicinity of basin rims, care must be exercised in interpreting surface compositions in these areas (Hawke *et al.*, 1979; Andre *et al.*, 1979a). Complications arise if the pyroclastics are difficult to detect, as would be the case if the deposits were either very old or originally thin and had been substantially mixed with subjacent material, or if the pyroclastics had a higher albedo than that which generally characterizes such deposits.

GEOLOGIC AND GEOCHEMICAL CHARACTERISTICS OF SPECIFIC EAST LIMB AND FARSIDE ANOMALIES

Several geochemical anomalies on the east limb and farside were chosen for detailed analysis based on geologic interest, degree and strength of anomaly, and data availability (Fig. 1). While most of the anomalies had been previously identified, some are described here for the first time.

Region near Langemak crater

Hubbard *et al.* (1978) pointed out the rather striking variations in Mg/Si and Al/Si intensity ratios which occur in the vicinity of Langemak crater. The lowest

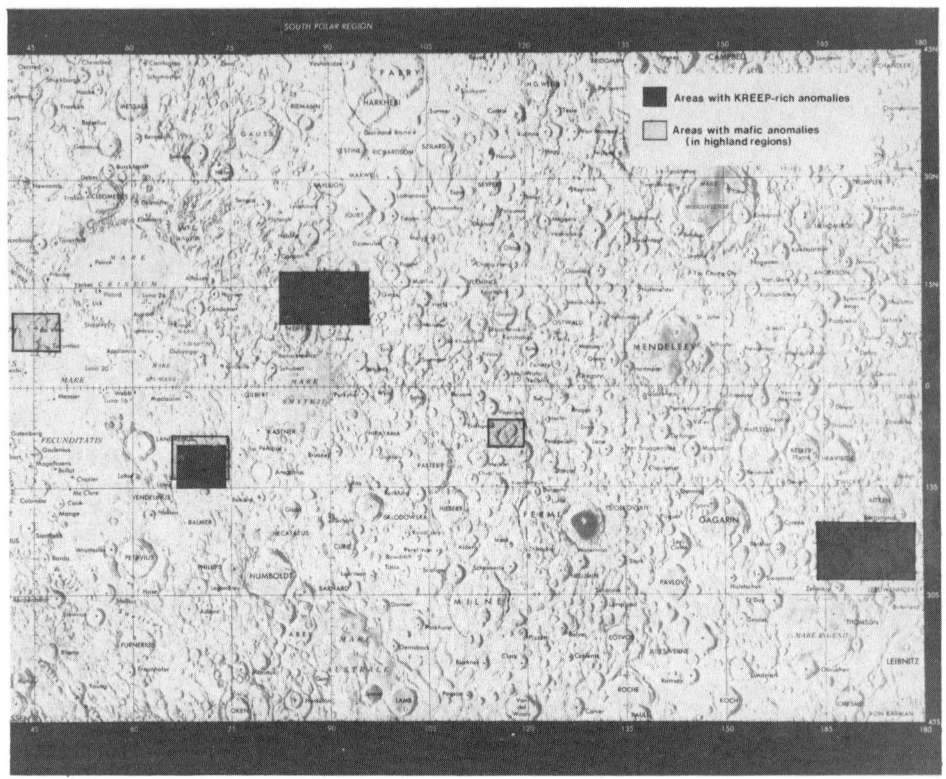

Fig. 1. Location map of geochemically anomalous regions discussed in text.

average Al/Si intensity ratios (~1.0) and the highest average Mg/Si intensity ratios (~0.8) are found around 115°E. In addition, Clark *et al.* (1978) noted that an adjacent highlands area immediately northeast of Pasteur crater was characterized by the unusual combination of high Fe and high Al/Si values. Examination of the most recent versions of the Fe and Ti distribution maps (Davis, 1979) showed that relatively high Fe and Ti values are associated with the Langemak region. The abundance of Ti is quite striking, with Ti content ranging between 2.4% and 3.5% along the northern edge of the ground track between 110°E and 125°E. No apparent radioactivity anomaly has been identified in this general region.

Important clues to understanding the source of geochemical anomalies have come from the work of Schultz and Spudis (1979), who correlated high Mg/Si intensity ratios with the dark- haloed craters in the Langemak region. The existence of these dark-haloed craters with associated high Fe, Ti, and Mg/Si values strongly suggest the presence of a buried ancient basalt layer. The occurrence of other dark-haloed craters in the region demonstrates that this unit is not limited to the immediate vicinity of Langemak and may, in fact, underlie some or most of the Imbrian and Nectarian light plains units located in this area (Wilhelms and El-Baz, 1977). The correspondence of an area of geochemical anomalies with an area of probable ancient mare volcanic activity suggests that early volcanism may be an important process in the production of regional anomalies elsewhere on the lunar surface.

Additional support for magmatic activity in the region comes from the work of Dvorak and Phillips (1978) concerning the Bouguer gravity anomalies associated with lunar craters. These workers found that in contrast to post-Imbrian craters, which have negative Bouguer anomalies, the Imbrian age craters which display evidence of post-impact magmatic activity (floor fractures, dark floor units, etc.) exhibit no Bouguer gravity anomalies. They suggest that magmatic activity may be responsible for part or all of the net zero mass anomalies associated with Imbrain age craters. It is of interest to note that no Bouguer gravity anomaly is associated with Pasteur which is a pre-Nectarian crater west of Langemak. The absence of a Bouguer gravity anomaly may be due to the presence of either a now covered extrusive unit in the crater floor or an intrusive body beneath the crater, and is consistent with the occurrence of an episode of volcanic activity in the region under study.

Mare Marginis region

The results of Th deconvolution modeling studies by Haines *et al.* (1978) allowed the identification of a region north of Mare Smythii with Th concentrations well above average highland levels. This region is dominated by Mare Marginis and the mare-flooded crater Neper, but also contains highland units including light plains deposits in Babcock crater and just east of Marginis (Fig. 2). While Haines *et al.* (1978) reported a Th concentration of 3.4 ppm for this region as a whole, they also noted that when the model was modified by decoupling the Mare Marginis-Neper and highland portions, the resulting fit was equally good. Using this alternate model, the resulting highland Th concentration was 2.3 ppm and the mare concentration was 5.4 ppm. The high Th concentrations are clearly associated with the mare material and hence there appears to be a mare basalt unit on the eastern limb with a Th concentration of at least 3.4 ppm and perhaps as high as 5.4 ppm; both values are considerably in excess of those associated with typical Apollo mare samples.

The high Th content of the Mare Marginis basalts suggests that they are intermediate in composition between mare and KREEP basalts and may be similar to the Apollo 17 KREEPy basalt described by Ryder *et al.* (1977). Estimates of Apollo 17 KREEPy basalt Th concentrations based on reported K_2O values range from ~4 to 6 ppm and are comparable to those for the Mare Marginis-Neper region. In addition, the MgO values reported by Andre *et al.* (1979a) for Neper (~8–9% are similar to the average composition of Apollo 17 KREEPy basalt (~8%).

Balmer crater region

Another area with anomalously high Th concentrations (~4.0 ppm) was identified by Haines *et al.* (1978) just north of Balmer crater (68°E, 15°S; see Figs. 3 and

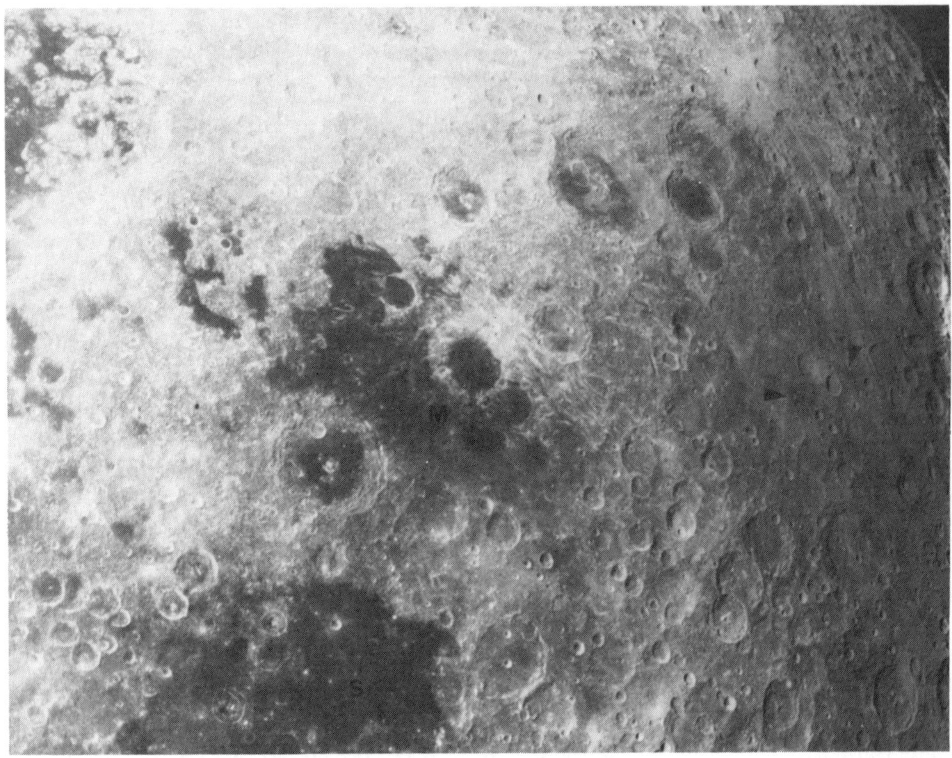

Fig. 2. Region on eastern limb showing Marc Marginis (M), a high Th region with possible KREEPy basalt affinities. Light plains on right contain numerous dark halo craters (arrows) indicating presence of covered ancient mare basalts. Mare Smythii (S) is at 0°N, 80°E. (AS 17-152-23327).

4). This Th high closely corresponds to an area of Imbrian-age light plains containing a cluster of dark-haloed craters (Schultz and Spudis, 1979) which suggest the presence of a buried basaltic unit. The orbital geochemistry data presented by Davis (1979) show that at least the northern portion of this plains unit exhibits relatively high FeO and TiO_2 values (TiO_2 = 2.3%–3.0%; FeO = 7.2%–12.4%). In addition, Andre *et al.* (1979b) noted that the Imbrian plains southeast of Langrenus exhibit Mg/Si values 16% higher than surrounding units.

While the Imbrian-age plains and the associated geochemical anomalies may be related to a genuine episode of highland volcanism as suggested by Haines *et al.* (1978), the presence of dark-haloed craters on the plains units suggests an alternate explanation. It is possible that the covered basalt is the source of the high Th material. The area is surrounded (within 250 km) by five major impact structures of Imbrian or younger age which could have contributed to the thin cover of highland material. The subjacent basaltic material would have been mixed with crater material during ejecta emplacement and by subsequent vertical mixing by small impacts which penetrated the thin highland debris surface units.

If this explanation is correct, the basaltic unit must have had Th concentrations approaching those found in the Apollo 14 and 15 KREEP basalts. In addition, this KREEP-rich unit have been emplaced by early in the Imbrian Period and may be pre-Imbrian in age.

Terrain north of Taruntius crater

An expanse of low-lying terrain north and northeast of Taruntius crater (Fig. 5) has been mapped as plains and smooth terra material of Imbrian age (Wilhelms, 1972). However, recent remote sensing data presented by Bielefeld *et al.* (1978) suggest that these units have a mare affinity. These workers defined a number of compositional units in the Crisium-Northern Fecunditatis region on the basis of the natural clustering of Al/Si ratios, Mg/Al ratios, and albedo values. The area north of Taruntius is included in a unit which has cluster centroid values of 0.48 and 0.50, respectively, for Al/Si concentration ratio and Mg/Al concentration ratio. Elsewhere in the Crisium region, the same unit generally corresponds to maria surfaces (Mare Undarum, Mare Spumans, and the mare flooded floor of Firmicus crater) or a narrow mare-dominated mixed unit near mare-highland contacts. Hence, mare material might be expected to be associated with the deposits north of Taruntius.

Other geologic and orbital geochemical data support the suggestion that the highland units under consideration have a mare affinity. The Mg/Si intensity ratio variation map presented by Hubbard and co-workers (Frontispiece, 1978) shows that major portions of these units exhibit high values (Mg/Si >0.9) more commonly associated with mare basalt units. In addition, the normal albedo data of Pohn and Wildey (1970) show that these deposits exhibit albedo values largely between 0.096 and 0.108, a range not typical of the circum-Crisium highlands but only slightly above that which characterizes most of Mare Fecunditatis and Mare Crisium. In Whitaker's 0.61–0.37 μm color difference photograph (E. A. Whiaker, pers. comm., 1976), the deposits appear slightly less "red" than the adjacent highlands to the north, suggesting a compositional difference. The deposit morphology is also consistent with an extrusive igneous origin. The deposits exhibit a smooth, generally flat surface, embay more rugged highland terrain, and exhibit what appears to be a mare-type ridge (8°N, 48.5°E).

In order to further investigate the composition of the possible volcanic units north of Taruntius, mixing model calculations (see Hawke and Ehmann, 1976) were performed using the geochemical data of Bielefeld *et al.* (1978) and FeO values inferred fromAl_2O_3 concentrations using the method described by Miller *et al.* (1974). The results show that the composition of the region can best be modeled as a mixture of 46% highlands material (unit B of Bielefeld *et al.*, 1978) and 54% mare material (unit H of Bielefeld *et al.*, 1978). Models using other endmember compositions yielded fits almost as good, but the results were essentially the same: a major component of mare basalt is required to model the composition of the units north of Taruntius.

Fig. 3. View looking westward of lunar eastern limb showing regional setting of Balmer light plains (beneath arrow which points to a dark-halved crater). (AS 17-152-23293).

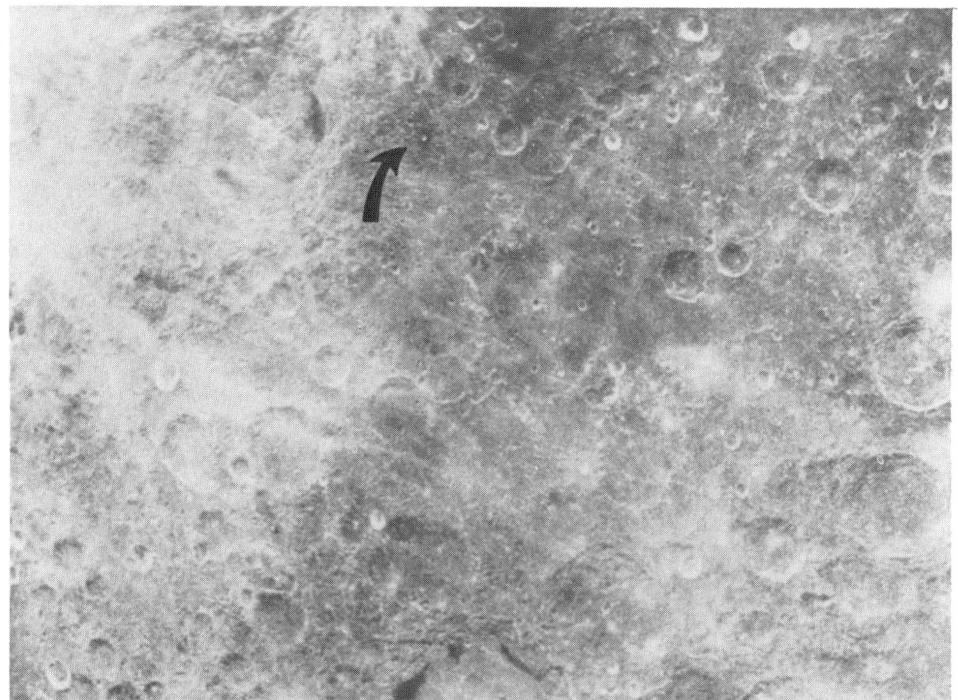

Fig. 4. Light plains region near the crater Balmer (20°S, 70°E) displaying dark halo craters (arrow) indicating the presence of a buried mafic unit. This region also has a high Th content. (AS 17-152-23332).

In summary, the evidence strongly suggests that the plains and smooth terra units north of Taruntius are volcanic units emplaced as fluid flows with some possible pyroclastic activity. The present surface composition is intermediate between mare basalts and highland material. Either the units represent early deposits of mare basalt which have been contaminated by lateral transport of highland material or, if highlands contamination has been minimal, a less mare-like initial composition.

Van de Graaff region

The region near Van de Graaff crater (Fig. 6) has long been noted for its abundance of associated geochemical and geophysical anomalies. The region is just within the proposed outermost rim of the large South Pole-Aitken basin and is also characterized by the occurrence of numerous unusual geologic features including a bright swirl belt and a unique surface unit composed of grooves and mounds (Stuart-Alexander, 1978).

Several major geochemical anomalies have been located in the Van de Graaff region. The region contains relatively high Th, K, and Fe concentrations and low

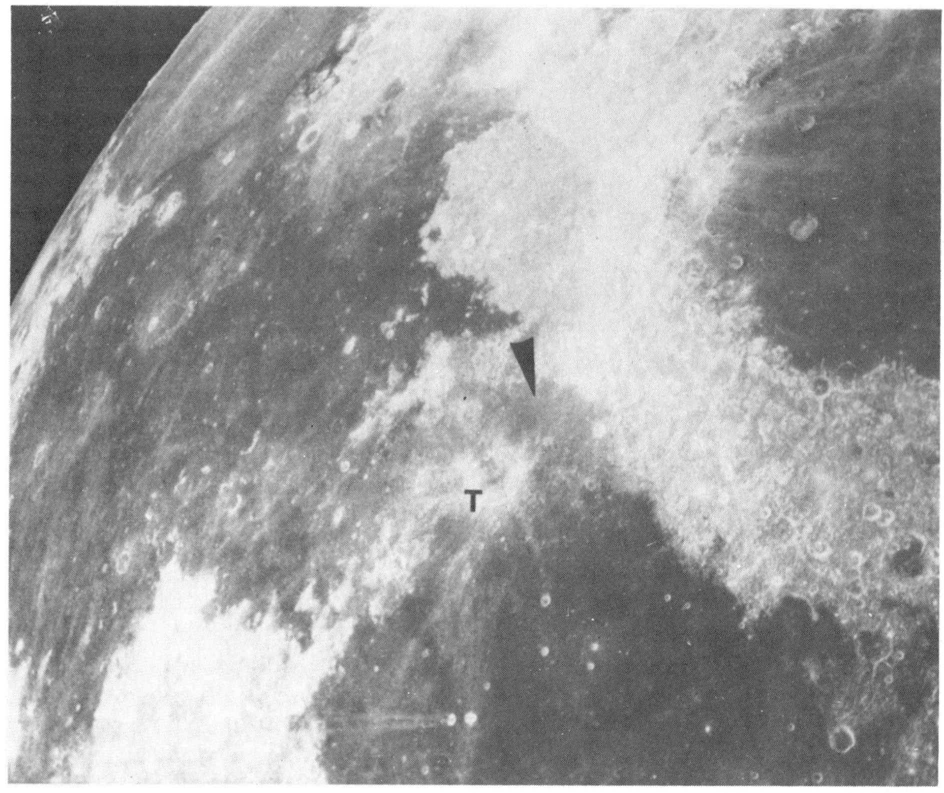

Fig. 5. Regional view of Crisium—Fecunditatis area showing moderate albedo plains (arrow) north of the crater Taruntius (T). (AS 17-152-23329).

Ti abundances (Metzger *et al.*, 1977; Bielefeld *et al.*, 1976; Davis, 1979). Attempts have been made to relate the unusual abundances of certain elements to the presence of specific deposits and rock types (e.g., low-K Fra Mauro basalt, granitic material, mare basalt). A basic question is whether the chemical anomalies are related (1) to material excavated from some depth in the farside crust by the South Pole-Aitken basin or other basins in the vicinity (Schonfeld, 1977; Davis, 1979), (2) to a local episode of extrusive igneous activity, or (3) to exotic material transported from some distant source with Imbrium basin being most often suggested (e.g., Stuart-Alexander, 1978).

Several major objections can be raised to the hypothesis that the geochemical anomalies are related to material transported from the Imbrium basin target site. This hypothesis rests on the interpretation of the grooved terrain (unit Ig, Stuart-Alexander, 1978) as Imbrium basin material concentrated in this region resulting from convergence of Imbrium ejecta at the antipodal point (Moore *et al.*, 1974; Stuart-Alexander, 1978). This interpretation is by no means generally accepted, and Schultz and Gault (1975) have proposed that this terrain was formed as a result of extensive mass wasting caused by the convergence of Imbrium seismic waves beneath the antipodal point. The observations that the Van de Graaff region exhibits high radioactivity and that the Imbrium event ejected KREEP-rich material enriched in radioactive elements have been used to support the Imbrium ejecta hypothesis. However, recent studies have shown that earlier ideas concerning the amount and extent of KREEP distribution by Imbrium basin were overestimates (Hawke and Head, 1978; Metzger *et al.*, 1979). Finally, as pointed out by Stuart-Alexander (1978), the grooved terrain does not extend under the easternmost 10° of the radioactivity anomaly. This observation makes any direct correlation between the grooved terrain and high radioactivity values highly suspect.

The results of our analysis suggest that the geochemical anomalies in the Van de Graaff region are due to volcanic activity. A major objection to the volcanic hypothesis has been the limited areal extent of mare-like basalt deposits in the region for which the geochemical data were obtained. There are four deposits of

Fig. 6. Area near the crater Van de Graaff (25°S, 170°E), a radioactivity anomaly on the lunar farside. (portion of LO II-33M).

mare basalt directly under the ground track but these make up less than 20 percent of the region (Stuart-Alexander, 1978). A partial explanation for this problem comes from consideration of the response characteristics of the Apollo gamma-ray spectrometer. Because the directional response of the spectrometer was essentially isotropic, the elemental concentrations observed from any given position were a composite of all concentrations within its field of view, which extended from horizon to horizon (Haines *et al.*, 1978; Metzger *et al.*, 1979). Although the instrument responded most efficiently to elements at the nadir and less efficiently the elements at greater distances, the spectrometer did respond to elements off the ground track (see Fig. 2 of Haines *et al.*, 1978). We suggest that at least some of the geochemical anomalies in the Van de Graaff region are due to the influence of major deposits of mare basalt just south of the ground track. For example, the highest Fe values (9%) in the region as determined by Davis (1979) are just north of Mare Ingenii.

Another perhaps more important factor may involve thinly covered basaltic units. As discussed above, the region is antipodal to Imbrium basin and any pre-Imbrian volcanic deposits could easily have been covered by the extensive mass-wasting caused by seismic waves associated with Imbrium (Schultz and Gault, 1979). Dark-haloed craters have been identified in the region and suggest the presence of buried basaltic units (Schultz and Spudis, 1979). At least some of the light plains deposits along the ground track may consist of thinly covered basalts buried either at the time of the Imbrium event or later in the Imbrian period. The presence of major magnetic anomalies in the region which do not correlate with maria (Hood *et al.*, 1978) strongly suggests that igneous activity was not strictly limited to areas where mare basalt is seen on the surface.

In order to further investigate the possibility that the occurrence of basalt is responsible at least in part for the geochemical anomalies in the Van de Graaff region, chemical mixing model calculations were performed on the composition of the total region as determined by the Apollo gamma-ray spectrometer. The latest values for TiO_2, FeO, MgO, K_2O, and Th concentrations were used (Bielefeld *et al.*, 1976; Metzger *et al.*, 1977; Davis, 1979). A wide variety of lunar rock type compositions was employed in the effort to model the composition of the region. The best fit obtained to date was for a mixture of anorthositic gabbro (61%), low-K Fra Mauro basalt (5%), and a basaltic component (34%), similar in composition to the Apollo 17 KREEPy basalt described by Ryder *et al.* (1977) as being intermediate in composition between mare and KREEP basalts. Attempts to model the regional composition with mixtures of KREEP, mare basalts, granites, and felsites with typical highland components (anorthositic gabbro, etc.) were either unsuccessful or yielded unsatisfactory standard errors.

In summary, the results of these mixing calculations together with the evidence discussed above, suggest that the geochemical anomalies in the Van de Graaff region are related to episodes of extrusive igneous activity. These basalts are unlike the typical mare basalts returned by the Apollo missions but appear similar to the igneous-textured Apollo 17 KREEPy basalts. Most are of Imbrian age but some may have been emplaced in pre-Imbrian time.

CONCLUSIONS AND IMPLICATIONS

We conclude that the east limb and farside geochemical anomalies discussed in this paper are related to episodes of extrusive igneous activity which produced deposits with a variety of ages and compositions. Some of the associated volcanic deposits (e.g., Mare Marginis) exhibit albedos and surface morphologies similar to those of the nearside maria and are post-Imbrian in age. Others are now thinly covered by highland debris and are commonly represented by light plains units which exhibit clusters of dark-haloed craters. These deposits are thought to be the products of an early phase of volcanic activity and were largely emplaced in pre-Imbrian time (Schultz and Spudis, 1979). The volcanic deposits associated with the geochemical anomalies discussed above generally correlate with regional topographic lows, areas of relatively thin crust, and structural features such as basin rings (Spudis, 1979). These associations are consistent with an extrusive igneous origin. Apparently, eastern limb and farside volcanism have been more extensive both in space and in time than has previously been thought.

The compositions of these volcanic deposits have important implications for the petrologic, geochemical, and thermal evolution of the lunar interior. Little evidence was found for a genuine episode of highland volcanism. The geochemical anomalies appear to be related to basaltic deposits with either mare affinities (e.g., terrain north of Taruntius) or that are intermediate in composition between KREEP and mare basalts (e.g., Mare Marginis, Van de Graaff). These basalts of intermediate composition may be similar to the Apollo 17 KREEPy basalts (Ryder *et al.*, 1977), which have been dated as 4.01 b.y. (Compston *et al.*, 1975). If the basalts in Mare Marginis are chemically similar to these Apollo 17 KREEPy basalts, it is significant that the Marginis deposits were emplaced at much later time (age <3.65 b.y.), (Boyce and Johnson, 1978; Boyce, 1979). In addition, the age data presented by Boyce and Johnson (1978) indicate that the mare units within Marginis were emplaced over an interval of 1.15 b.y. (3.65 b.y. to 2.5 b.y). This evidence argues against the hypothesis that an early (~4.0 b.y.) moonwide episode of volcanism produced all deposits intermediate in composition between KREEP and mare basalt. The existence of such intermediate basalts on the eastern limb and farside supports previous suggestions that KREEP-like volcanism was not limited to a small area of the lunar nearside (Haines *et al.*, 1978; Spudis, 1979).

Finally, the results of this study strongly suggest that thinly buried volcanic deposits are the source of certain geochemical anomalies on the eastern limb and farside and apparently can exert a strong influence on regional surface chemistry. Caution should be exercised in interpreting the surface compositions of what appear to be highland terrains which exhibit a high concentration of dark-haloed impact craters (see Fig. 6 of Schultz and Spudis, 1979). Inclusion of such terrains in attempts to establish average global and regional highlands compositions could lead to serious errors.

Acknowledgments—This work was carried out at the Hawaii Institute of Geophysics, University of Hawaii under NASA Grant NSG 7312 which is gratefully acknowledged. Thanks are extended to the National Space Sciences Data Center for many of the photographs used in this study. The attendance of PDS at the Lunar Highlands conference was supported in part by a grant from Sigma Xi and by NASA Grant NSG 7429. Helpful reviews were given by Ted Maxwell and R. A. De Hon. Useful comments and discussions were provided by P. Clark, D. Sellers, P. Schultz, R. Clark. N. Hubbard, and G. Ryder.

REFERENCES

Andre C. G., Maxwell T. A., El-Baz F., and Adler I. (1979b) Chemical diversity of lunar light plains from orbital x-ray data (abstract). In *Papers Presented to the Conference on the Lunar Highlands Crust,* p. 1–3. Lunar and Planetary Institute, Houston.

Andre C. G., Wolfe R. W., and Adler I. (1979a) Are early magnesium-rich basalts widespread on the moon? *Proc. Lunar Planet. Sci. Conf. 10th,* p. 1739–1751.

Bielefeld M. J., Reedy R. C., Metzger A. E., Trombka J. I., and Arnold J. R. (1976) Surface chemistry of selected lunar regions. *Proc. Lunar Sci. Conf. 7th,* p. 2661–2676.

Bielefeld M. J., Widley R. L., and Trombka J. I. (1978) Correlation of chemistry with normal albedo in the Crisium region. In *Mare Crisium: The View from Lunar 24* (R. B. Merrill and J. J. Papike, eds.), p. 33–42. Pergamon, N.Y.

Boyce J. M. (1979) Correlation of remote sensing data and crater density ages of flow units in the far eastern lunar maria. NASA TM-80339, p. 82–83.

Boyce J. M. and Johnson D. A. (1978) Ages of flow units in the far eastern maria and implications for basin-filling history. *Proc. Lunar Planet. Sci. Conf. 9th,* p. 3275–3283.

Clark P. E., Eliason E., Andre C. G., and Adler I. (1978) A new color correlation method applied to XRF Al/Si ratios and other lunar remote sensing data. *Proc. Lunar Planet. Sci. Conf. 9th,* p. 3015–3027.

Compston W., Foster J. J., and Gray C. M. (1975) Rb-Sr ages of clasts from within Boulder 1, Station 2, Apollo 17. *The Moon* **14,** 445–462.

Davis P. A. (1980) Iron and titanium distribution on the moon from orbital gamma-ray spectrometry with implications for crustal evolutionary models. *J. Geophys. Res.* In press.

Dvorak J. and Phillips R. J. (1978) Lunar Bouguer gravity anomalies: Imbrian age craters. *Proc. Lunar Planet. Sci. Conf. 9th,* p. 3651–3668.

Frontispiece (1978) *Proc. Lunar Planet. Sci. Conf. 9th,* Vol. 3, Plate 3.

Haines E. L., Etchegaray-Ramirez M. I., and Metzger A. E. (1978) Thorium concentrations in the lunar surface. II: Deconvolution modeling and its application to the regions of Aristarchus and Mare Smythii. *Proc. Lunar Planet. Sci. Conf. 9th,* p. 2985–3013.

Hawke B. R. and Ehmann W. D. (1976) Mixing model studies of the Apollo 17 regolith (abstract). In *Lunar Science VII,* p. 351–353. The Lunar Science Institute, Houston.

Hawke B. R. and Head J. W. (1977) Pre-Imbrian history of the Fra Mauro region and Apollo 14 sample provenance. *Proc. Lunar Sci. Conf. 8th,* p. 2741–2761.

Hawke B. R. and Head J. W. (1978) Lunar KREEP volcanism: Geologic evidence for history and mode of emplacement. *Proc. Lunar Planet. Sci. Conf. 9th,* p. 3285–3309.

Hawke B. R., MacLaskey D., McCord T. B., Adams J. B., Head J. W., Pieters C. M., and Zisk S. H. (1979) Multispectral mapping of the Apollo 15—Apennine region: The identification and distribution of regional pyroclastic deposits. *Proc. Lunar Planet. Sci. Conf. 10th,* p. 2995–3015.

Head J. W. (1974) Orientale multi-ringed basin interior and implications for the petrogenesis of lunar highland samples. *The Moon* **11,** 327–356.

Head J. W. and McCord T. B. (1978) Imbrian-age highland volcanism on the moon: The Gruithuisen and Mairan domes. *Science* **199,** 1433–1436.

Hood L. L., Russell C. T., and Coleman P. J. Jr. (1978) The magnetization of the lunar crust as deduced from orbital surveys. *Proc. Lunar Planet. Sci. Conf. 9th,* p. 3057–3078.

Hubbard N. J., Vilas F., and Keith J. E. (1978) From Serenity to Langemak: A regional chemical setting for Mare Crisium. In *Mare Crisium: The View from Luna 24* (R. B. Merrill and J. J. Papike, eds.), p. 13–32. Pergamon, N.Y.

Metzger A. E., Haines E. L., Parker R. E. and Radocineki R. (1977) Thorium concentrations in the lunar surface. I: Regional values and crustal content. *Proc. Lunar Sci. Conf. 8th,* p. 949–999.

Metzger A. E., Haines E. L., Etchegaray-Ramirez M. I., and Hawke B. R. (1979) Thorium concentrations in the lunar surface: III. Deconvolution of the Apennius region. *Proc. Lunar Planet. Sci. Conf. 10th,* p. 1701–1718.

Miller M. D., Pacer R. A., Ma M.-S., Hawke B. R., Lockhart G. L., and Ehmann W. D. (1974) Compositional studies of the lunar regolith at the Apollo 17 site. *Proc. Lunar Sci. Conf. 5th,* p. 1079–1086.

Moore H. J., Hodges C. A., and Scott D. H. (1974) Multi-ringed basins—illustrated by Orientale and associated features. *Proc. Lunar Sci. Conf. 5th,* p. 71–100.

Pohn H. and Wildey R. L. (1970) A photoelectric-photographic study of the normal albedo of the moon. *U.S. Geol. Survey Prof. Paper 599-E.* 20 pp.

Ryder G. Stoeser D. B., and Wood J. A. (1977) Apollo 17 KREEPy basalt: A rock type intermediate between mare and KREEP basalt. *Earth Planet. Sci. Lett.* **35,** 1–13.

Ryder G. and Taylor G. J. (1976) Did mare-type volcanism commence early in lunar history? *Proc. Lunar Sci. Conf. 7th,* p. 1741–1755.

Ryder G. and Wood J. A. (1977) Serenitatis and Imbrium impact melts: Implications for large-scale layering in the lunar crust. *Proc. Lunar Sci. Conf. 8th,* p. 655–668.

Schonfeld E. (1977) Comparison of orbital chemistry with crustal thickness and lunar sample chemistry. *Proc. Lunar Sci. Conf. 8th,* p. 1149–1162.

Schonfeld E. and Bielefeld M. J. (1978) Correlation of dark mantle deposits with high Mg/Al ratios. *Proc. Lunar Planet. Sci. Conf. 9th,* p. 3037–3048.

Schultz P. H. and Gault D. E. (1975) Seismic effects from major basin formations on the Moon and Mercury. *The Moon* **12,** 159–177.

Schultz P. H. and Spudis P. D. (1979) Evidence for ancient mare volcanism. *Proc. Lunar Planet. Sci. Conf. 10th,* p. 2899–2918.

Spudis P. D. (1978) Composition and origin of the Apennine Bench Formation. *Proc. Lunar Planet. Sci. Conf. 9th,* p. 3379–3394.

Spudis P. D. (1979) *The extent and duration of lunar highland volcanism.* NASA TM-80339, p. 270–272.

Stuart-Alexander D. E. (1978) Geologic map of the central far side of the moon. U.S. Geol. Survey Misc. Geol. Inv. Map I-1047.

Wilhelms D. E. (1972) Geologic map of the Taruntius quadrangle of the moon. U.S. Geol. Survey Misc. Geol. Inv. Map I-703.

Wilhelms D. E. and El-Baz F. (1977) Geologic map of the east side of the moon. U.S. Geol. Survey Misc. Geol. Inv. Map I-948.

Papike, J.J. and Merrill, R.B., eds.
Proc. Conf. Lunar Highlands Crust (1980), p. 483-499
Printed in the United States of America

Infrared and radar signatures of lunar craters: Implications about crater evolution*

T. W. Thompson[1], J. A. Cutts[1], R. W. Shorthill[2], and S. H. Zisk[3]

[1] Planetary Science Institute, Science Applications, Incorporated, Pasadena, California 91101
[2] University of Utah Research Institute, Salt Lake City, Utah 84108
[3] NEROC Haystack Observatory, Westford, Massachusetts 01886

Abstract—Lunar craters from different geological epochs have distinctive signatures in ground-based infrared eclipse imaging and radar imaging at 3.8 and 70 cm wavelengths. Craters have been classified as B (bright) or F (faint) depending on the contrast between the signal level from the crater floor and the average terrain background.

For large (diam. ≥ 10 km) terra craters, the IR/radar signatures are a function of crater age, supporting the hypothesis that lunar surface processes gradually reduce bright signatures to average terra background. The youngest craters have infrared temperatures and radar echo strengths which are enhanced relative to their backgrounds. In somewhat older craters, the infrared temperature enhancements have faded, but the high radar echo strengths remain. In still older craters, either the 3.8 cm enhancement or the 70 cm enhancement have disappeared. Finally, the oldest craters have no enhancements at any wavelength.

The evolution of IR/radar signatures depends upon crater size. Comparisons of size-frequency distributions of different crater types show that craters with both radar and infrared enhancements are only half as abundant as Copernican and Eratosthenian craters at 10 km diameter but are twice as abundant at 100 km diameter. Craters with any form of infrared or radar enhancement are only two thirds as abundant as Imbrian and younger craters at 10 km diameter, but are 50% more abundant at 100 km diameter. Thus, some larger radar enhanced craters are pre-Imbrian. The path of evolution is also size dependent: in larger craters (diam. ≥ 30 km) the 3.8 cm enhancement disappears first; in smaller craters, the 70 cm enhancement disappears first.

Some geological models for evolution of IR/radar signatures of crater floors have been examined. The simplest model involves the formation and subsequent gardening of an impact melt layer in the crater floor. This model accounts for the evolutionary path of larger craters (≥ 30 km) but not smaller craters. Quantitative models of the evolution of rock populations in regoliths and of the interaction of microwaves with regoliths are needed to better understand crater evolutionary processes.

I. INTRODUCTION

The infrared and radar signatures of lunar craters contain information about surface and subsurface rock populations. Infrared (10 micron) eclipse temperatures are controlled by surface rocks, whereas radar echo strengths are controlled by both surface and subsurface rocks (Thompson *et al.*, 1974). The infrared and

* PSI Contribution 124

radar signatures of craters change with time as the pristine crater is modified by surface processes. Thompson *et al.* (1974) reviewed the various combinations of infrared and radar signatures for 51 selected craters. They showed that the youngest craters have enhanced infrared and radar signatures and that enhanced radar echoes at the longest radar wavelength are the last enhanced signature to disappear. The objective of this work is to study intermediate states in the evolution of infrared and radar signatures of craters.

Whereas the earlier work of Thompson et al. (1974) was based upon only 51 craters, these hypotheses about lunar crater evolution can now be tested with a much larger data set of Thompson *et al.* (1979). In particular, craters with enhanced infrared and radar signatures were cataloged for the portion of the earthside lunar hemisphere covered by the LAC (Lunar Aeronatical Chart) maps. A previous study of this data set showed mare-terra differences for smaller lunar craters (diam. < 12 km) associated with depth of the megaregolith, the debris layer overlying the lunar terrae (Thompson *et al.*, 1979). In this study, larger (diam. > 12 km) craters on the terra were compared with crater age statistics of Wilhelms *et al.* (1978) to test a hypothesis of crater evolution originally proposed by Thompson *et al.* (1974). Mare-terra differences in infrared and radar crater signatures were also investigated to corroborate the findings from the comparison of terra craters and their photogeological ages.

II. CATALOG OF RADAR AND INFRARED SIGNATURES OF LUNAR CRATERS

A catalog of infrared and radar crater signatures was assembled by Thompson *et al.* (1979) using data from several sources. The infrared data was originally obtained during the lunar eclipse of December 19, 1964, using the Kottamia 74-inch telescope of the Helwan Observatory, Egypt (Shorthill, 1973). The 3.8 cm radar data were obtained in a series of observations from 1966 to 1970 using the planetary radar at Haystack Observatory (Zisk *et al.*, 1974). The 70 cm radar data were obtained between 1966 and 1969 using the 430 MHz radar at the Arecibo Observatory, Puerto Rico (Thompson, 1974). Surface resolution for these observations was 1–3 km for the 3.8 cm radar, 5–10 km for the 70 cm radar, and 14–30 km for the infrared observations. This resolution, which is the image size of a point target, is approximately one-half of the line-pair resolution.

This catalog covers about one-half of the earth-visible surface. Craters on the limb beyond the area of the available LAC charts were ignored. There are no 70 cm radar data for these areas and the infrared resolution is reduced. Also, as described by Thompson *et al.* (1979), the original data were sorted by distance from the subearth point into the Central, Middle, and Outer Ring areas shown in Fig. 1. Only the data for the Middle and Outer Rings have been used in our investigations because of the poor radar data in the Central Area.

This catalog contains 386 craters with diameters greater than 12 km. Crater names, positions, and diameters were taken from the Lunar and Planetary Laboratory (LPL) catalogs of Arthur *et al.* (1963, 1964, 1965, and 1966). Craters were classified as mare craters or terra craters based upon their settings in the USGS Lunar Geological Maps. Craters in both the irregular and circular (basin) mare were classified as mare craters; all other craters were classified as terra craters.

Radar and infrared signatures for these craters were measured from quantized video displays of the data. For larger craters with diameters several times the resolution cell size, the infrared and radar strengths were measured from the interior of the crater. These larger craters generally display uniform responses across their interiors. The radar data show topographic effects associated with the

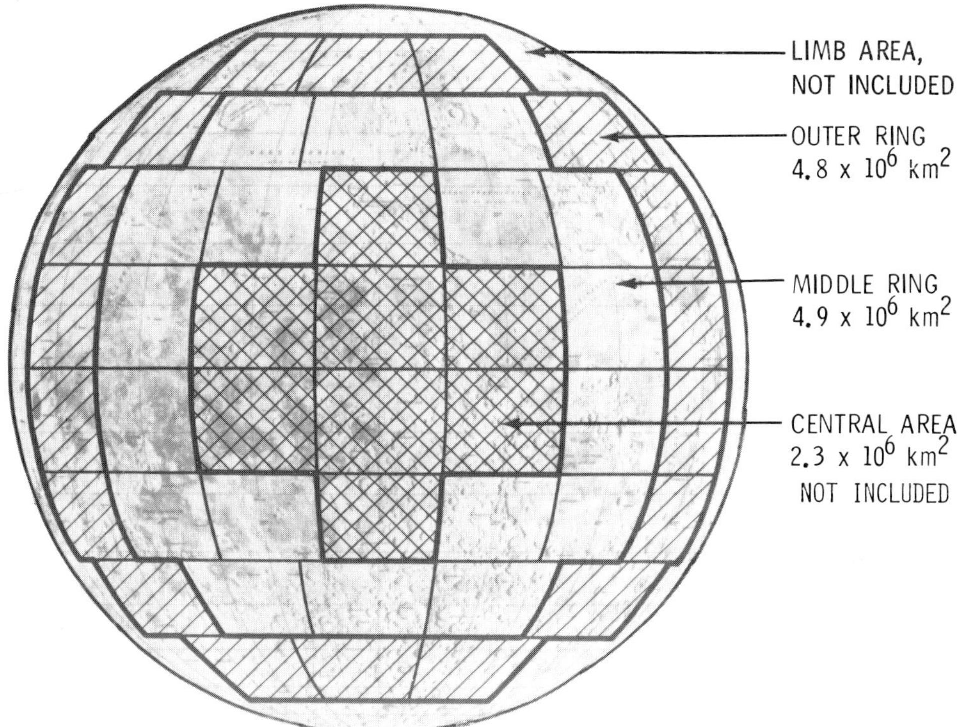

Fig. 1. Area for infrared radar statistics of this study. Middle and Outer Ring Areas were combined; Central and Limb Areas were ignored (see Thompson *et al.*, 1979).

rims since areas tilted toward the radar have stronger echoes and areas tilted away from the radar have weaker echoes. For smaller craters with diameters of a few resolution cell sizes and less, the infrared and radar strengths were measured for the center of spots which appear in the image. The infrared and radar strengths were measured relative to the brightness of the surrounding terrain.

To simplify the statistical analysis of the radar and infrared signatures, signal strengths were classified as bright or faint. Radar bright craters backscattered at least twice the power of the mean surface; infrared bright craters were warmer than the mean surface by 10°K or more. Faint craters did not meet these criteria and consequently had low contrasts relative to the adjoining terrains. A binary code, XYZ, was used to describe the infrared and radar signature of each crater in the catalog. Code X takes the value B when the crater is infrared bright and F when it is faint. Codes Y and Z were similarly assigned based on the 3.8 cm and 70 cm radar signals. (For example, a FBF crater has bright 3.8 cm radar echo with faint infrared and 70 cm radar strengths.) There are eight possible combinations. One crater type, the FFF crater, has not yet been cataloged for the terra since it was not pertinent to the earlier study (Thompson *et al.*, 1979). However, a comparison of the statistics for the infrared and radar bright craters with the data on Wilhelms *et al.* (1978) indicates that FFF craters on the terra should be ten times more abundant than all infrared and radar bright craters. Thus, about 2900 FFF craters should occur in our study area.

Histograms for the various classes of craters with enhanced infrared and radar signatures are given in Fig. 2. The surface conditions for these various combinations of infrared and radar signatures are those given in Table 1. We now consider what these crater populations imply about the origin of these craters and the evolution of craters from one class to another in response to various lunar surface properties.

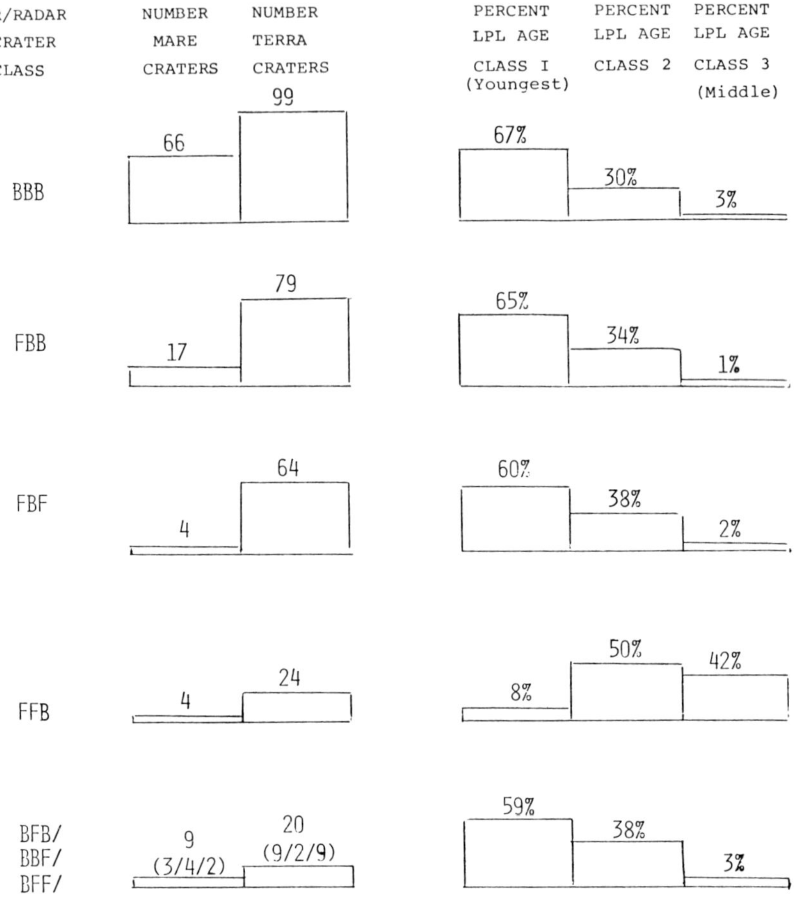

Fig. 2. Histograms for the occurrence of infrared and radar craters with diameters greater than 12 km in the Central and Outer Ring areas of Fig. 1. The FBF and FFB craters occur almost exclusively in the terra; the BFB, BBF, BFF craters occur rarely. The number of FFF craters is estimated to be 2,900 for the terra and zero for mare. Mare area was 4.4×10^6 km^2; terra area was 5.4×10^6 km^2.

III. STATISTICS AND AGES OF RADAR-INFRARED CRATER CLASSES

The eight crater classes defined by various combinations of the infrared and radar signatures have varied populations. Most craters in the maria are either BBB or FBB craters, while in the terra the BBB, FBB, FBF, FFB, and FFF craters are abundant. To understand this, it is useful to correlate the radar infrared crater classes with age signatures based on photogeologic analysis. Figure 2 shows histograms of the occurrence of the various anomaly types as a function of a photogeological degradation state from the LPL catalogs. These data are consistent with the hypothesis that BBB craters and FFF craters are end members of an evolutionary sequence.

Possible evolutionary paths

A framework of possible evolutionary paths for the radar-infrared crater signatures is shown in Fig. 3. This framework makes few assumptions about the mechanisms of evolution; it only requires that craters form first as BBB craters and progressively degrade to FFF craters through several possible intermediate states as the regolith at the location of the crater evolves back towards the local average for that part of the moon.

We can gain a crude idea of what evolutionary paths are most important from the frequency of occurrence of the different anomaly types on the moon (Fig. 4).

Table 1. Surface conditions implied by various combinations or infrared and radar signatures.

Anomaly index (IR–3.8 cm–70 cm)	Implied surface conditions
B—	Excess surface rocks with sizes larger than 10 cm.
-B-	Excess rocks with sizes of 1 to 40 cm on the surface or buried no deeper than 1.2 meters.
—B	Excess rocks with sizes of 20 cm to 7.0 m on the surface or buried no deeper than 20 meters.
BBB	Excess surface rocks with sizes of 1 cm and larger.
FBB	Average surface rocks Excess buried rocks with sizes of 1 to 40 centimeters within 1.1 m of surface. Excess buried rocks with sizes of 20 cm to 7.0 meters within 20 m of surface.
FBF	Average surface rocks Excess buried rocks with sizes 1 to 40 cm within 1.1 m of surface.
FFB	Average surface rocks Excess buried rocks with sizes of 20 cm to 7.0 m within 20 m of surface.
BBF	Excess surface rocks with sizes 1 cm to 20 cm. Average surface rocks with sizes 20 cm to 7.0 m.
BFB	Average surface rocks with sizes 1 cm to 40 cm. Excess surface rocks with sizes 40 cm to 7.0 m.
BFF	Average surface rocks with sizes 1 cm to 7.0 m. Excess surface rocks with sizes 7.0 m and larger.
FFF	Average surface, no excess surface and subsurface rocks.

Taken from Table 1 of Thompson, *et al*. (1979)

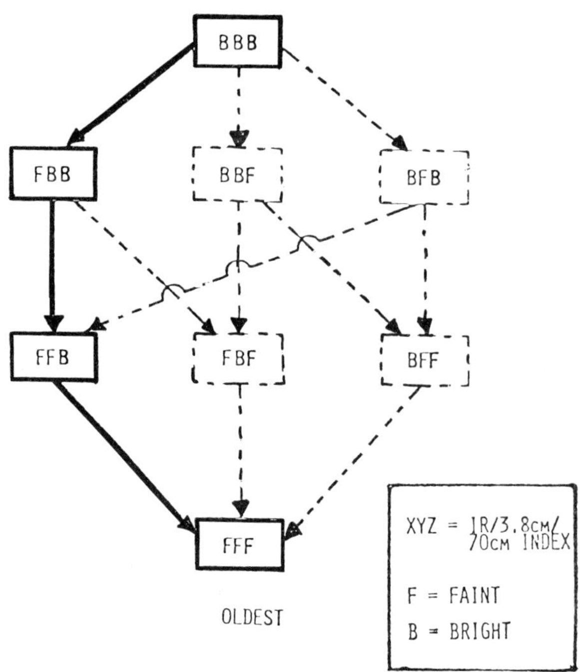

Fig. 3. Model for evolution of infrared and radar craters. This assumes that the youngest craters are BBB, the oldest are FFF and the intermediate cases lose one signature at a time. An expected evolution shown with the solid line assumes that the infrared signature disappears first, the 3.8 cm radar signature disappears second, and 70 cm radar signature disappears last.

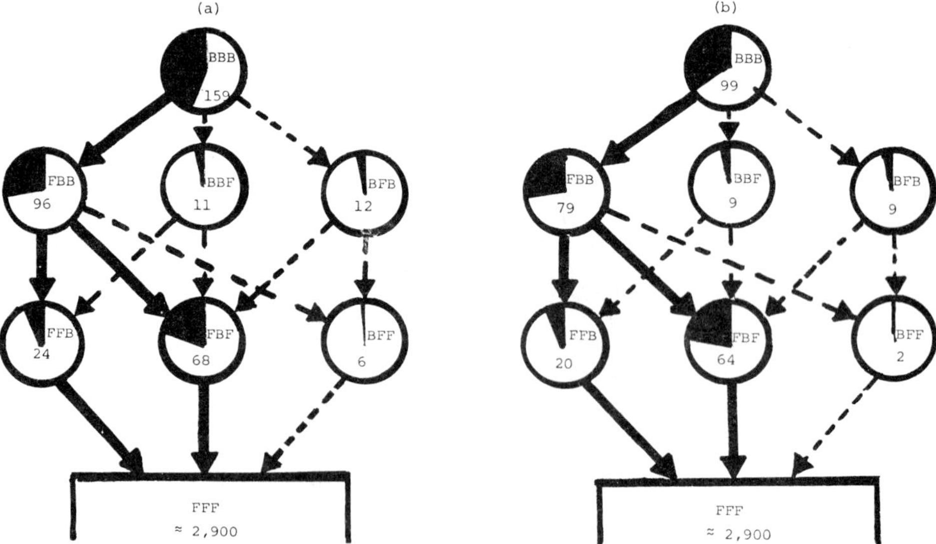

Fig. 4. Observed infrared radar crater populations plotted on evolution model paths shown in Fig. 3. (a) All mare and terra craters; (b) Terra craters only. Note that the FBB Craters evolve to both FBF and FFB craters.

There are 99 BBB, 79 FBB, 64 FBF, 30 FFB, 9 BFB, 9 BFF, and 2 BBF craters in the 5.4 × 10^6 km^2 of terra in the Central and Outer Ring areas of Fig. 1. Figures 2 and 4 show that the BBB, FBB, FBF and FFB craters occur frequently while the BBF, BFB, and BFF occur infrequently. These crater abundances suggest that BBB craters initially evolve into FBB craters; these FBB craters further evolve by two distinct paths to FBF or FFB craters; these FBF and FFB craters in turn finally evolve to FFF craters.

Time scale

In order to establish a time scale for the evolutionary paths illustrated in Figs. 3 and 4, and to estimate how long a crater remains in a particular class, we need to relate the populations of radar-infrared classes to a geologic time scale. Wilhelms *et al.* (1978) have recently classified large lunar craters by geologic age and constructed size-frequency distributions for craters produced in four time intervals: Copernican plus Erathosthenian, Imbrian, Nectarian and pre-Nectarian. By comparing these size-frequency distributions with those for radar-infrared bright crater classes we can search for a correspondence between crater class and age. Here we emphasize the craters on the terra since the radar-infrared populations indicate that craters from all stages in the evolution from BBB to FFF are preserved there (Fig. 4). Also, the terra background signal level in the radar and infrared has little variation across the moon and this simplifies the interpretation of the radar-infrared crater evolution.

In order to compare different crater populations we use relative size-frequency plots (Crater Analysis Group, 1979), where a function R is plotted versus crater diameter. The quantity R is $(\bar{D})^3 N/A\,(D_{max}-D_{min})$, where $\bar{D}$ is the geometric mean of crater diameters, N is the number of craters, A is the area, and D_{max}, D_{min} are the maximum and minimum crater diameters in a size bin. A crater population which has a cumulative distribution proportional to (crater diameter)$^{-2}$ and a differential distribution proportional to (crater diameter)$^{-3}$ plots as a horizontal line in a log (R) versus log (D) plot. Similarly, a crater population which has a cumulative distribution proportional to (crater diameter)$^{-3}$ and a differential population proportional to (crater diameter)$^{-4}$ has a slope of -1 in a log (R) versus log (D) plot.

In Fig. 5(a) we compare the diameter-frequency distribution of BBB craters on the terra with the Copernican and Eratosthenian craters. The two populations are generally similar at smaller diameters, suggesting that most small BBB craters are Copernican and Eratosthenian in age. However, BBB craters at larger diameter are more abundant than Copernican and Eratosthenian craters, implying that some of these larger BBB craters on the terra are Imbrian in age. In Fig. 5(b) we compare the size-frequency distributions of all infrared and radar bright craters on the terra with Imbrian and younger craters. The two populations have different slopes, indicating that the radar-infrared bright craters are again more numerous at larger sizes. Thus, some of the larger radar bright craters are pre-Imbrian.

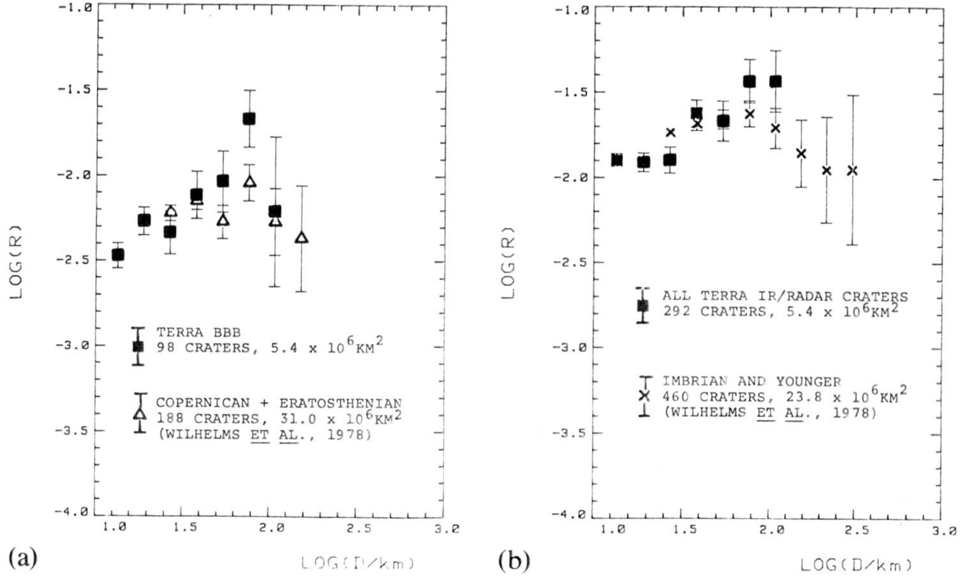

Fig. 5a. Relative size-frequency distributions for terra BBB and (Copernican & Eratosthenian) craters of Wilhelms *et al*. (1978).
Fig. 5b. Relative size-frequency distributions for all terra infrared/radar craters and Imbrian and younger craters of Wilhelms *et al*. (1978).

A more detailed insight into the age distributions of our radar-infrared bright craters as a function of the more populous crater classes is found in Figs. 6 and 7. The diameter-frequency distributions on the terra of BBB, FBB and FFB craters have positive slopes in this size range. The FBF crater distribution, in contrast, has a negative slope indicating that smaller craters dominate this population.

To investigate the slopes of these crater distributions, we assumed that the crater populations could be approximated by a linear relationship on a log (R)−log (D) plot. We used least squares methods to estimate crater densities at 10 km and 100 km diameters (Table 2). Crater data of Wilhelms *et al*. (1978) were similarly fitted to characterize cratering in the Nectarian and later ages. A chi-squared estimate of goodness-of-fit to the assumed straight line distributions was also computed. All fits were good except for the Nectarian and younger distribution which shows a marked deviation from the straight line relationship. This behavior was also noted by Wilhelms *et al*. (1978), who concluded that the impacting population in the Nectarian times was deficient in smaller bodies. Finally, in order to conveniently compare the various crater populations, the crater densities at 10 km and 100 km diameter were expressed as percentages of the Imbrian and younger values.

The data shown in Table 2 suggest that smaller craters evolve faster than larger craters. Note that terra BBB and FBB craters at 10 km diameter occur as frequently as Copernican and Eratosthenian craters. In contrast, terra BBB craters at 100 km are twice as abundant as the Copernican and Eratosthenian craters and must therefore comprise a number of Imbrian craters also. All infrared and radar craters at 10 km diameters comprise 65 percent of post-Imbrian craters; this implies that one-third of these smaller post-Imbrian craters have evolved to FFF craters. In contrast, all infrared and radar craters at 100 km diameter are 1.4 times as abundant as the Imbrian and younger craters; this implies that some larger radar bright craters are pre-Imbrian.

The data shown in Table 2 suggest that smaller craters and larger craters evolve to different classes. There are few FFB craters at 10 km diameter and few FBF craters at 100 km. This implies that smaller FBB craters on the terra evolve to FBF craters, while larger FBB craters evolve to FFB craters.

The general features of the data shown in Table 2 are corroborated by comparing the statistics of the terra and mare craters shown in Fig. 6. The preservation of some Imbrian craters in the 100 km diameter range as BBB craters implies that BBB populations on the Terra will exceed those on the maria; this is confirmed by the data in Fig. 6. However, at 10 km diameter, where only

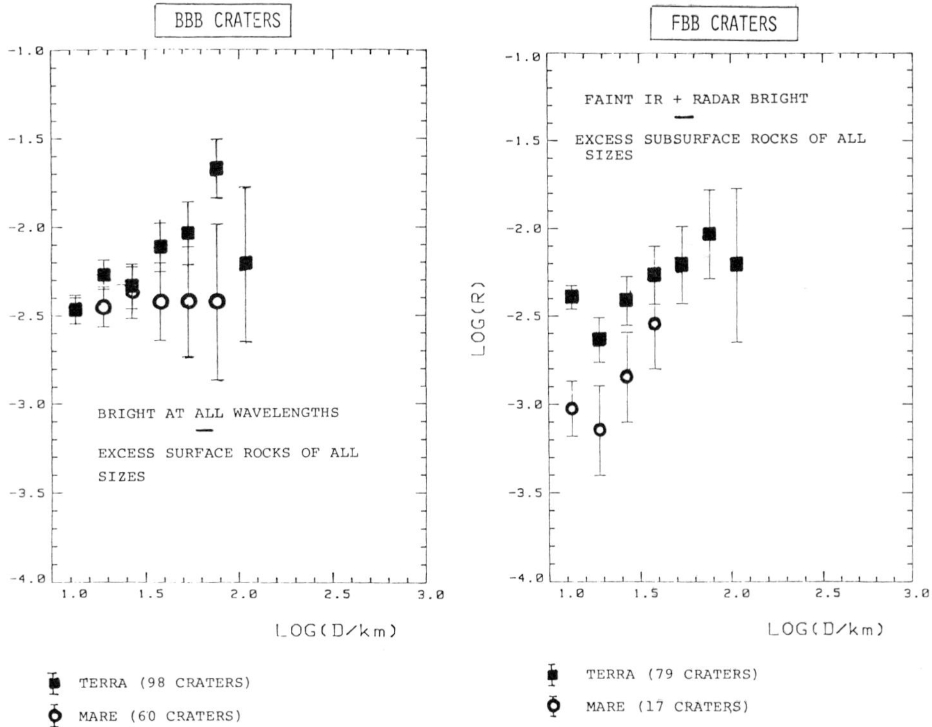

Fig. 6. Relative size-frequency distributions for the BBB and FBB craters. All of these craters except the Mare BBB show a size dependence different than the population of Imbrian and younger craters, which is approximately a horizontal line in these plots (Fig. 5a). Mare area was 4.4×10^6 km^2; terra area was 5.4×10^6 km^2.

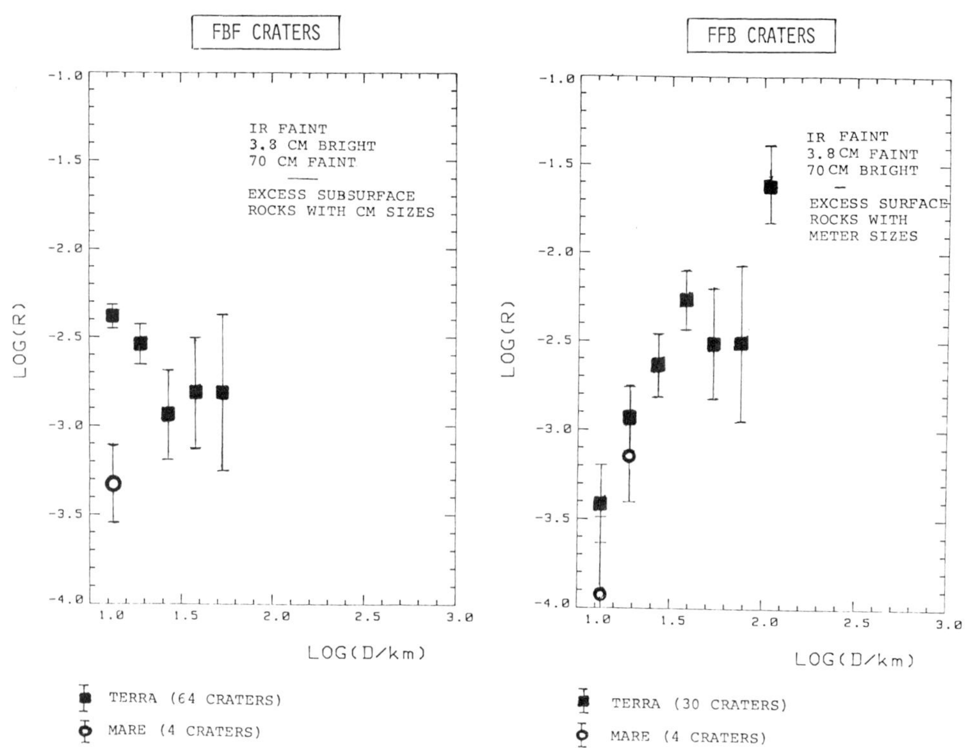

Fig. 7. Relative size-frequency distributions for the FBF and FFB craters. These craters occur primarily in the terra and have size dependences different than the population of Imbrian and younger craters, which would plot approximately as a horizontal line in these plots. Mare area was 4.4×10^6 km^2; terra area was 5.4×10^6 km^2.

Eratosthenian and younger craters are preserved as BBBs, we expect that BBB populations of mare and terra will be similar because all these BBBs formed since the maria. FBB craters are found on the maria with populations consistent with the hypothesis that smaller craters evolve faster than larger ones. All other crater classes are effectively absent on the mare. In addition, Fig. 8 shows that diameter distributions for all infrared and radar craters on the mare are nearly identical to the Copernican and Eratosthenian craters of Wilhelms *et al.* (1978). This suggests that there are few, if any, FFF craters on the mare with diameters greater than 12 km.

In summary, it appears that evolution of infrared and radar bright craters is size dependent. Smaller craters with diameters between 10 and 30 km tend to evolve from BBB to FBB to FBF to FFF craters. Larger craters with diameters larger than 30 km diameter tend to evolve from BBB to FBB to FFB to FFF craters. The ages associated with these crater classes are summarized in Table 3.

Table 2. Relative crater frequencies from linear least square fit to crater data.

Crater class	R(D) [R(Imbrian and Younger)]		R (D) × 10^4		Least square fit
	D=10 km	D=100 km	D=10 km	D=100 km	Chi-squared probability
Terra BBB	15.4%	66.6%	29.2	133.3	59.0%
Terra FBB	15.9%	26.4%	30.1	52.8	33.5%
Terra FBF	32.9%	1.4%	62.4	2.7	84.0%
Terra FFB	1.7%	57.8%	3.3	115.6	42.1%
Terra BBB+FBB +FBF+FFB	65.9%	152.2%	125.0	304.4	—
*Copernican & Eratosthenian	32.77%	31.6%	62.1	63.2	46.3%
*Imbrain & Younger	100.0%	100.0%	189.7	200.0	59.15%
*Nectarian & Younger	252.3%	375.7%	478.6	753.4	0.015%

* From Wilhelms *et al.* (1978)

DISCUSSION

To completely explain the formation and evolution of the infrared and radar signatures of craters is far beyond the scope of this paper. However, we will try to identify some possibly productive lines of future inquiry. Here we concentrate on explanations for (1) the typical behavior of large craters which have an evolution sequence BBB/FBB/FFB/FFF; (2) the typical behavior of small craters which evolve BBB/FBB/FBF/FFF; and (3) the faster rates of evolution of small craters. We will not discuss less common radar-infrared signatures.

First, it is useful to discuss how a typical terra region has evolved under the influence of impact gardening. Today, the terra has a uniform infrared and radar response, the end product of a deep gardening by impacting bodies. Except near very large craters which excavated bedrock beneath the megaregolith, the blocks now found at the surface and those mixed with finer debris to some depth are breccias and impact melts formed by the compaction and fusion of finer debris and smaller blocks by impacting events. Although the depth of the regolith continues to grow at a modest rate, the average size-frequency distribution of particles at shallow depths, less than a few tens of meters, has reached a steady state

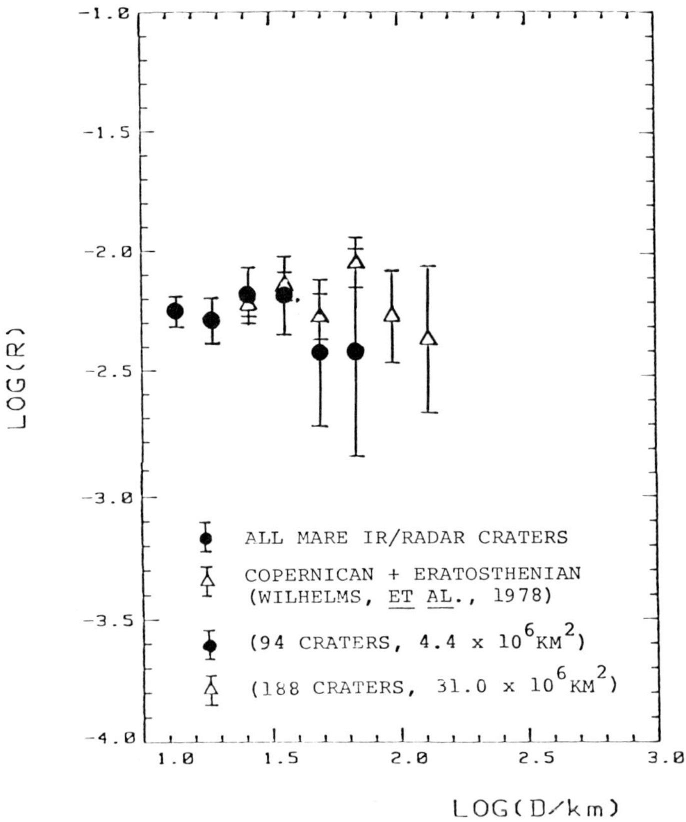

Fig. 8. Relative size-frequency distributions for all mare infrared/radar craters and (Copernican + Eratosthenian) craters of Wilhelms *et al.* (1978).

Table 3. Crater ages for IR-radar craters.

	Terra crater age	
Crater class	Smaller craters 10 km < diam. < 30 km	Larger craters 30 km < diam.
BBB	Copernican and Eratosthenian	Mostly Copernican and Eratosthenian; a few Imbrian
FBB		Imbrian
FBF	Imbrian	—
FFB	—	Mostly Imbrian; some pre-Imbrian

condition and the infrared and radar signatures of this surface have also reached steady state.

The formation of a large crater on the terra surface locally disturbs this steady state situation. An heterogeneous mix of blocks is excavated from depth and an impact melt is deposited on the floor of the pristine crater. Photogeological mapping and studies of the orbital infrared data show that the floors of large craters do not have uniform properties but are irregularly divided into units of distinct physical properties (Mendell and Low, 1975). We infer that older crater floors evolved from pristine floors primarily under the influence of impact gardening processes (Quaide and Oberbeck, 1975; Hörz *et al.*, 1975; Thompson *et al.*, 1974). However, endogenic processes may have been important, particularly for those craters that rim the maria and now constitute fractured floor craters (Schultz, 1976). Mass movement of debris from crater walls onto floors may have mantled the pristine surface (Mendell, 1976). Finally, ejecta from large basin-forming events such as Orientale and Imbrium have influenced the evolution of older craters (Moore *et al.*, 1974). All of these effects are potentially important in the evolution of the radar-infrared crater signatures.

How do the radar-infrared signatures of the floor of a pristine crater evolve under the action of impact gardening alone? Only some very qualitative ideas can be offered at present. First, initial conditions are important. If the floor is composed of a solidified impact melt then the signatures of the pristine surface and the subsequent evolution of the signatures may be different than if the floor is initially composed of debris. If the floor consists of a thin layer of impact melt overlying debris, still more complex patterns in the evolution of radar-infrared signatures can be expected.

Consider the evolution of a crater floor mantled with impact melt. The thickness of impact melt originally occupying the floor of large craters is dependent on crater diameter. Lange and Hawke (1979) show that the volume of impact melt generated is approximately (1.8×10^{-4}) $D^{3.4}$ km^3 where D is crater diameter in kilometers. Some of this melt is deposited beyond the crater rim. The melt that remains in the interior of the crater probably occupies only a fraction of the crater floor. To first order, the thickness of the melt deposit on the floor of a crater can be estimated by assuming that the melt volume described above occupies a disk with diameter D. Thus, the depth of impact melt on the floor of a crater is $(0.28)D^{1.4}$ meters. Table 4 shows that this melt thickness will vary from 7 meters for a 10 km crater to 177 meters for a 100 km crater.

The radar-infrared signature of the melt will change in a complex manner as an impact regolith develops on it. Initially, the high thermal inertia of exposed impact melt will give rise to a strong thermal infrared contrast as reported by Schultz and Mendell (1978) for young craters such as Aristarchus and Olbers A. Next, the thermal signature will decline in response to comminution of the uppermost layer by small scale impact cratering. As the regolith gets deeper, large blocks of material generated by deeper cratering events into the impact melt may dominate the surface particle size distribution and re-enhance the thermal IR signature. The re-enhancement of the thermal signature, if real, has not been

Table 4. Predicted time scale for decay of radar anomalies as a function of crater diameter.

Crater diameter	Melt thickness (in meters)	Crater† age (by)	Regolith† depth (H_{med}) (in meters)	$\left(\frac{H_{med}}{T_m}\right)$	$\left(\frac{H_{med}}{D_{3.8}}\right)^*$	$\left(\frac{H_{med}}{D_{70}}\right)^*$	Predicted signatures 3.8 cm	70 cm
		0.8	1.0–1.4	0.14	0.8	0.5	B	B
10	7.0	3.55	4.8–5.7	0.7	4.0	0.24	F	B
		3.95	18.0–25.8	2.6	15.0	0.9	F	F
		4.2	43.5–105.0	6.2	36.0	2.2	F	F
30	26.8	0.8	1.0–1.4	0.04	0.8	0.05	B	B
		3.55	4.8–5.7	0.18	4.0	0.24	F	B
		3.95	18.0–25.8	0.67	15.0	0.9	F	B
		4.2	43.5–105.0	1.60	36.0	2.2	F	F
		0.8	1.0–1.4	0.005	0.8	0.05	B	B
		3.55	4.8–5.7	0.03	4.0	0.24	F	B
100	176.7	3.95	18.0–25.8	0.10	15.0	0.9	F	B
		4.2	43.5–105.0	0.25	36.0	2.2	F	F

† Relationship of age to regolith determined from data of Shoemaker and Morris (1970), Shoemaker *et al.* (1970), Boyce (1976), Boyce and Johnson (1977). Crater ages are 0.8 b.y. (Copernicus), 3.55 b.y. (Mare Tranquillitatis at Apollo 11), 3.95 b.y. (Imbrian ejecta blanket) and 4.2 b.y. (Nectarian ejecta blanket). These times can be crudely referenced to the geological periods used in the text: the Nectarian period starts at 4.2 b.y., the Imbrian at 3.95 b.y. and the Copernican Eratosthenian at about 3 b.y. The range of regolith depths cited reflects uncertainties in the estimates. We have used minimum values in our calculations.

* Penetration depth ($D_{3.8}$ and D_{70}) at 3.8 cm and 70 cm are assumed to be 1.2 m and 20 m respectively.

resolved in the thermal signatures that we have investigated. With further regolith development the block population of the surface layers of the regolith will decline as new blocks from the substrate are excavated less frequently.

The radar scattering from pristine impact melt would be expected to be small initially but to increase rapidly as blocks are produced by impact cratering. As the regolith thickens, the decline in the block population at the surface will be followed by a decline just beneath the surface. In a very thick regolith, the material in the upper layers will be reduced to the steady state condition. Consequently, an early decline in the infrared signature will be followed by a decline in the 3.8 cm signatures and finally a fall in the 70 cm signature (see Fig. 3).

We expect that the evolution of the radar signature, the contrast between the crater floor and the surrounding terra, will depend in part on the thickness of the melt layer. In small craters, the depth of the regolith formed on the crater floor may exceed the impact melt thickness and will entirely destroy the coherent structure of the melt layer. In addition, Gault *et al.* (1974) show that material buried at depths significantly shallower than the median regolith depth will be turned over many times during the history of that surface. Consequently, rocks derived from the destruction of an impact melt layer by a deep regolith will be returned to the surface frequently where they are rapidly destroyed. As a result the radar signature from a small crater floor originally mantled by impact melt will disappear when the median regolith thickness is a few times the melt thickness.

In a larger crater, by contrast, the impact melt is so thick that it can survive impact gardening for all post-Imbrian to post-Nectarian time; thus, fresh rocks will continue to be broken away from the impact melt. However, those within 10 meters or so of the surface will be destroyed in the manner described above. Consequently, the radar signature will again disappear as the regolith deepens, but the time scale for the signature to disappear will depend on the radar penetration depth and not on the thickness of impact melt.

In Table 4 we have attempted to assemble these ideas into a crude prediction of the time scale for decay of the radar anomalies. Regolith depths have been estimated from crater diameter-frequency curves using a model relationship between the diameter (C_s) of change in slope of the curve (Boyce and Johnson, 1977) and the median regolith depth determined from various theoretical and experimental studies (Shoemaker *et al.*, 1970; Quaide and Oberbeck, 1975). The relationship between impact melt thickness (T_m), median regolith depth (H_{med}), and radar penetration depths at 3.8 cm and 70 cm ($D_{3.8}$ and D_{70}) is characterized by three ratios (H_{med}/T_m, $H_{med}/D_{3.8}$, and H_{med}/D_{70}). Radar signatures disappear when either H_{med}/T_m exceeds two, or the other ratios exceed unity. Craters as young as Copernicus will have very shallow regoliths and still preserve the 3.8 cm signatures. In the case of the 70 cm signatures, there is a dependence on size and age. The small craters are radar bright until the regolith development as reflected in the index (H_{med}/T_m) comminutes the layer of impact melt. For very large craters (H_{med}/T_m) is always less than unity and the radar penetration depth is important. Assuming (H_{med}/D_{70}) = 1 denotes the change from a bright crater to a faint crater, then there are still some large pre-Imbrian craters that are bright, in agreement with our observations.

The premature loss of the 70 cm signature and the accompanying extended survival of 3.8 cm signatures in the small FBF craters is not explained by the model just described. One possibility is that impact melt is not exposed on the crater floor. Lange and Hawke (1979) report that many impact craters in the size range 15 to 30 km show no photogeological evidence of impact melt in the crater floor. They cite evidence for mantling of the melt layer, by material derived from wall slumping. They also point out that in larger craters this mantling process is not as effective, although Mendell (1976) feels that wall slumping is an important floor modifier for all craters. Possibly, the debris layers that mantle the floors of the smaller craters are dominated by centimeter-sized particles which provide a strong 3.8 cm radar signature but no 70 cm signature.

It may also be possible to account for the occurrence of FBF craters purely in terms of impact gardening. Conceivably, the particle size distributions in some phases of regolith development are dominated by the small blocks resulting in an FBF signature. To seriously evaluate these various possibilities it is necessary to quantitatively model the evolution of particle sizes in all phases of regolith development using techniques such as those developed by Quaide and Oberbeck (1974), Housen *et al.* (1979), and Langevin and Arnold (1977). In addition, a model for predicting the diffuse radar scattering from fragmental debris layer is needed.

SUMMARY

A comparison of the statistics of the infrared and radar bright craters on the terra with the statistics for craters assigned to different geologic epochs, has provided insights into the evolution of rock populations in the first few meters of the lunar subsurface. The statistics for the infrared and radar craters are consistent with the following evolution: (I) Craters are formed as BBB (bright at all wavelength) craters with excess populations of surface and subsurface rocks. (II) BBB craters evolve to FBB (faint IR, bright radar) craters with average surface rocks and excess subsurface rocks of all sizes. (III) FBB craters evolve further to either FBF or FFB craters, which have enhanced radar signatures at only one radar wavelength. These FBF and FFB craters have excess subsurface rocks of either centimeter or meter sizes. (IV) FBF and FFB craters evolve finally to the FFF crater which is faint at all wavelengths and has no excess surface or subsurface rocks.

These infrared and radar signatures evolve with different rates and along different paths depending upon crater size. Smaller craters with diameters of 10 to 30 km evolve primarily to FBF craters, while larger craters with diameters of 30 km and larger evolve primarily to FFB craters. In addition, smaller craters evolve faster than larger craters. The association of these infrared and radar bright craters with various geologic epochs is given in Table 3.

The initial conditions and subsequent evolution of radar-infrared crater signatures of the crater floors depends upon a number of lunar surface processes, including impact melt generation, debris blanketing, impact gardening, and crater wall collapse. We investigated the initial formation and subsequent gardening on an impact melt as one possible evolution scenario. This yielded an evolution of BBB to FBB to FFB to FFF craters which was size-dependent and agreed with the crater ages derived from the large crater statistics. However, this failed to predict the evolution through the FBF class observed for smaller craters. Clearly, we need to develop more quantitative descriptions of the evolution of particle sizes in regoliths and of the radar signatures corresponding to the various stages of evolution. Further studies along these lines could enable us to use radar and infrared signatures as a tool for understanding how various surface processes affected the lunar highland crust.

Acknowledgments—The catalog of infrared and radar craters used here was originally compiled by Wm. James Roberts of the Planetary Science Institute. The generation of this catalog relied heavily upon the infrared and radar images of the moon generated by the NASA Apollo Experiment S-217. Ms. Amelia Vetrone generated the crater statistic plots. Reviews by Fred Hörz and Peter Schultz provided many improvements to this paper. The work reported here was funded by NASA Contract NASW 3205.

REFERENCES

Arthur D. W. G., Agnieray A. G., Horvath R. A., Wood C. A., and Chapman C. R. (1963) The system of lunar craters, Quadrant I. *Commun. Lunar and Planetary Lab., Univ. Arizona 2,* 71–78.

Arthur D. W. G., Agnieray A. G., Horvath R. A., Wood C. A., and Chapman C. R. (1964) The system of lunar craters, Quadrant II. *Commun. Lunar and Planetary Lab., Univ. Arizona 3,* 1–2.

Arthur D. W. G., Agnieray A. G., Pellicori R. H., Wood C. A., and Weller T. (1965) The system of lunar craters, Quadrant III. *Commun. Lunar and Planetary Lab., Univ. Arizona 3,* 61–62.

Arthur D. W. G., Pellicori R. H., and Wood C. A. (1966) The system of lunar craters, Quadrant IV. *Commun. Lunar and Planetary Lab., Univ. Arizona 4,* 1–2.

Boyce J. M. (1976) Ages of flow units in the lunar nearside maria based on Lunar Orbiter 14 photographs. *Proc. Lunar Sci. Conf. 7th,* p. 2717–2728.

Boyce J. M. and Johnson D. A. (1977) Ages of flow units in Mare Crisium based on crater data. *Proc. Lunar Sci. Conf. 8th,* p. 3495–3502.

Crater Analysis Group (1979) Standard techniques for presentation and analysis of crater size-frequency data. *Icarcus* **37,** 467–474.

Gault D. E., Hörz F., Brownlee D. E., and Hartung J. B. (1974) Mixing of the lunar regolith. *Proc. Lunar Sci. Conf. 5th,* 2365–2386.

Hörz F., Schneider E., Gault D. E., Hartung J. B., and Brownlee D. E. (1975) Catastrophic rupture of lunar rocks: A Monte Carlo simulation. *The Moon* **13,** 235–258.

Housen K. R. and Wilkening L. L., Chapman C. R., and Greenberg R. (1979) Asteroid regoliths. *Icarus* **39,** 317–351.

Lange M. A. and Hawke B. R. (1979) The generation of lunar impact melts: A comparison of theoretical and observational results (abstract). In *Papers Presented to the Conference on the Lunar Highlands Crust,* p. 98–100. Lunar and Planetary Institute, Houston.

Langevin Y. and Arnold J. R. (1977) The evolution of the lunar regolith. *Ann. Rev. Earth Planet. Sci. Lett.* **5,** 449–489.

Mendell W. W. (1976) Degradation of large, Period II lunar craters. *Proc. Lunar Sci. Conf. 7th,* p. 2703–2716.

Moore H. J., Hodges C. A., and Scott D. H. (1974) Multiring basins—illustrated by Orientale and associated features. *Proc. Lunar Sci. Conf. 5th,* p. 71–100.

Quaide W. and Oberbeck V. (1975) Development of the mare regolith: Some model considerations. *The Moon* **13,** 27–55.

Schultz P. H. (1976) Floor-fractured lunar craters. *The Moon* **15,** 241–273.

Schultz P. H. and Mendell W. (1978) Orbital infrared observations of lunar craters and possible implications for impact ejecta emplacement. *Proc. Lunar Planet. Sci. Conf. 9th,* p. 2857–2883.

Shoemaker E. M. and Morris E. C. (1970) Physical properties of the lunar regolith determined from Surveyor television observations. *Radio Sci.* **5,** 129–155.

Shoemaker E. M., Hait M. H., Swann G. A., Schleicher D. L., Schaber G. G., Sutton R. L., Dahlen D. H., Goddard E. N., and Waters A. C. (1970) Origin of the lunar regolith at Tranquility Base. *Proc. Apollo 11 Lunar Sci. Conf.,* p. 2399–2412.

Shorthill R. W. (1973) Infrared atlas charts of the eclipsal moon. *The Moon* **7,** 22–45.

Thompson T. W. (1974) Atlas of lunar radar maps at 70 cm wavelength. *The Moon* **10,** 51–85.

Thompson T. W., Masursky H., Shorthill R. W., Tyler G. L., and Zisk S. H. (1974) A comparison of infrared, radar and geologic mapping of lunar craters. *The Moon* **10,** 87–117.

Thompson T. W., Masursky H., Shorthill R. W., and Zisk S. H. (1979) Blocky craters: Implications about the lunar megaregolith. *Moon and Planets* **21,** 319–342.

Wilhelms D. E., Oberbeck V. R., and Aggarwal H. R. (1978) Size-frequency distributions of primary and secondary lunar impact craters. *Proc. Lunar Planet. Sci. Conf. 9th,* p. 3735–3762.

Zisk S. H., Pettengill G. H., and Catuna G. W. (1974) High resolution radar map of lunar surfaces of 3.8 cm wavelength. *The Moon* **10,** 17–50.

Lunar Sample Index

Index entries were compiled from information supplied by the authors, and refer only to opening pages of articles.

*For Apollo 16 sample numbers 60001 through 69941, see pages 42–49.

Subject Index

Index entries were compiled from key words supplied by the authors, and refer only to opening pages of articles.

Author Index